Weeds of California and Other Western States

Volume 1 • *Aizoaceae–Fabaceae*

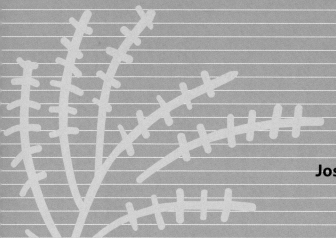

Joseph M. DiTomaso
Evelyn A. Healy

*Sponsored by
the California Weed Science Society*

University of California
Agriculture and Natural Resources

Publication 3488

For information about ordering this publication, contact
University of California
Agriculture and Natural Resources
Communication Services
6701 San Pablo Avenue, 2nd Floor
Oakland, California 94608-1239

Telephone 1-800-994-8849
(510) 642-2431
FAX (510) 643-5470
E-mail: danrcs@ucdavis.edu
Visit the ANR Communication Services Web site at http://anrcatalog.ucdavis.edu

Publication 3488

 This publication has been anonymously peer reviewed for technical accuracy by University of California scientists and other qualified professionals. This review process was managed by the ANR Associate Editor for Pest Management.

ISBN-13: 978-1-879906-69-3
ISBN-10: 1-879906-69-4
Library of Congress Control Number: 2006920353

©2007 by the Regents of the University of California
Division of Agriculture and Natural Resources
All rights reserved.

No part of this publication may be reproduced, stored in a retrieval system, or transmitted, in any form or by any means, electronic, mechanical, photocopying, recording, or otherwise, without the written permission of the publisher and the authors.

Photo credits are given in the captions. Cover photo by Joseph M. DiTomaso; design by Celeste Aida Marquiss. Illustrations by Evelyn A. Healy. Map on page 1808 reprinted from *The Jepson Manual*, J. Hickman, ed. (University of California Press, 1993), with permission from The Jepson Herbarium, © Regents of the University of California.

✺ Printed in Canada on recycled paper.

The University of California prohibits discrimination or harassment of any person on the basis of race, color, national origin, religion, sex, gender identity, pregnancy (including childbirth, and medical conditions related to pregnancy or childbirth), physical or mental disability, medical condition (cancer-related or genetic characteristics), ancestry, marital status, age, sexual orientation, citizenship, or status as a covered veteran (covered veterans are special disabled veterans, recently separated veterans, Vietnam era veterans, or any other veterans who served on active duty during a war or in a campaign or expedition for which a campaign badge has been authorized) in any of its programs or activities. University policy is intended to be consistent with the provisions of applicable State and Federal laws. Inquiries regarding the University's nondiscrimination policies may be directed to the Affirmative Action/Staff Personnel Services Director, University of California, Agriculture and Natural Resources, 1111 Franklin St., 6th Floor, Oakland, CA 94607-5201, (510) 987-0096. **For information about ordering this publication, telephone 1-800-994-8849.**

10m-pr-1/07-SB/CM

Contents
Volume 1

List of Tables xi
Acknowledgments xiii
About This Book 2
Format of Plant Descriptions 4
How To Identify Plants Using This Book 5
Lists of Species Covered in This Book 7
Shortcut Identification Tables 16
    Table 1. Parasitic plants 16
    Table 2. Succulent plants 16
    Table 3. Plants with papery sheath (ocrea) at base of leaf (not part of the petiole) 16
    Table 4. Plants very sticky (glandular) to the touch 17
    Table 5. Plants with square stems 18
    Table 6. Plants climbing on taller vegetation (viny) with and without tendrils 18
    Table 7. Plants with stellate, branched, or forked hairs 19
    Table 8. Plants with leaves whorled or appearing whorled at nodes 20
    Table 9. Plants that exude milky latex 20
    Table 10. Woody shrubs and trees 21
    Table 11. Emergent aquatic plants 22
    Table 12. Plants with prickles, spines, or thorns on leaves, stems fruits, flower heads, multiple parts, or entire plant 23
        A. Spines or prickles on leaves only 23
        B. Prickles, stinging hairs, or thorns on stem only 24
        C. Prickles and sharp structures on fruit only 24
        D. Prickles and sharp structures on flowerhead or inflorescence only 25
        E. Prickles, spines, or thorns on more than one plant part or entire plant 25
    Table 13. Plants with various types of compound or dissected leaves 26
        A. Two leaflets 26
        B. Three leaflets (trifoliolate) 26
        C. Palmately compound leaves 27
        D. Leaves once-pinnately compound or appearing once-pinnately compound .. 27
        E. Leaves twice-pinnately compound 29
        F. Leaves dissected or more than twice-pinnately compound 29
Key to the Common Mature Weedy Grasses of California 31
Vegetative Key to the Common Weedy Grasses of California 42
Weed Descriptions
Aizoaceae (Iceplant Family) 58
    Hottentot-fig *(Carpobrotus edulis)* 58
Amaranthaceae (Pigweed Family) 71
    Alligatorweed *(Alternanthera philoxeroides)* 71
    Tumble pigweed *(Amaranthus albus)* • **Prostrate pigweed** *(Amaranthus blitoides)* • **Low amaranth** *(Amaranthus deflexus)* 78
    Smooth pigweed *(Amaranthus hybridus)* • **Palmer amaranth** *(Amaranthus palmeri)* • **Redroot pigweed** *(Amaranthus retroflexus)* 90
Anacardiaceae (Sumac Family) 100
    Peruvian peppertree *(Schinus molle)* • **Brazilian peppertree** *(Schinus terebinthifolius)* 100
    Pacific poison-oak *(Toxicodendron diversilobum)* 105
Apiaceae (Carrot Family) 110
    Greater ammi *(Ammi majus)* • **Toothpick ammi** *(Ammi visnaga)* • **Wild celery** *(Cyclospermum leptophyllum)* 110
    Bur chervil *(Anthriscus caucalis)* • **Hedgeparsley** *(Torilis arvensis)* 117

Contents

Apiaceae (Carrot Family), *continued*
    Western waterhemlock (*Cicuta douglasii*) • Spotted waterhemlock
    (*Cicuta maculata*) 124
    Poison-hemlock (*Conium maculatum*) 132
    Wild carrot (*Daucus carota*) 137
    Fennel (*Foeniculum vulgare*) 142
    Floating pennywort (*Hydrocotyle ranunculoides*) • Lawn pennywort
    (*Hydrocotyle sibthorpioides*) 147
Apocynaceae (Dogbane Family) 151
    Big periwinkle (*Vinca major*) 151
Aquifoliaceae (Holly Family) 154
    English holly (*Ilex aquifolium*) 154
Araliaceae (Ginseng Family) 156
    English ivy (*Hedera helix*) 156
Arecaceae (Palm Family) 161
    Canary Island date palm (*Phoenix canariensis*) • Mexican fan palm
    (*Washingtonia robusta*) 161
Asclepiadaceae (Milkweed Family) 164
    Bladderflower (*Araujia sericifera*) 164
    Mexican whorled milkweed (*Asclepias fascicularis*) 167
Asteraceae (Sunflower Family) 172
    Common yarrow (*Achillea millefolium*) 172
    Russian knapweed (*Acroptilon repens*) 175
    Croftonweed (*Ageratina adenophora*) 179
    Annual bursage (*Ambrosia acanthicarpa*) • Giant ragweed (*Ambrosia trifida*) 182
    Mayweed chamomile (*Anthemis cotula*) 189
    Capeweed (*Arctotheca calendula*) 194
    Big sagebrush (*Artemisia tridentata*) 199
    Slender aster (*Aster subulatus*) 204
    English daisy (*Bellis perennis*) 208
    Hairy beggarticks (*Bidens pilosa*) 211
    Plumeless thistle (*Carduus acanthoides*) • Musk thistle (*Carduus. nutans*) • Italian
    thistle (*Carduus pycnocephalus*) • Slenderflower thistle (*Carduus tenuiflorus*) 217
    Smooth distaff thistle (*Carthamus baeticus*) • Woolly distaff thistle
    (*Carthamus lanatus*) 228
    Purple starthistle (*Centaurea calcitrapa*) • Diffuse knapweed (*Centaurea diffusa*) •
    Iberian starthistle (*Centaurea iberica*) • Spotted knapweed (*Centaurea biebersteinii*)
    • Squarrose knapweed (*Centaurea virgata*) 235
    Malta starthistle (*Centaurea melitensis*) • Yellow starthistle (*Centaurea solstitialis*) •
    Sicilian starthistle (*Centaurea sulphurea*) 250
    Pineappleweed (*Chamomilla suaveolens*) 261
    Rush skeletonweed (*Chondrilla juncea*) 264
    Chicory (*Cichorium intybus*) 268
    Canada thistle (*Cirsium arvense*) • Yellowspine thistle (*Cirsium ochrocentrum*) •
    Wavyleaf thistle (*Cirsium undulatum*) • Bull thistle (*Cirsium vulgare*) 272
    Hairy fleabane (*Conyza bonariensis*) • Horseweed (*Conyza canadensis*) 284
    Southern brassbuttons (*Cotula australis*) • Brassbuttons (*Cotula coronopifolia*) 291
    Smooth hawksbeard (*Crepis capillaris*) 295
    Common crupina (*Crupina vulgaris*) 299
    Artichoke thistle (*Cynara cardunculus*) 302
    Cape-ivy (*Delairea odorata*) 305
    Cutleaf burnweed (*Erechtites glomerata*) • Australian burnweed (*Erechtites minima*) .. 309

Contents

Asteraceae (Sunflower Family), *continued*
    Smallflower galinsoga (*Galinsoga parviflora*) • Hairy galinsoga
    (*Galinsoga quadriradiata*) 314
    Everlasting cudweed (*Gnaphalium luteo-album*) • Purple cudweed
    (*Gnaphalium purpureum*) 318
    Curlycup gumweed (*Grindelia squarrosa*) 325
    Common sunflower (*Helianthus annuus*) • Texas blueweed (*Helianthus ciliaris*) 329
    Spikeweed (*Hemizonia pungens*) 335
    Telegraphplant (*Heterotheca grandiflora*) 340
    Orange hawkweed (*Hieracium aurantiacum*) 343
    Virgate tarweed (*Holocarpha virgata*) 347
    Smooth catsear (*Hypochaeris glabra*) • Common catsear (*Hypochaeris radicata*) 353
    Poverty sumpweed (*Iva axillaris*) 357
    Willowleaf lettuce (*Lactuca saligna*) • Prickly lettuce (*Lactuca serriola*) •
    Bitter lettuce (*Lactuca virosa*) 360
    Oxeye daisy (*Leucanthemum vulgare*) 369
    Scotch thistle (*Onopordum acanthium*) • Illyrian thistle (*Onopordum illyricum*) •
    Taurian thistle (*Onopordum tauricum*) 372
    Bristly oxtongue (*Picris echioides*) 379
    Tansy ragwort (*Senecio jacobaea*) • Common groundsel (*Senecio vulgaris*) 382
    Blessed milkthistle (*Silybum marianum*) 391
    Lawn burweed (*Soliva sessilis*) 395
    Spiny sowthistle (*Sonchus asper*) • Annual sowthistle (*Sonchus oleraceus*) 398
    Small wirelettuce (*Stephanomeria exigua*) 404
    Feverfew (*Tanacetum parthenium*) • Common tansy (*Tanacetum vulgare*) 407
    Dandelion (*Taraxacum officinale*) 412
    Western salsify (*Tragopogon dubius*) • Common salsify (*Tragopogon porrifolius*) 416
    Spiny cocklebur (*Xanthium spinosum*) • Common cocklebur
    (*Xanthium strumarium*) 420
Boraginaceae (Borage Family) 426
    Coast fiddleneck (*Amsinckia menziesii* var. *intermedia*) • Menzies fiddleneck
    (*Amsinckia menziesii* var. *menziesii*) 426
    Houndstongue (*Cynoglossum officinale*) 430
    Pride of Madeira (*Echium candicans*) • Vipers bugloss (*Echium plantagineum*) .. 434
    Seaside heliotrope (*Heliotropium curassavicum*) • European heliotrope
    (*Heliotropium europaeum*) 440
    Broadleaf forget-me-not (*Myosotis latifolia*) 446
    Prickly comfrey (*Symphytum asperum*) • Common comfrey (*Symphytum officinale*). 450
Brassicaceae (Mustard Family) 455
    Yellow rocket (*Barbarea vulgaris*) 455
    Rapeseed mustard (*Brassica napus*) • Black mustard (*Brassica nigra*) • Birdsrape
    mustard (*Brassica rapa*) • Saharan mustard (*Brassica tournefortii*) 460
    Shepherd's-purse (*Capsella bursa-pastoris*) 471
    Hairy bittercress (*Cardamine hirsuta*) • Little bittercress (*Cardamine oligosperma*)... 474
    Lens-podded whitetop (*Cardaria chalepensis*) • Hoary cress (*Cardaria draba*) •
    Hairy whitetop (*Cardaria pubescens*) 478
    Blue mustard (*Chorispora tenella*) 486
    Lesser swinecress (*Coronopus didymus*) • Greater swinecress
    (*Coronopus squamatus*) 489
    Flixweed (*Descurainia sophia*) 493
    Dyer's woad (*Isatis tinctoria*) 497
    Field pepperweed (*Lepidium campestre*) • Clasping pepperweed
    (*Lepidium perfoliatum*) 500

Contents

Brassicaceae (Mustard Family), *continued*
    Perennial pepperweed *(Lepidium latifolium)* 511
    Wild radish *(Raphanus raphanistrum)* • Radish *(Raphanus sativus)* 516
    Austrian fieldcress *(Rorippa austriaca)* • Yellow fieldcress *(Rorippa sylvestris)* .. 521
    Sibara *(Sibara virginica)* 527
    Wild mustard *(Sinapis arvensis)* • Shortpod mustard *(Hirschfeldia incana)* 531
    Tumble mustard *(Sisymbrium altissimum)* • London rocket *(Sisymbrium irio)* •
    Hedge mustard *(Sisymbrium officinale)* 537
    Field pennycress *(Thlaspi arvense)* 549
Cannabaceae (Hemp Family) 552
    Marijuana *(Cannabis sativa)* 552
Caryophyllaceae (Pink Family) 555
    Mouseear chickweed *(Cerastium fontanum)* • Sticky chickweed *(Cerastium
    glomeratum)* • Common chickweed *(Stellaria media)* 555
    Babysbreath *(Gypsophila paniculata)* 564
    Dwarf pearlwort *(Sagina apetala)* • Corn spurry *(Spergula arvensis)* 567
    Bouncingbet *(Saponaria officinalis)* 576
    English catchfly *(Silene gallica)* 579
Chenopodiaceae (Goosefoot Family) 586
    Red orach *(Atriplex rosea)* • Australian saltbush *(Atriplex semibaccata)* 586
    Fivehook bassia *(Bassia hyssopifolia)* • Kochia *(Kochia scoparia)* 594
    Wild beet *(Beta vulgaris)* 600
    Common lambsquarters *(Chenopodium album)* • Nettleleaf goosefoot
    (Chenopodium murale) 602
    Mexicantea *(Chenopodium ambrosioides)* 612
    Halogeton *(Halogeton glomeratus)* 618
    Barbwire Russian thistle *(Salsola paulsenii)* • Russian thistle *(Salsola tragus)* •
    Spineless Russian thistle *(Salsola collina)* 622
Clusiaceae (St. Johnswort Family) 633
    Common St. Johnswort *(Hypericum perforatum)* 633
Convolvulaceae (Morningglory Family) 639
    Field bindweed *(Convolvulus arvensis)* 639
    Alkaliweed *(Cressa truxillensis)* 645
    Japanese morningglory *(Ipomoea nil)* • Tall morningglory *(Ipomoea purpurea)* .. 647
Cucurbitaceae (Squash Family) 654
    Citronmelon *(Citrullus lanatus)* 654
Cupressaceae (Cypress Family) 660
    Western juniper *(Juniperus occidentalis)* 660
Cuscutaceae (Dodder Family) 663
    Dodder *(Cuscuta* spp.*)* • Giant dodder *(Cuscuta reflexa)* 663
Cyperaceae (Sedge Family) 668
    Smallflower umbrella sedge *(Cyperus difformis)* • Tall flatsedge
    (Cyperus eragrostis) 668
    Yellow nutsedge *(Cyperus esculentus)* • Purple nutsedge *(Cyperus rotundus)* 674
    Green kyllinga *(Kyllinga brevifolia)* 679
Dennstaedtiaceae (Bracken Family) 682
    Western brackenfern *(Pteridium aquilinum)* 682
Dipsacaceae (Teasel Family) 685
    Common teasel *(Dipsacus fullonum)* • Fullers teasel *(Dipsacus sativus)* 685
Elaeagnaceae (Oleaster Family) 689
    Russian-olive *(Elaeagnus angustifolia)* 689
Equisetaceae (Horsetail Family) 693
    Field horsetail *(Equisetum arvense)* • Scouringrush *(Equisetum hyemale)* 693

Contents

Euphorbiaceae (Spurge Family) 698
    Turkey mullein *(Croton setigerus)* 698
    Leafy spurge *(Euphorbia esula)* • **Oblong spurge** *(Euphorbia oblongata)* • **Serrate spurge** *(Euphorbia serrata)* 701
    Caper spurge *(Euphorbia lathyris)* • **Petty spurge** *(Euphorbia peplus)* 710
    Spotted spurge *(Euphorbia maculata)* • **Ground spurge** *(Euphorbia prostrata)* • **Creeping spurge** *(Euphorbia serpens)* • **Nodding spurge** *(Euphorbia nutans)* 717
    Castorbean *(Ricinus communis)* 728
    Chinese tallowtree *(Sapium sebiferum)* 731
Fabaceae (Pea or Bean Family) 734
    Black acacia *(Acacia melanoxylon)* • **Kangaroothorn** *(Acacia paradoxa)* 734
    Camelthorn *(Alhagi maurorum)* 744
    Loco and milkvetch *(Astragalus spp.)* 748
    Scotch broom *(Cytisus scoparius)* • **French broom** *(Genista monspessulana)* • **Spanish broom** *(Spartium junceum)* 752
    Wild licorice *(Glycyrrhiza lepidota)* 763
    Everlasting peavine *(Lathyrus latifolius)* 766
    Birdsfoot trefoil *(Lotus corniculatus)* 770
    Yellow bush lupine *(Lupinus arboreus)* 774
    Black medic *(Medicago lupulina)* • **California burclover** *(Medicago polymorpha)* ... 776
    White sweetclover *(Melilotus albus)* • **Indian sweetclover** *(Melilotus indicus)* • **Yellow sweetclover** *(Melilotus officinalis)* 784
    Velvet mesquite *(Prosopis velutina)* • **Creeping mesquite** *(Prosopis strombulifera)* 790
    Black locust *(Robinia pseudoacacia)* 794
    Red sesbania *(Sesbania punicea)* 797
    Swainsonpea *(Sphaerophysa salsula)* 802
    Small hop clover *(Trifolium dubium)* • **Strawberry clover** *(Trifolium fragiferum)* • **Rose clover** *(Trifolium hirtum)* • **Crimson clover** *(Trifolium incarnatum)* • **White clover** *(Trifolium repens)* 805
    Gorse *(Ulex europaeus)* 820
    Common vetch *(Vicia sativa* ssp. *sativa)* • **Narrowleaf vetch** *(Vicia sativa* ssp. *nigra)* • **Winter vetch** *(Vicia villosa* ssp. *varia)* • **Hairy vetch** *(Vicia villosa* ssp. *villosa)* 823

Volume 2

Geraniaceae (Geranium Family) 835
    Broadleaf filaree *(Erodium botrys)* • **Redstem filaree** *(Erodium cicutarium)* • **Whitestem filaree** *(Erodium moschatum)* 835
    Cutleaf geranium *(Geranium dissectum)* • **Dovefoot geranium** *(Geranium molle)* .. 845
Iridaeae (Iris Family) 856
    Douglas iris *(Iris douglasiana)* • **Western blueflag iris** *(Iris missouriensis)* 856
Juncaceae (Rush Family) 862
    Toad rush *(Juncus bufonius)* • **Soft rush** *(Juncus effusus)* • **Spreading rush** *(Juncus patens)* 862
Lamiaceae (Mint Family) 868
    Henbit *(Lamium amplexicaule)* • **Purple deadnettle** *(Lamium purpureum)* 868
    White horehound *(Marrubium vulgare)* 873
    Pennyroyal *(Mentha pulegium)* 876
    Healall *(Prunella vulgaris* var. *vulgaris)* 882
    Mediterranean sage *(Salvia aethiopis)* • **Wand sage** *(Salvia virgata)* 885
Liliaceae (Lily Family) 890
    Three-corner leek *(Allium triquetrum)* • **Wild garlic** *(Allium vineale)* 890
    Onionweed *(Asphodelus fistulosus)* 897

Contents

Liliaceae (Lily Family), *continued*
    False garlic *(Nothoscordum gracile)* 900
    Deathcamas *(Zigadenus* spp.*)* 903
Lythraceae (Loosestrife Family).. 908
    Hyssop loosestrife *(Lythrum hyssopifolium)* • Purple loosestrife *(Lythrum salicaria)* 908
Malvaceae (Mallow Family) 915
    Velvetleaf *(Abutilon theophrasti)* 915
    Common mallow *(Malva neglecta)* • Bull mallow *(Malva nicaeensis)* •
    Little mallow *(Malva parviflora)* • High mallow *(Malva sylvestris)* 920
    Alkali mallow *(Malvella leprosa)* 931
Martyniaceae (Unicorn Plant Family) 935
    Devil's-claw *(Proboscidea lousianica)* • Yellow devil's-claw *(Proboscidea lutea)* 935
Molluginaceae (Carpetweed Family) 940
    Carpetweed *(Mollugo verticillata)* 940
Moraceae (Mulberry Family) 943
    Fig *(Ficus carica)* 943
Myoporaceae (Myoporum Family) 948
    Myoporum *(Myoporum laetum)* 948
Myrtaceae (Eucalyptus Family) 951
    Tasmanian blue gum *(Eucalyptus globulus)* 951
Onagraceae (Evening Primrose Family) 955
    Tall annual willowherb *(Epilobium brachycarpum)* • Fringed willowherb
    (Epilobium ciliatum) 955
    Scarlet gaura *(Gaura coccinea)* • Drummond's gaura *(Gaura drummondii)* •
    Wavyleaf gaura *(Gaura sinuata)* 962
    Hooker's eveningprimrose *(Oenothera elata)* • Cutleaf eveningprimrose
    (Oenothera laciniata) 971
Orobanchaceae (Broomrape Family) 978
    Cooper's broomrape *(Orobanche cooperi)* • Branched broomrape
    (Orobanche ramosa) 978
Oxalidaceae (Oxalis Family) 981
    Creeping woodsorrel *(Oxalis corniculata)* • Buttercup oxalis *(Oxalis pes-caprae)* .. 981
Papaveraceae (Poppy Family) 989
    Fumitory *(Fumaria officinalis)* 989
Phytolaccaceae (Pokeweed Family) 992
    Common pokeweed *(Phytolacca americana)* 992
Plantaginaceae (Plantain Family) 995
    Cutleaf plantain *(Plantago coronopus)* • Buckhorn plantain
    (Plantago lanceolata) • Broadleaf plantain *(Plantago major)* 995
Poaceae (Grass Family)... 1004
    Punagrass *(Achnatherum brachychaetum)* 1004
    Jointed goatgrass *(Aegilops cylindrica)* • Ovate goatgrass *(Aegilops ovata)* •
    Barb goatgrass *(Aegilops triuncialis)* 1008
    Creeping bentgrass *(Agrostis stolonifera)* 1016
    Meadow foxtail *(Alopecurus pratensis)* • Swamp pricklegrass *(Crypsis schoenoides)* .. 1021
    European beachgrass *(Ammophila arenaria)* 1027
    Sweet vernalgrass *(Anthoxanthum odoratum)* 1031
    Giant reed *(Arundo donax)* 1034
    Slender oat *(Avena barbata)* • Wild oat *(Avena fatua)* 1040
    Annual false-brome *(Brachypodium distachyon)* 1046
    Big quakinggrass *(Briza maxima)* • Little quakinggrass *(Briza minor)* 1049
    Rescuegrass *(Bromus catharticus)* 1053

Poaceae (Grass Family), *continued*
    Ripgut brome *(Bromus diandrus)* • Downy brome, Cheatgrass *(Bromus tectorum)* •
    Red brome *(Bromus madritensis)* 1056
    Soft brome *(Bromus hordeaceus)* • Japanese brome *(Bromus japonicus)* 1065
    Southern sandbur *(Cenchrus echinatus)* • Field sandbur *(Cenchrus incertus)* •
    Longspine sandbur *(Cenchrus longispinus)* 1078
    Feather fingergrass *(Chloris virgata)* 1083
    Jubatagrass *(Cortaderia jubata)* • Pampasgrass *(Cortaderia selloana)* 1090
    Bermudagrass *(Cynodon dactylon)* 1098
    Hedgehog dogtailgrass *(Cynosurus echinatus)* 1105
    Orchardgrass *(Dactylis glomerata)* 1108
    Smooth crabgrass *(Digitaria ischaemum)* • Large crabgrass *(Digitaria sanguinalis)*.1112
    Junglerice *(Echinochloa colona)* • Barnyardgrass *(Echinochloa crus-galli)* 1118
    Purple veldtgrass *(Ehrharta calycina)* 1129
    Goosegrass *(Eleusine indica)* 1136
    Quackgrass *(Elytrigia repens)* 1140
    Stinkgrass *(Eragrostis cilianensis)* • Little lovegrass *(Eragrostis minor)* • Tufted
    lovegrass *(Eragrostis pectinacea)* 1146
    Southwestern cupgrass *(Eriochloa acuminata)* 1162
    Tall fescue *(Festuca arundinacea)* 1167
    Nitgrass *(Gastridium phleoides)* • Annual junegrass *(Koeleria phleoides)* 1171
    Tanglehead *(Heteropogon contortus)* 1176
    Common velvetgrass *(Holcus lanatus)* 1179
    Mediterranean barley *(Hordeum marinum)* • Hare barley *(Hordeum murinum)*. 1182
    Bearded sprangletop *(Leptochloa fascicularis)* • Mexican sprangletop
    (Leptochloa uninervia)... 1190
    Italian ryegrass *(Lolium multiflorum)* • Perennial ryegrass *(Lolium perenne)*.... 1195
    Nimblewill *(Muhlenbergia schreberi)* 1201
    Blue panicgrass *(Panicum antidotale)* • Witchgrass *(Panicum capillare)* 1203
    Dallisgrass *(Paspalum dilatatum)* • Knotgrass *(Paspalum distichum)* 1211
    Kikuyugrass *(Pennisetum clandestinum)* • Crimson fountaingrass
    (Pennisetum setaceum) • Feathertop *(Pennisetum villosum)*.............................. 1219
    Harding grass *(Phalaris aquatica)* • Littleseed canarygrass *(Phalaris minor)* •
    Hood canarygrass *(Phalaris paradoxa)* 1230
    Smilograss *(Piptatherum miliaceum)* 1243
    Annual bluegrass *(Poa annua)*.. 1246
    Bulbous bluegrass *(Poa bulbosa)* 1250
    Kentucky bluegrass *(Poa pratensis)* 1253
    Rabbitfoot polypogon *(Polypogon monspeliensis)* 1262
    Arabian mediterraneangrass *(Schismus arabicus)* • Common
    mediterraneangrass *(Schismus barbatus)* 1270
    Giant foxtail *(Setaria faberi)* • Yellow foxtail *(Setaria pumila)* • Green foxtail
    (Setaria viridis).. 1275
    Shattercane *(Sorghum bicolor)* • Johnsongrass *(Sorghum halepense)* 1288
    Smooth cordgrass *(Spartina alterniflora)* 1297
    Smutgrass *(Sporobolus indicus)* 1306
    Medusahead *(Taeniatherum caput-medusae)* 1310
    Rattail fescue *(Vulpia myuros)* 1314
Polygonaceae (Buckwheat Family) 1319
    Swamp smartweed *(Polygonum amphibium)* • Pale smartweed
    (Polygonum lapathifolium) • Ladysthumb *(Polygonum persicaria)* 1319
    Common knotweed *(Polygonum arenastrum)* • Silversheath knotweed
    (Polygonum argyrocoleon) 1331

Contents

Polygonaceae (Buckwheat Family), *continued*
    Wild buckwheat (*Polygonum convolvulus*) 1337
    Japanese knotweed (*Polygonum cuspidatum*) • **Himalayan knotweed**
    (*Polygonum polystachyum*) • **Sakhalin knotweed** (*Polygonum sachalinense*) 1340
    Red sorrel (*Rumex acetosella*) 1347
    Curly dock (*Rumex crispus*) • **Broadleaf dock** (*Rumex obtusifolius*) •
    Fiddleleaf dock (*Rumex pulcher*) 1351
Portulacaceae (Purslane Family) 1363
    Redmaids (*Calandrinia ciliata*) 1363
    Miner's lettuce (*Claytonia perfoliata*) 1366
    Common purslane (*Portulaca oleracea*) • **Horse-purslane**
    (*Trianthema portulacastrum*) 1370
Primulaceae (Primrose Family) 1377
    Scarlet pimpernel (*Anagallis arvensis*) 1377
Ranunculaceae (Buttercup Family) 1381
    Larkspurs (*Delphinium* spp.) 1381
    Roughseed buttercup (*Ranunculus muricatus*) • **Creeping buttercup**
    (*Ranunculus repens*) • **Bur buttercup** (*Ranunculus testiculatus*) 1385
Rosaceae (Rose Family) 1400
    Biddy-biddy (*Acaena novae-zelandiae*) 1400
    Parney's cotoneaster (*Cotoneaster lacteus*) • **Silverleaf cotoneaster** (*Cotoneaster pannosus*) 1404
    English hawthorn (*Crataegus monogyna*) 1409
    Sulfur cinquefoil (*Potentilla recta*) 1413
    Dog rose (*Rosa canina*) • **Sweetbriar rose** (*Rosa eglanteria*) 1421
    Himalaya blackberry (*Rubus armeniacus*) • **Cutleaf blackberry** (*Rubus laciniatus*) • **Western raspberry** (*Rubus leucodermis*) • **Western thimbleberry** (*Rubus parviflorus*) • **Pacific blackberry** (*Rubus ursinus*) 1426
Rubiaceae (Madder Family) 1435
    Catchweed bedstraw (*Galium aparine*) • **Wall bedstraw** (*Galium parisiense*) •
    Field madder (*Sherardia arvensis*) 1435
Salicaceae (Willow Family) 1447
    Coyote willow (*Salix exigua*) • **Goodding's black willow** (*Salix gooddingii*) •
    Red willow (*Salix laevigata*) • **Arroyo willow** (*Salix lasiolepis*) 1447
Scrophulariaceae (Figwort Family) 1453
    Bellardia (*Bellardia trixago*) 1453
    Foxglove (*Digitalis purpurea*) 1456
    Sharppoint fluvellin (*Kickxia elatine*) 1460
    Dalmatian toadflax (*Linaria dalmatica*) • **Yellow toadflax** (*Linaria vulgaris*) 1463
    Yellow glandweed (*Parentucellia viscosa*) 1472
    Moth mullein (*Verbascum blattaria*) • **Common mullein** (*Verbascum thapsus*) .. 1475
    Water speedwell (*Veronica anagallis-aquatica*) 1482
    Corn speedwell (*Veronica arvensis*) • **Western purslane speedwell**
    (*Veronica peregrina*) • **Persian speedwell** (*Veronica persica*) 1486
Simaroubaceae (Quassia Family) 1499
    Tree-of-heaven (*Ailanthus altissima*) 1499
Solanaceae (Nightshade Family) 1503
    Chinese thornapple (*Datura ferox*) • **Jimsonweed** (*Datura stramonium*) •
    Sacred thornapple (*Datura wrightii*) 1503
    Black henbane (*Hyoscyamus niger*) 1510
    Manyflower tobacco (*Nicotiana acuminata*) • **Tree tobacco** (*Nicotiana glauca*) . 1513
    Wright groundcherry (*Physalis acutifolia*) • **Cutleaf groundcherry**
    (*Physalis angulata*) • **Lanceleaved groundcherry** (*Physalis lanceifolia*) •
    Tomatillo groundcherry (*Physalis philadelphica*) 1518

Solanaceae (Nightshade Family), *continued*
    American black nightshade *(Solanum americanum)* • Black nightshade *(Solanum nigrum)* • Hairy nightshade *(Solanum physalifolium)* 1528
    Horsenettle *(Solanum carolinense)* • Robust horsenettle *(Solanum dimidiatum)* • Silverleaf nightshade *(Solanum elaeagnifolium)* • Lanceleaf nightshade *(Solanum lanceolatum)* • White-margined nightshade *(Solanum marginatum)*... 1538
Tamaricaceae (Tamarisk Family) 1554
    Athel tamarisk *(Tamarix aphylla)* • Smallflower tamarisk *(Tamarix parviflora)* • Saltcedar *(Tamarix ramosissima)*.. 1554
Urticaceae (Nettle Family) 1562
    Stinging nettle *(Urtica dioica)* • Burning nettle *(Urtica urens)* 1562
Verbenaceae (Vervain Family) 1569
    Mat lippia *(Phyla nodiflora)*... 1569
    Tall vervain *(Verbena bonariensis)* • Seashore vervain *(Verbena litoralis)* 1572
Violaceae (Violet Family)... 1577
    English violet *(Viola odorata)* 1577
Viscaceae (Mistletoe Family) 1581
    Dwarf mistletoes *(Arceuthobium* spp.) • American or true mistletoes *(Phoradendron* spp.) • European mistletoe *(Viscum album)*............................... 1581
Zygophyllaceae (Caltrop Family) 1589
    African rue *(Peganum harmala)* 1589
    Puncturevine *(Tribulus terrestris)* 1592
    Syrian beancaper *(Zygopyllum fabago)* 1596
Appendix: Non-native Plants Rarely or Occasionally Naturalized in California..... 1601
Glossary 1661
Bibliography 1680
Index to Volumes 1 and 2.. 1741
Authors and Contributors 1807
Geographic Subdivisions of California 1808

List of Tables
Volume 1
Table 1. Parasitic plants 16
Table 2. Succulent plants 16
Table 3. Plants with papery sheath (ocrea) at base of leaf (not part of the petiole) 16
Table 4. Plants very sticky (glandular) to the touch 17
Table 5. Plants with square stems 18
Table 6. Plants climbing on taller vegetation (viny) with and without tendrils 18
Table 7. Plants with stellate, branched, or forked hairs 19
Table 8. Plants with leaves whorled or appearing whorled at nodes 20
Table 9. Plants that exude milky latex.. 20
Table 10. Woody shrubs and trees.. 21
Table 11. Emergent aquatic plants 22
Table 12. Plants with prickles, spines, or thorns on leaves, stems, fruits, flower heads, multiple parts, or entire plant 23
Table 13. Plants with various types of compound or dissected leaves......................... 26
Table 14. Iceplants and relatives (*Aptenia, Carpobrotus, Conicosia, Drosanthemum, Malephora,* and *Mesembryanthemum* spp.) 61
Table 15. *Amaranthus* species... 80
Table 16. Beggarticks (*Bidens* spp.)... 216
Table 17. *Carduus* species... 220
Table 18. Spiny-leaved thistles: Flowers pink, purple, or white, except where noted.. 227
Table 19. Purple-flowered knapweeds, starthistles, and common crupina (*Acroptilon, Centaurea,* and *Crupina* spp.) 244

List of Tables

Table 20. Yellow-flowered starthistles (*Centaurea* spp.) 258
Table 21. *Cirsium* species 283
Table 22. Cudweeds (*Gnaphalium* spp.) and licorice plant (*Helichrysum petiolare*) 324
Table 23. Common *Hemizonia* species with spine-tipped leaves and yellow anthers ... 339
Table 24. Tarweeds (*Hemizonia* and *Holocarpha* spp.) with yellow flower heads and soft-tipped leaves 352
Table 25. Lettuces (*Lactuca* spp.) 368
Table 26. Yellow-flowered mustards (*Brassica*, *Hirschfeldia*, and *Sinapis* spp.) 468
Table 27. Pepperweeds (*Lepidium* spp.) and hoary cress (*Cardaria draba*) 509
Table 28. *Sisymbrium* species 548
Table 29. Pearlworts, spurry, sandspurrys, and polycarp (*Sagina*, *Spergula*, *Spergularia*, and *Polycarpum* spp.) 575
Table 30. Catchflys and campions (*Silene* spp.) 585
Table 31. Morningglorys (*Ipomoea* spp.) 653
Table 32. Spurge (*Euphorbia* spp.) vegetative and reproductive characteristics 726
Table 33. *Acacia* species 742
Table 34. Brooms (*Cytisus*, *Genista*, *Spartium*, and *Retama* spp.) 762
Table 35. Clovers (*Trifolium* spp.) 818
Table 36. Vetches (*Vicia* spp.) 834

Volume 2

Table 37. Filarees (*Erodium* spp.) 844
Table 38. Geraniums (*Geranium* spp.) 855
Table 39. Mallows and sida (*Lavatera*, *Malva*, *Malvella*, *Modiola*, and *Sida* spp.) 930
Table 40. Woodsorrels (*Oxalis* spp.) 988
Table 41. Plantains (*Plantago* spp.) 1003
Table 42. Goatgrasses (*Aegilops* spp.) 1015
Table 43. Common oat species (*Avena* spp.) 1045
Table 44. Brome grasses (*Bromus* spp.) 1077
Table 45. Fingergrass, windmillgrass, and rhodesgrass (*Chloris* spp.) 1089
Table 46. Jubatagrass and pampasgrass (*Cortaderia* spp.) 1097
Table 47. Barnyardgrass and related *Echinochloa* species 1128
Table 48. Quackgrass (*Elytrigia repens*) and wheatgrasses (*Elytrigia* and *Pascopyrum* spp.) 1145
Table 49. Lovegrasses (*Eragrostis* spp.) 1161
Table 50. Barleys (*Hordeum* spp.) and cereal rye (*Secale cereale*) 1189
Table 51. Canarygrasses and relatives (*Phalaris* spp.) 1242
Table 52. Bluegrasses (*Poa* spp.) 1261
Table 53. *Polypogon* species 1269
Table 54. Foxtails (*Setaria* spp.) 1287
Table 55. Cordgrass (*Spartina* spp.) 1305
Table 56. Smartweeds (*Polygonum* spp.) 1330
Table 57. Docks (*Rumex* spp.) 1362
Table 58. Buttercups (*Ranunculus* spp.) 1399
Table 59. Blackberries and relatives (*Rubus* spp.) 1434
Table 60. Bedstraws (*Galium* spp.) and field madder (*Sherardia arvensis*) 1446
Table 61. Willows (*Salix* spp.) 1452
Table 62. Toadflaxes (*Linaria* spp.) without yellow flowers 1471
Table 63. Aquatic speedwells (*Veronica* spp.) 1485
Table 64. Terrestrial speedwells (*Veronica* spp.) 1498
Table 65. Groundcherries (*Physalis* spp.) 1527
Table 66. Nightshades (*Solanum* spp.) without prickles and with simple hairs 1537
Table 67. Nightshades (*Solanum* spp.) with spiny prickles (except *S. mauritianum*) and star-shaped hairs 1552

Acknowledgments

We would like to thank the many individuals for their help in the production of this book, including Pat Akers, Suzanne Albright, Dr. Lars Anderson, Dr. Debra Ayres, Bill Bailey, Dr. Joe Balciunas, Carl Bell, Robin Breckenridge, Brenda Brinton, Lori Brown, Dr. Chris Carmichael, Mary Carroll, David Chang, Peter Connor, Dr. Dave Cudney, Dr. Steve Dreistadt, Jenny Drewitz, Dr. Clyde Elmore, Dr. Steven Enloe, Ed Finley, Dr. Alison Fisher, Dr. Holly Forbes, Ed Fredrickson, Caroline Gibbs, Dr. Russ Hahn, Art Hazebrook, Dr. Jodie Holt, Dr. Fred Hrusa, Dr. Ruth Hufbauer, Steve Junak, Rick Keck, Barbara Keller, Mike Kelly, Jo Kitz, Butch Krebs, Dr. Tom Lanini, David Lile, Jan Lowrey, Dr. Dan Marcum, Dr. Mike Messler, Jessica Miller, Dr. Phil Motooka, Dr. Joe Neal, Ross O'Connell, Steve Orloff, Andrea Pickard, Carri Pirosko, Dr. Tim Prather, Dawn Rafferty, Dr. John Randall, Dr. Mark Renz, Dr. Barry Rice, Dr. Richard Riefner Jr., Ernie Roncoroni Jr., Ernie Roncoroni Sr., Geoff Sainty, Dr. Andy Sanders, Mary (Kitty) Schlosser, Steve Schoenig, Dr. Andy Senesac, Jake Sigg, Dr. James P. Smith, Dr. Dave Spencer, Alison Stanton, Dr. Scott Steinmaus, Dr. Don Strong, Rick Storre, Steve Tjosvold, Dr. Rick Uva, John Wade, Peter Warner, Dr. Cheryl Wilens, Rob Wilson, and Andrea Woolfolk. We would like to thank Dr. Bruce Baldwin for allowing us to use *The Jepson Manual* map of California. We are particularly indebted to Dr. Ellen Dean for assisting in the identification of many of our collections. Although the author photographed most of the images used in this text, we are very grateful to Suzanne Paisley for her photographic advice and assistance. Most important, we are fortunate to have had the opportunity to work with Jack Kelly Clark, who not only advised us on photographic equipment and techniques but also photographed many of the plant species and the seeds. We also thank Jim O'Brien for his beautiful seed photographs.

We would also like to thank the many reviewers for their helpful comments, including Harry Agamalian, Dr. Lars Anderson, Dr. Debra Ayres, Bill Bailey, Dr. Joe Balciunas, Carl Bell, Dr. Carla Bossard, Robin Breckenridge, Dr. John Brock, Dr. Matt Brooks, Mick Canevari, Vanelle Carrithers, David Chang, Mike Connor, Peter Connors, Kim Cooper, Dr. Dave Cudney, Jay Davidson, Dr. Ellen Dean, Dr. Steve Dewey, Deanne DiPietro, Morgan Doran, Jenny Drewitz, Dr. Tom Dudley, Celestine Duncan, Dr. Clyde Elmore, Dr. Steven Enloe, Dr. Jennifer Erskine, Dr. Steve Fennimore, Ed Finley, Dr. Albert Fischer, Dr. Alison Fisher, Dr. Ted Foin, Ed Fredrickson, Bill Frost, Holly George, Kerry Heise, Dr. Jim Hill, Dr. Jodie Holt, Dr. Fred Hrusa, Dr. Louise Jackson, Dr. Marie Jasieniuk, Phil Jenkins, Mike Kelly, Bruce Kidd, Guy Kyser, Don Lancaster, Dr. Tom Lanini, Michelle LeStrange, Laura Lee Lienk, Rachel Long, Dr. Dan Marcum, Bruce McArthur, Neil McDougald, Dr. Milt McGiffen, Dr. Richard Michelmore, Dr. Pat Minogue, Gene Miyao, Bob Mullen, Doug Munier, Mike Murray, Glenn Nader, Dr. Robert Norris, Oswaldo Ochoa, Ross O'Connell, Steve Orloff, Jeff Philips, Andrea Pickart, Carri Pirosko, Dr. Mike Pitcairn, Dr. Tim Prather, Dawn Rafferty, Dr. John Randall, Dr. Mark Renz, Dr. Kevin Rice, Dr. Barry Rice-Meyer, Dr. Mona Robison, John Roncoroni, Dr. Fred Ryan, Dr. Hank Seagall,

Acknowledgments

Jerry Schmierer, Dr. Anna Sher, Dr. Anil Shresthra, Jake Sigg, Dr. Lincoln Smith, Richard Smith, Dr. Dave Spencer, Alison Stanton, Dr. Scott Steinmaus, Dr. Barry Tickes, Steve Tjosvold, Joel Trumbo, Ron Vargas, Becky Waegell, Peter Warner, Ed Weber, Dr. Tom Whitson, Dr. Cheryl Wilens, Jack Williams, Rob Wilson, Bill Winans, Dr. Dale Woods, Steve Wright, Dr. John Yoder, and Dr. Jim Young.

We thank our parents for their inspiration and families and friends for their support, including Sue Webster and the CSs who assisted in the identification of buckbrush, wild buckwheat, and buckhorn plantain. Joe DiTomaso would like to thank his sons, Evan and John Paul, who accompanied him on so many photographing trips and whose reward is an occasional photograph of their hand holding a plant. Evelyn especially thanks Chris Withrow for his wonderful words of encouragement to his mother throughout the duration of the project, and Paul Teller for his ever-ready support during the frenetic culmination of the project in the final year.

In addition, we are most grateful to the California Weed Science Society for their substantial financial contribution and sponsorship of this book. We also thank the Center for Research in Pest Management, Pesticide Applicators Professional Association (PAPA), UC Division of Agriculture and Natural Resources, UC Statewide Integrated Pest Management Project, and Smith-Lever IPM Project at the University of California. We also thank the Weed Research and Information Center, as well as the California Department of Food and Agriculture, for their financial assistance in the development of the text. Finally, we thank the University of California Division of Agriculture and Natural Resources staff for working with the authors on the production and publication of this book.

J.M.D.
E.A.H.

Weeds of California and Other Western States

Volume 1

About This Book

Weeds of California and Other Western States is a two-volume practical guide to the identification and biology of over 700 terrestrial weeds. Although the book covers primarily weeds of California, many of these species also occur in other areas of the United States, particularly the western states. Their occurrence outside California is noted in the distribution section of the appropriate entry. This book is complemented by *Aquatic and Riparian Weeds of the West*, coauthored by Joseph M. DiTomaso and Evelyn A. Healy (ANR Publication 3421, 2003); there is some overlap, however, between the two books in emergent aquatic species of riparian and water use systems. Each species in this book was carefully researched to provide the most accurate and timely information possible.

The common names used in this book conform to those in *Composite List of Weeds*, to be published in 2007 by the Weed Science Society of America (WSSA). When a species was not included in the WSSA publication, we used the most widely recognized common name. The scientific names are those we considered the most widely accepted by current taxonomic treatments based on a variety of literature. Synonyms of both common and scientific names are included in the text and can also be found in the index. Five-letter Bayer codes (standardized abbreviation and computer code) were included when available from the *Composite List of Weeds*. We have also indicated when a particular species was listed on the federal or California state noxious weed list or on the California Invasive Plant Council list of non-native plants that threaten wildlands.

Rather than providing individual descriptions for each weed species, we grouped related species or morphologically similar species into a single description to allow for easier comparison. In addition to photographs and descriptions of the prominent weedy species, we have also included photographs and distinguishing characteristics of many morphologically similar or related species. In some cases these species are also weedy, although less common in distribution, but in other cases they are nonweedy native species. We have included some of the nonweedy natives to avoid potential misidentification and inadvertent damage to nontarget desirable species.

The book consists of 262 individual entries, including a full description of 451 species and another 361 plants compared as similar species, representing 63 plant families. There is at least one photograph of 737 plant species.

The main body of the book contains weed descriptions, which are presented in alphabetical order according to family, then genus and species. The family names (scientific and common) and the habitat where the species occur are indicated at the top of the page. The format of the text is similar to that in *Aquatic and Riparian Weeds of the West*. The text includes not only detailed descriptions of the morphological characteristics of each main species but also information on the distribution, habitat, postsenescence characteristics, propagation and phenology, management considerations that favor or discourage survival, and characteristics that allow distinguishing between other similar or related species (see "Format of Plant Descriptions"). The symbol ± is used to mean "about" and indicates variability in the given characteristic; numbers in parentheses, as in "12(15) cm wide," indicate occasional variability to the parenthetical number. Phrases or words in italics indicate important characteristics in the identification of that species. Where a particular genus or group contains many difficult-to-distinguish weedy species, we have included tables comparing important characteristics. Distribution information is according to *The Jepson Manual* (University of California Press, 1993), and a map showing these regions appears on page 1808.

A list of the species covered in the book is provided beginning on page 7. The list is organized alphabetically by family name, then scientific name and common name, and includes the page number where the description of the plant can be found. The book also contains shortcut identification tables for groups that share similar, unusual, or relatively uncommon characteristics (page 16). Two grass identification keys are provided: a key to all characteristics including inflorescences and reproductive parts (page 31) and a key to vegetative characteristics only (page 42).

An appendix of non-native plants rarely or occasionally naturalized in California is included in volume 2 (page 1601). The text employs certain technical botanical terms in describing the species; to assist in understanding these technical terms, we have included an illustrated glossary (page 1661). The index at the end of volume 2 (page 1741) covers both volumes and includes currently accepted common and scientific names as well as synonyms of common and scientific names. Measurements are given in metric units; aid in conversion can be found on the inside back cover of both volumes.

Although we have taken every precaution to ensure that the information provided is accurate and the identifications are correct, there may be instances where inaccuracies escaped our notice. Should a reader find an error in the text, please contact Joe DiTomaso at the Department of Plant Sciences, Robbins Hall, Mail Stop 4, University of California, Davis CA, 95616.

Format of Plant Descriptions

Each section of plant descriptions follows the format explained below. Note that every entry may not contain all categories of descriptions.

Plant category: Family name (Family common name)   Problem site(s)

Common name [*Scientific name* and author] [Bayer code]

NOXIOUS WEED LISTS: Federal Noxious Weed List, species included on the California Invasive Pest Plant Council (Cal-IPC) *California Invasive Plant Inventory* list (See http://cal-ipc.org), California Department of Food and Agriculture (CDFA) Noxious Weed List, or noxious weed lists from other western states.

SYNONYMS: Common and scientific names used in other texts or references but not considered as the commonly accepted names today.

GENERAL INFORMATION: Summary of the important aspects of the plant's life cycle, size, growth form, visual appearance, habitat, distribution, impact, method of introduction, and native environment and range. In cases where the plant can cause toxicity to humans or other animals, a brief description of the toxicology is included.

SEEDLING: Description of the vegetative characteristics of newly emerging perennials, as well as seedlings, including cotyledons and the first few leaves.

MATURE PLANT: Description of the vegetative characteristics of mature plants.

ROOTS AND UNDERGROUND STRUCTURES: Description of the root types and growth characteristics of annuals, biennials, and herbaceous and woody perennial species.

FLOWERS: Flowering season and a description of flowering stems, inflorescence, and individual flowers and structures associated with flowers. When available, pollination method, self-compatibility, and flower timing are included.

FRUITS AND SEEDS: Description of the size and type of fruit and seed.

SPIKELETS AND FLORETS: Description of the reproductive structures of grasses.

SPORE-BEARING STRUCTURES: Description of organisms that reproduce by spores, rather than flowers (e.g., horsetails, ferns).

POSTSENESCENCE CHARACTERISTICS: Description of the species either in the dormant stage or after the plant has died (annuals or biennials) or foliage has senesced (perennials).

HABITAT: Description of habitats where the species is typically found. This section may also include soil, nutrient, or exposure characteristics associated with the species.

ADDITIONAL ECOLOGICAL ASPECTS: Typically a comparison between the biology and ecology of the weedy species and native desirable species within the same habitat.

DISTRIBUTION: Detailed distribution in California, including maximum elevation in which the species is found. Other states were the plant is found, and when known, the worldwide distribution, are also given.

PROPAGATION AND PHENOLOGY: Description of the methods of reproduction, seed dispersal, germination requirements and conditions, seed survival and longevity, and early establishment characteristics and requirements.

MANAGEMENT FAVORING OR DISCOURAGING SURVIVAL: When available, cultural practices that can prevent establishment or management options that have proven effective or ineffective in controlling infestations. Because of the dynamic nature of weed control, specific recommendations are not included.

SIMILAR SPECIES: Description of species that resemble or are closely related to the main species described in the entry, with characteristics that allow separation. In some cases, these include detailed descriptions given only here, whereas in other cases they refer to similar species that are described in detail elsewhere in the text. Species described only here also include synonyms and the Bayer code (when available).

How To Identify Plants Using This Book

Weeds of California and Other Western States provides users with a number of resources to aid in identifying weeds in agricultural and nonagricultural areas. Although we expect that most users will leaf through the book until they see a photo that matches the plant of concern, there are other, perhaps faster, methods of identification. For example, if the plant of interest has an unusual characteristic, such as square stems, a glandular or sticky surface, spiny structures, or milky latex, it can be identified by finding these characteristics in the short-cut identification tables on pages 16–30. Using these tables narrows the number of choices and hopefully saves time in correctly identifying a plant.

We have also included two keys to help in identification of common weedy grasses. These include the "Key to the Common Mature Weedy Grasses of California" (page 31) and "Vegetative Key to the Common Weedy Grasses" (page 42). These keys are set up much like typical dichotomous keys, with two choices at each stage. Some steps in the keys end with a genus rather than with a particular species; in these cases, identification to species can be achieved by reading the individual entry for that genus. In many situations, we have included a plant in more than one place in the key. This was done to account for variability in the species or difficulty in separating a particular species based on that key step.

Some plant entries contain tables to assist in separating closely related species. These tables compare several characteristics, including life cycle and vegetative and reproductive characteristics. Rather than using standard dichotomous keys, readers should be able to use these tables to compare related species based on the characteristics that may be available. Identification to species should be relatively easy if the plant is at a stage where all characteristics are evident. However, if only a few vegetative characteristics are present, it is still possible to narrow the choices to one or two possible species.

Finally, volume 2 includes an index to the currently accepted common and scientific names to the plants discussed in both volumes, as well as synonyms (page 1741). The common names used in this book were those included in the most recent edition of the *Composite List of Weeds*, published by the Weed Science Society of America (2006 edition).

We hope that this combination of color photos, text descriptions, keys, tables, and the index will increase the accuracy and speed in the identification of weeds.

List of Species Covered in This Book

Species and genera are organized alphabetically by family and then genus within each family. Indented species and genera are described in the related species section of main entries. Nonindented species are more fully described in a main entry. See page numbers for species descriptions; volume numbers are given with page numbers (e.g., 2:1370) where required.

Species	Common name	Page
Volume 1		
Aizoaceae		58
Carpobrotus spp.	sea-fig, hottentot-fig	58
Trianthema portulacastrum	horse-purslane	2:1370
Aptenia cordifolia	baby sun rose	64
Conicosia pugioniformis	roundleaf iceplant	66
Drosanthemum floribundum	showy dewflower	68
Malephora crocea	coppery mesembryanthemum	69
Mesembryanthemum spp.	iceplants	69
Amaranthaceae		71
Alternanthera spp.	alligatorweed and chaff-flower	71
Amaranthus spp.	pigweeds and amaranths	78, 90
Anacardiaceae		100
Schinus spp.	peppertrees	100
Toxicodendron diversilobum	Pacific poison-oak	105
Rhus trilobata	skunkbrush	109
Apiaceae		110
Ammi spp.	ammis	110
Anthriscus caucalis	bur chervil	117
Cicuta spp.	waterhemlocks	124
Conium maculatum	poison-hemlock	132
Cyclospermum leptophyllum	wild celery	110
Daucus carota	wild carrot	137
Foeniculum vulgare	fennel	142
Hydrocotyle spp.	pennyworts	147
Torilis arvensis	hedgeparsley	117
Daucus pusillus	southwestern carrot	141
Pastinaca sativa	wild parsnip	131, 136
Scandix pecten-veneris	venus-comb	122
Torilis nodosa	knotted hedgeparsley	121
Apocynaceae		151
Vinca spp.	periwinkles	151
Aquifoliaceae		154
Ilex aquifolium	English holly	154
Araliaceae		156
Hedera spp.	ivys	156
Arecaceae		161
Phoenix canariensis	Canary Island date palm	161
Washingtonia robusta	Mexican fan palm	161
Asclepiadaceae		164
Araujia sericifera	bladderflower	164

List of Species

Species	Common name	Page
Asclepias spp.	milkweeds	167
Asteraceae		172
Achillea millefolium	common yarrow	172
Acroptilon repens	Russian knapweed	175
Ageratina adenophora	croftonweed	179
Ambrosia spp.	bursage and ragweeds	182
Anthemis spp.	chamomiles	189
Arctotheca calendula	capeweed	194
Artemisia spp.	sagebrush, mugwort, and wormwoods	199
Aster subulatus var. *ligulatus*	slender aster	204
Bellis perennis	English daisy	208
Bidens spp.	beggarticks	211
Carduus spp.	thistles	217
Carthamus spp.	distaff thistles	228
Centaurea spp.	starthistles and knapweeds	235, 250
Chamomilla suaveolens	pineappleweed	261
Chondrilla juncea	rush skeletonweed	264
Cichorium intybus	chicory	268
Cirsium spp.	thistles	272
Conyza spp.	fleabanes and horseweed	284
Cotula spp.	brassbuttons	291
Crepis spp.	hawksbeards	295
Crupina vulgaris	common crupina	299
Cynara cardunculus	artichoke thistle	302
Delairea odorata	Cape-ivy	305
Erechtites spp.	burnweeds	309
Galinsoga spp.	galinsogas	314
Gnaphalium spp.	cudweeds	318
Grindelia spp.	gumweeds and gumplants	325
Helianthus spp.	sunflowers	329
Hemizonia spp.	spikeweed and tarweeds	335
Heterotheca grandiflora	telegraphplant	340
Hieracium spp.	hawkweeds	343
Holocarpha virgata	virgate tarweed	347
Hypochaeris spp.	catsears	353
Iva axillaris	poverty sumpweed	357
Lactuca spp.	lettuces	360
Leucanthemum spp.	oxeye daisy and chrysanthemums	369
Onopordum spp.	thistles	372
Picris echioides	bristly oxtongue	379
Senecio spp.	ragworts and groundsels	382
Silybum marianum	blessed milkthistle	391
Soliva sessilis	lawn burweed	395
Sonchus spp.	sowthistles	398
Stephanomeria exigua	small wirelettuce	404
Tanacetum spp.	feverfew and common tansy	407
Taraxacum officinale	dandelion	412
Tragopogon spp.	salsifys	416

Species	Common name	Page
Xanthium spp.	cockleburs	420
Baccharis salicifolia	seepwillow	2:1452
Calendula arvensis	field marigold	197
Cnicus benedictus	blessed thistle	233
Dimorphotheca sinuata	African daisy	198
Dittrichia graveolens	stinkwort	350
Erigeron spp.	fleabanes	206
Euryops subcarnosus	sweet resinbush	390
Helichyrum petiolare	licoriceplant	324
Leontodon taraxacoides	lesser hawkbit	415
Scolymus hispanicus	golden thistle	234
Urospermum picroides	prickly goldenfleece	381
Boraginaceae		426
Amsinckia spp.	fiddlenecks	426
Cynoglossum officinale	houndstongue	430
Echium spp.	echiums, vipers bugloss, pride of Madeira	434
Heliotropium spp.	heliotropes	440
Myosotis spp.	forget-me-nots	446
Symphytum spp.	comfreys	450
Borago officinalis	common borage	453
Lappula squarrosa	European sticktight	448
Brassicaceae		455
Barbarea spp.	rockets and wintercresses	455
Brassica spp.	mustards	460
Capsella bursa-pastoris	shepherd's-purse	471
Cardamine spp.	bittercresses	474
Cardaria spp.	hoary cress and whitetops	478
Chorispora tenella	blue mustard	486
Coronopus spp.	swinecresses	489
Descurainia spp.	flixweed and tansymustard	493
Hirschfeldia incana	shortpod mustard	531
Isatis tinctoria	dyer's woad	497
Lepidium spp.	pepperweeds	500, 511
Raphanus spp.	radishes	516
Rorippa spp.	fieldcresses and yellowcress	521
Sibara virginica	sibara	527
Sinapis spp.	mustards	531
Sisymbrium spp.	rockets and mustards	537
Thlaspi arvense	field pennycress	549
Arabidopsis thaliana	mouse-ear cress	530
Malcolmia africana	Malcolm stock	488
Cannabaceae		552
Cannabis sativa	marijuana	552
Caryophyllaceae		555
Cerastium spp.	chickweeds	555
Gypsophila spp.	babysbreaths	564
Sagina spp.	pearlworts	567

List of Species

Species	Common name	Page
Saponaria officinalis	bouncingbet	576
Silene spp.	catchflys and campions	579
Spergula arvensis	corn spurry	567
Stellaria spp.	chickweeds and starworts	555
Arenaria serpyllifolia	thymeleaf sandwort	563
Polycarpum tetraphyllum	four-leaved polycarp	574
Spergularia spp.	sandspurrys	572
Chenopodiaceae		586
Atriplex spp.	oraches and saltbushes	586
Bassia hyssopifolia	fivehook bassia	594
Beta vulgaris	wild beet	600
Chenopodium spp.	Mexicantea, lambsquarters, and goosefoots	602, 612
Halogeton glomeratus	halogeton	618
Kochia scoparia	kochia	594
Salsola spp.	Russian thistles, glasswort, and saltwort	622
Monolepis nuttalliana	Nuttall povertyweed	610
Clusiaceae [= Hypericaceae]		633
Hypericum spp.	St. Johnsworts and hypericums	633
Convolvulaceae		639
Convolvulus spp.	bindweeds	639
Cressa truxillensis	alkaliweed	645
Ipomoea spp.	morningglorys	647
Calystegia occidentalis	western morningglory	644
Dichondra repens	dichondra	644
Cucurbitaceae		654
Citrullus lanatus	citronmelon	654
Cucumis spp.	smellmelon, paddymelon	656
Cucurbita foetidissima	buffalo gourd	657
Marah spp.	manroots and wild cucumbers	307, 659
Cupressaceae		660
Juniperus occidentalis var. *occidentalis*	western juniper	660
Cuscutaceae		663
Cuscuta spp.	dodders	663
Cyperaceae		668
Cyperus spp.	sedges, nutsedges, and flatsedges	668
Kyllinga brevifolia	green kyllinga	679
Dennstaedtiaceae		682
Pteridium aquilinum var. *pubescens*	western brackenfern	682
Dipsacaceae		685
Dipsacus spp.	teasels	685
Elaeagnaceae		689
Elaeagnus angustifolia	Russian-olive	689
Equisetaceae		693
Equisetum spp.	horsetails and scouringrushes	693
Euphorbiaceae		698

Species	Common name	Page
Croton setigerus (= *Eremocarpus setigerus*)	turkey mullein	698
Euphorbia spp.	spurges	701, 710, 717
Ricinus communis	castorbean	728
Sapium sebiferum	Chinese tallowtree	731
Fabaceae		734
Acacia spp.	kangaroothorn, acacias, and wattles	734
Alhagi maurorum	camelthorn	744
Astragalus spp.	locos and milkvetches	748
Cytisus spp.	brooms	752
Genista monspessulana	French broom	752
Glycyrrhiza lepidota	wild licorice	763
Lathyrus spp.	peavines and sweetpeas	766
Lotus spp.	trefoils and Spanish clover	770
Lupinus arboreus	yellow bush lupine	774
Medicago spp.	black medic, burclovers, and alfalfa	776
Melilotus spp.	sweetclovers	784
Prosopis spp.	mesquites	790
Robinia pseudoacacia	black locust	794
Sesbania spp.	sesbanias	797
Spartium junceum	Spanish broom	752
Sphaerophysa salsula	swainsonpea	802
Trifolium spp.	clovers	805
Ulex europaeus	gorse	820
Vicia spp.	vetches	823
Halimodendron halodendron	Russian salttree	746
Ononis alopecuroides	foxtail restharrow	782
Retama monosperma	bridal broom	761
Senna obtusifolia	sicklepod	800
Volume 2		
Geraniaceae		835
Erodium spp.	filarees	835
Geranium spp.	geraniums, little robin, and herb-robert	845
Iridaceae		856
Iris spp.	irises	856
Juglandaceae		1502
Juglans californica	California black walnut	1502
Juncaceae		862
Juncus spp.	rushes	862
Lamiaceae		868
Lamium spp.	henbit and deadnettles	868
Marrubium vulgare	white horehound	873
Mentha spp.	pennyroyal and mints	876
Prunella vulgaris	healall	882
Salvia spp.	sages	885
Trichostema lanceolatum	vinegarweed	880
Liliaceae		890
Allium spp.	leeks, onions, and garlics	890

List of Species

Species	Common name	Page
Asphodelus fistulosus	onionweed	897
Nothoscordum gracile	false garlic	900
Zigadenus spp.	deathcamases	903
Veratrum spp.	false-hellebores	906
Lythraceae		908
Lythrum spp.	loosestrifes	908
Malvaceae		915
Abutilon theophrasti	velvetleaf	915
Malva spp.	mallows	920
Malvella leprosa	alkali mallow	931
Hibiscus trionum	Venice mallow	919
Lavatera cretica	Cornish mallow	929
Modiola caroliniana	bristly mallow	929
Sida rhombifolia	arrowleaf sida	934
Martyniaceae		935
Proboscidea spp.	devil's-claws	935
Molluginaceae		940
Mollugo verticillata	carpetweed	940
Glinus lotoides	lotus sweetjuice	942
Moraceae		943
Ficus carica	fig	943
Maclura pomifera	Osage-orange	946
Morus alba	white mulberry	947
Myoporaceae		948
Myoporum laetum	myoporum	948
Myrtaceae		951
Eucalyptus globulus	Tasmanian blue gum	951
Oleaceae		692
Ligustrum lucidum	glossy privet	692
Olea europaea	olive	692
Onagraceae		955
Epilobium spp.	willowherbs and fireweed	955
Gaura spp.	gauras	962
Oenothera spp.	evening-primroses	971
Orobanchaceae		978
Orobanche spp.	broomrapes	978
Oxalidaceae		981
Oxalis spp.	woodsorrels and oxalises	981
Papaveraceae [= Fumariaceae]		989
Fumaria spp.	fumitorys	989
Phytolaccaceae		992
Phytolacca americana	common pokeweed	992
Plantaginaceae		995
Plantago spp.	plantains	995
Poaceae		1004
Achnatherum brachychaetum	punagrass	1004
Aegilops spp.	goatgrasses	1008
Agrostis spp.	bentgrasses	1016

List of Species

Species	Common name	Page
Alopecurus spp.	meadow foxtail and blackgrass	1021
Ammophila spp.	beachgrasses	1027
Anthoxanthum spp.	vernalgrasses	1031
Arundo donax	giant reed	1034
Avena spp.	oats	1040
Brachypodium distachyon	annual false-brome	1046
Briza spp.	quakinggrasses	1049
Bromus spp.	rescuegrass, cheat, and bromes	1053, 1056, 1065
Cenchrus spp.	sandburs	1078
Chloris spp.	fingergrasses, rhodesgrass, windmill-grasses	1083
Cortaderia spp.	jubatagrass and pampasgrass	1090
Crypsis spp.	pricklegrasses	1021
Cynodon spp.	bermudagrasses and stargrass	1098
Cynosurus spp.	dogtailgrasses	1105
Dactylis glomerata	orchardgrass	1108
Digitaria spp.	crabgrasses	1112
Echinochloa spp.	junglerice, barnyardgrasses, and watergrasses	1118
Ehrharta spp.	veldtgrasses	1129
Eleusine spp.	goosegrasses	1136
Elytrigia spp.	quackgrass and wheatgrasses	1140
Eragrostis spp.	stinkgrass and lovegrasses	1146
Eriochloa spp.	cupgrasses	1162
Festuca spp.	fescues	1167
Gastridium phleoides	nitgrass	1171
Heteropogon contortus	tanglehead	1176
Holcus spp.	velvetgrasses	1179
Hordeum spp.	barleys	1182
Koeleria phleoides	annual junegrass	1171
Leptochloa spp.	sprangletops	1190
Lolium spp.	ryegrasses	1195
Muhlenbergia schreberi	nimblewill	1201
Panicum spp.	panicgrass, witchgrass, panicum, and millet	1203
Paspalum spp.	dallisgrass, knotgrass, bahiagrass, and vaseygrass	1211
Pennisetum spp.	kikuyugrass, buffelgrass, fountaingrass, and feathertop	1219
Phalaris spp.	hardinggrass and canarygrasses	1230
Piptatherum miliaceum	smilograss	1243
Poa spp.	bluegrasses	1246, 1250, 1253
Polypogon spp.	polypogons and water-bent	1262
Schismus spp.	mediterraneangrasses	1270
Setaria spp.	foxtails and brittlegrass	1275
Sorghum spp.	shattercane and johnsongrass	1288
Spartina spp.	cordgrasses	1297

List of Species

Species	Common name	Page
Sporobolus spp.	smutgrass and dropseed	1306
Taeniatherum caput-medusae	medusahead	1310
Vulpia spp.	fescues	1314
Axonopus fissifolius	carpetgrass	1117
Dactyloctenium aegyptium	crowfootgrass	1088
Distichlis spicata	saltgrass	1104
Imperata brevifolia	satintail	1228
Miscanthus sinensis	eulaliagrass	1095
Pascopyrum smithii	western wheatgrass	1144
Phragmites australis	common reed	1038
Saccharum ravennae	ravennagrass	1039, 1095
Sclerochloa dura	hardgrass	1273
Secale cereale	cereal rye	1188
Stenotaphrum secundatum	St. Augustinegrass	1229
Stipa capensis	Mediterranean steppegrass	1006
Polygonaceae		1319
Polygonum spp.	ladysthumb, knotweeds, and smartweeds	1319, 1331, 1337, 1340
Rumex spp.	sorrels and docks	1347, 1351
Portulacaceae		1363
Calandrinia ciliata	redmaids	1363
Claytonia spp.	miner's lettuces	1366
Portulaca oleracea	common purslane	1370
Primulaceae		1377
Anagallis arvensis	scarlet pimpernel	1377
Ranunculaceae		1381
Delphinium spp.	larkspurs	1381
Ranunculus spp.	buttercups	1385
Rosaceae		1400
Acaena spp.	biddy-biddys	1400
Cotoneaster spp.	cotoneasters	1404
Crataegus monogyna	English hawthorn	1409
Potentilla spp.	cinquefoils	1413
Rosa spp.	roses	1421
Rubus spp.	blackberries, raspberries, and thimbleberry	1426
Prunus cerasifera	cherry plum	1412
Prunus ilicifolia	holly-leafed cherry	1:155
Pyracantha spp.	pyracanthas and firethorns	1407
Sanguisorba minor ssp. *muricata*	salad burnet	1420
Rubiaceae		1435
Galium spp.	bedstraws	1435
Sherardia arvensis	field madder	1435
Salicaceae		1447
Salix spp.	willows	1447
Populus alba	white poplar	1451
Scrophulariaceae		1453
Bellardia trixago	bellardia	1453

List of Species

Species	Common name	Page
Digitalis purpurea	foxglove	1456
Kickxia spp.	fluvellins	1460
Linaria spp.	toadflaxes	1463
Parentucellia spp.	glandweeds	1472
Verbascum spp.	mulleins	1475
Veronica spp.	speedwells	1482, 1486
Simaroubaceae		1499
Ailanthus altissima	tree-of-heaven	1499
Solanaceae		1503
Datura spp.	thornapples and jimsonweed	1503
Hyoscyamus niger	black henbane	1510
Nicotiana spp.	tobaccos	1513
Physalis spp.	groundcherrys	1518
Solanum spp.	nightshades, buffalobur, and horsenettles	1528, 1538
Tamaricaceae		1554
Tamarix spp.	tamarisks and saltcedar	1554
Urticaceae		1562
Urtica spp.	nettles	1562
Parietaria judaica	spreading pellatory	1567
Verbenaceae		1569
Phyla nodiflora	mat lippia	1569
Verbena spp.	vervains	1572
Violaceae		1577
Viola spp.	violets and pansy	1577
Viscaceae		1581
Arceuthobium spp.	dwarf mistletoes	1581
Phoradendron spp.	mistletoes	1581
Viscum album	European mistletoe	1581
Zygophyllaceae		1589
Peganum harmala	African rue	1589
Tribulus terrestris	puncturevine	1592
Zygophyllum fabago	Syrian beancaper	1596

Shortcut Identification Tables

Table 1. Parasitic plants

Species	Common name	Family	Vol:Page
Cuscuta spp.	dodders	Cuscutaceae	1:60
Orobanche spp.	broomrapes	Orobanchaceae	2:97
Arceuthobium spp.	dwarf mistletoes	Viscaceae	2:158
Phoradendron spp.	mistletoes	Viscaceae	2:158
Viscum album	European mistletoe	Viscaceae	2:158

Table 2. Succulent plants

Species	Common name	Family	Vol:Page
Aptenia cordifolia	baby sun rose	Aizoaceae	1:6
Carpobrotus spp.	hottentot-fig, sea-fig	Aizoaceae	1:5
Conicosia pugioniformis	narrowleaf iceplant	Aizoaceae	1:6
Drosanthermum floribundum	showy dewflower	Aizoaceae	1:6
Malephora crocea	coppery mesembryanthemum	Aizoaceae	1:6
Mesembryanthemum spp.	iceplants	Aizoaceae	1:6
Trianthema portulacastrum	horse-purslane	Aizoaceae	2:137
Cotula coronopifolia	brassbuttons	Asteraceae	1:29
Heliotropium curassavicum	seaside heliotrope	Boraginaceae	1:44
Bassia hyssopifolia (young plant)	fivehook bassia	Chenopodiaceae	1:59
Halogeton glomeratus	halogeton	Chenopodiaceae	1:61
Kochia scoparia (young plant)	kochia	Chenopodiaceae	1:59
Monolepis nuttalliana	Nuttall povertyweed	Chenopodiaceae	1:61
Salsola soda	glasswort	Chenopodiaceae	1:63
Portulaca oleracea	common purslane	Portulacaceae	2:137
Zygophyllum fabago	Syrian beancaper	Zygophyllaceae	2:159

Table 3. Plants with papery sheath (ocrea) at base of leaf (not part of the petiole)

Species	Common name	Family	Vol:Page
Spergula arvensis	corn spurry	Caryophyllaceae	1:50
Spergularia spp.	sandspurrys	Caryophyllaceae	1:57
Equisetum spp.	horsetails and scouringrushes	Equisetaceae	1:69
Polygonum spp.	smartweeds, knotweeds	Polygonaceae	2:131 2:133 2:133 2:134
Rumex spp.	docks, sorrels	Polygonaceae	2:134 2:135

Table 4. Plants very sticky (glandular) to the touch

Species	Common name	Family	Vol:Page
Ageratina adenophora (stems)	croftonweed	Asteraceae	1:179
Baccharis salicifolia	seepwillow	Asteraceae	2:1452
Conyza coulteri	Coulter's conyza	Asteraceae	1:288
Dittrichia graveolens	stinkwort	Asteraceae	1:350
Grindelia spp.	gumweeds, gumplants	Asteraceae	1:325
Hemizonia fitchii	Fitch's tarweed	Asteraceae	1:339
Hemizonia kelloggii	Kellogg's tarweed	Asteraceae	1:339, 1:350
Hemizonia lobbii	threeray tarweed	Asteraceae	1:350
Heterotheca grandiflora	telegraphplant	Asteraceae	1:340
Holocarpha virgata	virgate tarweed	Asteraceae	1:347
Onopordum tauricum	Taurian thistle	Asteraceae	1:372
Senecio elegans	purple ragwort	Asteraceae	1:389
Cerastium glomeratum	sticky chickweed	Caryophyllaceae	1:555
Silene conoidea	cone catchfly	Caryophyllaceae	1:584
Silene latifolia ssp. alba	white campion	Caryophyllaceae	1:582
Silene noctiflora	nightflowering catchfly	Caryophyllaceae	1:582
Spergula arvensis	corn spurry	Caryophyllaceae	1:567
Spergularia spp.	sandspurrys	Caryophyllaceae	1:572
Chenopodium ambrosioides	Mexicantea	Chenopodiaceae	1:612
Chenopodium botrys	Jerusalem-oak goosefoot	Chenopodiaceae	1:616
Chenopodium multifidum	cutleaf goosefoot	Chenopodiaceae	1:617
Chenopodium pumilio	clammy goosefoot	Chenopodiaceae	1:617
Glycyrrhiza lepidota	wild licorice	Fabaceae	1:763
Trichostema lanceolatum	vinegarweed	Lamiaceae	2:880
Proboscidea spp.	devil's-claws	Martyniaceae	2:935
Myoporum laetum (young leaves)	myoporum	Myoporaceae	2:948
Gaura parviflora	small-flowered gaura	Onagraceae	2:962
Oenothera spp. (most species)	evening-primroses	Onagraceae	2:971
Orobanche spp.	broomrapes	Orobanchaceae	2:978
Rosa canina (occasionally)	dog rose	Rosaceae	2:1421
Rosa eglanteria	sweetbriar rose	Rosaceae	2:1421
Rubus parviflora (inflorescence)	western thimbleberry	Rosaceae	2:1426
Bellardia trixago (inflorescence)	bellardia	Scrophulariaceae	2:1453
Linaria maroccana (inflorescence slightly)	Morocco toadflax	Scrophulariaceae	2:1463
Parentucellia spp.	glandweeds	Scrophulariaceae	2:1472
Verbascum blattaria (inflorescence)	moth mullein	Scrophulariaceae	2:1475
Verbascum virgatum (inflorescence)	purplestamen mullein	Scrophulariaceae	2:1475

Table 4. Plants very sticky (glandular) to the touch, cont.

Species	Common name	Family	Vol:P
Veronica perigrina (slightly)	western purslane speedwell	Scrophulariaceae	2:14
Veronica serpyllifolia (inflorescence slightly)	thymeleaf speedwell	Scrophulariaceae	2:14
Nicotiana acuminata var. multiflora	manyflower tobacco	Solanaceae	2:15
Nicotiana quadrivalvis	Indian tobacco	Solanaceae	2:15
Solanum marginatum	white-margined nightshade	Solanaceae	2:15
Solanum physalifolium	hairy nightshade	Solanaceae	2:15

Table 5. Plants with square stems

Species	Common name	Family	Vol:P
Bidens frondosa	devils beggarticks	Asteraceae	1:
Bidens pilosa	hairy beggarticks	Asteraceae	1:
Lamium spp.	henbit, purple deadnettle	Lamiaceae	2:
Marrubium vulgare	white horehound	Lamiaceae	2:
Mentha spp.	pennyroyal, mints	Lamiaceae	2:
Prunella vulgaris	healall	Lamiaceae	2:
Salvia spp.	sages	Lamiaceae	2:
Trichostema lanceolatum	vinegarweed	Lamiaceae	2:
Lythrum spp.	loosestrifes	Lythraceae	2:
Anagallis arvensis	scarlet pimpernel	Primulaceae	2:1
Galium spp. (young stage)	bedstraws	Rubiaceae	2:1
Parentucellia spp.	glandweeds	Scrophulariaceae	2:1
Parietaria judaica	spreading pellatory	Urticaceae	2:1
Urtica spp.	nettles	Urticaceae	2:1
Phyla nodiflora	mat lippia	Verbenaceae	2:1
Verbena spp.	vervains	Verbenaceae	2:1

Table 6. Plants climbing on taller vegetation (viny) with (*) and without tendrils

Species	Common name	Family	Vol:P
Vinca spp.	periwinkles	Apocynaceae	1:
Hedera spp.	ivys	Araliaceae	1:
Araujia sericifera	bladderflower	Asclepiadaceae	1:
Delairea odorata	Cape-ivy	Asteraceae	1:
Calystegia occidentalis	western morningglory	Convolvulaceae	1:
Convolvulus spp.	bindweeds	Convolvulaceae	1:
Dichondra repens	dichondra	Convolvulaceae	1:
Ipomoea spp.	morningglorys	Convolvulaceae	1:
Citrullus lanatus var. citroides*	citronmelon	Cucurbitaceae	1:
Cucumis spp.*	smellmelon, paddymelon	Cucurbitaceae	1:
Cucurbita foetidissima*	buffalo gourd	Cucurbitaceae	1:
Marah spp.*	manroots, wild cucumbers	Cucurbitaceae	1:307, 1:
Cuscuta spp.	dodders	Cuscutaceae	1:

Shortcut Identification Tables

ble 6. Plants climbing on taller vegetation (viny) with (*) and without tendrils, cont.

ecies	Common name	Family	Vol:Page
thyrus spp*	peavines, sweetpeas	Fabaceae	1:766
cia spp.*	vetches	Fabaceae	1:823
maria spp.	fumitorys	Papaveraceae	2:989
lygonum convolvulus	wild buckwheat	Polygonaceae	2:1337
sa multiflora	multiflora rose	Rosaceae	2:1421
bus armeniacus (= R. discolor)	Himalaya blackberry	Rosaceae	2:1426
bus laciniatus	cutleaf blackberry	Rosaceae	2:1426
lium aparine	catchweed bedstraw	Rubiaceae	2:1435
lanum dulcamara	bittersweet nightshade	Solanaceae	2:1528
lanum sisymbriifolium	sticky nightshade	Solanaceae	2:1538

ble 7. Plants with stellate, branched, or forked hairs

ecies	Common name	Family	Vol:Page
edera spp.	ivys	Araliaceae	1:156
abidopsis thaliana	mouse-ear cress	Brassicaceae	1:530
psella bursa-pastoris	shepherd's-purse	Brassicaceae	1:471
escurainia sophia	flixweed	Brassicaceae	1:493
alcolmia africana	Malcolm stock	Brassicaceae	1:488
bara virginica	sibara	Brassicaceae	1:527
roton setigerus (= Eremocarpus setigerus)	turkey mullein	Euphorbiaceae	1:698
entha suaveolens (lower leaf surface)	apple mint	Lamiaceae	2:876
butilon theophrasti	velvetleaf	Malvaceae	2:915
avatera cretica	Cornish mallow	Malvaceae	2:929
alva spp.	mallows	Malvaceae	2:920
alvella leprosa	alkali mallow	Malvaceae	2:931
rbascum speciosum	showy mullein	Scrophulariaceae	2:1475
rbascum thapsus	common mullein	Scrophulariaceae	2:1475
ysalis viscosa	grape groundcherry	Solanaceae	2:1518
lanum carolinense	horsenettle	Solanaceae	2:1538
lanum dimidiatum	robust nightshade	Solanaceae	2:1538
lanum elaeagnifolium	silverleaf nightshade	Solanaceae	2:1538
lanum lanceolatum	lanceleaf nightshade	Solanaceae	2:1538
lanum marginatum	white-margined nightshade	Solanaceae	2:1538
lanum mauritianum	woolly nightshade	Solanaceae	2:1538
lanum rostratum	buffalobur	Solanaceae	2:1538
lanum sisymbriifolium	sticky nightshade	Solanaceae	2:1538
hyla nodiflora (hairs forked)	mat lippia	Verbenaceae	2:1569

Table 8. Plants with leaves whorled or appearing whorled (*) at nodes

Species	Common name	Family	Vol:P
Asclepias fascicularis	Mexican whorled milkweed	Asclepiadaceae	1:1
Polycarpum tetraphyllum*	four-leaved polycarp	Caryophyllaceae	1:5
Spergula arvensis*	corn spurry	Caryophyllaceae	1:5
Spergularia spp.*	sandspurrys	Caryophyllaceae	1:5
Equisetum arvense (stem branches in a whorl)	field horsetail	Equisetaceae	1:6
Equisetum telmateia ssp. braunii (stem branches in a whorl)	giant horsetail	Equisetaceae	1:6
Acacia verticillata	star acacia	Fabaceae	1:7
Lythrum salicaria*	purple loosestrife	Lythraceae	2:9
Mollugo verticillata	carpetweed	Molluginaceae	2:9
Plantago indica*	psyllium	Plantaginaceae	2:9
Galium spp.	bedstraws	Rubiaceae	2:14
Sherardia arvensis	field madder	Rubiaceae	2:14
Linaria vulgaris*	yellow toadflax	Scrophulariaceae	2:14

Table 9. Plants that exude milky latex

Species	Common name	Family	Vol:P
Vinca spp.	periwinkles	Apocynaceae	1:1
Araujia sericifera	bladderflower	Asclepiadaceae	1:1
Asclepias spp.	milkweeds	Asclepiadaceae	1:1
Chondrilla juncea	rush skeletonweed	Asteraceae	1:2
Cichorium intybus	chicory	Asteraceae	1:2
Crepis spp.	hawksbeards	Asteraceae	1:2
Hieracium spp.	hawkweeds	Asteraceae	1:3
Hypochaeris spp.	catsears	Asteraceae	1:3
Lactuca spp.	lettuces	Asteraceae	1:3
Leontodon taraxacoides	lesser hawkbit	Asteraceae	1:4
Picris echioides	bristly oxtongue	Asteraceae	1:3
Scolymus hispanicus	golden thistle	Asteraceae	1:2
Sonchus spp.	sowthistles	Asteraceae	1:3
Stephanomeria exigua	small wirelettuce	Asteraceae	1:4
Taraxacum officinale	dandelion	Asteraceae	1:4
Tragopogon spp.	salsifys	Asteraceae	1:4
Urospermum picroides	prickly goldenfleece	Asteraceae	1:3
Euphorbia spp.	spurges	Euphorbiaceae	1:7, 1:710, 1:7
Sapium sebiferum	Chinese tallowtree	Euphorbiaceae	1:7
Ficus carica	fig	Moraceae	2:9
Maclura pomifera	Osage-orange	Moraceae	2:9
Morus alba	white mulberry	Moraceae	2:9

Table 10. Woody shrubs and trees

Species	Common name	Family	Shrub or Tree	Vol:Page
...us trilobata	skunkbrush	Anacardiaceae	Shrub	1:109
...hinus spp.	peppertrees	Anacardiaceae	Tree	1:100
...xicodendron diversilobum	Pacific poison-oak	Anacardiaceae	Shrub	1:105
...nca spp.	periwinkles	Apocynaceae	Shrub	1:151
...x aquifolium	English holly	Aquifoliaceae	Tree	1:154
...dera spp.	ivys	Araliaceae	Shrub	1:156
...oenix canariensis	Canary Island date palm	Arecidaceae	Shrublike	1:161
...ashingtonia robusta	Mexican fan palm	Arecidaceae	Shrublike	1:161
...temisia tridentata	big sagebrush	Asteraceae	Shrub	1:199
...accharis salicifolia	seepwillow	Asteraceae	Shrub	2:1452
...hium candicans	pride of Madeira	Boraginaceae	Shrub	1:434
...hium pininana	pine echium	Boraginaceae	Shrub	1:434
...riplex semibaccata	Australian saltbush	Chenopodiaceae	Shrub	1:586
...lsola vermiculata	Mediterranean saltwort	Chenopodiaceae	Shrub	1:631
...ypericum canariense	Canary Island hypericum	Clusiaceae (= Hypericaceae)	Shrub	1:637
...niperus occidentalis var. cidentalis	western juniper	Cupressaceae	Tree	1:660
...aeagnus angustifolia	Russian-olive	Elaeagnaceae	Shrub/tree	1:689
...cinus communis	castorbean	Euphorbiaceae	Shrub	1:728
...pium sebiferum	Chinese tallowtree	Euphorbiaceae	Tree	1:731
...cacia spp.	wattles, acacias, kangaroothorn	Fabaceae	Shrub/tree	1:734
...hagi maurorum (= A. pseudalhagi)	camelthorn	Fabaceae	Shrub	1:744
...ytisus spp.	brooms	Fabaceae	Shrub	1:752
...enista monspessulana	French broom	Fabaceae	Shrub	1:752
...alimodendron halodendron	Russian salttree	Fabaceae	Shrub	1:746
...upinus arboreus	yellow bush lupine	Fabaceae	Shrub	1:774
...rosopis spp.	mesquites	Fabaceae	Shrub/tree	1:790
...etama monosperma (= Genista monosperma)	bridal broom	Fabaceae	Shrub	1:761
...obinia pseudoacacia	black locust	Fabaceae	Tree	1:794
...esbania punicea	red sesbania	Fabaceae	Shrub	1:797
...partium junceum	Spanish broom	Fabaceae	Shrub	1:752
...lex europaeus	gorse	Fabaceae	Shrub	1:820
...icus carica	fig	Moraceae	Tree	2:943
...aclura pomifera	Osage-orange	Moraceae	Tree	2:946
...orus alba	white mulberry	Moraceae	Tree	2:947

Table 10. Woody shrubs and trees, cont.

Species	Common name	Family	Shrub or Tree	Vol:
Myoporum laetum	myoporum	Myoporaceae	Shrub/tree	2:
Eucalyptus globulus	Tasmanian blue gum	Myrtaceae	Tree	2:
Ligustrum lucidum	glossy privet	Oleaceae	Tree	1:
Olea europaea	olive	Oleaceae	Shrub/Tree	1.
Phytolacca americana	common pokeweed	Phytolaccaceae	Shrub	2:
Polygonum cuspidatum	Japanese knotweed	Polygonaceae	Shrub	2:1
Polygonum polystachyum	Himalayan knotweed	Polygonaceae	Shrub	2:1
Polygonum sachalinense	Sakhalin knotweed	Polygonaceae	Shrub	2:1
Cotoneaster spp.	cotoneasters	Rosaceae	Shrub	2:1
Crataegus monogyna	English hawthorn	Rosaceae	Tree	2:1
Prunus cerasifera	cherry plum	Rosaceae	Shrub/tree	2:1
Pyracantha spp.	pyracanthas, firethorns	Rosaceae	Shrub	2:1
Rosa spp.	roses	Rosaceae	Shrub	2:1
Rubus spp.	blackberries, thimbleberry, raspberry	Rosaceae	Shrub	2:1
Populus alba	white poplar	Salicaceae	Tree	2:1
Salix spp.	willows	Salicaceae	Shrub/Tree	2:1
Ailanthus altissima	tree-of-heaven	Simaroubaceae	Tree	2:1
Nicotiana glauca	tree tobacco	Solanaceae	Shrub	2:1
Solanum lanceolatum	lanceleaf nightshade	Solanaceae	Shrub	2:1
Solanum marginatum	white-margined nightshade	Solanaceae	Shrub	2:1
Solanum mauritianum	woolly nightshade	Solanaceae	Shrub/Tree	2:1
Tamarix spp.	tamarisks, saltcedar	Tamaricaceae	Shrub/tree	2:1

Table 11. Emergent aquatic plants

Species	Common name	Family	Vol:P
Alternanthera philoxeroides	alligatorweed	Amaranthaceae	1
Cicuta spp.	waterhemlocks	Apiaceae	1:
Hydrocotyle ranunculoides	floating pennywort	Apiaceae	1:
Cotula coronopifolia	brassbuttons	Asteraceae	1:
Lepidium latifolium	perennial pepperweed	Brassicaceae	1:
Equisetum spp.	horsetails, scouringrushes	Equisetaceae	1:0
Sesbania punicea	rattlebush	Fabaceae	1:
Iris pseudacorus	yellowflag iris	Iridaceae	2:8
Juncus spp.	rushes	Juncaceae	2:8
Mentha pulegium	pennyroyal	Lamiaceae	2:8
Lythrum hyssopifolia	hyssop loosestrife	Lythraceae	2:9
Lythrum salicaria	purple loosestrife	Lythraceae	2:9
Lythrum tribracteatum	threebract loosestrife	Lythraceae	2:9

Shortcut Identification Tables

e 11. Emergent aquatic plants, cont.

:ies	Common name	Family	Vol:Page
obium ciliatum	fringed willowherb	Onagraceae	2:955
ndo donax	giant reed	Poaceae	2:1034
nochloa spp.	junglerice, barnyardgrasses, watergrasses	Poaceae	2:1118
ochloa spp.	sprangletops	Poaceae	2:1190
alum distichum	knotgrass	Poaceae	2:1211
aris aquatica	hardinggrass	Poaceae	2:1230
aris arundinacea	reed canarygrass	Poaceae	2:1230
gmites australis	common reed	Poaceae	2:1038
pogon monspeliensis	rabbitfoot polypogon	Poaceae	2:1262
rtina spp.	cordgrasses	Poaceae	2:1297
gonum spp. (many species)	smartweeds, ladysthumb	Polygonaceae	2:1319 2:1340
ex spp. (many species)	docks	Polygonaceae	2:1351
unculus flammula	greater creeping spearwort	Ranunculaceae	2:1385
unculus muricatus	roughseed buttercup	Ranunculaceae	2:1385
unculus repens	creeping buttercup	Ranunculaceae	2:1385
unculus sceleratus	crowfoot buttercup	Ranunculaceae	2:1385
x spp. (many species)	willows	Salicaceae	2:1447
nica anagallis-aquatica	water speedwell	Scrophulariaceae	2:1482
nica beccabunga	European speedwell	Scrophulariaceae	2:1482
nica catenata	chain speedwell	Scrophulariaceae	2:1482
nica serpyllifolia	thymeleaf speedwell	Scrophulariaceae	2:1486
arix spp.	tamarisks, saltcedar	Tamaricaceae	2:1554

e 12. Plants with prickles, spines, or thorns on leaves, stems, fruits, flower heads, tiple parts, or entire plant

ines or prickles on leaves only

cies	Common name	Family	Vol:Page
aquifolium	English holly	Aquifoliaceae	1:154
enix canariensis	Canary Island date palm	Arecidaceae	1:161
shingtonia robusta	Mexican fan palm	Arecidaceae	1:161
tuca saligna	willowleaf lettuce	Asteraceae	1:360
chus asper	spiny sowthistle	Asteraceae	1:398
ogeton glomeratus	halogeton	Chenopodiaceae	1:618
sola paulsenii	barbwire Russian thistle	Chenopodiaceae	1:622
sola tragus	Russian thistle, tumbleweed	Chenopodiaceae	1:622
cia verticillata	star acacia	Fabaceae	1:740
x europaeus	gorse	Fabaceae	1:820

Shortcut Identification Tables

Table 12. Plants with prickles, spines, or thorns on leaves, stems, fruits, flower heads, multiple parts, or entire plant, cont.

B. Prickles, stinging hairs, or thorns on stem only

Species	Common name	Family	Vol:P
Acacia farnesiana	sweet acacia	Fabaceae	1:
Acacia pardoxa	Kangaroothorn	Fabaceae	1:
Alhagi maurorum (= *A. pseudalhagi*)	camelthorn	Fabaceae	1:
Halimodendron halodendron	Russian salttree	Fabaceae	1:
Prosopis spp.	mesquites	Fabaceae	1:
Robinia pseudoacacia	black locust	Fabaceae	1:
Abutilon theophrasti	velvetleaf	Malvaceae	2:
Maclura pomifera	Osage-orange	Moraceae	2:
Crataegus monogyna	English hawthorn	Rosaceae	2:1
Pyracantha spp.	pyracanthas, firethorns	Rosaceae	2:1
Rosa spp.	roses	Rosaceae	2:1
Rubus armeniacus (= *R. discolor*)	Himalaya blackberry	Rosaceae	2:1
Rubus laciniatus	cutleaf blackberry	Rosaceae	2:1
Rubus leucodermis	western raspberry	Rosaceae	2:1
Rubus ursinus	Pacific blackberry	Rosaceae	2:1
Solanum lanceolatum (occasionally on stems and leaves)	lanceleaf nightshade	Solanaceae	2:1
Urtica urens	burning nettle	Urticaceae	2:1

C. Prickles and sharp structures on fruit only

Species	Common name	Family	Vol:P
Daucus spp.	carrots	Apiaceae	1:
Torilis spp.	hedgeparsleys	Apiaceae	1:
Ambrosia acanthicarpa	annual bursage	Asteraceae	1:
Ambrosia trifida	giant ragweed	Asteraceae	1:
Bidens spp.	beggarticks	Asteraceae	1:
Soliva sessilis	lawn burweed	Asteraceae	1:
Xanthium strumarium	common cocklebur	Asteraceae	1:
Cynoglossum officinale	houndstongue	Boraginaceae	1:
Lappula squarrosa	European sticktight	Boraginaceae	1:
Bassia hyssopifolia	fivehook bassia	Chenopodiaceae	1:
Cucumis myriocarpus	paddymelon	Cucurbitaceae	1:0
Marah spp.	manroots, wild cucumbers	Cucurbitaceae	1:307, 1:
Ricinus communis	castorbean	Euphorbiaceae	1:
Glycyrrhiza lepidota	wild licorice	Fabaceae	1:
Medicago arabica	spotted burclover	Fabaceae	1:
Medicago polymorpha	California burclover	Fabaceae	1:
Proboscidea spp.	devil's-claws	Martyniaceae	2:9
Cenchrus spp.	sandburs	Poaceae	2:10
Ranunculus arvensis	corn buttercup	Ranunculaceae	2:1

Shortcut Identification Tables

Table 12. Plants with prickles, spines, or thorns on leaves, stems, fruits, flower heads, multiple parts, or entire plant, cont.

Species	Common name	Family	Vol:Page
Ranunculus testiculatus	bur buttercup	Ranunculaceae	2:1385
Ranunculus muricatus	roughseed buttercup	Ranunculaceae	2:1385
Acaena spp.	biddy-biddys	Rosaceae	2:1400
Galium aparine	catchweed bedstraw	Rubiaceae	2:1435
Galium parisiense	wall bedstraw	Rubiaceae	2:1435
Datura spp.	thornapples, jimsonweed	Solanaceae	2:1503
Tribulus terrestris	puncturevine	Zygophyllaceae	2:1592

Prickles and sharp structures on flower head or inflorescence only

Species	Common name	Family	Vol:Page
Amaranthus spp.	pigweeds, amaranths	Amaranthaceae	1:78, 1:90
Centaurea calcitrapa	purple starthistle	Asteraceae	1:235
Centaurea diffusa	diffuse knapweed	Asteraceae	1:235
Centaurea iberica	Iberian starthistle	Asteraceae	1:235
Centaurea melitensis	Malta starthistle	Asteraceae	1:250
Centaurea solstitialis	yellow starthistle	Asteraceae	1:250
Centaurea virgata ssp. squarrosa	squarrose knapweed	Asteraceae	1:235
Centaurea sulphurea	Sicilian starthistle	Asteraceae	1:250
Xanthium strumarium	common cocklebur	Asteraceae	1:420

Prickles, spines, or thorns on more than one plant part or entire plant

Species	Common name	Family	Vol:Page
Carduus spp.	thistles	Asteraceae	1:217
Carthamus spp.	distaff thistles	Asteraceae	1:228
Cirsium spp.	thistles	Asteraceae	1:272
Cnicus benedictus	blessed thistle	Asteraceae	1:233
Cynara cardunculus	artichoke thistle	Asteraceae	1:302
Hemizonia fitchii	Fitch's tarweed	Asteraceae	1:339
Hemizonia parryi	Parry's tarweed	Asteraceae	1:339
Hemizonia pungens	spikeweed	Asteraceae	1:335
Lactuca serriola	prickly lettuce	Asteraceae	1:360
Lactuca virosa	bitter lettuce	Asteraceae	1:360
Onopordum spp.	thistles	Asteraceae	1:372
Picris echioides	bristly oxtongue	Asteraceae	1:379
Scolymus hispanicus	golden thistle	Asteraceae	1:234
Silybum marianum	blessed milkthistle	Asteraceae	1:391
Urospermum picroides	prickly goldenfleece	Asteraceae	1:381
Xanthium spinosum	spiny cocklebur	Asteraceae	1:420
Amsinckia menziesii	fiddlenecks	Boraginaceae	1:426
Borago officinalis	common borage	Boraginaceae	1:453
Symphytum spp.	comfreys	Boraginaceae	1:450

Shortcut Identification Tables

Table 12. Plants with prickles, spines, or thorns on leaves, stems, fruits, flower heads, multiple parts, or entire plant, cont.

Species	Common name	Family	Vol:P
Salsola paulsenii	barbwire Russian thistle	Chenopodiaceae	1:
Salsola tragus	Russian thistle	Chenopodiaceae	1:
Dipsacus spp.	teasels	Dipsacaceae	1:
Ulex europaeus	gorse	Fabaceae	1:
Rubus spp. (many species)	blackberries, raspberry, thimbleberry	Rosaceae	2:14
Solanum carolinense	horsenettle	Solanaceae	2:1
Solanum dimidiatum	robust horsenettle	Solanaceae	2:1
Solanum elaeagnifolium	silverleaf nightshade	Solanaceae	2:1
Solanum lanceolatum	lanceleaf nightshade	Solanaceae	2:1
Solanum marginatum	white-margined nightshade	Solanaceae	2:1
Solanum mauritianum	woolly nightshade	Solanaceae	2:1
Solanum rostratum	buffalobur	Solanaceae	2:1
Solanum sisymbriifolium	sticky nightshade	Solanaceae	2:1
Urtica spp.	nettles	Urticaceae	2:1

Table 13. Plants with various types of compound or dissected leaves

A. Two leaflets

Species	Common name	Family	Vol:P
Lathyrus spp.	peavines, sweetpeas	Fabaceae	1:7
Zygophyllum fabago	Syrian beancaper	Zygophyllaceae	2:15

B. Three leaflets (trifoliolate)

Species	Common name	Family	Vol:P
Rhus trilobata	skunkbrush	Anacardiaceae	1:
Toxicodendron diversilobum	Pacific poison-oak	Anacardiaceae	1:
Cytisus spp. (sometimes)	brooms	Fabaceae	1:7
Genista monspessulana	French broom	Fabaceae	1:7
Lotus spp. (can appear to have 5 with 2 stipules at base)	trefoils	Fabaceae	1:7
Medicago spp.	burclovers, black medic, alfalfa	Fabaceae	1:7
Melilotus spp.	sweetclovers	Fabaceae	1:7
Trifolium spp.	clovers	Fabaceae	1:8
Geranium robertianum	herb-robert	Geraniaceae	2:8
Oxalis spp.	woodsorrels, oxalises	Oxalidaceae	2:9
Ranunculus bulbosus	bulbous buttercup	Ranunculaceae	2:13
Ranunculus repens	creeping buttercup	Ranunculaceae	2:13
Potentilla norvegica	rough cinquefoil	Rosaceae	2:14
Rubus laciniatus	cutleaf blackberry	Rosaceae	2:14
Rubus leucodermis	western raspberry	Rosaceae	2:14
Rubus pensilvanicus	Pennsylvania blackberry	Rosaceae	2:14

Table 13. Plants with various types of compound or dissected leaves, cont.

Species	Common name	Family	Vol:Page
...bus ulmifolius	elmleaf blackberry	Rosaceae	2:1426
...bus ursinus	Pacific blackberry	Rosaceae	2:1426

Palmately compound leaves

Species	Common name	Family	Vol:Page
...shingtonia robusta (or appearing with age)	Mexican fan palm	Arecidaceae	1:161
...nnabis sativa	marijuana	Cannabaceae	1:552
...moea cairica	Cairo morningglory	Convolvulaceae	1:652
...pinus arboreus	yellow bush lupine	Fabaceae	1:774
...ranium purpureum	little robin	Geraniaceae	2:845
...ranium robertianum	herb-robert	Geraniaceae	2:845
...nunculus arvensis	corn buttercup	Ranunculaceae	2:1385
...nunculus sardous	hairy buttercup	Ranunculaceae	2:1385
...nunculus sceleratus	crowfoot buttercup	Ranunculaceae	2:1385
...tentilla recta	sulfur cinquefoil	Rosaceae	2:1413
...bus armeniacus (= R. discolor)	Himalaya blackberry	Rosaceae	2:1426
...bus pensilvanicus	Pennsylvania blackberry	Rosaceae	2:1426
...bus ulmifolius	elmleaf blackberry	Rosaceae	2:1426

Leaves once-pinnately compound or appearing once-pinnately compound

Species	Common name	Family	Vol:Page
...hinus spp.	peppertrees	Anacardiaceae	1:100
...stinaca sativa	wild parsnip	Apiaceae	1:131, 1:136
...rilis arvensis	hedgeparsley	Apiaceae	1:117
...oenix canariensis	Canary Island date palm	Arecidaceae	1:161
...nbrosia psilostachya	western ragweed	Asteraceae	1:188
...nthemis tinctoria	yellow chamomile	Asteraceae	1:192
...dens spp.	beggarticks	Asteraceae	1:211
...arthamus spp.	distaff thistles	Asteraceae	1:228
...entaurea spp.	starthistles, knapweeds	Asteraceae	1:235, 1:250
...otula australis	southern brassbuttons	Asteraceae	1:291
...otula mexicana	Mexican brassbuttons	Asteraceae	1:294
...rupina vulgaris	common crupina	Asteraceae	1:299
...ynara cardunculus	artichoke thistle	Asteraceae	1:302
...necio spp. (some species)	ragworts, groundsels	Asteraceae	1:382
...nacetum spp.	common tansy, feverfew	Asteraceae	1:407
...chium plantagineum	vipers bugloss	Boraginaceae	1:434
...arbarea spp.	wintercress, rockets	Brassicaceae	1:455
...rassica spp.	mustards	Brassicaceae	1:460
...apsella bursa-pastoris	shepherd's-purse	Brassicaceae	1:471

Table 13. Plants with various types of compound or dissected leaves, cont.

Species	Common name	Family	Vol:
Cardamine spp.	bittercresses	Brassicaceae	1
Coronopus spp.	swinecresses	Brassicaceae	1
Hirschfeldia incana	shortpod mustard	Brassicaceae	1
Lepidium spp. (many species)	pepperweeds	Brassicaceae	1
Raphanus spp.	radishes	Brassicaceae	1:
Sibara virginica	sibara	Brassicaceae	1.
Sinapis spp.	mustards	Brassicaceae	1.
Sisymbrium spp.	mustards, rockets	Brassicaceae	1:
Astragalus spp.	locos, milkvetches	Fabaceae	1:
Glycyrrhiza lepidota	wild licorice	Fabaceae	1:
Halimodendron halodendron	Russian salttree	Fabaceae	1:
Lotus spp.	trefoils	Fabaceae	1:
Ononis alopecuroides	foxtail restharrow	Fabaceae	1:
Prosopis spp.	mesquites	Fabaceae	1:
Robinia pseudoacacia	black locust	Fabaceae	1:
Senna obtusifolia	sicklepod	Fabaceae	1:
Sesbania spp.	sesbanias	Fabaceae	1:
Sphaerophysa salsula	swainsonpea	Fabaceae	1:
Vicia spp.	vetches	Fabaceae	1:
Erodium spp. (upper leaves on some)	filarees	Geraniaceae	2:
Geranium purpureum	little robin	Geraniaceae	2:
Geranium robertianum	herb-robert	Geraniaceae	2:
Plantago coronopus	cutleaf plantain	Plantaginaceae	2:
Acaena spp.	biddy-biddys	Rosaceae	2:1
Rosa spp.	roses	Rosaceae	2:1
Rubus laciniatus	cutleaf blackberry	Rosaceae	2:1
Sanguisorba minor ssp. *muricata*	salad burnet	Rosaceae	2:1
Ailanthus altissima	tree-of-heaven	Simaroubaceae	2:1
Verbena tenuisecta	moss vervain	Verbenaceae	2:1
Peganum harmala	African rue	Zygophyllaceae	2:1
Tribulus terrestris	puncturevine	Zygophyllaceae	2:1

Shortcut Identification Tables

Table 13. Plants with various types of compound or dissected leaves, cont.

Leaves twice-pinnately compound

Species	Common name	Family	Vol:Page
Ammi majus	greater ammi	Apiaceae	1:110
Cicuta spp.	waterhemlocks	Apiaceae	1:124
Pastinaca sativa	wild parsnip	Apiaceae	1:131, 1:136
Torilis spp.	hedgeparsleys	Apiaceae	1:117
Ambrosia artemisiifolia	common ragweed	Asteraceae	1:188
Ambrosia psilostachya	western ragweed	Asteraceae	1:188
Anthemis arvensis	corn chamomile	Asteraceae	1:192
Artemisia annua	annual wormwood	Asteraceae	1:202
Artemisia biennis	biennial wormwood	Asteraceae	1:202
Cotula australis	southern brassbuttons	Asteraceae	1:291
Cotula mexicana	Mexican brassbuttons	Asteraceae	1:294
Senecio jacobaea	tansy ragwort	Asteraceae	1:382
Soliva sessilis	lawn burweed	Asteraceae	1:395
Tanacetum spp.	common tansy, feverfew	Asteraceae	1:407
Capsella bursa-pastoris	shepherd's-purse	Brassicaceae	1:471
Coronopus spp.	swinecresses	Brassicaceae	1:489
Lepidium spp. (many species)	pepperweeds	Brassicaceae	1:500
Acacia baileyana	cootamundra wattle	Fabaceae	1:738
Acacia dealbata	silver wattle	Fabaceae	1:738
Acacia decurrens	green wattle	Fabaceae	1:739
Acacia elata	cedar wattle	Fabaceae	1:739
Acacia farnesiana	sweet acacia	Fabaceae	1:739
Acacia mearnsii	black wattle	Fabaceae	1:739
Acacia melanoxylon (young leaves)	black acacia	Fabaceae	1:734
Prosopis spp.	mesquites	Fabaceae	1:790
Erodium cicutarium	redstem filaree	Geraniaceae	2:835
Fumaria spp.	fumitorys	Papaveraceae (= Fumariaceae)	2:989
Verbena tenuisecta	moss vervain	Verbenaceae	2:1572
Peganum harmala	African rue	Zygophyllaceae	2:1589

Leaves dissected or more than twice-pinnately compound

Species	Common name	Family	Vol:Page
Ammi spp.	ammis	Apiaceae	1:110
Anthriscus caucalis	bur chervil	Apiaceae	1:117
Cyclospermum leptophyllum	wild celery	Apiaceae	1:110
Cicuta spp.	waterhemlocks	Apiaceae	1:124
Conium maculatum	poison-hemlock	Apiaceae	1:132
Daucus spp.	carrots	Apiaceae	1:137

Table 13. Plants with various types of compound or dissected leaves, cont.

Species	Common name	Family	Vol
Foeniculum vulgare	fennel	Apiaceae	1
Scandix pecten-veneris	venus-comb	Apiaceae	1
Torilis spp.	hedgeparsleys	Apiaceae	1
Achillea millefolium	common yarrow	Asteraceae	1
Ambrosia artemisiifolia	common ragweed	Asteraceae	1
Ambrosia psilostachya	western ragweed	Asteraceae	1
Anthemis cotula	mayweed chamomile	Asteraceae	1
Artemisia annua	annual wormwood	Asteraceae	1
Artemisia biennis	biennial wormwood	Asteraceae	1
Chamomilla suaveolens	pineappleweed	Asteraceae	1
Soliva sessilis	lawn burweed	Asteraceae	1
Descurainia sophia	flixweed	Brassicaceae	1
Lepidium perfoliatum	clasping pepperweed	Brassicaceae	1
Pteridium aquilinum	western brackenfern	Dennstaedtiaceae	1
Fumaria spp.	fumitorys	Papaveraceae (= Fumariaceae)	2
Ranunculus testiculatus	bur buttercup	Ranunculaceae	2

Key to the Common Mature Weedy Grasses of California

This key to 145 weedy mature grasses of California and other western states is based primarily on characteristics associated with the inflorescence and spikelets, but also some vegetative features. (For a key based on vegetative characteristics only, see page 42.) To use this key effectively a hand lens or stereoscope is often necessary. The initial key is to four easily recognized groups. Within each group, species or genera can be further identified. The volume and page number where the species is described is indicated after the scientific name. Illustrations (figures) are included on pages 55–56 to assist in separating characteristics or to clarify terminology. When all weedy species within a genus are covered in the text, the genus is followed by the spp. (species plural) designation. Several species described in this text were not included in the key because of their limited distribution and unlikelihood of being encountered. These species include:

Agrostis capillaris (colonial bentgrass)

Axonopus fissifolius (carpetgrass)

Cynodon plectostachyus (stargrass)

Cynodon transvalensis (African bermudagrass)

Heteropogon contortus (tanglehead)

Miscanthus sinensis (eulaliagrass)

Panicum antidotale (blue panicgrass)

Sclerochloa dura (hardgrass)

Stipa capensis (Mediterranean steppegrass)

Key to Groups

A. Inflorescence appearing as a spike or very condensed panicle with branches of inflorescence either absent or inconspicuous (fig. K.1). **Group 1**

A. Inflorescence not appearing as a spike or a very condensed panicle; if a panicle, the branches generally conspicuous

   B. Inflorescence a more open panicle with branches conspicuous (figs. K.2, K.3); main branches of inflorescence conspicuously branched at least twice. **Group 2**

   B. Inflorescence appearing umbel-like (all branches radiating from a central terminal point) or with branches in a raceme or whorl along the main stem; inflorescence branched only once from main axis

      C. Inflorescence branches umbel-like (sometimes with only two branches) or with at least top branches in an umbel arrangement (figs. K.4, K.7); may have one or two branches below umbel, or several branches in a whorl, below terminal umbel (fig. K.5). **Group 3**

      C. Inflorescence a raceme (branches alternating along a central stem) without a terminal umbel of branches (fig. K.6); each branch appears as a spike. **Group 4**

Group 1. Inflorescence a spike or very condensed panicle

Vol 2: Page

1. Bristles associated with the base of each spikelet (fig. K.9)

    2. Inflorescence purple, bristles fused at the base
        Pennisetum ciliare 1228

    2. Inflorescence green, bristles not fused
        Setaria spp. 1275

1. No bristles at the base of each spikelet, may be hairy or with spiny burs

    3. Spikelet enclosed within a spiny bur (fig. K.10)
        Cenchrus echinatus 1078
        Cenchrus incertus
        Cenchrus longispinus

    3. Spikelet not enclosed within a spiny bur

        4. Awn present on either glumes or lemma

            5. Awns ≥ 10 mm long

                6. Awns < 20 mm long

                    7. Spikelets facing one direction so as to appear asymmetrical (fig. K.8)
                          Cynosurus echinatus. 1105

                    7. Spikelets not all facing one direction, inflorescence more symmetrical

                        8. Spikelets 3 per node, two lateral spikelets sterile and smaller than central spikelet (fig. K.11)
                              Hordeum marinum 1182

                        8. Spikelets 1 per node

                            9. Only 1 glume present, spikelets lateral to main stem axis (fig. K.12)
                                Lolium temulentum 1195

                            9. Glumes 2, spikelets flat against main stem axis

                                10. Fertile florets 1 per spikelet
                                  Polygonum maritimus 1262

                              10. Fertile florets more than 1 per spikelet

                                  11. Awn bent and twisted (fig. K.13)
                                    Bromus alopecuros 1065

                                  11. Awn straight or somewhat diverging but not bent and twisted

                                      12. Ligule > 1 mm long
                                        Brachypodium distachyon. 1046

                                      12. Ligule ≤ 1 mm long
                                        Vulpia spp. 1314

Key to Mature Weedy Grasses

Vol 2: Page

6. Awns ≥ 20 mm long

    13. More than 1 awn per glume or lemma on lower spikelets of inflorescence
        Aegilops ovata 1008
        Aegilops triuncialis

    13. Only 1 awn per glume or lemma on any spikelet

        14. Spikelets 2 or 3 per node
            Hordeum jubatum............................ 1182

            15. Spikelets 3 per node, lemma 8–14 mm long (excluding awn)
                Hordeum murinum 1182

            15. Spikelets 2 per node, lemma 5–8 mm long (excluding awn)
                Taeniatherum caput-medusae 1310

        14. Spikelets 1 per node

            16. Spikelets sunken into inflorescence stem (fig. K.14), glumes lacking a bristly ciliate margin
                Aegilops cylindrica.......................... 1008

            16. Spikelets not sunken into inflorescence stem, glumes with a bristly ciliate margin
                Secale cereale 1188

5. Awns < 10 mm long

    17. Spikelets facing one direction so as to appear asymmetrical (fig. K.8)
        Cynosurus spp. 1105

    17. Spikelets not all facing one direction, inflorescence more symmetrical

        18. Fertile florets more than 1 per spikelet

            19. Only 1 glume present, spikelets lateral to main stem axis (fig. K.12)
                Lolium multiflorum 1195
                Lolium rigidum
                Lolium temulentum

            19. Glumes 2, spikelets flat against main stem axis

                20. Awns on lemma or glume < 2 mm long
                    Elytrigia repens 1140, 1171
                    Koeleria phleoides

                20. Awns on lemma or glume > 2 mm long

                    21. Ligule ≥ 1 mm long

                        22. Inflorescence usually 2–6 cm long, spikelet 20–35 mm long, glumes and lemma glabrous to scabrous
                            Brachypodium distachyon..................... 1046

                      22. Inflorescence usually 5–16 cm long, spikelet 12–22 mm long, glumes and lemma usually soft hairy
                            Bromus hordeaceus.......................... 1065

                    21. Ligule < 1 mm long

Key to Mature Weedy Grasses

                          Vol 2: Page

   23. Annual, < 0.7 m tall
     Vulpia spp............................... 1314

   23. Perennial, ≥ 0.7 m tall
     Elytrigia repens 1140

18. Fertile florets 1 per spikelet

   24. Spikelets 3 per node, two lateral spikelets sterile and smaller than central spikelet (fig. K.11)
     Hordeum marinum 1182

   24. Spikelets 1 per node

     25. Spikelet ≤ 3 mm long (excluding awn)

      26. Rhizomatous perennial, leaves 2-ranked (fig. K.19), ligule < 1 mm long
       Muhlenbergia schreberi..................... 1201

      26. Annual, leaves not 2-ranked, ligule > 2 mm long
       Polypogon maritimus 1262
       Polypogon monspeliensis

     25. Spikelet > 3 mm long (excluding awn)

      27. Foliage densely soft hairy on blade and sheath
       Holcus spp. 1179

      27. Foliage may have hairs but not densely soft hairy on blade and sheath

       28. Fertile floret surrounded by 2-awned dark brown and hairy sterile lemmas (fig. K.15)
        Anthoxanthum spp. 1031

       28. Fertile floret not surrounded by 2-awned dark brown and hairy sterile lemmas

        29. Glumes glabrous
         Gastridium phleoides 1171

        29. Glumes hairy
         Alopecurus spp. 1021

4. Awn not present on either glumes or lemma

  30. Fertile florets more than 1 per spikelet

   31. Spikelets facing one direction so as to appear asymmetrical (fig. K.8)
     Cynosurus cristatus 1105

   31. Spikelets not all facing one direction, inflorescence more symmetrical

     32. Only 1 glume present, spikelets lateral to main stem axis (fig. K.12)
      Lolium perenne......................... 1195
      Lolium temulentum

     32. Glumes 2, spikelets flat against main stem axis

      33. Perennial, ≥ 0.5 m tall
       Elytrigia spp. 1140

Vol 2: Page

33. Annual or perennial, < 0.5 m tall

    34. Leaves 2-ranked (fig. K.19), rhizomatous perennial
        Distichlis spicata 1104

    34. Leaves not 2-ranked, annual
        Schismus spp. 1270

30. Fertile florets 1 per spikelet

    35. Growing on coastal sand dunes, rhizomatous perennial
        Ammophila spp. 1027

    35. Not found on coastal sand dunes, not a rhizomatous perennial

        36. Longest glume < 2 mm long
            Sporobolus indicus 1306

        36. Longest glume ≥ 2 mm long

            37. Ligule ciliate (fig. K.26), < 1 mm long, generally low growing, < 0.4 m tall
                Crypsis spp. 1021

            37. Ligule membranous (fig. K.27), > 3 mm long, erect plants, > 40 cm tall
                Phalaris spp. 1230

Group 2. Inflorescence an open panicle

1. Reedlike grass, ≥ 2 m tall

    2. Long brown hairs at base of leaf blade, 2 spikelets per node, spikelet < 8 mm long, fertile floret 1 per spikelet
        Saccharum ravennae 1039, 1095

    2. Base of leaf blade lacking long brown hairs, 1 spikelet per node, spikelet ≥ 8 mm long, fertile floret > 2 per spikelet

        3. Lemma with long silky hairs < 8 mm at base, collars and sheath hairy, glume 10–13 mm long
            Arundo donax 1034

        3. Lemma glabrous, collars and sheath glabrous (margin of sheath may be ciliate hairy); glume 3–7 mm long
            Phragmites australis 1038

1. Not a reedlike grass, < 2 m tall

    4. Spikelets facing one direction so as to appear asymmetrical, panicle condensed (fig. K.8)

        5. Inflorescence 10–25 cm long, lemma awn 3–12 mm long
            Cynosurus echinatus 1105

        5. Inflorescence 1–4 cm long, lemma awn ~1 mm long
            Dactylis glomerata 1108

                                                                                             Vol 2: Page

4. Spikelets not all facing one direction, inflorescence more symmetrical

    6. Bristles associated with the base of each spikelet (fig. K.9)
            Pennisetum villosum 1219

    6. Bristles not associated with the base of each spikelet

        7. Inflorescence a large open panicle $\geq$ 30 cm long on stalks $\geq$ 2 m long
                Cortaderia spp. 1090

        7. Inflorescence generally < 30 cm long and on stalks < 2 m long

            8. Spikelets consist of small bulblets or plantlets (fig. K.16)
                    Poa bulbosa 1250

            8. Spikelets not small bulblets or plantlets

                9. Both glumes and lemmas awnless

                    10. Fertile florets 1 per spikelet

                      11. Lemma < 3 mm long

                          12. Spikelets > 3.5 mm long
                              Panicum miliaceum 1203

                        12. Spikelets $\leq$ 3.5 mm long

                            13. Ligule ciliate at least on upper ⅓
                                Panicum capillare........................1203
                                Panicum dichotomiflorum

                        13. Ligule membranous

                            14. Ligule < 2 mm long

                                15. Immature spikelets awned, awns becoming deciduous with age
                                    Piptatherum miliaceum........1243

                                15. Immature spikelets not awned
                                    Polypogon viridis 1262

                            14. Ligule $\geq$ 2 mm long
                                Agrostis gigantea 1016, 1262
                                Agrostis stolonifera
                                Polypogon viridis

                    11. Lemma $\geq$ 3 mm long

                      16. Ligule $\leq$ 3 mm long

                          17. Ligule ciliate on upper half
                              Panicum miliaceum 1203

                        17. Ligule membranous
                            Ehrharta calycina........................1129
                            Ehrharta erecta

                    16. Ligule > 3 mm long

Key to Mature Weedy Grasses

                                                                       Vol 2: Page

    18. Ligule ciliate on top (fig. K.28)
        Sorghum bicolor..........................1288
        Sorghum halepense

    18. Ligule membranous
        Phalaris aquatica.........................1230
        Phalaris arundinacea

10. Fertile florets more than 1 per spikelet

    19. Fertile florets 2 per spikelet, glume > 15 mm long
        Avena sativa.............................1040

    19. Fertile florets > 2 per spikelet, glumes ≤ 15 mm long

        20. Lemma with 3 conspicuous veins (1 on the keel)

            21. Ligule ≥ 2 mm long
                Leptochloa uninervia........................1190

            21. Ligule < 2 mm long
                Eragrostis spp. 1146

        20. Lemma with more than 3 conspicuous veins (fig. K.13)

            22. Spikelets < 10 mm long

                23. Leaf blades with prow- or boat-shaped tips (fig. K.20)
                      Poa spp......... 1246, 1250, 1253

                23. Leaf blades lacking prow- or boat-shaped tips

                    24. Ligule ciliate (fig. K.26), < 1 mm long
                          Schismus spp. 1270

                    24. Ligule membranous (fig. K.27), > 3 mm long
                          Briza minor 1049

            22. Spikelets ≥ 10 mm long

                25. Ligule ≥ 3 mm long
                    Briza maxima...............1049

                25. Ligule < 3 mm long

                    26. Sheath closed for more than ½ the length
                        (figs. K.23, 24)
                          Bromus briziformis 1065
                          Bromus inermis
                          Bromus secalinus

                    26. Sheath open for more than ½ the length
                        (figs. K.21, K.22)
                          Festuca spp..................1167

9. Glumes or lemmas with conspicuous awns

    27. Fertile florets 1 per spikelet

        28. Awns of glumes or lemma > 8 mm

            29. Awns of glume or lemma ≥ 15 mm
                Achnatherum brachychaetum 1004

Vol 2: Page

29. Awns of glume or lemma < 15 mm

  30. Awn straight or nearly so
    Polypogon maritimus 1262
    Polypogon monspeliensis

  30. Awn bent (fig. K.13)
    Sorghum spp. 1288

28. Awns of glumes or lemma ≤ 8 mm

  31. Awns of glumes or lemma < 5 mm

    32. Foliage densely soft hairy on blade and sheath
      Holcus spp. 1179

    32. Foliage may have hairs but not densely soft hairy on blade and sheath

      33. Spikelet ≥ 8 mm long
        Ehrharta longiflora 1129

      33. Spikelet < 8 mm long

        34. Ligule ≥ 1 mm

          35. Panicle open, very diffuse (fig. K.2)

            36. Ligule > 2 mm long
              Agrostis avenacea 1016

            36. Ligule ≤ 2 mm long
              Piptatherum miliaceum 1243

          35. Panicle condensed or ascending, not diffuse (fig. K.3)

            37. Ligule ≥ 2 mm long, condensed panicle
              Polypogon spp. 1262

            37. Ligule < 2 mm long, generally open panicle
              Piptatherum miliaceum 1243

        34. Ligule < 1 mm

          38. Spikelet ≥ 4 mm long
            Muhlenbergia schreberi 1201

          38. Spikelet < 4 mm long
            Piptatherum miliaceum 1243

  31. Awns of glumes or lemma ≥ 5 mm

    39. Spikelet > 8 mm long
      Ehrharta longiflora 1129

    39. Spikelet ≤ 8 mm long

      40. Inflorescence very open and diffuse (fig. K.2)
        Agrostis avenacea 1016

Key to Mature Weedy Grasses

- **40.** Inflorescence a condensed panicle or with all branches ascending (fig. K.3)
  - **41.** Lemma > 3 mm long
    Sorghum bicolor 1288
  - **41.** Lemma ≤ 3 mm long
    Polypogon australis 1262
    Polypogon maritimus
    Polypogon monspeliensis
- **27.** Fertile florets more than 1 per spikelet
  - **42.** Awn of glume or lemma generally > 18 mm long
    - **43.** Awn bent and twisted (fig. K.13)
      Avena barbata 1040
      Avena fatua
      Avena sterilis
    - **43.** Awn straight or somewhat divergent but not bent or twisted
      - **44.** Longest glume < 10 mm long
        Vulpia myuros 1314
      - **44.** Longest glume ≥ 10 mm long
        Bromus diandrus 1056
        Bromus sterilis
        Bromus madritensis
  - **42.** Awn of glume or lemma generally ≤ 18 mm long
    - **45.** Longest glume ≥ 15 mm long
      Avena sativa 1040
    - **45.** Longest glume < 15 mm long
      - **46.** Awn of glume or lemma < 5 mm long
        - **47.** Spikelet (not including awn) ≥ 15 mm long
          Bromus catharticus 1053, 1065
          Bromus briziformis
          Bromus secalinus
          Bromus inermis
        - **47.** Spikelet (not including awn) < 15 mm long
          - **48.** Lemma with 3 conspicuous veins (1 is keel) (fig. K.17)
            Leptochloa fascicularis 1190
          - **48.** Lemma with more than 3 conspicuous veins
            - **49.** Spikelet ≤ 5 mm long
              Koeleria phleoides 1171
            - **49.** Spikelet > 5 mm long
              - **50.** Perennial
                Festuca spp. 1167

                                                               Vol 2: Page

50. Annual

        Vulpia bromoides 1314

    46. Awn of glume or lemma > 5 mm long

        51. Sheath closed for more than ½ the length (figs. K.23, K.24)

            Bromus alopecuros 1056, 1065
            Bromus arenarius
            Bromus hordeaceus
            Bromus japonicus
            Bromus madritensis
            Bromus secalinus
            Bromus tectorum

        51. Sheath open for more than ½ the length (figs. K.21, K.22)

            Vulpia spp...................1314

Group 3. Inflorescence branches umbel-like

                                                                      Vol 2: Page

1. Branches of inflorescence less than 3 (fig. K.7)

    2. Fertile florets more than 2 per spikelet

        Eleusine tristachya 1136

    2. Fertile florets 1 per spikelet

        3. Glume and sterile lemma glabrous

            Paspalum notatum 1211

        3. Glume and sterile lemma puberulent to long hairy at tip

            Paspalum distichum 1211

1. Branches of inflorescence more than 3

    4. Fertile florets more than 1 per spikelet

        5. Glumes lacking an awn, ligule membranous throughout (fig. K.25)

            Eleusine spp. 1136

        5. Upper glume with awn 1.5–2 mm long, ligule membranous below and ciliate hairy on top (fig. K.28)

            Dactyloctenium aegyptium 1088

    4. Fertile florets 1 per spikelet

        6. Spikelet awned

            Chloris spp..............................1083

        6. Spikelet lacking awn

            7. Terminal umbel with additional branches below

                Digitaria spp............................1112

            7. Terminal umbel without branching below

                Cynodon dactylon.........................1098

Key to Mature Weedy Grasses

Group 4. Inflorescence a raceme

Vol 2: Page

1. No ligule present at junction between leaf blade and sheath
Echinochloa spp. 1118

1. Ligule present (either membranous or ciliate hairy) at junction between leaf blade and sheath

    2. Awns on either glume or lemma > 3 mm long

        3. Lemma awn 4–15 mm long, inflorescence < 8 cm long, spikelet 20–35 mm long
Brachypodium distachyon. 1046

        3. Lemma awn 1–5 mm long, inflorescence > 10 cm long, spikelet 6–12 mm long
Leptochloa fascicularis 1090

    2. Awns on either glume or lemma absent or ≤ 3 mm long

        4. Fertile florets more than 1 per spikelet

            5. Stoloniferous and rhizomatous perennial, leaves 2-ranked (fig. K.19)
Distichlis spicata 1104

            5. Annual, leaves not 2-ranked
Leptochloa spp. 1090

        4. Fertile florets 1 per spikelet

            6. Cup-like structure surrounding the base of spikelet (fig. K.18)
Eriochloa spp. 1162

            6. No cup-like structure surrounding the base of spikelet

                7. Inflorescence with less than 3 branches
Paspalum distichum 1211
Paspalum notatum

                7. Inflorescence with 3 or more branches

                    8. Spikelets glabrous

                        9. Ligule membranous (fig. K.27), > 2 mm long
Ehrharta erecta 1129

                        9. Ligule ciliate hairy (fig. K.26), < 0.5 mm long
Sporobolus spp. 1306

                  8. Spikelets hairy

                    10. Spikelets 2 or 3 per node, annual, ligule generally < 3 mm long
Digitaria spp.. 1112

                    10. Spikelets 1 per node, perennial, ligule generally ≥ 3 mm long
Paspalum dilatatum 1211
Paspalum urvillei

Vegetative Key to the Common Weedy Grasses of California

This key to 142 weedy grasses of California and other western states is based on vegetative characteristics other than reproductive structures. (For a key to all characteristics, see page 1:31.) Proper identification of grasses using vegetative characteristics usually requires the use of magnification, either a hand lens or stereoscope. The initial key is to two easily recognized groups. In some cases, the key may suggest more than one species within a genus or multiple species in different genera. Refer to the specific entries for more information on how to separate these species. The volume and page number where the species is described is indicated after the scientific name. Illustrations (figures) are included on pages 55–56 to assist in separating characteristics or to clarify terminology. When all weedy species within a genus are covered in the text, the genus is followed by the spp. (species plural) designation. Several species described in this text were not included in the key because of their limited distribution and unlikelihood of being encountered. These species include:

> *Ammophila breviligulata* (American beachgrass)
> *Axonopus fissifolius* (carpetgrass)
> *Crypsis vaginiflora* (African pricklegrass)
> *Cynodon transvalensis* (African bermudagrass)
> *Heteropogon contortus* (tanglehead)
> *Miscanthus sinensis* (eulaliagrass)
> *Panicum antidotale* (blue panicgrass)
> *Poa nemoralis* (wood bluegrass)
> *Sclerochloa dura* (hardgrass)
> *Setaria sphacelata* (African bristlegrass)
> *Sporobolus vaginiflorus* (poverty dropseed)
> *Stipa capensis* (Mediterranean steppegrass)

Key to Groups

Vol 2: Page

A. Ligule absent (fig. K.25)
                 Echinochloa spp. 1118

A. Ligule present
    B. Ligule ciliate hairy throughout or for at least one-half the length (fig. K.26)
        Group 1

    B. Ligule membranous throughout or for at least one-half the length (fig. K.27)
        Group 2

Vegetative Key to Common Weedy Grasses

Group 1. Ligule ciliate hairy throughout or for at least one-half the length

1. Ligule > 2 mm long   **Vol 2: Page**

    2. Ligule ≥ 4 mm long, perennial

        3. Young leaf folded in bud (fig. K.21)
Chloris gayana 1083

        3. Young leaf rolled in bud (fig. K.22)

            4. Base of blade with dense cluster of long (5–10 mm) brown hairs, plant ≥ 2 m tall
Saccharum ravennae 1039, 1095

            4. Base of blade without dense cluster of long brown hairs, plants < 2 m tall
Sorghum spp. 1288

    2. Ligule < 4 mm long, annual or perennial

        5. Blades glabrous most of the length or with short scabrous hairs

            6. Leaf blade < 1 cm wide

                7. Marine grass, associated with estuaries
Spartina spp. 1297

                7. Not associated with marine environments

                    8. Annual
Panicum dichotomiflorum 1203

                    8. Perennial

                          9. Bunchgrass, young leaves rolled in bud (fig. K.22)
Achnatherum brachychaetum 1004

                          9. Stoloniferous, young leaves folded in bud (fig. K.21)
Chloris spp. 1083

            6. Leaf blade ≥ 1 cm wide

                10. Large bunchgrass
Cortaderia spp. 1090

                10. Annual or spreading grass, can be clump forming

                    11. Marine grass, associated with estuaries, < 1.5 m tall
Spartina spp. 1297

                    11. Not associated with marine environments, if marine then ≥ 1.5 m tall

                        12. Reedlike grass, ≥ 2 m tall
Phragmites australis 1038

                        12. Not a reedlike grass, < 2 m tall
Panicum antidotale or *Panicum dichotomiflorum* 1203

Vegetative Key to Common Weedy Grasses

Vol 2: Page

5. Blades hairy throughout
  Panicum capillare or *Panicum miliaceum*. 1203

1. Ligule < 2 mm long

  13. Ligules ≥ 1 mm

   14. Marine grass, associated with estuaries, < 1.5 m tall
     Spartina spp. 1297

   14. Not associated with marine environments, if marine then ≥ 1.5 m tall

    15. Reedlike grass, ≥ 2 m tall
      Phragmites australis. 1038

    15. Not a reedlike grass, < 2 m tall

     16. Young leaf folded in bud, stem flattened, perennial (figs. K.21, K.24)

      17. Plants hairy

       18. Hairs only at base of blade and/or on collar
         Cenchrus incertus. 1078

       18. Hairs more widespread than only on base of blade and collar

        19. Annual or sometimes perennial, but not stoloniferous or a bunchgrass
          Cenchrus spp. 1078

        19. Perennial, either stoloniferous or a bunchgrass

         20. Stoloniferous, typically low growing, except when climbing up other vegetation
           Pennisetum clandestinum 1219

         20. Bunchgrass, not low growing
           Pennisetum villosum. 1078

      17. Plants generally glabrous

       21. Low growing annual
         Cenchrus spp. 1078

       21. Erect perennial
         Chloris spp. 1083
         or *Eragrostis curvula* 1146

     16. Young leaf rolled in bud, stems flattened or round (figs. K.22, K.23, K.24)

      22. Leaf blade hairy throughout or at least on base of blade

       23. Blade with ciliate hairy margin (fig. K.30), perennial bunchgrass
         Pennisetum setaceum or *Pennisetum villosum*. 1219

       23. Blade without ciliate hairy margin

        24. Blade and sheath surface hairy

         25. Long hairs (2–4 mm) only at base of blade and top of sheath
           Pennisetum ciliare 1228

Vol 2: Page

25. Long stiff hairs at 90° angle to surface on most of foliage, not just base of blade and top of sheath
Panicum capillare or *Panicum miliaceum*......... 1203

24. Blade or sheath hairy or glabrous, but not both hairy

  26. Sheath with hairy margin

    27. Blades hairy on upper surface

      28. Annual
Setaria faberi............................1275

      28. Perennial bunchgrass
Pennisetum villosum.....................1219

    27. Blades glabrous

      29. Annual
Setaria verticillata........................1275

      29. Perennial bunchgrass
Pennisetum villosum.....................1219

  26. Sheath without hairy margin

    30. Blade with long sparse hairs at base of upper surface

      31. Low growing, < 0.5 m tall
Cenchrus echinatus......................1078

      31. Erect, ≥ 0.5 m tall
Setaria gracilis or *Setaria pumila*...........1275

    30. Blade hairy throughout upper surface

      32. Stem round, blade 10–20 mm wide, upper blade surface short hairy throughout, plant to 2 m tall
Setaria faberi............................1275

      32. Stem usually flattened, blade 3–12 mm wide, upper blade surface glabrous or occasionally hairy, plant to 1 m tall
Setaria verticillata........................1275

22. Blade glabrous

  33. Blade ≤ 2 mm wide
Achnatherum brachychaetum..................1004

  33. Blade > 2 mm wide

    34. Ligule ciliate above and membranous at base (fig. K.28)

      35. Stem flattened, base without knotlike hard swellings

        36. Midvein not conspicuously whitish
Setaria verticillata or *Setaria viridis*............1275

        36. Midvein conspicuously whitish
Panicum dichotomiflorum......................1203

Vegetative Key to Common Weedy Grasses

Vol 2: Page

    35. Stem round, short rhizome, base with knotlike hard swellings
        Setaria gracilis............................1275

  34. Ligule ciliate to base

    37. Sheath margin ciliate hairy (fig. K.31)
        Setaria viridis.............................1275

    37. Sheath margin not ciliate hairy

      38. Midvein conspicuously whitish
        Panicum dichotomiflorum......................1203

      38. Midvein not conspicuously whitish

        39. Annual
          Eragrostis minor........................1146

        39. Perennial bunchgrass
          Setaria gracilis...........................1275

13. Ligules generally < 1 mm long

  40. Young leaf folded in bud (fig. K.21)

    41. Blade < 2 mm wide
        Schismus spp. 1270

    41. Blade ≥ 2 mm wide

      42. Stem flattened

        43. Perennial, rhizomes and stolons present

          44. Collar hairy
            Cynodon dactylon...........................1098

          44. Collar glabrous
            Chloris truncata............................1083

        43. Annual or perennial, no stolons or rhizomes present

          45. Annual, nodes swollen, ligule ciliate to base
            Cenchrus longispinus........................1078

          45. Perennial, nodes not swollen, ligule membranous at base and ciliate hairy above
            Eleusine tristachya..........................1136

      42. Stem round
        Eragrostis curvula............................1146

  40. Young leaf rolled in bud (fig. K.22)

    46. Ciliate hairs on sheath margin (fig. K.31)

      47. Perennial bunchgrass, round stems, leaf blades with ciliate hairy margin (fig. K.30)
        Pennisetum setaceum........................1219

      47. Annual, flattened or round stem, leaf blade without ciliate hairy margin
        Setaria viridis..............................1275

Vol 2: Page

46. No ciliate hairs on sheath margin

    48. Long hairs (2–4 mm) present within 2–4 cm of blade base on upper surface (fig. K.32)

        49. Perennial bunchgrass, stems round
            Pennisetum ciliare 1228

        49. Annual, stems flattened

            50. Low growing, < 0.5 m tall, blade without prominent midvein
                Cenchrus echinatus 1078
                or *Schismus* spp. 1270

            50. Erect, ≥ 0.5 m tall, blade with prominent midvein
                Setaria pumila 1275

    48. Glabrous or hairy, if hairy then hairs only near collar or on sheath and/or entire blade

        51. Blade with dense soft hairs

            52. Hairs short (puberulent), < 0.5 mm long
                Eriochloa contracta 1162

            52. Hairs longer, ≥ 0.5 mm long
                Crypsis schoenoides 1021

        51. Blade glabrous or sparsely hairy, hairs may be present near collar or on sheath

            53. Stem flattened, ligule ciliate hairy above and membranous at base (fig. K.28)
                Chloris virgata 1083

            53. Stem flattened or round, ligule ciliate to base

                54. Perennial, ligule generally < 0.5 mm

                    55. Hairs on collar margin absent or < 1 mm long
                        Sporobolus indicus 1306

                    55. Hairs on collar margin 1–3 mm long
                        Eragrostis lehmanniana 1146
                    Annual or perennial, ligule generally ≥ 0.5 mm

                56. Blade with glands on margin, plants with foul odor (fig. K.33)
                    Eragrostis cilianensis or *Eragrostis minor* 1306

                56. Blade without glands on margin

                    57. Typically low growing and spreading
                        Crypsis spp. 1021

                    57. Erect and ascending
                        Eragrostis spp. 1146
                        or *Eriochloa acuminata* 1162

Group 2. Ligule membranous throughout or for at least one-half the length

Vol 2: Page

1. Sheath closed (fused) for at least one-half the length (figs. K.23, K.24)
    2. Ligule about 0.5 mm long
        Aegilops cylindrica 1008
    2. Ligule > 0.5 mm long
        3. Ligule subconical in its attachment to the blade base, unevenly attached to stem
            Briza minor 1049
        3. Ligule attached to blade base at an even level and evenly attached to stem
            4. Young blade folded in bud, stem flattened (fig. K.21, K.24)
                5. Blade ≤ 4 mm wide with boat-shaped tip
                    Poa spp. 1246, 1250, 1253
                5. Blade generally > 4 mm wide and lacking boat-shaped tip
                    Dactylis glomerata 1108
            4. Young blade rolled in bud, stem round (figs. K.22, K.23)
                Bromus spp. 1053, 1056, 1065
1. Sheath open (not fused) for more than one-half the length
    6. Young blade folded in bud (fig. K.21)
        7. Ligule ≤ 1 mm long
            8. Blade with ciliate hairs on margin, often near base of blade (fig. K.30)
                9. Annual, ligule > 0.5 mm long
                    10. Blade and sheath surface hairy
                        Koeleria phleoides 1171
                    10. Blade and sheath surface not hairy or very sparsely long hairy, sheath whitish at base
                        Eleusine indica 1136
                9. Perennial, ligule ≤ 0.5 mm long
                    Cynodon spp. 1098
            8. Blade without ciliate hairs on margin
                11. Ligule ciliate on top of membranous base (fig. K.28)
                    12. Rhizomatous and stoloniferous
                        Cynodon spp. 1098
                    12. Not rhizomatous or stoloniferous
                        Eleusine tristachya 1136
                11. Ligule membranous throughout, may be jagged on top but not ciliate

Vegetative Key to Common Weedy Grasses

Vol 2: Page

13. Blade often ≤ 4 mm wide with boat-shaped tips in *Poa* (fig. K.20)
  Cynosurus cristatus 1105
  or *Poa nemoralis* 1253

13. Blade > 4 mm wide without boat-shaped tips

  14. Annual, sheath whitish at base
    Eleusine indica 1136

  14. Perennial, sheath not whitish at base

    15. Stem flattened, blade to 30 cm long, 4–10 mm wide, glabrous, with short rhizomes
      Paspalum notatum 1211

    15. Stem round, blades 6–10 cm long, < 5 mm wide, glabrous to slightly pubescent, no rhizomes
      Cynosurus cristatus 1105

7. Ligule ≥ 1 mm long

  16. Ligule < 2 mm long

    17. Long hairs (1–2 mm) on collar margin
      Ehrharta calycina 1129

    17. Collar margin glabrous

      18. Small auricle sometimes present (fig. K.34), whitish midvein on back of blade, blade lacking boat-shaped tip
        Lolium temulentum 1195

      18. No auricles or white midvein on back of blade, blade with (fig. K.20) or without boat-shaped tip

        19. Sheath glabrous, blade tips boat-shaped (fig. K.20)
          Poa spp. 1246, 1250, 1253

        19. Sheath hairy, lacking boat-shaped tips
          Koeleria phleoides 1171

  16. Ligule ≥ 2 mm long

    20. Blade with boat-shaped tip (fig. K.20)

      21. Low growing, ≤ 0.1 m tall
        Sclerochloa dura 1273
        or *Poa annua* 1246

      21. More erect, > 0.1 m tall
        Poa spp. 1246, 1250, 1253

    20. Blade without boat-shaped tip

      22. Tip of membranous ligule not ciliate hairy
        Dactylis glomerata 1108

      22. Tip of membranous ligule ciliate hairy
        Chloris gayana 1083

6. Young blade rolled in bud (fig. K.22)

Vegetative Key to Common Weedy Grasses

Vol 2: Page

23. Tall reedlike grass, ≥ 2 m tall
    24. Blade > 1 cm wide, hairy or glabrous, ligule < 3 mm long
        25. Ligule ciliate on top, culm ≤ 1.5 cm wide (fig. K.28)
            Phragmites australis 1038
        25. Ligule membranous throughout, culm > 1.5 cm wide, collar long (> 4 mm) hairy
            Arundo donax 1034
    24. Blade ≤ 1 cm wide, plant glabrous, ligule ≥ 3 mm long
        Phalaris arundinacea 1230

23. Not a tall reedlike grass, < 2 m tall
    26. Ligule ≤ 1 mm long
        27. Ligule with ciliate hairs on top of membranous base (fig. K.28)
            28. Stems round, blade 1–4 mm wide and 2-ranked
                Distichlis spicata 1104
            28. Stems flattened, blade > 4 mm wide and not 2-ranked
                29. Low growing, blade margin ciliate hairy at base (fig. K.30), collar white
                    Dactyloctenium aegyptium 1088
                29. Erect, no ciliate hairs on blade margin
                  30. Annual
                      Chloris virgata. 1083
                  30. Creeping perennial
                      Muhlenbergia schreberi. 1201
        27. Ligule membranous throughout, may be jagged at top but not ciliate hairy
            31. Auricle usually present at collar (fig. K.34)
                32. Annual, blade 1–3 mm wide and soft hairy
                    Taeniatherum caput-medusae 1310
                32. Annual or perennial, blade 2–14 mm wide and glabrous or hairy
                  33. Foliage completely glabrous, even on collar, but blade may have scabrous hairs
                    34. Bunchgrass or rhizomatous perennial, > 1 m tall
                        Elytrigia spp. 1140
                    34. Annual or perennial without obvious rhizomes, ≤ 1 m tall
                        Lolium spp. 1195
                        Festuca spp. 1167
                        or *Hordeum jubatum* (usually without auricle). ... 1182
                33. Foliage hairy or at least with hairs on collar
                  35. Annual, ≤ 0.5 m tall
                    Aegilops spp. 1008

Vegetative Key to Common Weedy Grasses

Vol 2: Page

    35. Perennial, > 0.5 m tall
        Elytrigia spp. 1140

31. Auricle absent at collar

  36. Blade usually 1–3 mm wide, annual

    37. Tuft of hairs at collar only
        Anthoxanthum aristatum 1004

    37. Glabrous or hairy on blade and/or sheath, but no hairs on collar

      38. Ligule $\geq$ 0.5 mm
        Koeleria phleoides. 1171
        Taeniatherum caput-medusae ... 1310
        or *Lolium perenne*. 1195

      38. Ligule < 0.5 mm
        Vulpia spp. 1314

  36. Blade usually $\geq$ 3 mm wide, annual or perennial

    39. Creeping perennial with 2-ranked blades (fig. K.19)
        Muhlenbergia schreberi. 1201

    39. Annual or erect perennial without 2-ranked blades

      40. Blade hairy, not just near collar
        Aegilops spp. 1008
        or *Hordeum jubatum*. 1182

      40. Blade glabrous throughout

        41. Stem flattened

          42. Blade shiny
            Lolium perenne 1195

          42. Blade not shiny
            Chloris virgata 1083

        41. Stem round

          43. Blade usually < 10 cm long, sometimes hairy in *Hordeum jubatum*
            Hordeum jubatum 1182
            Cynosurus cristatus 1105
            or *Lolium* spp. 1195

          43. Blade usually > 10 cm long, always glabrous

            44. Blade shiny, annual or perennial, but not a bunchgrass
              Lolium spp. 1195

            44. Blade not shiny, perennial bunchgrass
              Piptatherum miliaceum. 1243

Vegetative Key to Common Weedy Grasses

Vol 2: Page

26. Ligule > 1 mm long

  45. Ligule < 3 mm long (sometimes to 4 mm in *Holcus lanatus*)

    46. Auricle present at collar (fig. K.34)

      47. Perennial

        48. Blade not shiny, collar ciliate hairy
Anthoxanthum odoratum 1031

        48. Blade shiny, collar glabrous

          49. Veins prominent, generally ≥ 0.9 m tall, auricles often with ciliate margin
Festuca arundinacea 1167

          49. Veins not prominent, generally < 0.9 m tall, auricles lacking or without ciliate margin
Lolium perenne 1195

      47. Annual

        50. Auricle long (≥ 2 mm), usually wraps around stem, blade hairy
Hordeum spp. 1182

        50. Auricle short (< 2 mm), does not wrap around stem, blade glabrous and shiny
Lolium spp. 1195

    46. Auricle not present at collar

      51. Ligule ciliate hairy on top and membranous below (fig. K.28)

        52. Perennial, ciliate hairy on collar margin
Anthoxanthum odoratum 1031

        52. Annual, hairy on blade or sheath

          53. Hairs on sheath both long (1 mm) and short, not at 90° angle to surface
Brachypodium distachyon. 1046

          53. Hairs on sheath and blade > 1 mm long and at 90° angle to surface (fig. K.29)
Panicum capillare or *Panicum miliaceum* 1203

      51. Ligule membranous throughout, may be jagged at top but not ciliate hairy

        54. Ligule hairy on back, plant densely soft hairy
Holcus spp. 1179

        54. Ligule not hairy on back, plant not densely soft hairy

          55. Nodes dramatically swollen with long tuft of hairs at 90° angle to surface at nodes, sheath margin ciliate hairy
Paspalum distichum 1211

Vol 2: Page

55. Nodes not dramatically swollen or with long tufted hairs, sheath margin generally not ciliate hairy

   56. Sheath long hairy, hairs at 90° angle to surface

      57. Plant ≥ 1 m tall, not long hairy on blade
            Secale cereale 1188

      57. Plant < 1 m tall, long hairs on blade at 90° angle to surface
            Digitaria sanguinalis 1112

   56. Sheath lacking long hairs at 90° angle to surface, or with soft or very sparse hairs

      58. Stem flattened
            Digitaria ischaemum 1112
            or *Lolium perenne*............ 1195

      58. Stem round

        59. Collar with ciliate hairs on margin

          60. Collar whitish or purplish
              Ehrharta spp................ 1129

          60. Collar yellowish or green
              Anthoxanthum spp. 1031

        59. Collar without ciliate hairs on margin

          61. Annual, ≤ 40 cm tall
              some *Polypogon* spp.......... 1262
              Gastridium phleoides 1171
              Koeleria phleoides............ 1171
              or *Hordeum marinum* 1182

          61. Perennial or annual in *Lolium* spp., > 40 cm tall
              Agrostis spp. 1016
              Alopecurus pratensis.......... 1021
              Piptatherum miliaceum....... 1243
              Lolium spp. or some
              Polypogon spp................ 1262

45. Ligule ≥ 3 mm long

   62. Ligule > 10 mm long
      Ammophila arenaria 1027

   62. Ligule ≤ 10 mm long

      63. Ligule ciliate on top and membranous below (fig. K.28)
            Sorghum spp................................ 1288

      63. Ligule membranous throughout, may be jagged at top but not ciliate hairy

         64. Stem flattened

Vol 2: Page

65. Sheath and collar hairy, hairs on sheath at 90° angle to surface
Paspalum urvillei 1211

65. Sheath not very hairy and hairs not at 90° angle to surface

    66. Some hairs at base of blade on collar, perennial
Paspalum dilatatum. 1211

    66. Blade glabrous, annual
Leptochloa spp. 1090
or *Sorghum bicolor* 1288

64. Stem round

    67. Collar uneven in attachment to culm

        68. Sheath margin membranous (whitish), annual
Cynosurus echinatus. 1105
or *Briza minor* 1049

        68. Sheath margin not obviously membranous, perennial bunchgrass
Phalaris aquatica 1230

    67. Collar even in attachment to culm

        69. Longest blades mostly < 15 cm long

            70. Plant ≥ 0.6 m tall

                71. Perennial, rhizomatous and/or stoloniferous
Agrostis spp. 1016
or *Polypogon viridis* 1262

                71. Annual or perennial, not rhizomatous or stoloniferous
Polypogon spp. 1262, 1230
or *Phalaris* spp. (annual species)

            70. Plant < 0.6 m tall

                72. Stoloniferous perennial
Agrostis spp. 1016
or *Polypogon viridis* 1262

                72. Annual or perennial, not stoloniferous
Alopecurus myosuroides 1021
Gastridium phleoides 1171
Agrostis avenacea 1016
Polypogon spp. 1262
or *Briza minor* 1049

        69. Longest blades mostly ≥ 15 cm long

            73. Annual

                74. Base of blade with ciliate hairs on margin (fig. K.30)
Avena barbata or *Avena fatua* . . 1040

Vegetative Key to Common Weedy Grasses

                                                               Vol 2: Page

74. Base of blade without ciliate hairs on margin

    75. Minute hairs on back of ligule
           Polypogon monspeliensis 1262

    75. No hairs on back of ligule

        76. Blade generally $\leq$ 4 mm wide
                Leptochloa spp. 1190

        76. Blade generally > 4 mm wide

            77. Ligule mostly < 5 mm long

                78. Plant < 0.75 m tall
                      Briza maxima 1049

                78. Plant $\geq$ 0.75 m tall
                      Avena spp........1040
                      or *Sorghum bicolor* 1288

            77. Ligule mostly $\geq$ 5 mm long
                Phalaris spp......1230
                Avena spp. 1040
                or *Sorghum bicolor* 1288

73. Perennial

    79. Blade $\leq$ 3 mm wide
        Agrostis avenacea 1016

    79. Blade > 3 mm wide

        80. Blade minutely scabrous, plant generally $\leq$ 1 m tall
            Polypogon interruptus or *Polypogon imberbis* 1262

        80. Blade glabrous, plant generally > 1 m tall
            Phalaris arundinacea or *Phalaris aquatica* 1230

Grass Key Figures

Fig. K.1

Fig. K.2

Fig. K.3

Fig. K.4

Fig. K.5

Fig. K.6

Fig. K.7

Fig. K.8

Fig. K.9

Fig. K.10

Grass Key Figures, continued

Fig. K.11   Fig. K.12   Fig. K.13   Fig. K.14   Fig. K.15

Fig. K.16   Fig. K.17   Fig. K.18   Fig. K.19   Fig. K.20

Fig. K.21   Fig. K.22   Fig. K.23   Fig. K.24   Fig. K.25

Fig. K.26   Fig. K.27   Fig. K.28   Fig. K.29   Fig. K.30

Fig. K.31   Fig. K.32   Fig. K.33   Fig. K.34

Weed Descriptions

Volume 1 • *Aizoaceae–Fabaceae*

Hottentot-fig [*Carpobrotus edulis* (L.) N.E. Br.] [Cal-IPC: High]

SYNONYMS: freeway iceplant; iceplant; sea-fig; *Carpobrotus edule* (L.) Bolus; *Mesembryanthemum edule* L.

GENERAL INFORMATION: Mat-forming or trailing *shrub* to 50 cm tall, with elongate succulent leaves and showy yellow or pink flowers. Hottentot-fig is extensively planted along highways as an ornamental and to prevent soil erosion. It has escaped cultivation in many coastal areas, where it especially thrives in dune communities. Hottentot-fig displaces native dune species, and large infestations change the ecology of the community by increasing soil organic matter, encouraging invasion by other non-native species. Plants trap more sand than native dune species and generally stabilize dune communities unnaturally. Plants also die when completely buried. In time, an increased amount of organic matter in the sandy soil can promote invasion by other weed species that otherwise would not be able to inhabit dune soils. Native to South Africa.

SEEDLINGS: Seldom encountered, but most frequent on disturbed soils. Observations suggest that herbivory accounts for high seedling mortality.

MATURE PLANT: Stems woody near base, trailing, mostly to 3 m long, *root at the nodes*. Individual clones can grow to about 50 m in diameter. Leaves sessile, *opposite*, elongate, *succulent, 6–10 cm long, 1–1.5 cm wide, sharply triangular in cross-section, widest below middle,* slightly curved, outer angle of tip often ser-

Hottentot-fig (*Carpobrotus edulis*) infestation in a sand dune.   J. M. DiTomaso

Hottentot-fig *(Carpobrotus edulis)* with yellow flowers. J. M. DiTomaso

rate, glabrous but not covered with whitish bloom (glaucous). Leaf pairs fused at the base.

ROOTS AND UNDERGROUND STRUCTURES: Roots fibrous, dense, horizontal and vertical, mostly concentrated in top 30 cm of soil. *Stems develop adventitious roots at the nodes.*

FLOWERS: Primarily April–October, sometimes nearly year-round, but peaks in spring. Flowers solitary at stem tip, stalked, *8–10 cm in diameter,* with *numerous pink or yellow (aging pink) linear petals.* Stamens numerous. Sepals 5, succulent, 3–4 cm, unequal, largest two are leaflike and *sharply triangular in cross-section.* Ovary inferior, *chambers 10–12, with seeds attached to the inner surface of the outer wall* (parietal placentation). Insect-pollinated.

FRUITS AND SEEDS: *Capsules berrylike,* large, fleshy, edible, persist on plants for months, turn yellow with age. Seeds numerous per fruit, remain in fruit until fruits are consumed by animals or decompose.

POSTSENESCENCE CHARACTERISTICS: Plants turn reddish when stressed or dying. Plants can also be reddish in areas with nitrogen deficiency.

HABITAT: Coastal scrub, grassland, chaparral, bluffs, dunes, and other sandy coastal sites. Cultivated as a landscape ornamental in coastal regions and inland in mild winter areas. Does not tolerate cold winter climate.

DISTRIBUTION: Naturalized throughout coastal California, Channel Islands, mostly to 100 m; Florida.

PROPAGATION/PHENOLOGY: Reproduces *vegetatively by stem fragments* and by seed. Fruits are consumed by and primarily disperse with animals such as deer, rabbits, and rodents. Seeds survive ingestion by animals, and those that pass through an animal's gut germinate more readily than seeds from intact fruits. Fruits that are not eaten become hard, forcing seeds to remain dormant until fruits decompose, usually within 3 years. Seeds mostly germinate in fall after first significant rain. In grassland, seedlings compete poorly with grasses, but individuals that establish can spread rapidly by vegetative means. In foredune and dune scrub areas, seedling establishment appears to be limited primarily by herbivory. Plants appear to grow actively year-round.

MANAGEMENT FAVORING/DISCOURAGING SURVIVAL: Soil disturbance favors survival. Manual removal can help to control hottentot-fig. However, plants and stem fragments left on the ground after being pulled out can survive and reroot. Seeds are killed by exposure to a temperature of at least 105°C for 5 minutes, but the succulent nature of the foliage generally makes burning ineffective.

SIMILAR SPECIES: Refer to table 14 (p. 61) for a comparison of distinguishing characteristics among iceplants and relatives. In inland regions, these species are desirable ornamentals and generally do not escape cultivation. However, they can be invasive in sensitive coastal habitats.

Sea-fig [*Carpobrotus chilensis* (Molina) N.E., synonyms: *Carpobrotus aequilateralus* auct. non (Haw.) N.E. Br.; *Mesembryanthemum chilense* Molina] [Cal-IPC:

Hottentot-fig (*Carpobrotus edulis*) with purplish flower.   J. M. DiTomaso

Table 14. Iceplants and relatives (*Aptenia, Carpobrotus, Conicosia, Drosanthemum, Malephora,* and *Mesembryanthemum* spp.)

Species	Habit, life cycle	Leaf shape	Leaf length (cm)	Flower color; diameter (cm)	Fruit type; chamber number; dehiscence	Other
Aptenia cordifolia baby sun rose	prostrate perennial with a woody base	ovate to heart-shaped	1–3	rose-magenta to purple; 1	capsule; 4; opens when moistened	sepals 4, unequal; leaves covered with fine papillae; axile placentation
Carpobrotus chilensis sea-fig	trailing or matlike perennial, roots at the nodes	elongate, cross-section rounded-triangular	4–7	bright pink or magenta; 3–5	berrylike; 8–10; does not open	sepals 5, unequal; leaves opposite, usually with a whitish bloom (glaucous); parietal placentation
Carpobrotus edulis Hottentot-fig	trailing or matlike perennial, roots at the nodes	elongate, cross-section triangular	6–10	pink, or yellow aging to pink; 8–10	berrylike; 10–12; does not open	sepals 5, unequal; leaves opposite, outer angle of tip and larger sepals often serrate; parietal placentation; hybridizes with sea-fig
Conicosia pugioniformis roundleaf iceplant	prostrate to ascending perennial with a single, vertical underground stem and annual above ground stems	linear, cross-section triangular	mostly 15–20	shiny pale yellow; 5–8	capsule with a conical top; 10–20; opens while dry	sepals 5, slightly unequal; stem leaves alternate; basal rosette of leaves; leaves covered with minute papillae; parietal placentation
Drosanthemum floribundum showy dewflower	matlike shrub, older stems root at the nodes	narrow linear, cross-section rounded to rounded-triangular	mostly 1–1.5	bright pink; 1.5–2	capsule; 4–6	sepals usually 5, equal; stigmas 4–6; leaves covered with glistening papillae; parietal placentation
Malephora crocea coppery mesembryanthemum	prostrate perennial, sometimes roots at the nodes	narrow linear, cross-section round to rounded-triangular	2–6	orange with purple underneath or all purple; < 5	capsule; 8(–11); opens when moistened	sepal 4–6, unequal; stigmas 8 or more; leaves with a whitish bloom (glaucous); parietal placentation
Mesembryanthemum crystallinum crystalline iceplant	trailing annual, biennial, or short-lived perennial, stems forked	ovate to spatula-shaped, lower ± heart-shaped	2–10(20)	white, aging to pink; 0.7–1	capsule; 5; opens when moistened	sepals 5, equal; leaves covered with glistening papillae; axile placentation
Mesembryanthemum nodiflorum slender-leaved iceplant	prostrate to ascending annual, biennial, or short-lived perennial, stems branched at base	linear, cross-section round	1–2	white, aging to yellow; ± 0.5	capsule; 5; opens when moistened	sepals 5, equal; leaves covered with glistening papillae; axile placentation

Hottentot-fig

Hottentot-fig (*Carpobrotus edulis*) seeds. J. O'Brien

Moderate] is a mat-forming or trailing *shrub* that closely resembles Hottentot-fig. Leaves are mostly 4–7 cm long, rounded-triangular in cross-section, and have a smooth outer angle near the tip. Sea-fig is distinguished by having bright pink or magenta flowers *3–5 cm in diameter, ovaries and fruits with 8–10 chambers*, and leaves and larger sepals that are *rounded-triangular in cross-section* and that have a *smooth outer angle*. However, sea-fig hybridizes with Hottentot-fig, and some plants and populations have a combination of characteristics mak-

Sea-fig (*Carpobrotus chilensis*) plant. J. M. DiTomaso

ing identification to species impossible. Hybrid plants appear to be as invasive as Hottentot-fig. Sea-fig inhabits dunes, grassland, and coastal scrub throughout coastal California and the northern Channel Islands. Sea-fig also occurs in Oregon. It is still inconclusive whether sea-fig is native or non-native in California. Although it is widely thought that it is probably native to South Africa and may have been introduced to California from Chile, other evidence

Baby sun rose (*Aptenia cordifolia*) plant. J. M. DiTomaso

Roundleaf iceplant (*Conicosia pugioniformis*) plant. J. M. DiTomaso

Hottentot-fig

Roundleaf iceplant (*Conicosia pugioniformis*) flower.   J. M. DiTomaso

Roundleaf iceplant (*Conicosia pugioniformis*) seeds.   J. K. Clark

suggests that it is a California native. Historical records show that coastal Native American tribes used sea-fig in their cultural practices.

Baby sun rose [*Aptenia cordifolia* (L.f.) N.E. Br., synonyms: heartleaf iceplant; *Mesembryanthemum cordifolium* L. f.] is an uncommon prostrate *perennial*, with stems to about 0.6 m long, succulent *heart-shaped leaves* 1–3 cm long that are covered with minute papillae, and solitary *rose-magenta to purple flowers* ± 1 cm *wide* in the leaf axils. Unlike other members of the iceplant family, baby sun rose has *4 ovary chambers* with *seeds attached to the internal dividing membrane or septa within the ovary (axile placentation)*. In addition, leaves are stalked, flowers have *4 unequal sepals ± 5 mm long*, and *petals ± 3 mm long*. Fruits are *capsules* about 1–1.5 cm long that open at the top by *4 valves*. Baby sun rose has escaped cultivation as an ornamental and inhabits disturbed places and coastal

Showy dewflower (*Drosanthemum floribundum*) infestation. J. M. DiTomaso

Showy dewflower (*Drosanthemum floribundum*) flowers. J. M. DiTomaso

Coppery mesembryanthemum (*Malephora crocea*) plant.   J. M. DiTomaso

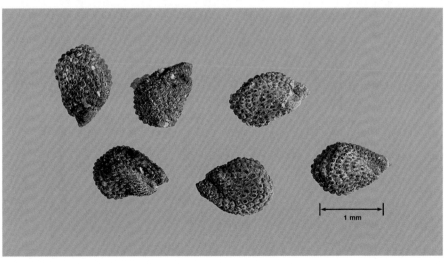

Coppery mesembryanthemum (*Malephora crocea*) seeds.   J. K. Clark

wetland margins in the Central Coast, South Coast, and Channel Islands, to 100 m. It also occurs in Oregon and Florida. Native to South Africa.

Roundleaf iceplant [*Conicosia pugioniformis* (L.) N.E. Br., synonyms: narrow-leaved iceplant; *Mesembryanthemum pugioniforme* L.] [Cal-IPC: Limited] is a short-lived *perennial*, with a *single perennial underground stem* to about 30 cm long, annual aboveground stems that are prostrate to ascending, succulent leaves about 15–20 cm long, and shiny yellow flowers 5–8 cm in diameter that have an unpleasant scent. Leaves are rounded-triangular in cross-section. Roundleaf iceplant is further distinguished by usually having a *basal rosette of leaves* and *alternate stem leaves* that are often densely clustered near the stem tip. In addition, *capsules* have conical tops that eventually open by 10–20 valves, the tips of which remain erect. Fruit valves do not open when moistened. Roundleaf iceplant inhabits dunes and other sandy places along the Central Coast, to 100 m. It is especially invasive on dunes in Santa Barbara and San Luis Obispo Counties. Native to South Africa.

Crystalline iceplant (*Mesembryanthemum crystallinum*) in flower.   J. M. DiTomaso

Crystalline iceplant (*Mesembryanthemum crystallinum*) infestation on a sand dune.   J. M. DiTomaso

Hottentot-fig

Showy dewflower [*Drosanthemum floribundum* (Haw.) Schwant., synonyms: rosy iceplant; *Mesembryanthemum floribundum* Haw.] is an uncommon mat-forming *shrub*, with *bright pink flowers* and *opposite linear leaves*. Leaves are 1–1.5 cm long, ± 2.5 mm wide, weakly triangular to round in cross-section, rounded and *broadest near the tip, slightly curved*, and covered with glistening papillae. Flowers are 1.5–2 cm in diameter and have *equal sepals* about 3–4 mm long, *4–6 stigmas*, and an *ovary with 4–6 chambers* and seeds that are attached to a membrane on the outer wall (parietal placentation). Capsules are *persistent*

Crystalline iceplant (*Mesembryanthemum crystallinum*) papillae on stem and petiole.
J. M. DiTomaso

and *open at the top by 4–6 winged valves*. Older stems often root at the nodes. Showy dewflower inhabits disturbed coastal sites in the North, Central, and South Coast, and Channel Islands, to 35 m. Native to South Africa.

Coppery mesembryanthemum [*Malephora crocea* (Jacq.) Schwan, synonyms: coppery mesemb; *Mesembryanthemum croceum* Jacq.] is a prostrate *shrub*, with succulent slender leaves 2–6 cm long that are opposite, sessile and slightly fused at the base, rounded-triangular to round in cross-section, covered with whitish bloom, and clustered on short branches. Stems are thick, pale, corky, and sometimes root at the nodes. Flowers are usually less than 5 cm wide, have 4–6 sepal lobes, and generally have *orange petals that are purple underneath*; but sometimes petals are purple on both surfaces. Coppery mesembryanthemum is distinguished by having ovaries and *persistent capsules* with 8(–11) *chambers* and *seeds attached to the inner surface of the outer wall*. Capsule valves are winged and open like lids when moistened. Coppery mesembryanthemum inhabits coastal bluffs and wetland margins in the Central Coast, South Coast, and southern Channel Islands, to 50 m. Native to South Africa.

Crystalline iceplant [*Mesembryanthemum crystallinum* L., synonyms: common iceplant; iceplant; *Gasoul crystallinum* (L.) Rothm.][MEKCR] [Cal-IPC: Moderate Alert] is a common trailing *annual, biennial*, sometimes short-lived perennial, with alternate or opposite leaves that are *ovate to spatula- or heart-shaped*, mostly 2–10 cm long, and *covered with glistening papillae*. Stems are mostly forked. Flowers have *5 sepals, 5 ovary chambers*, and are about *1 cm wide* with *white petals that age to pink*. Ovaries and capsules have seeds that are *attached to a membrane or septa near the center of the fruit* (axile placentation).

Crystalline iceplant (*Mesembryanthemum crystallinum*) seeds.   J. K. CLARK

Slender-leaved iceplant (*Mesembryanthemum nodiflorum*) in flower.   J. M. DiTomaso

Capsules open by 5 valves when moistened with water. Crystalline iceplant inhabits coastal bluffs and other disturbed sites throughout coastal California and the Channel Islands, to 100 m. It has also naturalized to some extent in Arizona and Pennsylvania, Mexico, and the Mediterranean. Native to South Africa.

Slender-leaved iceplant [*Mesembryanthemum nodiflorum* L., synonyms: slender-leaf iceplant; *Gasoul nodiflorum* (L.) Rothm.] is an uncommon *annual, biennial,* or sometimes short-lived perennial, with prostrate to ascending stems that are branched near the base, sessile *cylindrical leaves 1–2 cm long that are round in cross-section,* and *white flowers about 0.5 cm wide* that age to yellow. Like crystalline iceplant, slender-leaved iceplant has leaves that are covered with glistening papillae and flowers that have *5 sepals, 5 ovary chambers,* and *axile placentation.* Capsules open by 5 valves when moistened with water. Slender-leaved iceplant inhabits coastal bluffs, salty flats in the interior of the coastal slopes, and saline wetland margins in the San Francisco Bay region, South Coast, and Channel Islands, to 100 m. It also occurs in Arizona and Oregon, Mexico, Australia. Native to South Africa.

Alligatorweed [*Alternanthera philoxeroides* (C. Martius) Griseb][ALRPH] [Cal-IPC: High Alert] [CDFA: A]

SYNONYMS: *Achyranthes philoxeroides* (Mart.) Standl.; *Alternanthera paludosa* Bunbury; *Alternanthera philoxerina* Suess.

GENERAL INFORMATION: Noxious *aquatic to terrestrial perennial*, with *opposite leaves* and horizontal to ascending stems to 1 m long that root at the nodes. The aquatic form has *hollow, floating,* emergent and submerged stems, while the terrestrial form has solid stems. Plants are typically rooted in soil in shallow water and can form dense, interwoven floating mats that extend over the surface of deeper water. Mats can become dense enough to support the weight of a person and *can break away and colonize new sites.* Floating mats disrupt the natural ecology of a site by reducing light penetration and crowding out native species. Serious infestations can create anoxic, disease- and mosquito-breeding conditions. Native to South America. The alligatorweed flea beetle *(Agasicles hygrophila)*, stem borer moth *(Vogtia malloi)*, and alligatorweed thrips *(Amynothrips andersoni)* have been released as biocontrol agents in the southeastern United States. These insects can effectively control infestations of alligatorweed, but have not established in California.

SEEDLINGS: Seedlings are seldom encountered since viable seed is rarely produced.

MATURE PLANT: Foliage herbaceous. Stems simple or branched, glabrous or with 2 opposing lengthwise rows of hairs. *Leaves opposite*, pairs nearly equal at

Alligatorweed *(Alternanthera philoxeroides)*, occasionally found on dry areas. D.O. CLARK

Alligatorweed

Alligatorweed (*Alternanthera philoxeroides*) habit in a canal   D.O. CLARK

each node, narrowly lanceolate to obovate, *4–11 cm long*, 1–3 cm wide, margins smooth, glabrous, surface waxy, sessile or with narrow winged petioles to 1 cm long that clasp the stem at the base.

ROOTS AND UNDERGROUND STRUCTURES: Stems stolonlike, root at the nodes. Floating plants have shorter, finer roots than plants rooted in soil. Stem fragments with a node often develop into new plants.

FLOWERS: June–October. Fragrant. *Spikes headlike*, about 1–2 cm in diameter, terminal and axillary, on *stalks 4–9 cm long*. Flowers and bracts *pearly white, glabrous*. Petals lacking. Sepals 5, separate, 5–7 mm long. Stamens 5, opposite sepals, alternate with 5 longer sterile stamens (staminodia). Ovary 1-chambered, contains 1 ovule.

FRUITS AND SEEDS: Utricles membranous, do not open to release the single seed. Seeds smooth, disk-shaped to flattened wedge-shaped. Mature fruits seldom encountered.

POSTSENESCENCE CHARACTERISTICS: Dead stems fall over and contribute to formation of the mat. Mild frost kills leaves but not stems. Severe frosts kill emergent stems but not submerged or buried parts.

HABITAT: Shallow water, wet soils, ditches, marshes, pond margins, slow-moving watercourses. Tolerates saline conditions to 10% salt by volume. Requires a warm summer growing season. Tolerates cold winters, but cannot survive prolonged freezing temperatures.

DISTRIBUTION: Uncommon. San Joaquin Valley (Tulare, Kings Cos.), Southwestern region (Los Angeles, San Diego, Riverside, and probably San Bernardino Cos.), and possibly elsewhere, to 200 m. Southeastern states, including Texas; Central America.

PROPAGATION/PHENOLOGY: *Reproduces vegetatively from stems and stem fragments.* Each node or fragment with a node is capable of developing into a new plant. Alligatorweed typically has a rapid growth rate and is highly competitive with other aquatic vegetation. Plants seldom inhabit water deeper than 2 m.

Alligatorweed (*Alternanthera philoxeroides*) stem segment with adventitious roots.
CDFA INTEGRATED PEST CONTROL PROGRAM

Alligatorweed (*Alternanthera philoxeroides*) node and flower cluster.
CDFA INTEGRATED PEST CONTROL PROGRAM

Alligatorweed

Alligatorweed *(Alternanthera philoxeroides)* seedlings surrounded by mosquitofern *(Azolla* sp.*)*.
CDFA INTEGRATED PEST CONTROL PROGRAM

Mat chaff-flower *(Alternanthera caracasana)* infestation in turf.
J. M. DITOMASO

Mat chaff-flower (*Alternanthera caracasana*) plants.   J. M. DiTomaso

Mat chaff-flower (*Alternanthera caracasana*) shoot section.   J. M. DiTomaso

Seeds rarely develop, and those that do are usually not viable.

MANAGEMENT FAVORING/ DISCOURAGING SURVIVAL: Plants grow best under eutrophic conditions. Mechanical removal without careful removal of all plant parts can facilitate spread. Buried stems can regenerate from depths to 30 cm.

SIMILAR SPECIES: Mat chaff-flower [*Alternanthera caracasana* Kunth, synonyms: washerwoman; *Achyranthes peploides* (Humb. & Bonpl. ex J.A. Schultes) Britt. & Wilson; *Alternanthera peploides* (Humb. & Bonpl. ex J.A. Schultes) Urban; *Alternanthera pungens* Kunth; the name *Alternanthera repens* (L.) Kuntze has been misapplied in some older references][ALRPE] is a *prostrate terrestrial perennial* with stems to 0.5 m long and flower heads that are similar to alligatorweed. Besides its terrestrial habit, mat chaff-flower is distinguished by having *sessile flower heads, leaves 1–3 cm long*, and *spine-tipped sepals*. In addition, flower heads are *ovoid to short cylindrical*, and 5–15 mm long. Stems and flower clusters usually have long wavy hairs. Leaves are ovate to weakly diamond-shaped. Mature plants usually have a thick woody vertical root. Mat chaff-flower grows on drier soils of waste places, roadsides, vacant lots, fields, and turf in the Southwestern region, to 150 m. It also occurs in Arizona, New Mexico, Texas, and most southeastern states. Native to Central and South America.

Mat chaff-flower (*Alternanthera caracasana*) in flower J. M. DiTomaso

Alligatorweed

Mat chaff-flower (*Alternanthera caracasana*) seeds, with and without calyx.   J. K. CLARK

Swamp smartweed [*Polygonum amphibium* L. var. *emersum* Michaux] is an *aquatic perennial with rhizomes* that may be confused with alligatorweed. Swamp smartweed is distinguished by having *alternate leaves* with *fused, sheathing stipules* (ocrea) and *pink flowers*. Refer to the entry for **Swamp smartweed, Pale smartweed, and Ladysthumb** (p. 1319, vol. 2) for more information.

Tumble pigweed [*Amaranthus albus* L.][AMAAL]

Prostrate pigweed [*Amaranthus blitoides* S. Wats.] [AMABL]

Low amaranth [*Amaranthus deflexus* L.]

SYNONYMS: Tumble pigweed: tumbleweed; white pigweed; *Amaranthus albus* L. var. *pubescens* (Uline & Bray) Fern.; *Amaranthus graecizans* auct. non L.; *Amaranthus graecizans* L. var. *pubescens* Uline & Bray; *Amaranthus pubescens* (Uline & Bray) Rydb.

Prostrate pigweed: creeping amaranth; mat amaranth; spreading pigweed; *Amaranthus blitoides* S. Wats. var. *crassior* Jeps.; *Amaranthus graecizans* auct. non L.

Low amaranth: largefruit amaranth; perennial pigweed; *Amaranthus argentinus* Speg.; *Amaranthus viridus* auct. non L.; *Euxolus deflexus* Raf. Plants with wrinkled seeds have been referred to as *Amaranthus gracilis* Desf.

GENERAL INFORMATION: Refer to table 15 (p. 80) for a comparison of distinguishing characteristics among *Amaranthus* species. Under certain conditions, tumble pigweed and prostrate pigweed foliage can accumulate levels of free nitrates that are *toxic* to cattle when ingested in quantity.

Tumble pigweed (*Amaranthus albus*) plant.

J. M. DiTomaso

Tumble pigweed: Bushy summer *annual* to about 0.75 m tall. Tumble pigweed is a widespread weed throughout North America, Eurasia, and southern Australia. Native to tropical America.

Prostrate pigweed: Prostrate summer *annual* with stems to about 0.75 m long. Prostrate pigweed is a ruderal native of the western United States. It is typically

Tumble pigweed (*Amaranthus albus*) flower clusters in leaf axils.   J. K. CLARK

Tumble pigweed • Prostrate pigweed • Low amaranth

Table 15. *Amaranthus* species

Amaranthus spp.	Habit	Inflorescence	Monoecious or dioecious*	Female flower sepals	Other	Distribution
A. albus tumble pigweed	erect, bushy	flowers axillary	monoecious	—	leaves light green, margins wavy; stems firm, usually pale green or whitish; plants become tumbleweeds	common throughout C
A. arenicola sandhills amaranth	erect, bushy	long terminal spike with spikelike branches at the base	dioecious	spoon-shaped, tips rounded to slightly notched, ± with a minute nipplelike point	bracts below all flowers 1–2.5 mm long; main spike to 60 cm long, 1–2 cm wide; upper stems mostly glabrous	very uncommon; Sacrame Valley, Centra and South Co
A. blitoides prostrate pigweed	prostrate	flowers axillary	monoecious	—	leaves dark green, margins not or barely wavy, sometimes white-lined; stems fleshy, often reddish	common throughout m of CA
A. deflexus low amaranth	prostrate or low and spreading	terminal spikelike panicle	monoecious	—	fruits fleshy, ± inflated, do not open to release the seed; panicles mature pinkish brown	occasional an locally comm Sacramento Valley, Centra western and Southwestern regions
A. hybridus smooth pigweed	erect	terminal panicle with spikelike branches	monoecious	tapered to a point, ± slightly pinched in	bracts below flowers 2–4 mm long; upper stems mostly glabrous to pubescent	occasional; Central Valley Central-weste and Southwes regions
A. palmeri Palmer amaranth	erect	long terminal spike ± with spikelike branches at the base	dioecious	spoon-shaped, tips rounded to slightly notched, ± with a minute nipplelike point	bracts below all flowers 2.5–6 mm long; main spike to 50 cm long, ~ 1–2 cm wide; upper stems mostly glabrous	common; San Joaquin Valley Central-weste region, deserts
A. powellii Powell amaranth	erect	terminal panicle with spikelike branches	monoecious	tapered to a point, ± slightly pinched in	bracts below flowers mostly 4–8 mm long; upper stems mostly glabrous	uncommon; Central Valley, Central-weste and Southwes regions
A. retroflexus redroot pigweed	erect	terminal panicle with spikelike branches	monoecious	spoon-shaped, tips rounded to slightly notched, ± with a minute nipplelike point	bracts below flowers mostly 2.5–4.5 mm long; upper stems pubescent to densely hairy	common; Cen Valley, Central-wester and Southwes regions
A. spinosus spiny amaranth	low, bushy	terminal spikelike panicle	monoecious	—	lower stems have 2 spines ~ 1–2 cm long at most nodes	very uncomm South Coast

Note: *Dioecious: male and female flowers develop on separate plants; monoecious: male and female flowers develop the same plant (male flowers sometimes develop only at the tips of panicles or spikes).

Tumble pigweed • Prostrate pigweed • Low amaranth

Tumble pigweed (*Amaranthus albus*) seedling. J. M. DiTomaso

Tumble pigweed (*Amaranthus albus*) seeds, with and without calyx. J. K. Clark

Tumble pigweed • Prostrate pigweed • Low amaranth

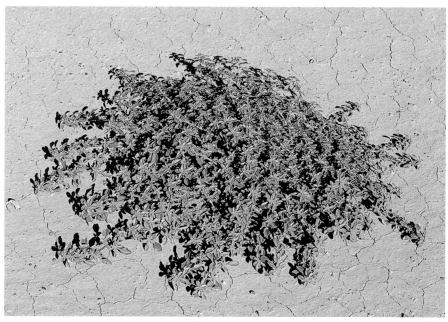

Prostrate pigweed (*Amaranthus blitoides*) plant.   J. M. DiTomaso

Prostrate pigweed (*Amaranthus blitoides*) flowering stem.   J. M. DiTomaso

weedy in dry disturbed places and is an introduced weed throughout much of Europe and the Mediterranean region.

Low amaranth: Prostrate or spreading summer *annual, biennial, or short-lived perennial*, with stems to about 0.5 m long. California plants most often appear to be annual or biennial. European and Australian plants are usually perennial. Native to South America.

Prostrate pigweed (*Amaranthus blitoides*) seedling.   J. M. DiTomaso

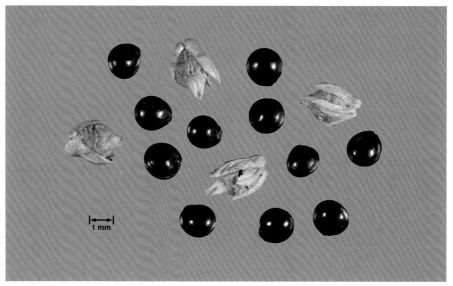

Prostrate pigweed (*Amaranthus blitoides*) seeds, with and without calyx.   J. K. Clark

Tumble pigweed • Prostrate pigweed • Low amaranth

Low amaranth (*Amaranthus deflexus*) infestation in an agricultural field.　　J. M. DiTomaso

SEEDLINGS: Leaves alternate.

Tumble pigweed: Cotyledons narrowly lanceolate, tips slightly rounded, 3–9 mm long, often reddish below, glabrous. First leaf oval to ovate, tip indented (emarginate), midvein and lower surface often reddish. Fifth and later leaf margins slightly wavy, often appear rough (scabrous) with a hand lens.

Prostrate pigweed: Cotyledons narrowly lanceolate to linear oblong, tips ± rounded, 4–10 mm long, 0.75–1.5 mm wide, glabrous. First leaf oval to ovate, tip ± indented, upper surface ± glossy, midvein and lower surface often reddish. Fifth and later leaf margins ± translucent appear rough with a hand lens.

Low amaranth: Cotyledons narrowly lanceolate, ~ 4–5 mm long, tip acute, 1–2 mm wide, stalk below (hypocotyl) often reddish. First leaf ovate-diamond-shaped, tip indented, ~ 3–7 mm long, 2.5–4 mm wide, ± with a few hairs.

MATURE PLANT: Leaves alternate. Stems typically highly branched.

Tumble pigweed: Plants *bushy*. *Stems erect to ascending*, whitish to pale green. Foliage glabrous to sparsely covered with minute hairs. Leaves elliptic to obovate, mostly 1–7 cm long, light green, sometimes early-deciduous, veins conspicuous, tapered to a stalk 0.5–5 cm long. Margins often slightly wavy or ruffled. Lower surfaces sometimes reddish. Axillary leaves smaller, persistent.

Prostrate pigweed: Plants *prostrate-spreading*. Stems mostly prostrate, often pink or red-tinged. Foliage mostly glabrous. Leaves elliptic to ovate or obovate, 0.5–4 cm long, often crowded, dark glossy green, sometimes early-deciduous, veins conspicuous, tapered to a stalk 0.5–3 cm long. Margins often

Tumble pigweed • Prostrate pigweed • Low amaranth

Low amaranth *(Amaranthus deflexus)* flowering stem. J. M. DiTomaso

slightly wavy, often white-lined. Axillary leaves smaller, persistent.

Low amaranth: Plants *low, spreading*. Stems usually *prostrate to decumbent*. Foliage sparse to moderately covered with short soft, somewhat crinkled hairs. Leaves ovate to diamond-shaped, 0.5–5 cm long, dull grayish green, tapered to a stalk 0.5–4 cm long. Margins often irregularly tightly rolled under.

ROOTS AND UNDERGROUND STRUCTURES: Tumble pigweed, prostrate pigweed: Taproot generally shallow. Lateral roots usually spreading.

Low amaranth: Taproot often deep. Plants frequently break at the crown when pulled by hand. New shoots can regrow from the upper portion of the remaining taproot and crown.

FLOWERS: *Male and female flowers develop separately on the same plant* (monoecious). Petals lacking.

Tumble pigweed: June–October. Flowers *axillary, inconspicuous*. Male/female flower sepals 3.

Prostrate pigweed: July–November. Flowers *axillary, inconspicuous*. Male/female flower sepals 4–5.

Low amaranth: May–November. Flower clusters *spikelike, terminal, conspicuous*, usually pinkish green or pinkish brown. Male flower sepals 2–3. Female flower sepals 2.

FRUITS AND SEEDS: Fruits contain 1 glossy seed.

Tumble pigweed: Capsules 1.5–2 mm long, open like a lid (circumscissile). Seeds lens-shaped, ± 1 mm in diameter, dark reddish brown.

Prostrate pigweed: Capsules 2–3 mm long, open like a lid (circumscissile). Seeds lens-shaped, 1–2 mm in diameter, margin acute, glossy black.

Low amaranth: Utricles *fleshy, inflated*, ovoid, 2.5–3 mm long, *do not open to release the seed*. Seeds oval lens-shaped, ± 1 mm long, glossy dark reddish brown to black.

Low amaranth (*Amaranthus deflexus*) seedling. J. M. DiTomaso

Tumble pigweed • Prostrate pigweed • Low amaranth

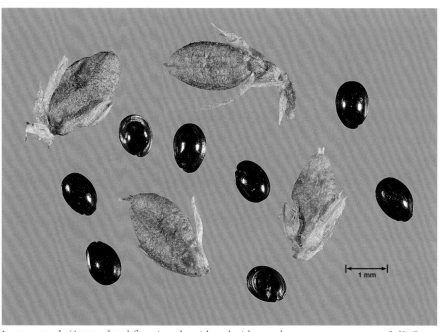

Low amaranth (*Amaranthus deflexus*) seeds, with and without calyx. J. K. Clark

Spiny amaranth (*Amaranthus spinosus*) plant. J. M. DiTomaso

Tumble pigweed • Prostrate pigweed • Low amaranth

Spiny amaranth (*Amaranthus spinosus*) flowering stem with spines.   J. M. DiTomaso

Spiny amaranth (*Amaranthus spinosus*) foliage.   J. M. DiTomaso

HABITAT: Disturbed places, roadsides, fields, gardens, landscaped areas, waste places, field and vegetable crops, orchards, vineyards, turf (prostrate pigweed), urban sites.

Low amaranth: Often found in urban environments in southern California, most frequently on compacted soils or in pavement cracks.

DISTRIBUTION: Tumble pigweed: Common throughout California to 2200 m. All contiguous states.

Prostrate pigweed: Throughout California, except the Great Basin, to 2200 m. Nearly all contiguous states.

Low amaranth: Central-western region and Southwestern region, to 650 m. Most central and eastern states.

PROPAGATION/PHENOLOGY: *Reproduce by seed.* Seeds of all species germinate in spring as the temperature warms.

Tumble pigweed: At maturity, main stems of tumble pigweed become brittle and detach at ground level under windy conditions. Most seeds disperse as the dried bushes tumble with the wind. Exposure to light increases seed dormancy. Ingestion by sheep enhances germination. Plants use the C4 photosynthetic pathway.

Prostrate pigweed, low amaranth: The biology of these species is poorly understood. Most seeds probably fall near the parent plant, but some disperse to greater distances with water, mud, soil movement, and agricultural and other human activities, and as seed and feed contaminants.

MANAGEMENT FAVORING/DISCOURAGING SURVIVAL: Cultivation, summer annual crops that use effective herbicides, and manual removal of plants as needed before seed develops can control *Amaranthus* species. Tumble pigweed and prostrate pigweed tolerate certain herbicides.

SIMILAR SPECIES: Spiny amaranth [*Amaranthus spinosus* L][AMASP] is a low, bushy summer *annual* with reddish stems to about 1 m long and *terminal spike-like flower clusters*. Plants also have axillary flower clusters. Spiny amaranth is readily distinguished from other amaranth species by having *2 rigid spines 0.8–2 cm long at most lower nodes*. Spiny amaranth is a very uncommon weed of roadsides, waste places, and other disturbed habitats in the South Coast and western Mojave Desert, to 700 m. It is more common in the eastern United States and Hawaii. Native to tropical America.

Spiny amaranth (*Amaranthus spinosus*) seeds, with and without calyx.　　　J. K. CLARK

Smooth pigweed [*Amaranthus hybridus* L.][AMACH]
Palmer amaranth [*Amaranthus palmeri* S. Wats.] [AMAPA]
Redroot pigweed [*Amaranthus retroflexus* L.][AMARE]

SYNONYMS: Smooth pigweed: amaranth pigweed; green amaranth; red amaranth; rough pigweed; slender pigweed; slim amaranth; spleen amaranth; *Amaranthus chlorostachys* Willd.; *Amaranthus hypochondriacus* L.; *Amaranthus incurvatus* Tim. ex Gren. & Godr.; *Amaranthus patulus* Bertol.; *Amaranthus quitensis* Kunth

Smooth pigweed (*Amaranthus hybridus*) plant.   J. M. DiTomaso

Smooth pigweed (*Amaranthus hybridus*) flowering stem. J. M. DiTomaso

Smooth pigweed (*Amaranthus hybridus*) seedling. J. M. DiTomaso

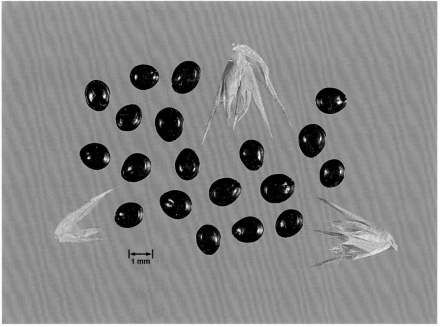
Smooth pigweed (*Amaranthus hybridus*) seeds, with and without calyx. J. K. Clark

Smooth pigweed • Palmer amaranth • Redroot pigweed

Palmer amaranth *(Amaranthus palmeri)* plant.
J. M. DiTomaso

Palmer amaranth *(Amaranthus palmeri)* male flowering stem.
J. M. DiTomaso

Palmer amaranth *(Amaranthus palmeri)* foliage
J. M. DiTomaso

Palmer amaranth: carelessweed

Redroot pigweed: carelessweed; Chinaman's greens; green amaranth; pigweed; redroot; redroot amaranth; rough pigweed; *Amaranthus retroflexus* L. var. *salicifolius* I.M. Johnston; *Amaranthus tricolor* L.

GENERAL INFORMATION: Erect summer *annuals* with *dense terminal panicles of inconspicuous flowers*. Refer to table 15 (p. 80) for a comparison of distinguishing features among *Amaranthus* species. Under certain conditions, smooth pigweed and redroot pigweed foliage can accumulate levels of free nitrates that are *toxic* to livestock when ingested. Plants contain the highest quantity of nitrate just prior to flowering and can also accumulate oxalates. However, livestock poisoning due to elevated nitrate or oxalate levels has not been reported in California. Both species are highly nutritious and have been suggested for use as forage by some scientists. Redroot pigweed and smooth pigweed can hybridize with some amaranth species, including certain cultivated species and **Powell amaranth** [*Amaranthus powellii* S. Wats.][AMAPO]. Such hybridization often makes identification to species difficult. Smooth pigweed is eaten as a green vegetable in parts of Africa and India and is native to the eastern United States, Mexico, Central America, and northern South America. Palmer amaranth is a ruderal native of the southwestern and south-central United States and northern Mexico. Redroot pigweed is a ruderal native of the central and eastern United States, southeastern Canada, and northeastern Mexico. It is state-listed as a secondary noxious weed in Minnesota.

SEEDLINGS: Leaves alternate.

Smooth pigweed, redroot pigweed: Cotyledons narrowly lanceolate, mostly 3–12 mm long, 0.75–3 mm wide, glabrous, lower surface often reddish. Stalk below (hypocotyl) often reddish. Leaves alternate. First leaf ovate, tip usually

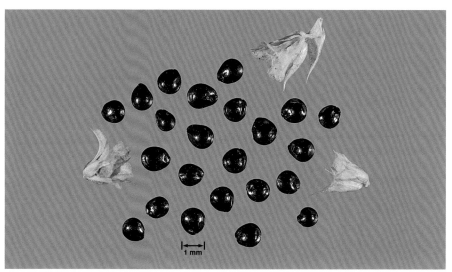

Palmer amaranth (*Amaranthus palmeri*) seeds, with and without calyx.   J. K. CLARK

Smooth pigweed • Palmer amaranth • Redroot pigweed

Redroot pigweed (*Amaranthus retroflexus*) plant.
J. M. DiTomaso

Redroot pigweed (*Amaranthus retroflexus*) seedling.
J. M. DiTomaso

Flower cluster of redroot pigweed (*Amaranthus retroflexus*).
J. K. Clark

Redroot pigweed (*Amaranthus retroflexus*) root.
J. K. Clark

Smooth pigweed • Palmer amaranth • Redroot pigweed

Redroot pigweed *(Amaranthus retroflexus)* flowering stem. J. M. DiTomaso

slightly indented (emarginate), lower surface and main veins often reddish. Redroot pigweed leaf stalks (petioles) usually hairy along the margins. Smooth pigweed leaf stalks glabrous to hairy.

MATURE PLANT: Leaves alternate, mostly 1.5–17 cm long, on stalks about 1–8 cm long. Lower leaves ovate to diamond-shaped. Upper leaves generally lanceolate.

Smooth pigweed: To 2.5 m tall. Upper stems mostly glabrous to pubescent with minute contorted hairs. Lower stems often reddish. Stems usually not as thick as those of redroot pigweed.

Palmer amaranth: To 2 m tall. Stems coarse. Upper stems usually nearly glabrous. Lower stems usually green.

Redroot pigweed: To 3 m tall. Stems coarse. Upper stems pubescent to densely covered with contorted hairs. Lower stems often reddish.

ROOTS AND UNDERGROUND STRUCTURES: Taproots thick, often shallow, upper portion often reddish in smooth pigweed and redroot pigweed.

FLOWERS: For species with panicles that consist of male and female flowers (monoecious), male flowers sometimes develop only at the branch tips. Male and female flower sepals usually 5. Bracts immediately below each flower acute, often spine-tipped. Mostly wind-pollinated. Self-compatible.

Smooth pigweed: June–November. Terminal panicles dense, sometimes flexuous, mostly 5–30 cm long, with ± spreading spikelike branches throughout, *consist of male and female flowers* (monoecious). Female flower sepals mostly 1.5–2.5 mm long. *Inner sepal tips mostly acute or slightly pinched-in and tapered to a point* (acuminate). Female and male flower bracts below spine-tipped, mostly 2–3.5 mm long.

Palmer amaranth: July–November. Terminal panicles *spikelike*, simple or with axillary spikes at the base, *consist only of male flowers or only of female flowers* (dioecious). *Main spike about 10–50 cm long,* mostly *0.7–2 cm wide*, often drooping. *Male and female flower outer sepal spine-tipped, longer than inner sepals. Inner female flower sepals spoon-shaped, mostly 2–4 mm long, tips rounded or notched, often with a minute abrupt nipplelike point* (mucronulate). Bracts

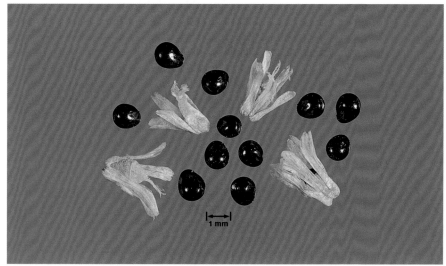

Redroot pigweed *(Amaranthus retroflexus)* seeds, with and without calyx. J. K. CLARK

Smooth pigweed • Palmer amaranth • Redroot pigweed

Powell amaranth (*Amaranthus powellii*) flowering stem.   J. M. DiTomaso

below rigid, *spine-tipped, 2.5–6 mm long*, some tips bent downward (reflexed) in fruit. Male flower inner sepals *tapered to a point, 2.5–6 mm long*. Bracts below spinelike, 2.5–6 mm long.

Redroot pigweed: June–November. Terminal panicles dense, usually stiff, mostly 5–20 cm long, 2–6 cm wide, with ascending spikelike branches throughout, *consist of male and female flowers* (monoecious). Female flower 2–3.3 mm long. *Inner sepal tips rounded, truncate, or notched, often with a minute abrupt nipplelike point.*

FRUITS AND SEEDS: Capsules ± 1.5 mm long, open like a lid to release 1 seed. Seeds lens-shaped, 1–1.5 mm in diameter, glossy, dark reddish brown to black.

POSTSENESCENCE CHARACTERISTICS: Dead stems with brown panicles may persist for a short period.

HABITAT: Cultivated fields, pastures, orchards, vineyards, gardens, landscaped areas, ditch banks, washes, roadsides, waste places, and other disturbed places. Smooth pigweed and redroot pigweed grow best on fertile soil. All thrive in open sunny places.

DISTRIBUTION: Smooth pigweed: Central Valley, Central-western region, Southwestern region, and possibly elsewhere, to 300 m. All contiguous states, except possibly Utah and Wyoming.

Palmer amaranth: Central Valley, Central-western region, deserts, to 1200 m. Throughout the southern two-thirds of the United States.

Redroot pigweed: Central Valley, Northwestern region, Central-western region, Southwestern region, Modoc Plateau, possibly elsewhere, to 2400 m. All contiguous states.

PROPAGATION/PHENOLOGY: Reproduce by seed. Seeds fall near the parent plant or disperse short distances with wind. Some seeds disperse greater distances with water, mud, soil movement, animals, agricultural and other human activities, and as seed and feed contaminants. Seeds germinate in spring as the temperature warms, and continue throughout summer when adequate moisture is available. Light and/or high temperature stimulate germination. Seeds between populations, within populations, and from the same plant vary in dormancy and germination requirements. Buried redroot pigweed seeds survive from about 3 years up to 40 years, depending on the biotype and environmental conditions. Smooth pigweed, redroot pigweed, and Powell amaranth use the C4 photosynthetic pathway. The photosynthetic pathway of Palmer amaranth is undocumented.

Powell amaranth (*Amaranthus powellii*) seedling. J. M. DiTomaso

MANAGEMENT FAVORING/DISCOURAGING SURVIVAL: Cultivation and manual removal of plants before seed develops can control *Amaranthus* species. Some redroot pigweed biotypes tolerate certain herbicides.

SIMILAR SPECIES: Refer to table 15 (p. 80) for a comparison of distinguishing features among *Amaranthus* species. **Powell amaranth** [*Amaranthus powellii* S. Wats., synonyms: *Amaranthus bouchonii* Thellung; *Amaranthus bracteosus* Uline & Bray; *Amaranthus viscidulus* Greene][AMAPO] is a ruderal native summer annual to 2 m tall. Powell amaranth has *panicles of male and female flowers* that are ± narrow and spikelike or resemble those of redroot pigweed and smooth pigweed. In addition, it has *inner sepals that are tapered to a point* like those of smooth pigweed. Unlike smooth pigweed, Powell amaranth has *female flower bracts 4–8 mm long* and *male flower bracts 3–8 mm long*. Powell amaranth is a widespread native that inhabits disturbed places. Occasionally it is weedy in crop fields, orchards, and vineyards, but it is less common than redroot pigweed, smooth pigweed, and Palmer amaranth. Powell amaranth occurs in the Central Valley, Sierra Nevada foothills, Central-western region, and Southwestern region, to 800 m, and is scattered throughout much of the United States.

Sandhills amaranth [*Amaranthus arenicola* I.M. Johnston, synonym: *Amaranthus torreyi* auct. non (Gray) Benth. ex S. Wats.][AMAAR] is an uncommon summer annual to 2 m tall, with a bushy habit and *dioecious panicles* similar to those of Palmer amaranth. Unlike Palmer amaranth, sandhills amaranth has *male flowers with the outermost sepal short-pointed and slightly longer than the other sepals* and *female flowers with an outermost sepal that is spoon-shaped with a rounded tip. Bracts of all flowers are mostly 1–2.5 mm long*. Sandhills amaranth sporadically inhabits sandy disturbed places in the Sacramento Valley, Central Coast, and South Coast, to 200 m. Native to the central and eastern United States.

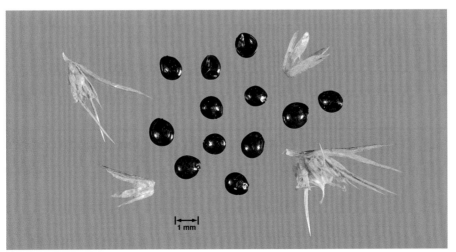

Powell amaranth (*Amaranthus powellii*) seeds, with and without calyx.   J. K. Clark

Peruvian peppertree [*Schinus molle* L.][SCIMO] [Cal-IPC: Limited]

Brazilian peppertree [*Schinus terebinthifolius* Raddi] [SCITE][Cal-IPC: Limited]

SYNONYMS: Peruvian peppertree: American pepper; California peppertree; escobilla; false pepper; molle de Peru; *Schinus angustifolius* Sessé & Moc.; *Schinus areira* L.; *Schinus bituminosus* Salisb.; *Schinus huigan* Molina; *Schinus occidentalis* Sessé & Moc.; others

Brazilian peppertree: Christmasberry; false pepper; faux poivrier; Florida holly; *Schinus antiarthriticus* Mart. ex March.; *Schinus mellisii* Engl.; *Schinus mucronulatus* Mart.; *Sarcotheca bahiensis* Turcz.

GENERAL INFORMATION: Evergreen *shrubs to trees* with *aromatic odd-pinnate-compound leaves*. The peppertrees are common landscape ornamentals that were introduced to California 100–200 years ago. Both species have escaped cultivation and become invasive in some areas.

Peruvian peppertree: More widespread than Brazilian peppertree in California, but appears to be less problematic. Peruvian peppertree is susceptible to black scale (*Saissetia oleae*), a detrimental pest of citrus. Plants can cause dermatitis in

Peruvian peppertree (*Schinus molle*) plant. J. M. DiTomaso

sensitive individuals. Fruits are used to make a drink in South America. Native to the riparian habitats of Peru.

Brazilian peppertree: This species is locally invasive in certain riparian areas of southern California and has aggressively colonized hundreds of thousands of acres in Florida. Brazilian peppertree foliage can be *toxic* to horses and cattle when ingested, and direct contact with the sap can cause contact dermatitis in sensitive individuals. Brazilian peppertree fruits are sometimes sold as "pink peppercorns" in the United States. However, ingestion of fruits in large quantities can cause severe digestive tract irritation in animals and humans. Native to the dry grasslands of southern Brazil.

SEEDLINGS: Brazilian peppertree seedlings tolerate shade.

MATURE PLANT: Leaves *alternate, odd-pinnate-compound*, glabrous, often slightly resinous, *aromatic*, especially when crushed. Stipules lacking.

Peruvian peppertree: To 18 m tall. Trunk bark scaly, light brown. Plants rounded in outline, with spreading branches. *Branchelets, twigs, and leaves drooping.* Leaves mostly 10–30 cm long. *Leaflets 15–59 per leaf*, sessile, narrowly lanceolate, about 3–6 cm long, mostly less than 1.25 cm wide. *Upper and lower surfaces light green.* Margins smooth to weakly toothed.

Brazilian peppertree: To 10 m tall, often multistemmed. Trunk bark coarse, gray-brown. Branches spreading to erect. *Branchlets, twigs, and leaves stiff,*

Peruvian peppertree (*Schinus molle*) flowers and foliage. J. M. DiTomaso

Peruvian peppertree (*Schinus molle*) fruit. J. M. DiTomaso

Brazilian peppertree *(Schinus terebinthifolius)* in flower. J. M. DiTomaso

spreading to erect. Leaves about 8–15 cm long. *Leaflets usually about 7 per leaf, leathery, sessile or short-stalked, elliptic to oblong, 2.5–7 cm long, 1–3 cm wide, upper surface slightly glossy and dark green, lower surface paler.* Margins smooth to toothed.

ROOTS AND UNDERGROUND STRUCTURES: Both species generally develop a shallow root system and can produce new shoots from roots. Roots of Peruvian peppertree are typically more widely spreading than those of Brazilian peppertree, but undamaged Brazilian peppertree is more likely to root sprout.

FLOWERS: June–August (Peruvian peppertree); September–November and sometimes March–May (Brazilian peppertree). *Male and female flowers develop on separate trees* (dioecious). Panicles consist of numerous greenish white flowers mostly 1–3 mm long. Petals and sepals 5. Stamens 10 or styles 3, fused at base. Insect-pollinated.

FRUITS AND SEEDS: Fruits *spherical, berrylike, 4–8 mm in diameter, leathery, pink to red, contain 1 seed*. Peruvian peppertree fruits mature in fall. Brazilian peppertree fruits mature primarily in late fall through winter.

HABITAT: Canyons, washes, slopes, riparian areas, fields, roadsides. Plants grow best where some soil moisture is available during the warm season. Both species tolerate poor soils, heat, air pollution, and drought.

Brazilian peppertree (*Schinus terebinthifolius*) in fruit.   J. M. DiTomaso

DISTRIBUTION: Peruvian peppertree: Central Valley, Central-western region, Southwestern region, Sierra Nevada foothills, and Tehachapi Mountains, to 700 m. Texas and Mexico.

Brazilian peppertree: Southern South Coast, to 200 m. Hawaii, Texas and Florida.

PROPAGATION/PHENOLOGY: *Reproduce by seed* and sometimes vegetatively from root sprouts. Fruits are readily consumed and dispersed by wildlife, particularly birds, or persist on female trees for up to ~ 8 months. Germination occurs anywhere from November to April, depending on environmental conditions. Seeds on the soil surface do not survive fire. Saplings grow rapidly under favorable conditions.

Peruvian peppertree: Seeds germinate more quickly than those of Brazilian peppertree.

Brazilian peppertree: Most seeds remain viable for less than 1 year after dispersal. Once established, seedlings exhibit high survival rates under many conditions. Brazilian peppertree seedlings are more drought tolerant than those of Peruvian peppertree.

MANAGEMENT FAVORING/DISCOURAGING SURVIVAL: Manually cutting or burning mature trees can stimulate root sprouting. After removal of mature plants, hand-pulling seedlings and removing root sprouts for 3 years or more can control infestations.

SIMILAR SPECIES: Peruvian and Brazilian peppertree are unlikely to be confused with other weedy tree species. However, at first glance, Brazilian peppertree may be mistaken for **toyon** [*Heteromeles arbutifolia* (Lindl.) Roem.], a *desirable native shrub* with red berrylike fruits in fall and winter. Toyon is distinguished by having *simple, nonaromatic leaves* that are similar in size, shape, color-, and texture to Brazilian peppertree leaflets. In addition, toyon fruits contain *3–6 seeds* and have a *crown of 5 small sepals at the apex.*

Pacific poison-oak [*Toxicodendron diversilobum* (Torr. & Gray) Greene][RHUDI]

SYNONYMS: poison oak; western poison oak; *Rhus diversiloba* Torr. & Gray; *Toxicodendron comarophyllum* Greene; *Toxicodendron dryophyllum* Greene; *Toxicodendron isophyllum* Greene; *Toxicodendron oxycarpum* Greene; *Toxicodendron radicans* (L.) Kuntze ssp. *diversilobum* (Torr. & Gray) Thorne; *Toxicodendron vaccarum* Greene

GENERAL INFORMATION: *Deciduous shrub* to 4 m tall, with compound leaves that typically consist of *3 leaflets*. Plants are sometimes treelike or vinelike with stems to 25 m long. Pacific poison-oak is a common native shrub of disturbed and undisturbed habitats. Although native plants are generally desirable in natural areas, Pacific poison-oak is considered one of the most hazardous plants in the western states. It can be problematic wherever people are likely to contact the plant, such as along trails in recreational areas or during brush removal around homes, along rights-of-way, fire breaks, construction sites, etc. All plant parts, except the pollen, have resin canals that contain the phenolic compound urushiol. Direct contact with bruised, broken, or insect-damaged plant parts, including dormant leafless stems, or contact with items such as tools, clothing, gloves, and pets that have had direct contact with plants can cause allergenic contact dermatitis in sensitive individuals. Smoke from burning plant material

Pacific poison-oak (*Toxicodendron diversilobum*) in open sun growing as a shrub.   J. M. DiTomaso

can cause severe respiratory irritation if inhaled. Sensitivity to Pacific poison-oak often increases with repeated exposure.

SEEDLINGS: Cotyledons obovate to oblong, short-stalked, glabrous, often slightly glossy. First leaves appear opposite, consist of 3 glabrous leaflets. Terminal leaflet lanceolate, much longer than the elliptic lateral leaflets.

MATURE PLANT: Highly variable. *Twigs glabrous to sparsely*

Pacific poison-oak (*Toxicodendron diversilobum*) in shade growing as a climbing vine.
J. M. DiTomaso

hairy, gray to reddish brown. Leaves alternate, *compound with 3(–5) leaflets*. Terminal leaflets 1–13 cm long, 1–8 cm wide, *base rounded or tapered with straight margins to a short, conspicuous stalk*. Lateral leaflets 1–7 cm long, 1–8 cm wide. Leaflet margins smooth, wavy, or slightly rounded-lobed. Upper surface glabrous or nearly glabrous, usually *slightly glossy*. Lower surface sparsely short-hairy.

ROOTS AND UNDERGROUND STRUCTURES: Roots and rhizomes woody, extensive. Rhizomes develop new shoots.

Pacific poison-oak (*Toxicodendron diversilobum*) leaf and flower cluster. J. K. Clark

Pacific poison-oak (*Toxicodendron diversilobum*) leaf turning red in fall. J. M. DiTomaso

Pacific poison-oak (*Toxicodendron diversilobum*) fruit. J. M. DiTomaso

FLOWERS: April–May. Male and female flowers develop on separate plants (dioecious). Panicles axillary, drooping, or spreading to erect. *Flower stalks 2–8 mm long.* Petals 5, yellowish green, 2–4 mm long. Sepals and stamens 5. Stigmas 3. Insect-pollinated.

FRUITS AND SEEDS: Berries *cream-colored or creamy white with dark striations*, turn brown with age, ± spherical, 1.5–6 mm in diameter, glabrous to sparsely stiff-hairy, contain 1 seed.

POSTSENESCENCE CHARACTERISTICS: Leaves deciduous, turn to shades of orange and red in fall.

HABITAT: Chaparral, oak woodlands, conifer and mixed conifer forests.

DISTRIBUTION: Throughout California, except Great Basin and southwestern edge of the Mojave Desert, to 1650 m. Oregon, Washington, and Nevada.

PROPAGATION/PHENOLOGY: *Reproduces by seed and vegetatively from rhizomes.* Seeds disperse primarily with animals, especially birds. Ingestion by birds facilitates germination by decreasing the dormancy period.

MANAGEMENT FAVORING/DISCOURAGING SURVIVAL: Mechanical removal of plants, including the root systems, is most effective when the soil is moist. Repeated grazing by sheep and/or goats will eventually kill the plant by exhausting the root carbohydrate reserves. This is most effective in small areas. Burning Pacific poison-oak does not kill the root system, and the smoke is hazardous to human health. Late-season systemic herbicide treatments to mature

Skunkbrush (*Rhus trilobata*) foliage resembles Pacific poison-oak, but lacks a terminal petiolule.

J. M. DiTomaso

Pacific poison-oak

Skunkbrush *(Rhus trilobata)* flowers.
J. M. DiTomaso

Skunkbrush *(Rhus trilobata)* foliage turning red resembles Pacific poison-oak.
J. M. DiTomaso

plants can also be effective. Protective clothing, including washable cotton gloves over plastic gloves, can help prevent exposure to the resin that causes contact dermatitis.

SIMILAR SPECIES: Skunkbrush or squaw bush [*Rhus trilobata* Torr. & Gray, synonyms: *Schmaltzia trilobata* (Nutt. ex Torr. & Gray) Small; *Rhus aromatica* Aiton ssp. *trilobata* (Nutt. ex Torr. & Gray) W.A. Weber] is a *native shrub* that is not considered a weed but is often confused with Pacific poison-oak. Unlike Pacific poison-oak, skunkbrush has *red berrylike fruits that are covered with short glandular hairs, terminal flower clusters with stiff branches, sessile flowers, current year's twigs moderate to densely covered with short hairs* (pubescent), *leaflets with a dull upper surface,* and *terminal leaflet bases that are sessile and tapered with concave margins.* Skunkbrush occurs on slopes, in washes, chaparral, and woodlands throughout California, except the Great Basin, to 2200 m. It is also found in all other western states, except Washington.

Greater ammi [*Ammi majus* L.][AMIMA]
Toothpick ammi [*Ammi visnaga* (L.) Lam.][AMIVI]
Wild celery [*Cyclospermum leptophyllum* (Pers.) Sprague ex Britt. & E. Wilson][APULE]

SYNONYMS: Greater ammi: Bishop's flower; Bishop's weed; lace flower; large bullwort; Queen Anne's lace

Toothpick ammi: bisnaga; toothpickweed; *Apium visnaga* (L.) Crantz; *Carum visnaga* (L.) Koso-Pol.; *Daucus visnaga* L.; *Selinum visnaga* (L.) Krause; *Visnaga daucoides* Gaertn.

Wild celery: marsh-parsley; wild parsley; *Apium leptophyllum* (Pers.) F. Muell. ex Benth.; *Apium tenuifolium* (Moench) Thellung ex Hegi; *Cyclospermum ammi* Lag.; *Sison ammi* L.; *Pimpinella leptophylla* Pers. Cyclospermum leptophyllum appears to be the original spelling and *Ciclospermum* is an orthographic variant used in many floras.

GENERAL INFORMATION: *Glabrous* plants with *dissected leaves* and *umbels of white flowers.*

Greater ammi (*Ammi majus*) plant.
J. M. DiTomaso

Greater ammi (*Ammi majus*) infestation.
J. M. DiTomaso

Greater ammi, toothpick ammi: Erect winter *annuals or biennials* to about 1 m tall, with compound umbels. Under certain conditions, greater ammi can accumulate levels of nitrates that are *toxic* to livestock. Greater ammi seeds contain furocoumarins and can cause photosensitization in livestock and fowl when consumed in quantity. Both species are cultivated for seeds, which are used medicinally in some countries. Toothpick ammi rarely causes photosensitization in livestock. Greater ammi tolerates certain herbicides and can be difficult to control in carrot fields. Native to Eurasia.

Greater ammi (*Ammi majus*) in fruit. J. M. DiTomaso

Greater ammi (*Ammi majus*) umbel.
J. M. DiTomaso

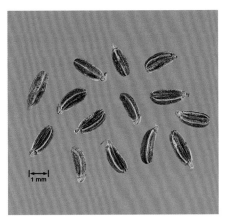

Greater ammi (*Ammi majus*) seeds.
Jim O'Brien

Wild celery: Erect to decumbent winter or summer *annual* to 0.6 m tall, with compound and simple umbels. Wild celery occurs nearly worldwide. It is considered to be introduced to the United States, but its origin is unclear.

SEEDLINGS: Glabrous.

Toothpick ammi (*Ammi visnaga*) infestation. J. M. DiTomaso

Toothpick ammi (*Ammi visnaga*) flowering umbel. J. M. DiTomaso

Greater ammi: Cotyledons linear, mostly 4–12 mm long, 0.5–1.5 mm wide. First leaf ± ovate, irregularly shallow-lobed and toothed or deeply 3-lobed, mostly 5–20 mm long, 4–14 mm wide. Second leaf often compound with 3 leaflets, mostly 5–20 mm long, on a stalk about 1–2 cm long. Central leaflet much larger than lateral leaflets.

Toothpick ammi: Cotyledons linear, 7–20 mm long, 1–2 mm wide. First leaf mostly 5–12 mm long, 6–8 mm wide, dissected 2 times into narrow segments, on a stalk ± 1 cm long.

Wild celery: Cotyledons linear, mostly 4–10 mm long, ± 1 mm wide. First leaf deeply 3-lobed, long-stalked.

MATURE PLANT: Leaves alternate, *dissected 1–3 times.*

Greater ammi: Leaves 6–20 cm long, on stalks 1–5 cm long. Lower leaflets or segments lanceolate, 1–1.5 cm long, margins serrate. Upper leaf segments linear.

Toothpick ammi: Leaves 5–20 cm long, on stalks ± 1 cm long. Leaf segments linear to threadlike, 0.5–3.5 cm long.

Toothpick ammi (*Ammi visnaga*) in fruit.
J. M. DiTomaso

Toothpick ammi (*Ammi visnaga*) seedling.
J. M. DiTomaso

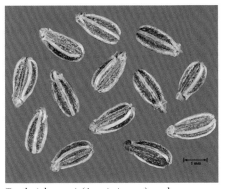

Toothpick ammi (*Ammi visnaga*) seeds.
J. K. Clark

Wild celery: Leaves 3–10 cm long. Upper leaves ± sessile. Lower leaves on stalks 2–12 cm long. Leaf segments linear to threadlike, 0.3–1.5 cm long.

ROOTS AND UNDERGROUND STRUCTURES: Taprooted, with fibrous laterals.

FLOWERS: Petals 5, white. Stamens 5. Styles 2. Ovary inferior, 2-chambered. Insect-pollinated.

Greater ammi, toothpick ammi: May–August. Umbels compound. Primary umbel bracts leaflike, *pinnate-dissected into linear segments*. Petal tips 2-lobed.

Wild celery *(Cyclospermum leptophyllum)* plant.
J. M. DiTomaso

Wild celery *(Cyclospermum leptophyllum)* flowers and fruiting stem.
J. M. DiTomaso

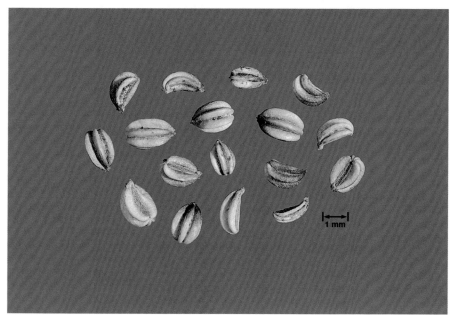

Wild celery *(Cyclospermum leptophyllum)* seeds.
J. O'Brien

Greater ammi • Toothpick ammi • Wild celery

Wild celery (*Cyclospermum leptophyllum*) fruiting stem. J. M. DiTomaso

Wild celery (*Cyclospermum leptophyllum*) seedling. J. M. DiTomaso

Sepal lobes 5, minute. Greater ammi: Base of primary umbels *slender*. Primary rays mostly 20–60, 2–7 cm long, slightly rough to touch (scabrous). Toothpick ammi: Base of primary umbels thick, *disklike*. Primary rays mostly 60–100, 2–5 cm long, glabrous.

Wild celery: April–August. Umbels compound and sometimes simple. Primary and secondary *umbel bracts lacking*. *Primary umbel rays 1–3*, less than 2 cm long. Flowers minute. Petal tips ± acute, not curved inward. Sepal lobes lacking.

FRUITS AND SEEDS: Fruits (schizocarps) *oblong to ovoid*, slightly compressed laterally (perpendicular to the central wall), ribbed, glabrous. Fruits have 1 conspicuous oil tube between each rib.

Greater ammi: Primary umbel rays open, spreading. Fruits mostly 1.5–2 mm long, ribs narrow.

Toothpick ammi: Primary umbel rays curved inward, form a *dense, bird's nest–like umbel*. Fruits mostly 2–2.5 mm long, ribs narrow.

Wild celery: Fruits 1–3 mm long and wide, ± 5-angled, ribs thick.

POSTSENESCENCE CHARACTERISTICS: Greater ammi and toothpick ammi have fruits that remain attached to the umbels, and flowering stems can persist into winter.

HABITAT: Disturbed places, fields, roadsides, railroad tracks, gardens, orchards, field and vegetable crops, and associated ditches and margins. Wild celery also occurs in turf and landscaped areas. Greater ammi usually inhabits moist sites.

DISTRIBUTION: Greater ammi: North Coast, southern North Coast Ranges, San Francisco Bay region, Central Valley, southern South Coast Ranges, South Coast, and possibly elsewhere, to about 1000 m. Oregon, Arizona, Texas, Louisiana, South Dakota, Missouri, Pennsylvania, and southern states.

Toothpick ammi: Sacramento Valley, northern Sierra Nevada foothills, southern San Francisco Bay region, Central Coast, and South Coast, to about 1000 m. Oregon, Texas, North Carolina, Florida, Alabama, and Pennsylvania.

Wild celery: Scattered throughout California to 150 m. Oregon, Nevada, Arizona, New Mexico, and southern and eastern states.

PROPAGATION/PHENOLOGY: *Reproduce by seed.* Seeds fall near parent plants or disperse to greater distances with water, soil movement, agricultural and landscape machinery and other human activities, and possibly as seed or feed contaminants.

MANAGEMENT FAVORING/DISCOURAGING SURVIVAL: Cultivation before seeds mature can control all species.

SIMILAR SPECIES: Wild carrot [*Daucus carota* L.] and **poison-hemlock** [*Conium maculatum* L.] somewhat resemble greater ammi. Unlike greater ammi, wild carrot has *hairy foliage* and *bristly fruits*. Poison-hemlock is typically a stout *biennial* to 3 m tall and is distinguished by *musty-scented foliage*, especially when crushed, and stems that are usually *purple-spotted or purple-streaked*. In addition, primary umbels of poison-hemlock have a *few small, simple bracts at the base*. Refer to the entries for **Wild carrot** (p. 137) and **Poison-hemlock** (p. 132) for more information.

Bur chervil [*Anthriscus caucalis* Bieb. [ANRCA]
Hedgeparsley [*Torilis arvensis* (Huds.) Link][TOIAR] [Cal-IPC: Moderate]

SYNONYMS: Bur chervil: *Anthriscus neglecta* Boiss. & Reut. var. *scandix* (Scop.) Hyl.; *Anthriscus scandicina* (Weber ex Wigg.) Mansf.; *Anthriscus vulgaris* Pers.; *Caucalis scandicina* Wigg.; *Heteromorpha arborescens* Cham. & Schltdl. var. *abyssinica* (A. Rich) H. Wolff

Hedgeparsley: bur parsley; spreading hedgeparsley; upright hedgeparsley; *Caucalis arvensis* Huds.; *Caucalis capensis* Lam.; *Torilis africana* (Thunb.) Spreng; *Torilis japonica* (Houtt.) DC.

GENERAL INFORMATION: Erect winter or summer *annuals* to 1 m tall, with *pinnate-dissected leaves* and *bristly fruits*. Native to Eurasia.

SEEDLINGS: Stalk below cotyledons (hypocotyl) darker than cotyledons.

Bur chervil: Cotyledons oblong-lanceolate, 7–15 mm long, 1.5–3 mm wide, glabrous, on stalks shorter than the cotyledons. First leaf once or twice deeply pinnate-lobed, about 0.6–1 cm long, on a short-hairy stalk mostly 1–2 cm long. Leaf margins and lower veins often sparsely hairy.

Bur chervil (*Anthriscus caucalis*) plant.
J. M. DiTomaso

Bur chervil (*Anthriscus caucalis*) fruiting umbel.
J. M. DiTomaso

Bur chervil (*Anthriscus caucalis*) foliage and flowering umbels. J. M. DiTomaso

Bur chervil (*Anthriscus caucalis*) seedling.
J. M. DiTomaso

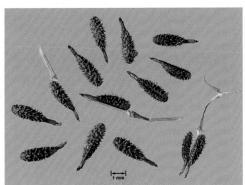

Bur chervil (*Anthriscus caucalis*) seeds. J. O'Brien

Hedgeparsley: Cotyledons linear-oblong, about 10–20 mm long, 1–2 mm wide, glabrous, on stalks shorter than the cotyledons. First and subsequent leaves twice deeply pinnate-lobed, hairy, mostly 1.5–2 cm long, on hairy stalks about 3–5 cm long.

MATURE PLANT: Leaves alternate, *mostly pinnate-dissected 2–3 times*, on stalks 2–8 cm long.

Bur chervil: Leaves sparse to moderately bristly-hairy, 5–15 cm long. Ultimate segments regularly deeply lobed nearly all the way to the midvein, lobes ± oblong and slightly narrowed at the base.

Hedgeparsley: Leaves sparsely covered with short flattened (appressed) hairs, 5–12 cm long. Ultimate segments lanceolate with margins coarsely toothed part way to the midvein.

ROOTS AND UNDERGROUND STRUCTURES: Taprooted, with fibrous lateral roots.

FLOWERS: March–July. Umbels compound, lowest bracts usually lacking. Ultimate umbel bractlets few to several, entire, 2–5 mm long. Petals 5, white. Stamens 5. Ovary inferior, 2-chambered. Styles 2.

Bur chervil: Umbels alternate, open. Primary rays mostly 3–6, about 1–2.5 cm long. Ultimate umbel bractlets lanceolate, reflexed backward. Sepal lobes lacking.

Hedgeparsley: Umbels opposite and terminal, open, on stalks (peduncles) longer than the leaves. Primary rays 2–10. Ultimate umbel bractlets linear. Sepal lobes 5, triangular, or lacking.

FRUITS AND SEEDS: Fruits (schizocarps) consist of 2 parts (mericarps) that eventually separate.

Bur chervil: Fruits *lanceolate, tip long-tapered to a point, ± 4 mm long, covered with minutely hook-tipped bristles.*

Hedgeparsley: Fruits generally *oblong, 3–5 mm long, ribbed, ribs covered with minutely barbed, hook-tipped bristles.* Inner faces of fruit halves as bristly as the outer parts.

Hedgeparsley *(Torilis arvensis)* plant.
J. M. DiTomaso

Hedgeparsley *(Torilis arvensis)* flowering stem.
J. M. DiTomaso

Hedgeparsley (*Torilis arvensis*) flowering umbels. J. M. DiTomaso

Hedgeparsley (*Torilis arvensis*) fruiting umbel with leaf. J. M. DiTomaso

Hedgeparsley (*Torilis arvensis*) seedling.
J. M. DiTomaso

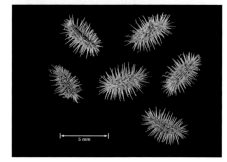
Hedgeparsley (*Torilis arvensis*) seeds. J. O'Brien

POSTSENESCENCE CHARACTERISTICS: Dead plants generally do not persist.

HABITAT: Disturbed sites, roadsides, fields, woodlands. Occasionally in orchards and vineyards. Bur chervil usually inhabits shady places.

DISTRIBUTION: Bur chervil: Throughout California, except Great Basin and deserts, to at least 400 m and possibly up to 1500 m. Oregon, Washington, Idaho, Wyoming, and Arizona. Scattered in eastern United States.

Hedgeparsley: Throughout California, except Great Basin and deserts, to 1600 m. Oregon, Washington, Idaho, Utah, and eastern half of United States, except far northern states and Florida.

PROPAGATION/PHENOLOGY: *Reproduce by seed.* Fruits fall near the parent plant or disperse to greater distances with water, mud, and by clinging to animals, to the shoes and clothing of humans, and to vehicle tires. Most seeds germinate after the first fall rains in areas with mild winters and in spring where winters are more severe.

MANAGEMENT FAVORING/DISCOURAGING SURVIVAL: Cultivation or hand-removal before fruits develop can control these species.

SIMILAR SPECIES: Knotted hedgeparsley [*Torilis nodosa* (L.) Gaertner] [TOINO] is an *annual* to 0.5 m tall that closely resembles hedgeparsley. Knotted hedgeparsley is distinguished by having a *spreading habit, umbel stalks (peduncles) that are shorter than the leaves, headlike umbels,* and *fruit halves with tubercled inner faces.* Knotted hedgeparsley inhabits turf and disturbed places

Knotted hedgeparsley *(Torilis nodosa)* plant.
J. M. DiTomaso

Knotted hedgeparsley *(Torilis nodosa)* flowering stem.
J. M. DiTomaso

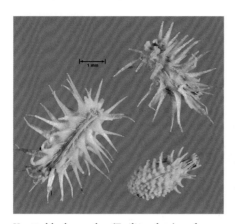

Knotted hedgeparsley *(Torilis nodosa)* seeds.
J. O'Brien

Venus-comb *(Scandix pecten-veneris)* plant.
J. M. DiTomaso

throughout northern and central California, except the Great Basin and east of the Sierra Nevada, to 1500 m. Native to Eurasia.

Venus-comb [*Scandix pecten-veneris* L., synonyms: Venus' needle; shepherds needle] [SCAPV] is an erect to spreading *annual* to 0.5 m tall with ± hairy foliage. Unlike bur chervil and hedgeparsley, Venus-comb has *elongated bristly fruits* with a *body 6–15 mm long* and *1–2 mm wide* and a *beak 20–70 mm long*. In addition, leaves are *finely dissected mostly 3–4 times into threadlike segments*. Venus-comb inhabits grasslands, waste places, roadsides, and other disturbed sites. It is scattered throughout much of California, except the Great Basin and desert regions, to 1000 m. It is most common along the North Coast and in the San Francisco Bay region. Native to the Mediterranean region.

Venus-comb *(Scandix pecten-veneris)* flowering umbel.
J. M. DiTomaso

Bur chervil, hedgeparsley, Venus-comb, and knotted hedgeparsley can resemble some native members of the carrot family, including southwestern carrot [*Daucus pusillus*] and California hedgeparsley [*Yabea macrocarpa*]. However, both these native species have leaflike bracts subtending the inflorescence and the four non-natives either lack or have inconspicuous leaflike bracts.

Venus-comb (*Scandix pecten-veneris*) fruiting umbel.   J. M. DiTomaso

Western waterhemlock [*Cicuta douglasii* (DC.) J.M. Coult. & Rose][CIUDO]

Spotted waterhemlock [*Cicuta maculata* L. vars. *angustifolia* Hook. and *bolanderi* (S.Wats.) Mulligan][CIUMC]

SYNONYMS: Western waterhemlock: *Cicuta californica* A. Gray; *Cicuta maculata* L. var. *californica* (A. Gray) Boivin

Spotted waterhemlock: common waterhemlock; poison parsley; spotted cowbane; spotted parsley; *Cicuta bolanderi* S. Wats. in part; *Cicuta occidentalis* Greene in part; *Cicuta virosa* L. var. *maculata* J.M. Coult. & Rose

GENERAL INFORMATION: Erect, native *perennials* of *wet habitats*, with *glabrous foliage* and a *thickened tuberous stem base that is horizontally partitioned into narrow hollow chambers* and leaves that are *1–2(–3) times pinnate-*

Western waterhemlock (*Cicuta douglasii*) plants in ditch.   J. M. DiTomaso

Western waterhemlock (*Cicuta douglasii*) foliage. J. M. DiTomaso

compound. Both species contain cicutoxin, an unsaturated aliphatic alcohol that is a potent nerve toxin, and are *extremely toxic* to livestock and humans when small quantities are ingested. All plant parts are poisonous, but roots are the most toxic part. Plants are usually most toxic in spring and fall. Human poisonings often occur when waterhemlock is mistaken for **wild parsnip** [*Pastinaca sativa* L.] or other edible plant and the roots are eaten as a vegetable. Livestock poisonings most often occur in spring when new foliage emerges from the fleshy roots. Dried plant material does not lose toxicity. Western and spotted waterhemlock are infrequent native species and are usually considered weeds only when they occur in places that are accessible to livestock or children.

SEEDLINGS: Cotyledons linear, glabrous.

MATURE PLANT: Foliage *glabrous*. Stems hollow except at nodes. Leaf stalks sheath stems.

Western waterhemlock: Stems to 3 m tall. Leaves 1–3 times pinnate compound. Most leaflets linear to lanceolate, 1–15 cm long, margins nearly smooth to sharp-serrated. Some lower leaflets are often deeply lobed into 2–3 lanceolate segments.

Spotted waterhemlock: Stems to 1.5 m tall, often streaked with purple. Leaves 1–2 times pinnate compound. Leaflets lanceolate, 2–10 cm long, ± coarsely sharp-serrated.

ROOTS AND UNDERGROUND STRUCTURES: *Stem bases thick, tuberous, often buried, horizontally partitioned into narrow hollow chambers. Cut surfaces of the* stem base exude an oily orangish yellow fluid that oxidizes to reddish brown and has a strong parsniplike scent. Young stem bases can be solid, but appear layered with yellowish lines on a white matrix. Below the stem base is usually a cluster of thick tuberlike storage roots or taproots and fibrous roots. Sometimes there is only 1 taproot or tuberlike root. The root cluster is often tuberlike early in the season and more taprootlike later in the season as energy reserves become depleted. Toward the end of the season before the root cluster disappears, it produces an overwintering horizontal rootstock from which new shoots and roots grow the following spring. The rootstock is usually above ground in western waterhemlock and underground in spotted waterhemlock.

FLOWERS: June–September. Umbels compound. Primary umbels usually lack bracts, rays mostly 15–30, spreading. Secondary umbels often have inconspicuous narrow bractlets. Petals 5, white or greenish, tips narrow. Sepal lobes 5, minute. Stamens 5. Styles 2. Insect-pollinated.

FRUITS AND SEEDS: Fruits (schizocarps) glabrous, slightly compressed laterally (perpendicular to central walls), with low, often pale, *corky ribs* and a *conspicuous oil tube between each pair of visible ribs*. The *fruit axis is divided to the base*.

Western waterhemlock: Fruits nearly round, with ribs that are much wider than the intervals between, 2–4 mm long.

Spotted waterhemlock: Fruits ovate, with ribs less than or equal to the width of the intervals between, 3–4 mm long.

POSTSENESCENCE CHARACTERISTICS: Aboveground parts die at the end of the season. Old stems with fruits and chambered base may persist into winter.

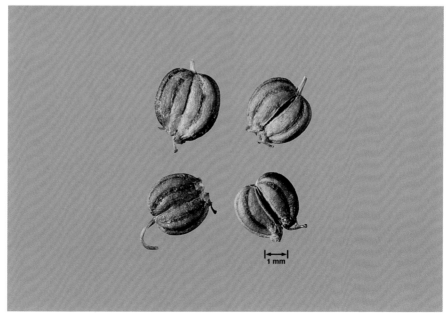

Western waterhemlock (*Cicuta douglasii*) seeds. J. O'Brien

Western waterhemlock • Spotted waterhemlock

Spotted waterhemlock (*Cicuta maculata* var. *angustifolia*) plant. J. M. DiTomaso

Spotted waterhemlock (*Cicuta maculata* var. *angustifolia*) flowering umbels. J. M. DiTomaso

Spotted waterhemlock (*Cicuta maculata* var. *angustifolia*) leaf. J. M. DiTomaso

Spotted waterhemlock (*Cicuta maculata* var. *angustifolia*) chambered root.
J. M. DiTomaso

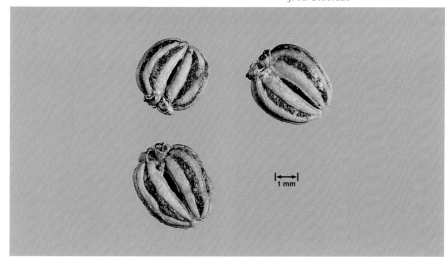

Spotted waterhemlock (*Cicuta maculata* var. *angustifolia*) seeds.
J. O'Brien

Wild parsnip (*Pastinaca sativa*) plant.   J. M. DiTomaso

Wild parsnip *(Pastinaca sativa)* flowering umbels. J. M. DiTomaso

Wild parsnip *(Pastinaca sativa)* rosette. J. M. DiTomaso

HABITAT: Wet places, often in water, many plant communities.

Western waterhemlock: Freshwater habitats.

Spotted waterhemlock: Var. *angustifolia* inhabits freshwater sites. Var. *bolanderi* usually inhabits salt marshes.

DISTRIBUTION: Western waterhemlock: North, Central, and South Coast; usually higher elevations of the Cascade Range and Sierra Nevada; and Great Basin, to 2500 m. Oregon, Washington, Idaho, Montana, and Nevada.

Spotted waterhemlock: Var. *angustifolia*: Great Basin, and San Bernardino Mountains, at 1500–2100 m. Western and central United States. Var. *bolanderi*: southern Sacramento Valley (Suisun Marsh) and Central and South Coast, to 200 m. Washington, Arizona, New Mexico, and southern and central United States.

PROPAGATION/PHENOLOGY: *Reproduce by seed* and vegetatively from overwintering rootstocks. Fruits disperse primarily with water and mud. Fruits float until the corky ribs become water-saturated. Seeds survive longer under moist conditions, up to ~ 3 years under moist field conditions. Overwintering rootstock can disperse with flooding. Plants typically do not flower until the second season or later.

SIMILAR SPECIES: Poison-hemlock [*Conium maculatum* L.][COIMA] is most easily distinguished by having *unpleasantly musty-scented foliage, purple-spotted or purple-streaked stems* that *lack* the tuberous horizontally chambered base, and fruits with *inconspicuous oil tubes between the prominent wavy ribs*. **Wild parsnip** [*Pastinaca sativa* L.][PAVSA] is a *biennial* with *nearly glabrous to short-hairy foliage* that *often grows in dry habitats*. Unlike the waterhemlocks, wild parsnip *lacks* the tuberous horizontally chambered stem base and has *fruits 4–6 mm wide* that are *greatly flattened front to back* (parallel to the central walls) with *narrow-winged marginal ribs*. In addition, wild parsnip has *yellow or orange flowers* and leaves that are pinnate-compound 1–2 times, often with some leaflets deeply lobed. Refer to the entry for **Poison-hemlock** (p. 132) for more information.

Wild parsnip (*Pastinaca sativa*) seeds. J. O'BRIEN

Poison-hemlock [*Conium maculatum* L.][COIMA] [Cal-IPC: Moderate]

SYNONYMS: California or Nebraska fern; carrot fern; deadly hemlock; poison parsley; poison stinkweed; snake-weed; spotted hemlock; wode thistle

GENERAL INFORMATION: Erect *biennial* (sometimes annual or short-lived perennial) to 3 m tall, with *large triangular, dissected compound leaves* and usually with *purple-spotted or purple-streaked stems*. Plants exist as a large basal rosette of leaves during the first year. Crushed foliage has a musty odor that is often described as similar to mouse excrement. Poison-hemlock contains piperidine alkaloids, and all plant parts are *highly toxic* to humans and animals when ingested. Alkaloid concentrations vary depending on plant location, environmental conditions, and stage of growth. Roots are the least toxic. Mature foliage and green seeds contain the highest concentration of total alkaloids, but new vegetative growth contains the greatest proportion of the most toxic alkaloid. Hence, many livestock poisonings occur in spring and fall when seeds germinate and new foliage emerges. Domesticated animals most susceptible to poisoning are horses, cattle, and goats. Symptoms of poisoning appear soon after ingestion and include nervousness, trembling, knuckling at the fetlock joints, uncoordinated gait (especially of the hind legs), dilated pupils, coldness of the limbs or body, weak and slow heartbeat, coma, and death from respiratory paralysis. Cattle, pigs, and goats that ingest small amounts of poison-hemlock

Poison-hemlock (*Conium maculatum*) infestation. J. M. DiTomaso

Poison-hemlock

Poison-hemlock (*Conium maculatum*) plant.
J. M. DiTomaso

Poison-hemlock (*Conium maculatum*) leaf.
J. M. DiTomaso

during a critical phase of the first trimester pregnancy may deliver deformed young. However, most animals avoid eating poison-hemlock when suitable forage is available. The toxic principles are volatile and long-term inhalation of the vapors is also poisonous. Volatilization accounts for the loss of toxicity of dried or well-heated plant material. Handling plants can cause contact dermatitis in sensitive individuals. Poison-hemlock is a state-listed noxious weed in Colorado, Idaho, Iowa, Nevada, Ohio, Oregon, and Washington. It is designated a facultative wetland indicator plant in all western states, except Arizona and New Mexico, where it is considered an obligate wetland species. Poison-hemlock is susceptible to ringspot virus, carrot thin leaf virus, and alfalfa and celery mosaic viruses. It was introduced from Europe in the 1800s as a fernlike garden ornamental.

SEEDLINGS: Cotyledons narrowly lanceolate-elliptic, 7–20 mm long, 2–6 mm wide, glabrous, light green, veiny below, on stalks that lengthen as development progresses. Stalk below cotyledons (hypocotyl) often purple-tinged. Leaves alternate, glabrous, 2–3 times pinnately lobed and/or dissected, sometimes slightly glossy. Stalks long, often purple-tinged, somewhat stipulelike, sheathing at the base. Crushed foliage has a musty odor.

MATURE PLANT: Foliage *glabrous. Stems typically purple-spotted or purple-streaked*, but sometimes only reddish or barely streaked, hollow except at the nodes. Leaves generally triangular, *2–3 times pinnate-compound, mostly 15–30 cm long*. Leaflets 1–2-pinnate-lobed or dissected. Lower leaf stalks expanded, somewhat stipulelike, sheathing at the base. Upper leaves nearly sessile.

ROOTS AND UNDERGROUND STRUCTURES: Taproot long, *thick, fleshy*, white, sometimes branched. Lateral roots often numerous.

FLOWERS: April–July. Umbels compound, often slightly rounded on top. Main stalk (peduncle) 2–8 cm long with inconspicuous lanceolate bracts below the secondary axes (pedicels). Pedicels usually 10–20, 1.5–5 cm long. Bractlets

Poison-hemlock (*Conium maculatum*) stem with purple blotches. J. M. DiTomaso

Poison-hemlock (*Conium maculatum*) seedling. J. M. DiTomaso

below pedicels mostly 1.5–2 mm long, resemble bracts. Flowers white, small, lack sepal lobes.

FRUITS AND SEEDS: Fruits (schizocarps) grayish brown, *slightly flattened laterally* (perpendicular to central walls), *ovate to nearly round, 2–3 mm wide*, with *conspicuously wavy longitudinal ribs. Oil tubes between ribs lacking.* Fruits ultimately separate into 2 halves (mericarps), each of which contains 1 seed.

POSTSENESCENCE CHARACTERISTICS: Flowering stems senesce in late summer or fall and can persist with a few fruits still attached well into winter or early spring.

HABITAT: Roadsides, pastures, fields, ditches, riparian areas, cultivated fields, waste places, and other disturbed, often moist sites.

DISTRIBUTION: Throughout California, except deserts and Modoc Plateau, to 1000 m. All contiguous states, except possibly Florida and Mississippi.

PROPAGATION/PHENOLOGY: *Reproduces by seed.* Most seeds fall near the parent plant, but some may disperse to greater distances with water, soil movement, animals, and human activities. Seed dispersal is prolonged and occurs from late summer through winter. After dispersal, most seeds can germinate almost immediately if conditions are favorable, but a small proportion remains dormant. Dormant seeds require a period of high summer and/or low winter temperatures before they can germinate. Nondormant seeds that do not disperse to a favorable environment can become dormant. Germination does not require light and in California typically commences with the first fall rains through early spring. Seedlings establish rapidly, especially on disturbed sites with bare soil. Seeds survive up to about 3 years under field conditions.

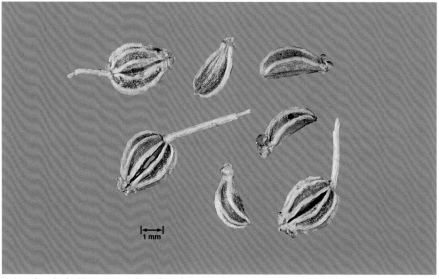

Poison-hemlock (*Conium maculatum*) seeds.   J. O'BRIEN

MANAGEMENT FAVORING/DISCOURAGING SURVIVAL: Plants do not regenerate when hand-pulled or cut below the crown. Removing plants before seeds mature every year will eventually deplete the seedbank. Repeated cultivation prevents poison-hemlock establishment in agricultural fields. Repeated mowing can eventually control it where cultivation is not possible. The European palearctic moth (*Agonopterix alstremeriana*), whose larvae feed only on poison-hemlock, was accidentally introduced to the United States in the early 1970s and has since become widespread in the western United States. Large populations of the moth can control poison-hemlock.

SIMILAR SPECIES: Wild parsnip [*Pastinaca sativa* L.][PAVSA] is a glabrous to minutely hairy *biennial*, sometimes annual or short-lived perennial, to about 2 m tall. Unlike poison-hemlock, wild parsnip has *yellow or orange flowers* and *fruits 4–6 mm wide that are strongly flattened front to back in cross-section* (parallel to central walls), *narrowly winged on the margins*, and have *1 oil tube between each rib pair*. In addition, crushed foliage has a parsniplike scent, stems lack purple spots and are conspicuously angled and grooved, and the leaves are 1-pinnate compound with each leaflet pinnate-lobed or toothed. Wild parsnip inhabits disturbed places throughout California, except the Great Basin and deserts, to about 1000 m. Wild parsnip is the weedy form of the cultivated parsnip and has an edible root. Native to Eurasia. See pages 129–131 for photos.

Western waterhemlock [*Cicuta douglasii* (DC.) J.M. Coult. & Rose][CIUDO] and **spotted waterhemlock** [*Cicuta maculata* L. vars. *angustifolia* Hook. and *bolanderi* (S.Wats.) Mulligan][CIUMC] are highly toxic native *perennials* of *wet habitats*, with *glabrous foliage*, *a thickened tuberous stem base that is horizontally partitioned into narrow hollow chambers*, and leaves that are *1–2(–3) times pinnate-compound*. The waterhemlocks also *lack purple spots and streaking on the stems* and have *slightly flattened, wingless fruits less than 4 mm wide*, with *corky ribs* and a *conspicuous oil tube between each pair of visible ribs*. Refer to the entry for **Western waterhemlock and Spotted waterhemlock** (p. 124) for more information.

Wild carrot [*Daucus carota* L.][DAUCA]

SYNONYMS: bird's-nest; devil's plague; Queen Anne's lace; *Carota sativa* Rupr.; *Caucalis carota* Crantz; *Caucalis daucus* Crantz

GENERAL INFORMATION: Erect *annual, biennial, or short-lived perennial* to 1.2 m tall, with *dissected leaves* and *compound umbels of white flowers*. Broken foliage or roots have a *carrot scent*. Plants exist as rosettes of long-stalked leaves until flowering stems develop at maturity. A purple-rooted biotype originating in Afghanistan is thought to be the progenitor of cultivated carrot. Wild carrot readily hybridizes with cultivated carrot and is susceptible to the pests and diseases that affect cultivated carrot. As livestock forage in pastures, wild carrot foliage has a nutritive value similar to that of legumes. Wild carrot is state-listed noxious weed in Iowa, Minnesota, Washington, and Ohio. Introduced from Eurasia and North Africa as a garden vegetable.

SEEDLINGS: Cotyledons linear, ± 20 mm long, 1 mm wide, glabrous. Root generally white. Leaves alternate, glabrous or sparsely hairy on margins and lower veins. Stalks long, usually with short bristly hairs in rows. First leaf dissected into 3 lobes. Lobes ± ovate in outline, pinnate-lobed.

MATURE PLANT: Stems *bristly-hairy*, branched, hollow, ridged. Rosette and stem leaves alternate, nearly glabrous to bristly-hairy, ± fernlike, about 5–15 cm long, finely dissected 3–4 times, ultimate segments ± linear to lanceolate.

ROOTS AND UNDERGROUND STRUCTURES: Taproots ± thick, carrot-shaped,

Wild carrot (*Daucus carota*) infestation.   J. M. DiTomaso

carrot-scented, yellowish white, sometimes deep, with fibrous lateral roots.

FLOWERS: May–September. Umbels compound, top convex to flat in flower, with *conspicuous finely pinnate-dissected bracts. Bract segments elongate, linear to threadlike.* Flowers white, except *central flower sometimes dark red to purplish.* Petals 5, tips unequally 2-lobed. Sepal lobes 5 or lacking. Stamens 5. Ovary inferior, 2-lobed. Out-crossing and self-fertile. Insect-pollinated.

Wild carrot (*Daucus carota*) plants. J. M. DiTomaso

Wild carrot (*Daucus carota*) flowering and fruiting umbels. J. M. DiTomaso

Wild carrot

FRUITS AND SEEDS: Umbels first become concave as fruits mature, then flat, with primary rays mostly 3–7.5 cm long. Fruits (schizocarps) ± ovoid, *slightly compressed front to back* (parallel to central walls), mostly 3–4 mm long, *broadest at the middle*, with 5 slender, bristly, longitudinal ribs. Slender ribs 5, bristly. *Winged ribs 4, lined with barb-tipped bristles.* Fruits later separate into 2 parts (mericarps).

Wild carrot (*Daucus carota*) seedling. J. M. DiTomaso

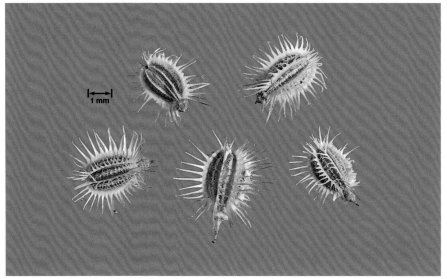

Wild carrot (*Daucus carota*) seeds. J. K. Clark

Southwestern carrot (*Daucus pusillus*) plant. J. M. DiTomaso

POSTSENESCENCE CHARACTERISTICS: Old flower stems with a few fruits sometimes persist for a few months.

HABITAT: Disturbed places, roadsides, fields, pastures, orchards, and vegetable crops. Usually grows on sandy or gravelly soils.

DISTRIBUTION: Throughout California, especially in the northern half of the state and coastal regions, except Great Basin and deserts, to 1500 m. All contiguous states.

PROPAGATION/PHENOLOGY: *Reproduces by seed.* Most fruits fall near the parent plant, but some disperse to greater distances with water, mud, as seed and feed contaminants, and by clinging to animals, to the shoes and clothing of humans, and to vehicle tires and machinery. Some seeds survive ingestion by pigeons, pheasants, and possibly some other animals. Most seeds germinate in spring, but smaller flushes often occur in summer and fall. Seeds survive several years under field conditions. Seeds submersed in water can survive up to 2 years. Buried seed can survive up to 20 years.

MANAGEMENT FAVORING/DISCOURAGING SURVIVAL: Disturbance at intervals greater than 1 year appears to enhance survival. Annual or more frequent cultivation can eliminate wild carrot. Mowing while plants are in full flower can prevent seed production. Some biotypes tolerate certain herbicides.

SIMILAR SPECIES: Southwestern carrot [*Daucus pusillus* Michx., synonym: American wild carrot][DAUPU] is a native *annual* to 0.9 m tall that resembles wild carrot. Compared to wild carrot, southwestern carrot has *umbel bract segments that are short and linear to lanceolate*, and the *oblong fruits are broadest below the middle*. In addition, the central flower of an umbel is always white in southwestern carrot, and the leaves are typically smaller and more finely dissected than wild carrot leaves. Primary rays of umbels are mostly 0.4–4 cm long. Southwestern carrot is a desirable component of the vegetation in natural areas, but it is occasionally weedy on disturbed sites and turf. Southwestern carrot inhabits grassland (including serpentine grassland), coastal scrub, chaparral, pastures, roadsides, and other disturbed places, especially on the coast, throughout much of California, except deserts and Great Basin, to 1500 m. It also occurs in Arizona, Idaho, New Mexico, Oregon, Washington, and most south-central and southern states.

Greater ammi [*Ammi majus* L.] is a *winter annual or biennial* to 1 m tall with lower leaves and umbels that also resemble those of wild carrot. Unlike wild carrot, greater ammi has *glabrous foliage, upper leaves pinnate-dissected into linear or threadlike segments*, and *smooth fruits*. Refer to the entry for **Greater ammi** (p. 110) for more information.

Southwestern carrot (*Daucus pusillus*) flowering umbel. J. M. DiTomaso

Fennel [*Foeniculum vulgare* Mill.][FOEVU][Cal-IPC: High]

SYNONYMS: anise; aniseed; sweet anise; sweet or common fennel; *Anethum foeniculum* L.; *Foeniculum foeniculum* (L.) Karst.; *Foeniculum officinale* All.; *Ligusticum foeniculum* Crantz

GENERAL INFORMATION: *Aromatic perennial* to 3 m tall, with *finely dissected leaves* and *flat-topped umbels of small yellow flowers*. Foliage and seeds have a *strong licorice or anise scent*, especially when crushed. Fennel invades grasslands, riparian areas, and other natural communities, particularly in coastal regions of central and southern California. Quite common along roadsides. Established plants are competitive, and soil disturbance facilitates the development of dense stands, which can exclude native vegetation in some areas. Birds and rodents consume the seeds, and feral pigs relish the roots. Fennel is also sometimes weedy in agronomic crops. Australian farmers have observed that tomatoes and beans grow very poorly when associated with fennel, suggesting that plants may have allelopathic properties. Fennel is rated as a noxious weed

Fennel (*Foeniculum vulgare*) plant.   J. M. DiTomaso

Fennel (*Foeniculum vulgare*) leaf. J. K. CLARK

Fennel (*Foeniculum vulgare*) foliage at base of plant. J. M. DiTomaso

in some regions of Australia. Unlike the weedy form, cultivated varieties are seldom invasive. Some varieties are cultivated as a spice or vegetable, for an essential oil used to flavor foods, and in some countries, for medicinal purposes. Native to southern Europe.

SEEDLINGS: Cotyledons linear, mostly 2–4 cm long, glabrous, on slightly

Fennel (*Foeniculum vulgare*) plant in the vegetative stage. J. K. CLARK

Fennel (*Foeniculum vulgare*) flowering umbel. J. M. DiTomaso

sheathing stalks 2–3 cm long. First leaves alternate, ovate-triangular, twice pinnate-dissected into linear segments, about 1–2 cm long, on slightly sheathing stalks about 1–4 cm long.

MATURE PLANT: Stems numerous from the crown, branched, faintly striated, glabrous, covered with a waxy bloom (glaucous), solid with white pith, appear jointed at the nodes. Leaves to about 40 cm long, mostly 30–40 cm wide, *finely pinnate-dissected into numerous threadlike segments*. Leaf stalks 7–15 cm long, base expanded and sheathing the stem.

Fennel (*Foeniculum vulgare*) seedling. J. M. DiTomaso

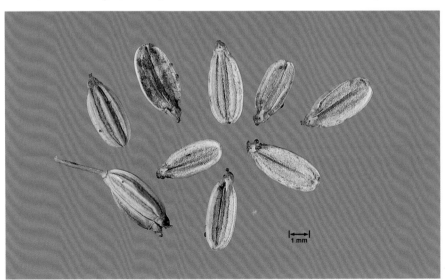

Fennel (*Foeniculum vulgare*) seeds. J. K. Clark

ROOTS AND UNDERGROUND STRUCTURES: Taproot deep, thick, woody, branched, with many lateral roots and a prominent crown.

FLOWERS: May–October (November). Umbels compound, flat-topped, with about 5–40 unequal rays, lack bracts or bractlets. *Flowers yellow*, small. Sepal lobes lacking.

FRUITS AND SEEDS: Fruits oblong, mostly 3–4 mm long, slightly laterally flattened (parallel to central plane), glabrous, with nearly equal, acute ribs. Fruits separate into 2 halves (mericarps) at maturity.

POSTSENESCENCE CHARACTERISTICS: Many flower stems or upper portions of flower stems die at the end of the dry season in late fall or early winter. Some apparently dead stems can produce new foliage at the nodes the following season. Dead flower stems with umbels of persistent fruits often persist for at least a year.

HABITAT: Open disturbed sites, roadsides, slopes, fields, grasslands, coastal scrub, riparian and wetland areas, and agronomic crops. Tolerates drought and frost. Grows in many soil types.

DISTRIBUTION: Most common throughout low elevation areas of California, except Great Basin and desert regions. Much more common along the coast in southern California. To 1800 m, but generally below 350 m.

PROPAGATION/PHENOLOGY: *Reproduces by seed* and sometimes vegetatively from root or crown fragments. Seed production is usually prolific. Seeds disperse with water, soil movement, animals, human activities, and as a seed contaminant. Most seeds germinate in fall through spring during the wet season, but germination can occur nearly year-round when conditions are favorable. Optimal temperature for germination is 16°–23°C. Seeds appear to survive several years under field conditions, but quantitative information is lacking. Flowering typically begins when plants are about 1.5–2 years of age, but plants can occasionally flower in first year. Fragmentation of roots and crowns may occur during flooding events, mudslides, or agricultural operations. New shoots grow from the crown and lower portions of overwintering stems in mid-winter to early spring.

MANAGEMENT FAVORING/DISCOURAGING SURVIVAL: Manually removing individual plants, including the roots, is an efficient means of control on sites with limited numbers of plants. One or an occasional cutting, mowing, or grazing does not kill the roots and facilitates seed dispersal if plants are in fruit. Cutting 3 to 4 times a year for a few years will eventually deplete the carbohydrate reserves and kill the plant. Fall burning followed by chemical treatment of new foliage during the growth period for 2 years can control larger stands. Minimizing soil disturbance helps to reduce the spread of fennel.

SIMILAR SPECIES: Fennel is easily distinguished from other weedy members of the carrot family by having *large finely dissected leaves with threadlike segments, yellow flowers*, and a distinctive *anise or licorice scent*.

Floating pennywort [*Hydrocotyle ranunculoides* L.f.] [HYDRA]

Lawn pennywort [*Hydrocotyle sibthorpioides* Lam.] [HYDSI]

SYNONYMS: Floating pennywort: floating marshpennywort; *Hydrocotyle batrachioides* DC.; *Hydrocotyle natans* Cirillo; *Hydrocotyle ranunculoides* L.f. vars. *ranunculoides* and *adoensis* Urb.

Lawn pennywort: lawn marshpennywort; *Hydrocotyle americana* L.; *Hydrocotyle confusa* H. Wolff; *Hydrocotyle ranunculoides* L.f. var. *incisa* Blume; *Hydrocotyle rotundifolia* Roxb.; others

GENERAL INFORMATION: Variable species with *palmate-lobed leaves* and branched *creeping stems* that *root at the nodes*.

Floating pennywort: *Aquatic* or terrestrial *perennial*. Plants typically form dense, low-growing mats in shallow water or on wet soil near water. Occasionally, small colonies are free-floating. Floating pennywort is a widespread native of North America. In natural areas, colonies are usually considered a desirable component of aquatic ecosystems. Because of its creeping habit, floating pennywort can be a nuisance in irrigation and drainage ditches. Plants are sometimes sold as an aquatic or pond ornamental and have escaped cultivation in some regions. In Britain, floating pennywort has become a problematic weed of natural aquatic habitats, and in southern and western Australia, it is a government-listed noxious weed.

Floating pennywort (*Hydrocotyle ranunculoides*) in a swampy site.
CDFA Integrated Pest Control Program

Lawn pennywort: Low-growing *terrestrial perennial* of moist sites. Introduced from Asia, where it is sometimes eaten as a vegetable and used medicinally. Some biotypes have been described as separate species.

SEEDLINGS: Leaves alternate.

Lawn pennywort: Foliage glabrous. Cotyledons ovate, ± 2 mm long, tip rounded to slightly indented (emarginate). First few leaves nearly round to kidney-shaped, deeply lobed at the base, about 2–4 mm long, 3–5 mm wide, palmate-veined. Margins shallowly 5–7-lobed, lobes rounded-acute with tips slightly pinched-in. Stalks 1–3 mm long.

MATURE PLANT: Foliage *glabrous*. Stems root at most nodes. Leaves alternate, nearly *round or kidney-shaped, deeply lobed or cleft at the base*, palmate-veined. Stalks have *papery stipules* at the base and are *not sheathing*.

Floating pennywort: Foliage *fleshy*. Stems fragment easily. Leaves mostly 1–8 cm wide, deeply 3–7 lobed (cleft), sometimes with a reddish central spot (point of stalk attachment). Margins smooth to scalloped (crenate). Stalks (petioles) thick, mostly 5–35 cm long, sometimes reddish.

Lawn pennywort: Stems slender. Leaves usually 1–2 cm wide, shallowly 5–7 lobed. Margins finely scalloped. Stalks slender, about 1–5 cm long.

ROOTS AND UNDERGROUND STRUCTURES: Roots grow from most nodes and are shallow in soil substrate.

FLOWERS: Umbels *simple*, on stalks (peduncles) shorter than leaf stalks. Calyx (sepals as a unit) lacking or consisting of 5 minute lobes. Petals 5, greenish or

Floating pennywort (*Hydrocotyle ranunculoides*) in a wet meadow.   J. M. DiTomaso

Floating pennywort (*Hydrocotyle ranunculoides*) foliage.   J. M. DiTomaso

Floating pennywort (*Hydrocotyle ranunculoides*) inflorescence.   J. M. DiTomaso

yellowish white to purplish. Stamens 5. Ovary inferior, 2-chambered, with 2 separate styles at the apex.

Floating pennywort: March–August. Umbels dense, with about 5–10 flowers on *stalks* (pedicels) about 1–3 mm long. Umbel stalk about 1–6 cm long.

Lawn pennywort: Sometimes flowers nearly year-round under favorable conditions. Umbels headlike, with 3–10 *sessile flowers*. Umbel stalk about 0.5–1.5 cm long.

FRUITS AND SEEDS: Fruits (schizocarps) flattened laterally (perpendicular to the central wall), separate into halves (mericarps) at maturity. Each mericarp contains 1 seed.

Floating pennywort: Fruits on *stalks* about 1–6 cm long, elliptic to round, mostly 1–3 mm long, with slightly thickened, inconspicuous ribs.

Lawn pennywort: Fruits *sessile*, round, about 1–1.2 mm in diameter, with acute ribs.

POSTSENESCENCE CHARACTERISTICS: Floating pennywort: Foliage sometimes dies back after plants flower.

HABITAT: Floating pennywort: Pond and lake margins, marshes, slow streams, irrigation and drainage ditches.

Lawn pennywort: Turf, moist places. Tolerates some shade.

DISTRIBUTION: Floating pennywort: Scattered throughout California, except Great Basin and deserts, to 1500 m. Arizona, Oregon, Washington, and most southern and eastern states, except for those of the extreme northeastern region. South America.

Lawn pennywort: South Coast (Los Angeles Co.), to about 330 m. Many southern and eastern states.

PROPAGATION/PHENOLOGY: *Reproduce by seed and/or vegetatively from creeping stems and stem fragments.* Seeds and stem fragments disperse with water, substrate movement, animals, and human activities. In Britain, floating pennywort plants rooted in the substrate produce seed, while floating colonies primarily reproduce vegetatively. Floating mats sometimes support other vegetation.

SIMILAR SPECIES: Musky pennywort [*Hydrocotyle moschata* Forster f.] is a low-growing *perennial* with *creeping stems* that closely resembles lawn pennywort. Unlike lawn pennywort, musky pennywort has *hairy leaves* and umbels with about *10–20 nearly sessile flowers*. Musky pennywort is an uncommon turf weed in the South Coast (Los Angeles Co.) area to about 100 m. Native to New Zealand.

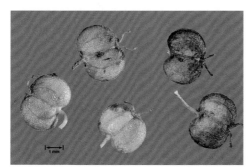

Floating pennywort *(Hydrocotyle ranunculoides)* fruit (schizocarps).
J. O'Brien

Lawn pennywort *(Hydrocotyle sibthorpioides)* seeds. Very similar to seeds of musky pennywort *(Hydrocotyle moschata)*.
J. K. Clark

Musky pennywort *(Hydrocotyle moschata)* plant with flowers and fruit.
J. M. DiTomaso

Big periwinkle [*Vinca major* L.][VINMA]
[Cal-IPC: Moderate]

SYNONYMS: bigleaf periwinkle; blue periwinkle; greater periwinkle; large periwinkle; *Vinca pubescens* D'Urville. *Vinca major* L. var. *variegata* Loud. is a variegated form with pale spotted or margined leaves.

GENERAL INFORMATION: Herbaceous *perennial* with *milky sap, trailing sterile stems* to about 1 m long, and ascending to erect flower-bearing stems to 0.5 m tall that develop *showy lavender-blue funnel-shaped flowers*. Big periwinkle is commonly cultivated as an ornamental ground cover, but it has escaped cultivation in many places. Under favorable conditions, plants spread invasively and can develop a dense ground cover that outcompetes other vegetation in natural areas. Some infestations around old homesteads have been present for many years. Big periwinkle contains numerous alkaloids and is used medicinally. Some plants in the dogbane family are extremely toxic, although poisoning due to the ingestion of big periwinkle is poorly documented. Native to central Europe.

SEEDLINGS: Rarely encountered. Cotyledons narrow elliptic, about 12–35 mm long, base long-tapered, glabrous. Leaves resemble those of mature plants, but are usually slightly smaller.

MATURE PLANT: Evergreen. Sterile stems typically arching, rooting at the tips. Leaves *opposite*, mostly glabrous, semiglossy, ovate, about *2–7 cm long, margins ciliate with minute hairs* (magnification required), on stalks to about 1 cm long.

Big periwinkle (*Vinca major*) infestation in a riparian area.   J. M. DiTomaso

ROOTS AND UNDERGROUND STRUCTURES: Perennial rootstock tough, ± woody.

FLOWERS: February–August, depending on location. Along the coast, plants can flower later into the summer. Flowers solitary in leaf axils, *lavender-blue*, rarely white, *funnel-shaped*, about 3–5 cm in diameter, on slender stalks (pedicels) 3–5 cm long. Corolla 5-lobed, lobes overlapping and pinwheel-like in bud. Sepals 5, slender, about 1 cm long, tips long-tapered, margins ciliate with minute hairs. Stamens 5, attached near top of corolla tube.

FRUITS AND SEEDS: Fruits and seeds rarely develop. Pods (follicles) 1 or 2 per flower, slender, curved, cylindrical, about 4–6 cm long, tip pointed, open on one side to release seeds. Seeds 1–5 per pod, glabrous, dark, nearly cylindrical, mostly 2.5–8 mm long, slightly compressed, grooved on one side, truncate at both ends.

HABITAT: Riparian sites, old homesteads, moist woodlands, and roadsides. More abundant along the coast. Grows best under moist shady conditions. Tolerates deep shade and poor soil. Foliage is susceptible to frost damage.

DISTRIBUTION: North Coast Ranges, northern Sierra Nevada foothills, Central Valley, San Francisco Bay region, South Coast Ranges, and South Coast, to 1500 m. Oregon, Washington, Idaho, Utah, Arizona, and much of the southern and eastern United States.

PROPAGATION/PHENOLOGY: *Reproduces vegetatively from trailing stems that root at the tips* and *stem fragments* and rarely by seed. Plants and stem fragments disperse with human activities, such as purposeful landscape planting and careless disposal of yard waste. Under favorable conditions, stem cuttings left on the ground can take root. In riparian areas, water currents can fragment stems and carry them downstream where they may root if lodged in a suitable place. Fruits with viable seeds rarely develop on cultivated and naturalized plants in California and elsewhere. However, seedlings have been observed near one population in Sunol, Alameda County. In England in the early 1960s, after decades of rarely developing fruit, several colonies of big periwinkle and common periwinkle [*Vinca minor* L.] produced numerous fruits with seeds during one particularly dry summer. Big periwinkle has a polyploid chromosome set ($2n = 92$), which is sometimes associated with low seed production.

MANAGEMENT FAVORING/DISCOURAGING SURVIVAL: Persistent manual removal can control the spread of big periwinkle. Leaves are covered with a waxy cuticle that can diminish the effectiveness of some herbicides.

SIMILAR SPECIES: Common periwinkle [*Vinca minor* L.] [VINMI] is an ornamental *perennial* that closely resembles big periwinkle. Common periwinkle has escaped cultivation in Washington, Utah, Arizona, and throughout most of the eastern half of the United States, but naturalized populations are not known to occur in California. Common periwinkle is distinguished by having *leaves about 4 cm long* or less, *flowers to about 2 cm in diameter*, and *leaf* and *sepal margins* that are *glabrous* (not ciliate with minute hairs). Like big periwinkle, common periwinkle rarely develops fruits.

Big periwinkle

Big periwinkle (*Vinca major*) foliage and flowers.　　　J. M. DiTomaso

Big periwinkle (*Vinca major*) flower.　　　J. M. DiTomaso

English holly [*Ilex aquifolium* L.]
[Cal-IPC: Moderate Alert]

SYNONYMS: Christmas holly

GENERAL INFORMATION: Evergreen *shrub or small tree* to 12 m tall, usually with *prickly leaves* and *red berries*. English holly has been cultivated since ancient times. In the United States, it is cultivated commercially for Christmas décor and as a landscape ornamental. Currently, there are many highly variable cultivars on the market. English holly has escaped cultivation and become invasive in certain areas of the moist coastal forests in California, Oregon, and Washington. The berries and leaves contain saponins that can cause digestive tract irritation when ingested. Most reported problems have involved children who ingested the berries. Native to Europe and western Asia.

SEEDLINGS: Cotyledons oblong, about 1–1.6 cm long, glabrous, leathery. Leaves alternate, glabrous, oblong-elliptic, ~ 1.5–2 cm long, margins prickly-toothed.

MATURE PLANT: Branches numerous, short, erect to spreading. Twigs glabrous or covered with minute hairs. Leaves alternate, ovate to ovate-oblong, mostly 2.5–6 cm long, glabrous, upper surface *glossy dark green*, lower surface lighter, dull, on short stalks. Margins mostly *spiny-toothed* and *wavy*, sometimes smooth; both types often occur on one plant. *Spiny teeth triangular, coarse, stiff, sharp, usually fewer than 20 per mature leaf*. Stipules small, deciduous.

ROOTS AND UNDERGROUND STRUCTURES: Main roots woody. Resistant to oak root fungus.

FLOWERS: May–June. Fragrant. Male and female flowers usually on separate plants (dioecious). Flowers small, in clusters on previous year's growth, fragrant. Petals 4, dull white, fused at base. Sepals 4, fused at base. Stamens 4. Ovary superior, ± 4-chambered, often with 1 ovule per chamber.

FRUITS AND SEEDS: Sepals persistent in fruit. Fruits (drupes) *berrylike, red*, ± spherical, mostly *7–8 mm in diameter*, smooth, contain 2–8 (usually 4) small 1-seeded nutlets. Fruits ripen in late summer/early fall.

HABITAT: Coastal forests, woodlands, riparian areas.

DISTRIBUTION: Western North Coast Ranges, San Francisco Bay region, and Central Coast, to 200 m. Expanding range. Oregon and Washington.

PROPAGATION/PHENOLOGY: *Reproduces by seed*. Birds and mammals consume and disperse a proportion of the fruits. Fruits left on the plant fall to the ground. Seeds have hard but permeable seed coats and dormant or immature embryos. Most seeds do not germinate for at least 2–3 years after dispersal, even with scarification and a cool, moist stratification period. English holly grows slowly, and natural regeneration is typically sparse. Female plants usually begin to produce seed at 5–12 years of age. Seed production is highest in trees more than 20 years old.

English holly

MANAGEMENT FAVORING/DISCOURAGING SURVIVAL: Manually removing individual plants before seed production begins can control the spread of English holly.

SIMILAR SPECIES: Holly-leafed cherry [*Prunus ilicifolia* (Nutt.) Walp.] is a desirable native *shrub or small tree* that is *not considered a weed*, but may be confused with English holly. Unlike English holly, holly-leafed cherry leaves usually have *more than 20 slender, weak spines per leaf* and *fruits 1.2–2.2 cm in diameter* that resemble those of commercial cherries. Holly-leafed cherry occurs in woodlands, chaparral, and canyons throughout the coastal regions and coastal mountain ranges, except for in the North Coast and northern North Coast Ranges.

English holly *(Ilex aquifolium)* stem with fruit.   J. M. DiTomaso

English ivy [*Hedera helix* L.][HEEHE][Cal-IPC: High]

SYNONYMS: *Hedera helix* L. ssp. *helix*. The systematics of *Hedera* are very confusing. *Hedera helix* is represented by approximately 500 cultivars, which has further complicated the identification of the species. The vast majority of *H. helix* cultivars are not considered invasive. There are over 13 species of *Hedera* worldwide, and many of them are difficult to distinguish morphologically. While most have not proven to be invasive, almost all the invasive populations in Washington and Oregon are probably *Hedera hibernica* (Kirchner) Bean [*Hedera helix* ssp. *hibernica* (Kirchner)

English ivy *(Hedera helix)* climbing on an oak tree. J. M. DiTomaso

McClintock]. The common name for *H. hibernica* is Irish or Atlantic ivy. Unfortunately, *Hedera hibernica is* typically sold as *Hedera helix*, and called English ivy. Juvenile *H. hibernica* can form large dense clonal patches. It has been speculated that the invasive ivies of California are likely also *H. hibernica*, but this has not yet been confirmed. *Hedera helix* patches generally are much smaller and less dense than those of *Hedera hibernica*. In addition, they are not as likely to reproduce by seed. Although the impact of *H. helix* in wildland for-

English ivy *(Hedera helix)* fruiting stem.
J. M. DiTomaso

English ivy *(Hedera helix)* leaf on a flowering stem.
J. M. DiTomaso

English ivy *(Hedera helix)* leaves.
J. M. DiTomaso

English ivy

ests of Oregon and Washington is considered to be fairly minor compared to *H. hibernica*, a few *H. helix* cultivars are invasive. *Hedera hibernica* and *H. helix* can be distinguished by subtle morphological traits, such as trichome orientation, color and leaf size. However, these traits are not always reliable under field conditions. One way that *H. hibernica* can be differentiated from *H. helix* is by comparing their chromosome number with flow cytometry. *Hedera helix* has a different chromosome number (2x) than *H. hibernica* (4x), but *H. hibernica* and *H. canariensis* have the same number and are both considered invasive. Unfortunately, this reliable lab technique cannot be used for field identification.

GENERAL INFORMATION: Vigorous *woody perennial* with *2 growth forms*. The juvenile form has *viny stems* to about 30 m long with leaves that are usually *3-lobed*. The adult reproductive form has *erect, shrubby stems* with *ovate to rhomboid (diamond-shaped) leaves*. English ivy is a common landscape ornamental of which there are numerous cultivars. It has escaped cultivation in many places, especially near the coast. English ivy grows over the natural vegetation in an area, including the trees, and eventually kills most resident plants by shading them out with its dense canopy of foliage. English ivy especially thrives in deciduous trees, which allow plants to receive more light and to continue upward growth during the winter months. Trees covered with ivy are more susceptible to wind damage from the extra weight. English ivy berries and leaves can be *toxic* to humans and cattle when ingested in quantity, and the sap can cause contact dermatitis in sensitive individuals. Native to Europe.

SEEDLINGS: Cotyledons ovate-oblong, mostly 1.5–2 cm long, bases rounded, tips slightly rounded, fairly leathery, glabrous, on stalks 2–3 mm long. Leaves

Algerian ivy *(Hedera canariensis)* leaves. J. M. DiTomaso

English ivy

Algerian ivy *(Hedera canariensis)* (right) and English ivy *(H. helix)* (left) leaves.   J. M. DiTomaso

alternate. First leaf weakly 3-lobed-ovate, about 10–18 mm long, slightly wavy, leathery, with star-shaped (stellate) hairs. Stalk covered with star-shaped hairs.

MATURE PLANT: Leaves evergreen, variable, alternate, leathery. Upper surfaces glabrous, often slightly glossy, usually *dark green*. Juvenile stems *vinelike*, develop aerial rootlets that enable stems to cling to objects such as trees and buildings. Juvenile leaves *palmately 3–5-lobed*, to 35 cm long, lobes shallow to deep. Adult reproductive stems erect, shrubby, lack aerial rootlets, nonclimbing. Adult leaves ovate to diamond-shaped, to 15 cm long. Leaf stalks and lower leaf surfaces of both forms are usually covered with grayish *star-shaped hairs* (stellate hairs) or sometimes glabrous. *Rays of star-shaped hairs 12 or less, needlelike, project at different angles, joined only at the base to form a central point.*

ROOTS AND UNDERGROUND STRUCTURES: Juvenile stems develop numerous adventitious roots at the nodes. Mature plants develop a thick woody base.

FLOWERS: Unlike most plants of the region, ivy flowers in the fall. Only the shrubby adult form develops flowers in racemes or panicles of simple umbels. Infestations in forest areas produce flowers high in the tree canopy. Sepals, petals, and stamens 5. Petals whitish or tinged green, mostly 3–5 mm long. Insect-pollinated.

FRUITS AND SEEDS: Fruits (drupes) berrylike, dark blue to black, about 5–10 mm in diameter, often covered with star-shaped hairs, contain a few seeds, typically 2. Fruits and stalks are usually covered with the same type of star-shaped hairs described for the leaf stalks. Fruits mature in spring. Each adult can produce tens of thousands of fruit each year.

HABITAT: Disturbed forests and woodlands, riparian areas. Requires some moisture year-round. Tolerates deep shade, but thrives in where plants receive some summer shade and direct winter sun.

DISTRIBUTION: Throughout California, except Great Basin and deserts, to

1000 m. Oregon, Washington, Arizona, Utah, and southern and eastern United States.

PROPAGATION/PHENOLOGY: *Reproduces vegetatively from juvenile stems* and *by seed*. Stem fragments of juvenile and adult plants left in contact with moist soil can regenerate into a new plant. Plants from adult fragments retain adult characteristics. In the horticultural industry, propagation is almost never by seed. The juvenile stage may last for 10 years or more before they reproduce by seed. Fruit production in adult plants can be high. Fruits are consumed and dispersed primarily by birds. Neglected horticultural ivy often fruits prolifically. Birds can carry the seeds from gardens and yards into nearby natural areas. An average of about 70% of the seeds are viable. Plants can live 100 years or more. One plant is reported to be over 400 years old.

MANAGEMENT FAVORING/DISCOURAGING SURVIVAL: English ivy is typically difficult to control. Manually removing plants can sometimes control problematic infestations and reduce spread by seed. Thick waxy cuticle on mature plants reduces the efficacy of systemic herbicides. Mechanical removal followed by herbicide treatment to the young recovering tissues can sometimes be effective.

SIMILAR SPECIES: Algerian ivy [*Hedera canariensis* Willd., synonym: *Hedera helix* L. ssp. *canariensis* (Willd) Coutinho; *Hedera canariensis* var. *algeriensis* B.-M.J. Auzende; *Hedera algeriensis* Hibberd] [Cal-IPC: High] is a closely related perennial that has escaped cultivation as an ornamental in some areas. In California, weedy populations of Algerian ivy have been mistakenly thought to be English ivy. As a result, until recently Algerian ivy was not included in most California floras. Algerian ivy and English ivy appear to hybridize with one another only in certain areas of their native range, and some taxonomists consider Algerian ivy to be a subspecies of English ivy. Some naturalized plants in California have an intermediate form and cannot be assigned to either species. Unlike pure forms of English ivy, pure forms of Algerian ivy have leaf stalks that are typically covered with yellowish *scalelike star-shaped hairs*. The star-shaped hairs have *12–22 rays that lie in one plane* and are *united at the base for about one-quarter of their length, forming a central shieldlike scale* (peltate). Hairs on growing stem tips are typically rust-colored. Flower stalks and fruits are also covered with the scalelike star-shaped hairs. In addition, leaf stalks are often noticeably reddish. Juvenile leaves are green or variegated and typically weakly 3-lobed. Compared to English ivy, Algerian ivy leaf lobes are generally more rounded, leaf veins are fainter, and stems are more coarse and root more shallowly at the nodes. Algerian ivy occurs primarily in disturbed woodland, forest, and riparian areas in the San Francisco Bay region, Central Coast, South Coast, eastern Transverse Ranges (San Gabriel and San Bernardino Mountains), and northwestern Peninsular Ranges (Santa Ana Mountains), to about 900 m. Native to southwestern Europe, northwestern Africa, and the Atlantic Islands, including the Azores and Canary Islands.

Canary Island date palm [*Phoenix canariensis* Chabaud][Cal-IPC: Limited]

Mexican fan palm [*Washingtonia robusta* H. Wendl.] [Cal-IPC: Moderate Alert]

SYNONYMS: Canary Island date palm: *Phoenix canariensis* Hort. ex Chabaud

Mexican fan palm: cotton palm; Washington fan palm; *Neowashingtonia robusta* (H. Wendl.) A. Heller; *Washingtonia filifera* (Linden ex Andre) H. Wendl. var. *robusta*

Canary Island date palm (*Phoenix canariensis*) plant fully grown.   J. M. DiTomaso

(H. Wendl.) Parish; *Washingtonia gracilis* Parish; *Washingtonia sonorae* S. Wats.

GENERAL INFORMATION: These palm trees have become problematic in natural riparian stream and river corridors, orchard crops, and as seedlings that volunteer in landscaped areas. Both are commonly cultivated as landscape ornamentals. In wildlands, invasive palms are most common in southern California. Populations are densest downstream from the source of invasion, which are typically residential areas. Canary Island date palm is native to the Canary Islands. Mexican fan palm is native to central Mexico, but not the northern mountain deserts. It has also invaded northwestern Mexico.

MATURE PLANT: Canary Island date palm: Trunk single to 25 m tall, with a diamond-like pattern when old leaves are removed. Leaves bright green, *pinnate-compound*, mostly 5–7 m long, stiff, arched, *persist on the trunk. Lower leaflets daggerlike.* Upper leaflets longitudinally folded upward.

Mexican fan palm: Trunk single to 30 m tall, *abruptly flared at the base*, with *leaf stalks that form a criss-cross pattern* unless removed. Leaves persistent, ± glossy green, *palmately veined* (can appear palmately compound when leaves are torn), about 1–2 m wide, split into narrow segments nearly half-way to the base and mostly 1–2 cm wide, with fibers along the margins of the segments. *Stalk margins lined with coarse teeth.*

ROOTS AND UNDERGROUND STRUCTURES: Roots fibrous, shallow.

FLOWERS: Flower clusters develop within the crown, shorter than leaves. Petals 3. Calyx lobes 3.

Mexican fan palm (*Washingtonia robusta*) plant fully grown. J. M. DiTomaso

Young Canary Island date palm (*Phoenix canariensis*). J. M. DiTomaso

Young Mexican fan palm (*Washingtonia robusta*). J. M. DiTomaso

Mexican fan palm (*Washingtonia robusta*) seedling.
J. M. DiTomaso

Canary Island date palm: *Male and female flowers develop on separate trees* (dioecious). Flowers yellowish. Stalks bright orange. Ovaries 3, separate.

Mexican fan palm: Flower clusters have main stem bracts that are longitudinally split and hang downward. Flowers bisexual, white. Ovary 3-lobed.

FRUITS AND SEEDS: Canary Island date palm: Fruits ovate to round, ± 2 cm long, brown, edible. Seeds oblong, mostly 8–12 mm long, with a *deep longitudinal groove on 1 side*.

Mexican fan palm: Fruits oblong, about *1 cm long or less*, black.

HABITAT: Landscaped areas, urban places, riparian streams (particularly near rural areas), and orchards.

DISTRIBUTION: Canary Island date palm: San Francisco Bay area and South Coast, to 1000 m. Florida.

Mexican fan palm: San Francisco Bay area, South Coast, and southern Sacramento Valley. Florida.

PROPAGATION/PHENOLOGY: *Reproduce by seed*. Birds ingest fruits and disperse the seeds with their droppings. Seeds are large and readily carried by winter rains from landscaped areas down storm drains into nearby creeks and rivers.

MANAGEMENT FAVORING/DISCOURAGING SURVIVAL: Manually removing seedlings can control both species.

SIMILAR SPECIES: Canary Island date palm and Mexican fan palm are the species most often encountered as volunteer seedlings in landscaped areas.

California fan palm [*Washingtonia filifera* (L. Linden) H.A. Wendl.] is native to the Sonoran Desert of southern California, as well as northern Baja California and southeastern Arizona. Although it has become naturalized in many locations on the coastal slope of southern California, as well as in the Kern River in Kern County and the Mojave Desert, it is not typically considered a weedy species in California. California fan palm has a more robust trunk than Mexican fan palm and is not markedly flared at the base. The trunk is more uniform throughout its length and lacks the criss-cross pattern.

Bladderflower [*Araujia sericifera* Brot.][CDFA list: B]

SYNONYMS: common moth-vine; cruel plant; moth catcher; *Araujia albens* (Martius) Don; *Araujia hortorum* Fourn; *Physianthus albens* Martius. *Araujia sericofera* Brot. is an orthographic error.

GENERAL INFORMATION: Fast-growing *perennial vine* with *milky juice*. Bladderflower has escaped cultivation as an ornamental and become a noxious pest in some regions of California, especially in areas where citrus is cultivated. Bladderflower generally thrives in citrus groves and can compete with trees for water, nutrients, and light. Plants grow extremely fast, and vines can grow over tree canopies within a couple of years. Individual branches of trees are sometimes killed by girdling vines. Significant infestations reduce fruit yields and interfere with tree maintenance. Bladderflower foliage and, to a lesser degree,

Bladderflower (*Araujia sericifera*) vine in flower.   J. M. DiTomaso

Bladderflower

Bladderflower (*Araujia sericifera*) flower.
J. M. DiTomaso

Bladderflower (*Araujia sericifera*) fruit cut open with immature seeds.
J. M. DiTomaso

fruits are reported to contain serotonin and other compounds that can cause nonfatal digestive tract irritation and neurological disturbances when ingested in sufficient quantity. However, toxicity problems have not been reported in North America. Plants also contain an enzyme in the sap that can dissolve skin and cause severe sores if not quickly washed off. Native to Central South America.

MATURE PLANT: Stems *twining*, slender, woody, sometimes branched, typically less than 12 m long, older growth glabrous or nearly glabrous, new growth covered with short white hairs. Leaves *opposite, narrowly triangular*, bases truncate to slightly lobed (cordate), 5–12 cm long, 2–6 cm wide, spaced 7–18 cm along stems, evergreen or partially deciduous in cooler climates. Upper surfaces glabrous, glossy dark green. Lower surfaces gray-green, minutely pitted, densely covered with minute hairs. Petioles 1.5–3 cm long.

ROOTS AND UNDERGROUND STRUCTURES: Under certain conditions, severed pieces of underground stems or crowns can produce new roots and shoots.

FLOWERS: Summer. Flower clusters (cymes) 2–10-flowered, develop from just below leaf axils. Flowers fragrant, waxy, white to pink, *2–3 cm long*, about 1–2 cm wide. Petals fused, 5-lobed, *bell- to funnel-shaped*. Sepals 5, fused near the base, green, erect, leaflike. Stamens fused into a *filament column with appendages* and an anther head. *Appendages separate, solid, margins convex.* Styles fused into a pistil head with 2 erect, elongate lobes. Insect-pollinated.

FRUITS AND SEEDS: Pods narrowly ovoid, *pendant*, 8–15 cm long, 4–5 cm in diameter, pale gray-green, open at maturity to release numerous seeds. Seeds dark brown to black, narrowly ovate to elliptic, 5–6 mm long, surface minutely reticulate and irregularly tubercled, with numerous silky white, deciduous hairs attached at the apex.

HABITAT: Orchards (especially citrus groves), landscaped areas, gardens, and disturbed sites. Tolerates poor, wet, or dry soils and light to moderate frost.

Bladderflower

Bladderflower (*Araujia sericifera*) fruit.   J. M. DiTomaso

DISTRIBUTION: Uncommon. North Coast Ranges (Mendocino and Sonoma Cos.), San Francisco Bay region, Central Valley (especially Sacramento and Fresno Cos.), South Coast Ranges, and South Coast region (especially Ventura, sw San Bernardino, nw Riverside, ne Orange Cos.), to 400 m. Florida.

PROPAGATION/PHENOLOGY: *Reproduces by seed* and *vegetatively from severed underground stems or crowns*. Stems can grow 6–9 m in one season. Seeds disperse with wind. Seed production is prolific, except in areas where temperatures drop below freezing in early fall. Seed viability is typically high (~ 90%), but longevity is undocumented. Plants usually produce seed the first season.

MANAGEMENT FAVORING/DISCOURAGING SURVIVAL: Improper disking can disperse underground stem and crown fragments of mature plants, which can develop into new vines under favorable conditions. Young seedlings do not tolerate light cultivation.

SIMILAR SPECIES: Bladderflower is unlikely to be confused with other weedy vines. There are a few nonweedy, native vines in the milkweed family (*Cynanchum, Matelea, Sarcostemma* species) that inhabit natural areas in the deserts. The native species are further distinguished by having flowers *less than 1 cm long*.

Mexican whorled milkweed [*Asclepias fascicularis* Decne.][ASCFA]

SYNONYMS: narrow-leaf milkweed; whorled milkweed; *Asclepias macrophylla* var. *comosa* Dur. & Hilg.; *Asclepias mexicana* auct. non Cav.

GENERAL INFORMATION: Erect native *perennial* to 1 m tall, with *milky white juice, leaves in whorls of 3 to 6*, and often with creeping roots. Under favorable conditions, dense colonies develop from the rhizomelike roots. In natural areas, native milkweeds are considered a desirable component of the plant community and are important insect-attracting plants. Monarch butterfly larvae feed solely on the foliage of certain species that make the larvae toxic to predators. However, all plant parts of milkweed species contain compounds that are *highly toxic* to livestock, poultry, and humans when ingested. **Western whorled milkweed**

Mexican whorled milkweed (*Asclepias fascicularis*) plant.
J. M. DiTomaso

Mexican whorled milkweed

Mexican whorled milkweed (*Asclepias fascicularis*) flower clusters and fruit. J. M. DiTomaso

[*Asclepias subverticillata* (Gray) Vail], which occurs in southwestern states other than California, and Mexican whorled milkweed are most often associated with livestock poisonings in North America. Both species contain a tentatively identified neurotoxin. Ingestion of 1/2–1% of an animal's weight in green foliage of Mexican whorled milkweed, less for western whorled milkweed, is often lethal. Toxicity symptoms in large animals can include colic, trembling, twitching, weakness, staggering, falling, inability to stand, profuse sweating, seizures with rigidly flexed limbs or paddling movements, and death from respiratory failure. Horses are especially susceptible to the toxic effects. In poultry symptoms include loss of muscular control, strange posturing, excitability, convulsions (often with backward flips), inability to stand, and breathing difficulties leading to death. Generally, animals avoid eating the bitter-flavored foliage unless more palatable forage is scarce, but dried plant material in hay remains toxic and is less easily avoided by livestock. Because of its toxicity and tendency to develop colonies from creeping roots, Mexican whorled milkweed can be problematic in pastures, rangeland, and crop fields.

SEEDLINGS: Cotyledons oblong, 4–15 mm long, 2–8 mm wide, glabrous, with a conspicuous midvein. First and subsequent leaves opposite, linear to narrowly lanceolate, about equal to or longer than the cotyledons, sometimes slightly folded upward at the midvein. Broken leaves and stems exude milky juice.

MATURE PLANT: Foliage usually glabrous, sometimes sparsely covered with minute hairs. *Leaves linear to narrowly lanceolate, in whorls of 3–5(6) at the nodes*, mostly 4–12 cm long, 0.3–1 cm wide, often folded upward along the midvein.

ROOTS AND UNDERGROUND STRUCTURES: Creeping roots generally shallow.

FLOWERS: June–September. Inflorescences umbel-like. Petal lobes 5, *greenish white to pale yellow*, often tinged purplish, about 4–5 mm long, reflexed downward. Sepals 5. Stamens 5, filaments fused into a column with 5 hoodlike structures near the top, anthers fused to a broad flat stigma head. Hoods broadly ovate, each with a slit from which a *slender incurved horn protrudes above the hoods* and anther or stigma head. Ovaries 2, separate. Styles 2, fused at the top into the stigma head. Insect-pollinated.

FRUITS AND SEEDS: Pods (follicles) *smooth, narrow*, mostly 6–9 cm long, tip tapered and slightly pinched-in (acuminate), on an erect stalk. Seeds ± oblong, strongly flattened, ± 6 mm long, with a tuft of deciduous silky hairs ± 3 cm long at the apex.

Mexican whorled milkweed (*Asclepias fascicularis*) fruit with dispersing seeds. J. M. DiTomaso

Mexican whorled milkweed (*Asclepias fascicularis*) seeds. J. O'Brien

Mexican whorled milkweed (*Asclepias fascicularis*) seedling. J. K. Clark

Showy milkweed (Asclepias speciosa) flowers.　　　　　　　　　J. M. DiTomaso

HABITAT: Grassland, pastures, roadsides, cultivated fields and margins, stream banks, and ditches. Grows in dry or wet places. Classified as a facultative wetland indicator species in California.

DISTRIBUTION: Throughout California, except North, Central, and South Coast and Baja California, to 2200 m. Oregon, Washington, Nevada, Idaho, and Utah.

PROPAGATION/PHENOLOGY: *Reproduces by seed* and *vegetatively from creeping roots*. Most seeds fall near the parent plant or disperse short distances with wind.

MANAGEMENT FAVORING/DISCOURAGING SURVIVAL: Cultivation or manual removal can control Mexican whorled milkweed.

SIMILAR SPECIES: Showy milkweed [*Asclepias speciosa* Torrey, synonyms: Greek milkweed; *Asclepias giffordii* Eastw.] is a native *perennial* to 1.2 m tall, with *opposite oval to oblong leaves 8–15 cm long on short stalks*. Flowers have rose-purple petal lobes with hairy backs and pinkish hoods that fade to yellowish. Unlike Mexican whorled milkweed and other native milkweeds, showy milkweed has *hoods that are longer than the anther or stigma head by at least half the length of the hoods* and *protruding horns that are much shorter than the hoods*. Showy milkweed lacks creeping roots and has foliage that is usually covered with soft woolly hairs. Pods are narrowly ovoid, densely covered with woolly hairs, and point downward on strongly reflexed stalks. Showy milkweed inhabits roadsides, fields, grassland, rangeland, woodlands, openings in coniferous forests, and other places, usually on dry, stony soils where little competing vegetation exists. It occurs throughout California to 1900 m. Showy milkweed contains cardiac glycosides and is fatally *toxic* to livestock when ingested at an amount of 1–2% of body weight. Symptoms include depression, digestive tract

Mexican whorled milkweed

Showy milkweed (*Asclepias speciosa*) plant.
J. M. DiTomaso

Showy milkweed (*Asclepias speciosa*) fruit.
J. M. DiTomaso

disturbances, respiratory difficulties, and sometimes death, typically without seizures or struggling. Like Mexican whorled milkweed, livestock usually avoid consuming showy milkweed if more palatable forage is available.

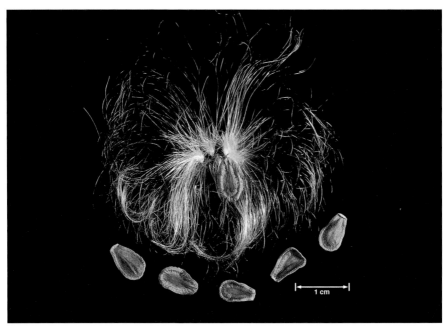

Showy milkweed (*Asclepias speciosa*) seed. J. O'Brien

Common yarrow [*Achillea millefolium* L.][ACHMI]

SYNONYMS: bloodwort; carpenter's weed; hierba de las cortaduras; milfoil; plumajillo; thousand leaf; *Achillea borealis* Bong. ssp. *arenicola* (A.A. Heller) Keck and ssp. *californica* (Pollard) Keck; *Achillea lanulosa* Nutt. ssp. *lanulosa* and ssp. *alpicola* (Rydb.) Keck; many others

GENERAL INFORMATION: Low, tufted *perennial* with flowering stems mostly to 1 m tall, *finely dissected leaves*, and *creeping rhizomes*. Plants exist as rosettes until flowering stems develop in late spring or summer. Common yarrow is a widespread native of North America and Eurasia and is usually considered a desirable component of the plant community in natural areas. However, it can be weedy in landscaped areas and turf due to its creeping habit and seedling

Common yarrow (*Achillea millefolium*) plant. J. M. DiTomaso

Common yarrow

Common yarrow (*Achillea millefolium*) flower heads. J. M. DiTomaso

Common yarrow (*Achillea millefolium*) foliage in turf. J. M. DiTomaso

Common yarrow (*Achillea millefolium*) seedling. J. M. DiTomaso

establishment. Common yarrow consists of a highly variable complex of polyploid biotypes. Cultivars with various flower colors are commonly grown as landscape ornamentals and occasionally as a low-maintenance replacement for turf. Plants are used medicinally in some places.

SEEDLINGS: Cotyledons ovate to oblong, about 3–4 mm long, ± 2 mm wide. Leaves alternate. First leaf finely dissected mostly 2 times into linear segments.

MATURE PLANT: Foliage nearly glabrous to soft-hairy. Leaves alternate, *aromatic when crushed, finely pinnate-dissected* 3 times into linear segments. Rosette and lower stem leaves 10–20 cm long. Upper leaves smaller, sessile, ± clasping the stem.

ROOTS AND UNDERGROUND STRUCTURES: Plants develop an extensive system of creeping rhizomes.

FLOWERS: March–September, depending on location. Flower clusters (corymbs) *flat-topped*. Heads usually consist of 3–8 *broad, white to pink ray flowers*

Common yarrow

and about 15–40 *white disk flowers*. Involucre (phyllaries as a unit) usually 3–5 mm in diameter. Phyllaries ovate, margins membranous, often hairy, in 3–4 unequal rows. Receptacle flat to convex, with membranous, *flat, chaffy bracts*. Self-incompatible. Insect-pollinated.

FRUITS AND SEEDS: Achenes narrowly oblong-lanceolate, ± 2 mm long, flattened, longitudinally ribbed, margins narrowly thick-winged, lack a pappus.

POSTSENESCENCE CHARACTERISTICS: Foliage remains green in winter, but sometimes dies during extended summer drought. New shoots grow from the rhizomes when growing conditions become favorable. Flowering stems die at the end of the season and can persist into winter.

HABITAT: Landscaped areas, turf, pastures, many plant communities, including grassland, alpine, semidesert, coastal bluffs, sand dunes, salt marshes. Tolerates extended drought.

DISTRIBUTION: Throughout California, except deserts, to 3500 m. All contiguous states, Canada, and Eurasia.

PROPAGATION/PHENOLOGY: *Reproduces vegetatively from creeping rhizomes* and *by seed*. Small creeping root fragments readily develop into new plants under favorable conditions. Root fragments disperse with soil movement. Achenes fall near the parent plant and disperse to greater distances with water, soil movement, and human activities such as landscape maintenance. Seed longevity, dormancy, and germination requirements vary with biotype and environment. Plants are winter-dormant or winter-active, depending on the biotype.

MANAGEMENT FAVORING/DISCOURAGING SURVIVAL: Plants tolerate some mowing. Manual removal of seedlings and mature plants, including creeping roots, can control common yarrow in landscaped areas and turf.

SIMILAR SPECIES: Common yarrow is distinguishable from other weedy species with finely dissected leaves by having *creeping rhizomes*.

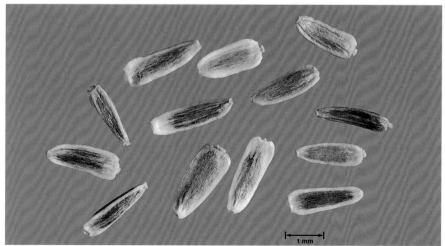

Common yarrow (*Achillea millefolium*) seeds. J. K. Clark

Russian knapweed [*Acroptilon repens* (L.) DC.]
[CENRE][Cal-IPC: Moderate][CDFA list: A]

SYNONYMS: hardheads; Russian starthistle; Turkestan thistle; *Centaurea picris* Pallas ex Willd.; *Centaurea repens* L.

GENERAL INFORMATION: Noxious *perennial* to 1 m tall, with clusters of pink, lavender-blue, or white flower heads and *dark creeping roots*. Plants appear to have allelopathic properties and are aggressively competitive. Russian knapweed can rapidly colonize an area and develop dense stands. Populations are often extremely long-lived due to extensive root systems. Stems die back after flowering in summer, and new shoots emerge in spring. Russian knapweed has been shown to accumulate high levels of zinc, and soils beneath its canopy can have 8 times the level of zinc as uninfested soils. It has been hypothesized that high soil zinc concentrations can inhibit the growth of more desirable species and interfere with restoration efforts. Like **yellow starthistle** [*Centaurea solstitialis* L.], Russian knapweed is *toxic* to horses and can cause nigropallidal encephalomalacia, or "chewing disease," when an animal ingests a certain amount of plant material over time. Generally, livestock will avoid grazing Russian knapweed because of its bitter taste. Besides California, Russian knapweed is a state-listed noxious weed in Arizona (restricted), Colorado, Hawaii, Idaho, Iowa (primary), Kansas, Minnesota (secondary), Montana (category 1), Nevada, New Mexico, North Dakota, Oregon (class B), South Dakota, Utah, Washington (class B), and Wyoming. Native to Central Asia.

Russian knapweed (*Acroptilon repens*) plant.   J. M. DiTomaso

Russian knapweed

Russian knapweed (*Acroptilon repens*) flowering stem.   J. M. DiTomaso

Russian knapweed (*Acroptilon repens*) seedling.   J. M. DiTomaso

SEEDLINGS: Uncommon in the field. Cotyledons ovate to spatulate and scurfy on the lower surface. First several rosette leaves elliptic to oblanceolate and covered with a white, powdery bloom. Margins entire. Subsequent rosette leaves irregularly 1-pinnately lobed with pronounced wavy margins.

MATURE PLANT: Stems erect, openly branched, *leafy*, and mostly covered with cobwebby gray hairs. Stem leaves alternate, *do not extend down the stem as wings*. Basal and lower stem leaves mostly oblong, 4–10 cm long, 1–2-pinnate-lobed or margins conspicuously wavy. *Upper stem leaves narrow lanceolate to linear*, entire or toothed, and 1–3 cm long. Leaves lack hairs or are covered with short to medium interwoven hairs.

ROOTS AND UNDERGROUND STRUCTURES: Creeping roots branch frequently at various depths to form an *extensive vertical and horizontal root system, dark brown to black at maturity*. Scale leaves alternate, small, narrow, appressed, clasping. Each scale leaf has a bud in its axil that is capable of producing a new shoot. Immature rhizomes paler, with longer, less-appressed scale leaves. Roots can grow from 1 m to several meters deep. New shoots and roots develop at various intervals to a depth 1.2 m or more. Root fragments as small as 2.5 cm can develop into a new plant from depths to 15 cm. Roots do not appear to associate with mycorrhizae.

FLOWERS: May–September. Flower head clusters panicle-like or flat-topped. Flower heads hemispheric, consist of about 30 white, pink, or lavender-blue *disk flowers interspersed with bristles* on the receptacle. Corollas about 15 mm long. Phyllaries (flower head bracts) in several overlapping rows, *ovate, base green, tip usually acute* (not comblike or spine-tipped), *tip margin broadly papery*. Primarily outcrossing.

FRUITS AND SEEDS: Achenes *white or pale gray*, obovoid, *3–4 mm long*, basal attachment scar barely lateral and *lacks a notch*. Pappus bristles unequal, white, about 6–10 mm long, lower portion minutely barbed, upper portion plumose, *early-deciduous*.

Russian knapweed (*Acroptilon repens*) immature bolted stem.   J. M. DiTomaso

POSTSENESCENCE CHARACTERISTICS: Old flower stems can persist for an extended period after senescence. Phyllaries and achenes remaining on old stems can aid species identification.

HABITAT: Fields, rangeland, cultivated sites, orchards, vineyards, roadsides, ditchbanks, and waste places. Grows on many soil types.

DISTRIBUTION: Throughout California, except for the wettest areas of the northwest and driest regions of the Great Basin and deserts, to 1900 m. Known populations occur in all counties except Calaveras, Del Norte, El Dorado,

Russian knapweed (*Acroptilon repens*) plant following flower senescence.   J. M. DiTomaso

Russian knapweed (*Acroptilon repens*) seeds.   J. O'Brien

Humboldt, Mariposa, Mendocino, Sierra, and Trinity. New populations have recently been reported from southern California. All western and central states.

PROPAGATION/PHENOLOGY: *Reproduces primarily by vegetative shoots from creeping roots.* Plants usually produce small quantities of viable seed. Seed heads mostly remain closed. Seeds fall near the parent plant or disperse with the seed head. Seeds generally germinate at temperatures from 0.5°–35°C, with an optimal range of about 20°–30°C, and light is not required. Scarification, fluctuating temperatures, and alternating light and dark periods enhance germination. Seed appears to survive about 2–3 years under field conditions.

MANAGEMENT FAVORING/DISCOURAGING SURVIVAL: Aboveground removal of plants, such as mowing, encourages new shoots to sprout from the extensive root system. Shallow cultivation can facilitate spread of rhizome fragments. However, repeated cultivation to about 30 cm deep over a period of about 3 years can kill much of the root system in the top 1 m of soil. Cultivation of dense competitive crops such as winter rye or wheat for a few years can significantly reduce Russian knapweed in crop fields, especially when the crop is harvested early for silage.

SIMILAR SPECIES: Other white, pink, and purple-flowered knapweeds (*Centaurea* species) and **common crupina** [*Crupina vulgaris* Cass.] are most easily distinguished by their *lack of dark creeping roots* and by phyllary and achene characteristics. Refer to table 18 (p. 227) for a comparison of the flowers of spiny-leaved species. In addition, only bearded creeper has *leaf margins with stiff hairs that are barbed at the tips* (glochidiate hairs) and flower receptacles with *flattened, scalelike, chaffy bracts.*

Croftonweed [*Ageratina adenophora* (Spreng.) King & H. Robins.][EUPAD][Cal-IPC: Moderate][Federal Noxious Weed]

SYNONYMS: catweed; eupatory; Maui pamakani; Mexican devil; sticky agrimony; sticky snakeroot; white thoroughwort; *Eupatorium adenophora* Spreng.; *Eupatorium adenophorum* Spreng.; *Eupatorium glandulosum* Kunth. non Michx.; *Eupatorium pasadense* McClat.

GENERAL INFORMATION: *Perennial or subshrub to 2 m tall, with opposite ovate-triangular leaves and nearly flat-topped clusters of white (sometimes pink) flower heads.* Croftonweed has escaped cultivation as an ornamental in

Croftonweed (*Ageratina adenophora*) plant.   J. M. DiTomaso

California and is especially invasive in mild coastal regions. Croftonweed can cause a fatal respiratory illness in horses when ingested over a period of several months to years. Symptoms develop more rapidly when animals ingest flowering plants. Sheep and goats are unaffected, and cattle generally avoid eating the plant. Croftonweed is a noxious weed in parts of Asia, coastal Australia, New Zealand, and Hawaii. Although it is not reported to have naturalized in Florida or North Carolina, croftonweed is a state-listed noxious weed in these states. The gall fly *Procecidochares utilis* has been successfully introduced as a biocontrol agent in Hawaii. However, the fly is not approved by the USDA for release in the contiguous states. Native to southern and central Mexico.

MATURE PLANT: Crown and lower stems woody. Upper stems *purplish*, often sparsely covered with minute, purplish, glandular hairs. *Leaves opposite, ovate to ovate-triangular*, with *3 main veins from the base, 4–10 cm long*, 2–9 cm wide, margins coarsely toothed, on a stalk about *2–4 cm long*. Lower surfaces sometimes purplish, often sparsely covered with very short, purplish glandular hairs.

ROOTS AND UNDERGROUND STRUCTURES: Crown often weakly woody. Main rootstock short, thick, pale yellowish, with a carrotlike scent when broken. Numerous lateral roots grow from the rootstock and extend peripherally up to about 1 m in all directions and downward to about 40 cm. New shoots grow from the crown and short rootstock, but not from the lateral roots. Lower stems and stem fragments that contact moist soil can develop adventitious roots.

FLOWERS: Nearly year-round. Flower head clusters (cymes) small, nearly flat-topped. Flower heads mostly *6–7 mm long*, consist of 10–60 *white to pink-tinged disk flowers* with *long exserted styles*. Phyllaries lanceolate, longitudinally ribbed, glandular, in 1–3 nearly equal series. Receptacles lack chaff.

Croftonweed (*Ageratina adenophora*) foliage. J. M. DiTomaso

Croftonweed

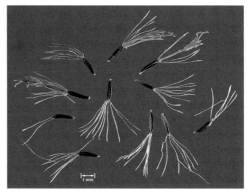

Croftonweed (*Ageratina adenophora*) flowering stems. J. M. DiTomaso

Croftonweed (*Ageratina adenophora*) seeds. J. O'Brien

FRUITS AND SEEDS: Achenes nearly cylindrical, ± 2 mm long excluding pappus, 5-angled or ribbed, black to brown. Pappus bristles few to several in a single row, fine, minutely barbed, white, deciduous.

HABITAT: Disturbed places, coastal canyons, riparian areas, scrub, and slopes. Grows best where moisture is available year-round and can tolerate both full sun and highly shaded areas.

DISTRIBUTION: San Francisco Bay region, Central Coast, western South Coast Ranges, South Coast, central and eastern Transverse Ranges (San Gabriel and San Bernardino Mountains), and south to the Mexican border in the Peninsular Range, to about 300 m.

PROPAGATION/PHENOLOGY: *Reproduces primarily by seed.* Mature plants can produce abundant quantities of apomictic seed (seed that develops without fertilization). Seeds disperse with wind, water, soil movement, human activities, and by clinging to animals. In Australia, only about 70–85% of seeds are viable. Seeds can germinate nearly year-round under favorable conditions, and light is required. Seedlings tolerate light shade and if damaged can regrow from the crown by 8 weeks of age. Most vegetative growth occurs in summer and fall.

MANAGEMENT FAVORING/DISCOURAGING SURVIVAL: Hand-removal can control small infestations provided that the crown and short rootsock is removed to prevent the growth of new shoots. Plants cut above the crown typically regenerate. Disturbance that creates openings of bare soil, including burning, can favor germination and establishment of croftonweed. Follow-up on managed populations is critical to avoid reinfestation from locally abundant seedbank.

SIMILAR SPECIES: There are three native *Ageratina* species in California, but only *A. herbacea* has opposite leaves like croftonweed. Unlike croftonweed, it is not weedy and grows in the eastern Mohave Desert mountains. It also has leaves *1.5–5.5 cm long on stalks to 1 cm long*, whereas croftonweed has leaves *4–10 cm long on stalks 2–4 cm long*.

Annual bursage [*Ambrosia acanthicarpa* Hook.][FRSAC]
Giant ragweed [*Ambrosia trifida* L.][AMBTR][CDFA list: B]

SYNONYMS: Annual bursage: annual burweed; flatspine burweed; sand-bur; *Franseria acanthicarpa* Hook.; *Franseria californica* Gand.; *Franseria montana* Nutt.; *Franseria palmeri* Rydb.; *Gaertneria acanthicarpa* Britt.

Giant ragweed: buffaloweed; great ragweed; kinghead; tall ragweed

GENERAL INFORMATION: These species typically colonize disturbed open sites. Pollen of *Ambrosia* species is a major cause of allergies in the summer and fall months.

Annual bursage: Widespread native summer annual to 1.5 m tall. Inhabits many natural plant communities and is not considered a pest under most circumstances. However, it can become problematic in agricultural fields, forestry regeneration sites, and other disturbed sites.

Giant ragweed: Erect summer annual typically to 2 m tall, rarely to 6 m in moist fertile soils. Giant ragweed populations can be noxious for a period but seldom persist in California. Native to the central and eastern United States, where it is often problematic in agricultural fields and drainage areas.

SEEDLINGS: Annual bursage: Cotyledons oblong to elliptic (about 0.6–1.4 cm long). Subsequent leaves deeply pinnate-lobed, sparsely covered with short white or gray bristly hairs.

Giant ragweed: Cotyledons round, ovate, or oblong, thick, sometimes slightly indented at the tips, mostly 2–4 cm long, 1–1.6 cm wide. First leaves ovate to lanceolate, slightly coarse-lobed. Subsequent leaves opposite, coarsely 3-lobed. First and subsequent leaves moderately covered with stiff hairs.

MATURE PLANT: Annual bursage: Foliage covered with white to gray, short, bristly hairs. Leaves gray-green, often opposite near the stem bases, *alternate on the*

Annual bursage (*Ambrosia acanthicarpa*) plant.
J. M. DiTomaso

Annual bursage (*Ambrosia acanthicarpa*) flowering stem.
J. M. DiTomaso

Annual bursage • Giant ragweed

Annual bursage (*Ambrosia acanthicarpa*) foliage.
J. M. DiTomaso

Annual bursage (*Ambrosia acanthicarpa*) seedling.
J. M. DiTomaso

Annual bursage (*Ambrosia acanthicarpa*) seeds.
J. O'Brien

upper stems, to 8 cm long and 7 cm wide, *pinnate-lobed 2 times, primary lobes deep.*

Giant ragweed: Stems coarse, single or branched, woody at the base, longitudinally black-lined, covered with soft to bristly hairs. Leaves *opposite, broad, palmately 3 to 5-lobed, 6–35 cm long*, margins finely serrate, surfaces sparsely covered with minute stiff hairs.

ROOTS AND UNDERGROUND STRUCTURES: Both species generally have a short taproot, with many fibrous roots.

FLOWERS: Heads small, greenish, composed of staminate (male) or pistillate (female) disk flowers. Staminate and pistillate heads are separate on a single plant (monoecious). Terminal spikes consist of nodding staminate heads 2–5 mm in diameter. Pistillate heads are clustered in the leaf axils below the spikes. Phyllaries of staminate heads fused, cuplike, with the 3 longest lobes *blackish along the midveins.* Phyllaries of pistillate heads fused, persistent, enclose a single ovary, become a bur in fruit. Wind-pollinated.

Annual bursage: August–November.

Giant ragweed (*Ambrosia trifida*) plant. J. Neal

Giant ragweed (*Ambrosia trifida*) flowering stem. J. M. DiTomaso

Giant ragweed (*Ambrosia trifida*) seedling. J. M. DiTomaso

Giant ragweed (*Ambrosia trifida*) leaf. J. M. DiTomaso

Giant ragweed (*Ambrosia trifida*) seeds. J. O'Brien

Annual bursage • Giant ragweed

Common ragweed (*Ambrosia artemisiifolia*) plant. J. M. DiTomaso

Common ragweed (*Ambrosia artemisiifolia*) flowering stem. J. M. DiTomaso

Common ragweed (*Ambrosia artemisiifolia*) seedlings. J. M. DiTomaso

Common ragweed (*Ambrosia artemisiifolia*) foliage. J. M. DiTomaso

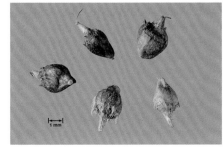

Common ragweed (*Ambrosia artemisiifolia*) seeds. J. O'Brien

Giant ragweed: June–September.

FRUITS AND SEEDS: Hardened phyllaries tightly enclose a single achene to form a bur. Burs ± obovoid.

Annual bursage: Burs highly variable, often golden brown, body 4–8 mm long, typically with *6–30 scattered spines. Spines sharp-pointed, flattened, straight, 2–5 mm long.*

Giant ragweed: Burs brown to gray, 6–12 mm long, with a *thick blunt*

Western ragweed (*Ambrosia psilostachya*) plants. J. M. DiTomaso

Western ragweed (*Ambrosia psilostachya*) flowering stem. J. M. DiTomaso

Western ragweed (*Ambrosia psilostachya*) seedling. J. M. DiTomaso

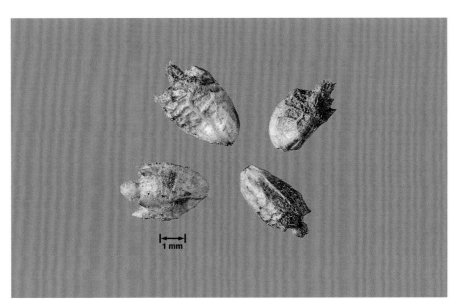

Western ragweed (*Ambrosia psilostachya*) seeds. J. O'Brien

beak at the apex, body usually 5–8-ribbed. Each rib terminates with a short, thick tooth or vestigial spine such that the beak is surrounded by a crown of 5–8 short blunt teeth. Most seed matures August–October.

POSTSENESCENCE CHARACTERISTICS: Rigid stems with fruits can persist into the winter months.

HABITAT: Annual bursage: Dry slopes, sandy flats and alluvial plains, grasslands, foothill woodlands, coastal sage scrub, disturbed sites, agricultural fields, and forestry sites. Commonly grows on dry or moist sandy soils.

Giant ragweed: Disturbed sites, agricultural fields, roadsides, and drainage areas. Grows best on fertile, moist soils.

DISTRIBUTION: Annual bursage: North Coast Ranges, Sierra Nevada, Great Basin, Southwestern region, and Mojave and Sonoran deserts, to 2200 m. All western states, most central states.

Giant ragweed: Uncommon. San Joaquin Valley (especially San Joaquin, Contra Costa, Madera Cos.), Central Coast (especially Monterey Co.), South Coast region (especially Orange Co.), and Modoc Plateau (Lassen Co.), to 100 m. All contiguous states, except possibly Nevada.

PROPAGATION/PHENOLOGY: *Reproduce by seed.*

Annual bursage: Burs disperse by clinging to agricultural machinery, shoes, clothing, and to animals. The biology of this species is poorly understood.

Giant ragweed: Most burs fall near the parent plant, but some can disperse long distances with water, animals, and human activities. Burs are rarely consumed by animals. An average-size plant produces roughly 275 seeds. Newly matured seed is usually dormant and requires a cold, moist period to germinate. Seeds typically germinate at temperatures between 8°–41°C, optimum 10°–24°C. Most germination in the field occurs early to mid-spring and at soil depths to 16 cm, optimal 2 cm. Germination rarely occurs on the soil surface. Seedlings emerging from shallow depths are most likely to survive. Plants effectively compete with other species for light.

MANAGEMENT FAVORING/DISCOURAGING SURVIVAL: Giant ragweed cut in mid- to late summer can still recover and produce seed. Cultivation to prevent seed production can help control infestations.

SIMILAR SPECIES: Common ragweed [*Ambrosia artemisiifolia* L., synonyms: annual ragweed; low ragweed; short ragweed; small ragweed; Roman ragweed] [AMBEL] is an *annual* that closely resembles annual bursage. It is native to the eastern states and is less common in California than it is there. Some botanists recognized 3 varieties of which only *Ambrosia artemisiifolia* L. var. *elatior* (L.) Descourtils occurs in the western states. Common ragweed is most easily distinguished by having staminate flower heads with green phyllaries that *lack a black midvein* and burs that are similar to those of giant ragweed but considerably *smaller* (2–4 mm long). It inhabits disturbed sites in northwestern California, eastern Sacramento Valley, South Coast, and low regions of the eastern Sierra Nevada, to 650 m. Common ragweed occurs in all contiguous states. It is a state-listed noxious weed in Illinois, Minnesota (secondary), and Oregon (class B).

Western ragweed [*Ambrosia psilostachya* D.C., synonyms: Cuman ragweed; perennial ragweed; *Ambrosia californica* Rydb.; *Ambrosia coronopifolia* Torrey & Gray; *Ambrosia cumanensis* auct. non Kunth] [AMBPS] is a widespread erect, native *perennial* that is sometimes weedy in orchards and vineyards. Its foliage is similar to that of annual bursage, but its burs are shaped like those of giant ragweed. Unlike annual bursage and giant ragweed, western ragweed can *reproduce vegetatively* from *creeping roots*, and leaves are mostly *1-pinnate-divided*, with irregularly toothed margins. Plants generally flower from July to November. Western ragweed commonly inhabits roadsides and dry fields throughout California, except some regions in the Mojave Desert and Great Basin, to 1000 m. It also occurs in most contiguous states, except a few eastern states.

Mayweed chamomile [*Anthemis cotula* L.][ANTCO]

SYNONYMS: dillweed; dog daisy; dog-fennel; dog's chamomile; fetid chamomile; hogs fennel; mather; manzanillo; mayweed; stinking chamomile; stinking daisy; stinkweed; white stinkweed; *Anthemis foetida* Lam.; *Maruta cotula* (L.) DC.; *Maruta foetida* Cass.

GENERAL INFORMATION: Spreading to erect winter or summer *annual* to about 0.5 m tall, with *finely dissected leaves* and *daisylike flower heads*. Crushed foliage has a *strong unpleasant scent*. Plants exist as rosettes until flowering stems develop at maturity. Mayweed chamomile is a state-listed noxious weed in Colorado and is a secondary noxious weed in Tasmania, Australia. Native to Europe.

SEEDLINGS: Cotyledons oval, 2.5–8 mm long, fused at the base, glabrous, wither early. First leaf pair opposite, pinnate-dissected, nearly glabrous to hairy. Subsequent leaves alternate, pinnate-dissected 2 times, nearly glabrous to hairy.

MATURE PLANT: Flowering stems branched. Leaves alternate, sparsely pubescent, *finely dissected 2–3 times into linear or nearly linear segments*, mostly 2–6 cm long.

ROOTS AND UNDERGROUND STRUCTURES: Taproot short, thick, with fibrous laterals.

FLOWERS: April–August. Heads 1–2.5 cm in diameter, *consist of marginal white ray flowers and central yellow disk flowers*. Ray flowers 10–15, with corollas 6–9 mm long. Phyllaries green, tips acute, equal in 2 overlapping rows, margins

Mayweed chamomile (*Anthemis cotula*) infestation. J. M. DiTomaso

Mayweed chamomile

Mayweed chamomile (*Anthemis cotula*) plant. J. M. DiTomaso

Mayweed chamomile (*Anthemis cotula*) flowering stems. J. M. DiTomaso

Mayweed chamomile (*Anthemis cotula*) seeds. J. O'Brien

membranous and pale brown. Receptacle conical with persistent chaffy bracts 2–3 mm long on the upper half.

FRUITS AND SEEDS: Achenes narrowly wedge-shaped, 1–2 mm long, 10-ribbed, ribs minutely tubercled. *Pappus consists of a minute scalelike crown.*

POSTSENESCENCE CHARACTERISTICS: Dead stems with flower head and leaf remnants can persist into winter.

HABITAT: Disturbed places, fields, roadsides, coastal dunes, chaparral, woodlands, waste places, agricultural crops, orchards, vineyards, landscaped areas, and gardens.

DISTRIBUTION: Northwestern region, Central-western region, central Sierra

Mayweed chamomile

Corn chamomile (*Anthemis arvensis*) stem and foliage. J. M. DiTomaso

Corn chamomile (*Anthemis arvensis*) flower head. J. M. DiTomaso

Corn chamomile (*Anthemis arvensis*) seedling. J. M. DiTomaso

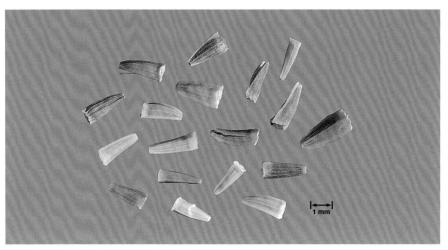

Corn chamomile (*Anthemis arvensis*) seeds. J. K. Clark

Nevada, Central Valley, South Coast, Transverse Ranges, and Peninsular Ranges, to 2000 m. All contiguous states.

PROPAGATION/PHENOLOGY: *Reproduces by seed.* Most seeds fall near the parent plant, but some can disperse to greater distances with water, mud, soil movement, human activities, and as a feed and seed impurity. Seeds can germinate nearly year-round under favorable conditions. In California, most seeds germinate in fall after the first rain of the season or in spring. Some seeds are

Yellow chamomile (*Anthemis tinctoria*) plant.
J. M. DiTomaso

Yellow chamomile (*Anthemis tinctoria*) seedling.
J. M. DiTomaso

Yellow chamomile (*Anthemis tinctoria*) seeds.
J. K. Clark

hard-coated and can survive up to 25 years under field conditions. Scarification enhances germination. Germination does not require light.

MANAGEMENT FAVORING/DISCOURAGING SURVIVAL: Cultivation or mowing as needed before seeds develop can control mayweed chamomile.

SIMILAR SPECIES: Corn chamomile [*Anthemis arvensis* L., synonym: *Anthemis arvensis* L. var. *agrestis* (Wallr.) DC.][ANTAR] is an *annual,* with stems usually more than 0.2 m tall. Corn chamomile closely resembles mayweed chamomile. Unlike mayweed chamomile, corn chamomile has flower head *receptacles with chaffy bracts throughout, smooth achenes,* and *foliage that does not have a strong, unpleasant scent when crushed.* Corn chamomile has escaped cultivation as a garden ornamental and inhabits disturbed places, fields, and roadsides in the Klamath Ranges, western North Coast Ranges, and central Sierra Nevada, to 1000 m. In California, corn chamomile is not as widespread as mayweed chamomile, but it is expected to expand its range. Native to Europe.

Yellow chamomile [*Anthemis tinctoria* L., synonyms: golden marguerite; *Cota tinctoria* (L.) J. Gay][ANTTI] is a *perennial* to 0.3 m tall, with *yellow flower heads, pinnate-lobed leaves,* and foliage that *lacks a strong scent when crushed.* Leaves are 2–5 cm long and lobes are toothed to shallow-lobed. Flower heads are ± 2.5 cm in diameter and have hemispheric receptacles with persistent, narrow chaffy bracts ± 2 mm long throughout. Yellow chamomile has escaped cultivation as a garden ornamental and inhabits roadsides, fields, and disturbed sites in the eastern North Coast Ranges, Cascade Range foothills, and northern Sierra Nevada, to 2000 m. The flower heads are sometimes used to produce a yellow dye. Native to Europe.

In its vegetative state, **pineappleweed** [*Chamomilla suaveolens* (Pursh) Rydb., synonym: *Matricaria matricarioides* (Less.) Porter] resembles mayweed chamomile. Pineapple-weed is distinguishable by the *sweet pineapple-like fragrance of crushed foliage* and by having *flower heads that lack conspicuous ray flowers*. Refer to the entry for **Pineappleweed** (p. 261) for further information.

Yellow chamomile (*Anthemis tinctoria*) flower heads. J. M. DiTomaso

Capeweed [*Arctotheca calendula* (L.) Levyns]
[Cal-IPC: Moderate Alert][CDFA list: A (fertile form only)]

SYNONYMS: cape dandelion; cape gold; *Arctotheca calendulacea* (R. Br.) Lewin; *Arctotis calendula* L.; *Cryptostemma calendula* (L.) Druce

GENERAL INFORMATION: Rosette-forming *perennial* to 0.3 m tall, with showy yellow or yellow with purple flower heads and *creeping stolons*. Capeweed is often cultivated as an ornamental groundcover. Plants in the horticultural trade are a sterile clone and reproduce strictly vegetatively from stolons. However, seed-bearing types have been introduced into California and have escaped cultivation in some areas. Seed-bearing plants typically colonize open sites with exposed soils. Sterile plants occasionally escape cultivation and can spread locally from the creeping stolons, but are much less likely to invade new sites than are seed-bearing plants. Capeweed is perennial in areas with a mild frost-free Mediterranean climate, such as coastal California. Seed-bearing types are annual elsewhere, including southern Australia, where capeweed is an abundant pasture weed, and South Africa. Certain capeweed populations in Australia have developed resistance to bipyridylium herbicides. Handling plants can cause contact dermatitis in sensitive individuals. Native to South Africa.

SEEDLINGS: Cotyledons spoon-shaped, glabrous. First leaves appear opposite, narrowly oblong, margins deeply lobed, lobes broadly acute to rounded. Terminal lobe larger than lateral lobes, tip usually rounded. Upper surface sparse to moderately covered with white hairs. *Lower surface densely covered*

Capeweed (*Arctotheca calendula*) infestation of infertile type.　　　　J. M. DiTomaso

Capeweed (*Arctotheca calendula*) young sprout of infertile type.   J. M. DiTomaso

Capeweed (*Arctotheca calendula*) plant of infertile type.   J. M. DiTomaso

Capeweed (*Arctotheca calendula*) flower and foliage of fertile type.   J. M. DiTomaso

Capeweed (*Arctotheca calendula*) seedling of fertile type.   J. M. DiTomaso

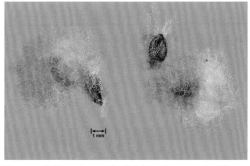

Capeweed (*Arctotheca calendula*) seeds.   J. O'Brien

with white, woolly hairs. Subsequent leaves alternate, similar to first leaves.

MATURE PLANT: Rosette leaves mostly 5–25 cm long, 2–6 cm wide, generally oblanceolate, *deeply pinnate-lobed 1–2 times*, lobes irregularly toothed to lobed, teeth and lobe apices acute or weakly acute and often tipped with a short bristle, upper surface glabrous to moderately covered with fine white cobweblike hairs, *lower surface densely covered with white woolly hairs*. Leaves on flowering stems 0–few, alternate, much reduced, sessile, clasping stem, pinnate-lobed to nearly entire.

ROOTS AND UNDERGROUND STRUCTURES: *Stolons often vigorously creeping, root at the nodes.* One plant can spread vegetatively to cover an area of up to 18 m² in 1–2 years.

FLOWERS: Most of year, peaking March–June. Heads ± 5 cm in diameter, solitary on hairy stalks mostly 15–20 cm long. Receptacle flat, *lacks chaffy bracts*.

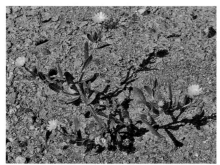

Field marigold *(Calendula arvensis)* plant.
J. M. DiTomaso

Field marigold *(Calendula arvensis)* flowering and fruiting heads.
J. M. DiTomaso

Field marigold *(Calendula arvensis)* seeds and bracts that enclose seeds.
J. O'Brien

African daisy (*Dimorphotheca sinuata*) plant.
J. M. DiTomaso

African daisy (*Dimorphotheca sinuata*) flower heads.
J. M. DiTomaso

Phyllaries strongly overlapping in 3–6 unequal rows, green, covered with woolly hairs, margins membranous, tip reflexed backward. Ray flowers sterile, fewer than 20, corollas 15–25 mm long, pale *yellow* in upper half, sometimes darker yellow below, usually *purple or greenish at the base*. In sterile plants, disk flowers are also yellow. However, disk flowers in fertile plants are *dark purplish*, numerous. *Pappus scales 6–8, ± 1 mm long*.

FRUITS AND SEEDS: Achenes ovate, flattened, mostly 2–4 mm long, longitudinally 3–5-ribbed, laterally wrinkled, mostly *hidden by a dense ball of long, pale, pinkish brown to brown woolly hairs*.

POSTSENESCENCE CHARACTERISTICS: Dead plants generally do not persist for long.

HABITAT: Disturbed coastal and urban sites. Tolerates drought. Grows best on well-drained soils. Plants tolerate light frost but are not hardy.

DISTRIBUTION: North Coast and San Francisco Bay region, to 250 m. Naturalized populations are uncommon.

PROPAGATION/PHENOLOGY: *Reproduces by seed and/or vegetatively from stolons*. The biology of capeweed is poorly understood. Seeds disperse with human activities, animals, and probably wind. Seedlings tolerate dry conditions.

MANAGEMENT FAVORING/DISCOURAGING SURVIVAL: Manual removal of plants, including stolons, can control small populations. In pastures, mowing and close grazing can increase the dominance of capeweed. Shallow cultivation can eliminate seedlings, but if stolons have developed, cultivation will increase vegetative spread.

SIMILAR SPECIES: Field marigold [*Calendula arvensis* L.] is a finely *glandular-hairy annual* mostly to 15 cm tall, with alternate *lanceolate leaves* and yellow flower heads ± 4 cm wide that are *nodding at maturity*. Field marigold has *phyllaries in 2–3 nearly equal rows, achenes that lack a pappus*, and foliage and achenes *without woolly hairs*. Ray achenes are ± 3 mm long, *curve strongly*

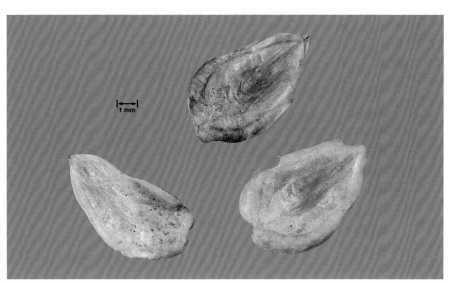

African daisy (*Dimorphotheca sinuata*) seeds.　　　　　　　　　　J. O'BRIEN

inward to nearly form a circle, and have *many sharp prickles on the back*. Disk flowers are staminate and do not develop fruit. Receptacles lack chaffy bracts. Field marigold primarily inhabits disturbed urban and coastal sites in the North Coast, Central Coast, and western South Coast Ranges, to 200 m. Field marigold is expected to expand its range in California. Native to Central Europe and the Mediterranean region.

African daisy [*Dimorphotheca sinuata* DC., synonym: glandular cape marigold] is a *glandular-hairy annual* to 0.3 m tall, with alternate entire to pinnate-lobed leaves that *lack woolly hairs* and have orange to yellow flower heads 3–7 cm wide. Ray flowers sometimes have violet bases or tips, and disk flowers are sometimes violet-tipped. Unlike capeweed, African daisy has *nearly equal phyllaries in 1 row* and *achenes that lack a pappus and woolly hairs*. In addition, ray achenes are *tubercled, 3-angled*, and mostly 4–5 mm long. Disk achenes are *smooth, winged, flattened*, and 6–7 mm long. Receptacles lack chaffy bracts. Lowers leaves taper to a stalklike base, and upper leaves are sessile. African daisy has escaped cultivation as an ornamental and occasionally inhabits roadsides and other disturbed sites in the San Joaquin Valley, western South Coast Ranges, South Coast, and Peninsular Ranges, to 1000 m. It has also naturalized in Oregon. African daisy seed is sometimes included in wildflower seed mixes. Native to South Africa.

Big sagebrush [*Artemisia tridentata* Nutt.][ARTTR]

SYNONYMS: basin big sagebrush; *Seriphidium tridentatum* (Nutt.) W.A. Weber. Four subspecies of big sagebrush are recognized in California.

GENERAL INFORMATION: Several subspecies of big sagebrush occur in California and the West. *A. tridentata* ssp. *tridentata* (big sagebrush), *A. tridentata* ssp. *vaseyana* (Rydb.) Beetle (mountain big sagebrush), and *A. tridentata* ssp. *wyomingensis* Beetle & A. Young (Wyoming big sagebrush) are the most common and they can hybridize. Wyoming big sagebrush is the smallest of the three (< 40 cm tall) and has a very narrow inflorescence (< 3 cm wide). Mountain big sagebrush is generally smaller (< 1 m tall) than big sagebrush (< 2 m tall) and has a narrower inflorescence (< 10 cm) with the inflorescence branches erect. In contrast, big sagebrush has a wider inflorescence (5–15 cm) and the branches are generally spreading. All big sagebrush subspecies are *aromatic shrubs* with persistent *gray-green narrowly wedge-shaped leaves*. Big sagebrush is a common dominant native shrub of cold desert and some high desert regions in the western United States. It is an important component of the plant community and is generally not considered a weed problem. In the past, overgrazing degraded arid and semiarid shrublands and decreased stands of native grasses within the shrubland matrix has led to an increase in the density of big sagebrush. Over the past several years, grazing practices have changed and overgrazing is not as common. Today, long-term fire suppression is considered the cause of dense sagebrush stands and reduced sage grouse habitat. Control of big sagebrush may be necessary for crop or forage production, and stand thinning may be required to increase wildlife habitat or release understory vegetation.

Big sagebrush (*Artemisia tridentata*) in rangeland. J. M. DiTomaso

Big sagebrush

Big sagebrush (*Artemisia tridentata*) plant in bloom. J. M. DiTomaso

Big sagebrush (*Artemisia tridentata*) young plants. J. M. DiTomaso

Big sagebrush (*Artemisia tridentata*) seeds. J. O'Brien

MATURE PLANT: Tall, erect, heavily branched shrub with trunk thick and bark shaggy. Leaves alternate, usually narrowly wedge-shaped, 1–4(6) cm long, tips often *bluntly 3-lobed*, sometimes 5-lobed, gray-green, densely short-hairy, often in axillary clusters.

ROOTS AND UNDERGROUND STRUCTURES: Main roots woody, generally deep.

FLOWERS: July–October. Heads mostly 2–2.5 mm in diameter, erect, consist of ~ 4–6 glandular or hairy yellow disk flowers. Phyllaries oblanceolate to obovate, hairy, margins membranous, in several unequal rows. Receptacles conical, usually lack chaffy bracts.

FRUITS AND SEEDS: Achenes elliptic to obovoid, 1–2 mm long, hairy or glandular. Pappus lacking or reduced to a minute crown of scales.

HABITAT: Generally found in valleys and slopes in cold desert and high desert regions. Basin big sagebrush is predominant on plains, valleys, and canyon bottoms below 2500 m on deep sandy loam, alluvial, and occasionally alkaline soils. Wyoming big sagebrush is found on plains, valleys, and slopes below 2700 m on shallow gravelly soils. Mountain big sagebrush occurs in valleys and slopes between 800 and 3200 m on well-drained soils in plant communities ranging from sagebrush-grass to aspen or spruce-dominated forests.

DISTRIBUTION: Great Basin, Mojave Desert, Transverse Ranges, South Coast, southern San Joaquin Valley, southwestern South Coast Ranges, and arid high elevation areas of the Cascade Range and Sierra Nevada, from 300 m to 3000 m or more. All states west of eastern Wyoming.

PROPAGATION/PHENOLOGY: *Reproduces by seed.* Seeds typically mature and

Biennial wormwood (*Artemisia biennis*) plant.
J. M. DiTomaso

Biennial wormwood (*Artemisia biennis*) flowering stem.
J. M. DiTomaso

Biennial wormwood (*Artemisia biennis*) immature plant.
J. M. DiTomaso

Biennial wormwood (*Artemisia biennis*) rosette.
J. M. DiTomaso

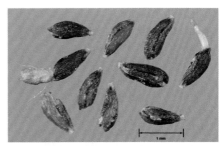

Biennial wormwood (*Artemisia biennis*) seeds.
J. K. CLARK

Annual wormwood (*Artemisia annua*) plant.
J. M. DITOMASO

Annual wormwood (*Artemisia annua*) seeds.
J. K. CLARK

disperse from late September through winter. Most seeds fall near the parent plant, can germinate immediately with favorable conditions, and typically survive for less than 1 year. Fire or disturbance appears to enhance seed germination.

MANAGEMENT FAVORING/DISCOURAGING SURVIVAL: Limiting grazing pressure in arid and semiarid rangeland can prevent an increase in big sagebrush density. Although most subspecies of big sagebrush cut below the crown or burned do not root-sprout, mountain big sagebrush [*A. tridentata* ssp. *vaseyana*] shrubs readily resprout after fire.

SIMILAR SPECIES: Several species of *Artemisia* can be found in California. Of these, only a few are occasionally considered weedy.

Annual wormwood [*Artemisia annua* L., synonyms: sweet wormwood; sweet annie][ARTAN] is an uncommon erect *annual* to 2 m tall, with *glabrous green foliage* that is *sweetly aromatic* when crushed, leaves that are *pinnate-dissected 2–3 times*, and *nodding flower heads*. Unlike California mugwort, annual wormwood has *pinnate-dissected leaves* and *lacks rhizomes*. It is cultivated for its essential oils and as a garden ornamental, but has escaped cultivation in some places. Annual wormwood inhabits disturbed sites in the Central Valley to 1000 m. Native to Europe.

Biennial wormwood [*Artemisia biennis* Willd.][ARTBI] is an erect *annual or biennial* to 2 m tall that resembles annual wormwood. In North America, biennial wormwood is typically annual. Unlike annual wormwood, biennial wormwood has *unscented or nearly unscented foliage* and *erect flower heads*. Leaves are pinnate-dissected 2 times. Biennial wormwood grows in open disturbed moist places in noncrop areas of the western North Coast Ranges, San Joaquin Valley,

Big sagebrush

California mugwort (*Artemisia douglasiana*) flowering stems. J. M. DiTomaso

California mugwort (*Artemisia douglasiana*) clump. J. M. DiTomaso

California mugwort (*Artemisia douglasiana*) foliage. J. M. DiTomaso

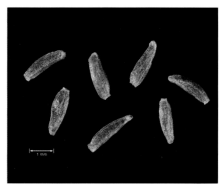

California mugwort (*Artemisia douglasiana*) seeds. J. O'Brien

Central-western region, South Coast, western Transverse Ranges, San Bernardino Mountains, and Great Basin, to 2200 m. It is rapidly expanding range in the north-central states where it is primarily a crop weed. Native to Europe.

California mugwort [*Artemisia douglasiana* Besser] [ARTDO] is a herbaceous native *perennial* to 2.5 m tall with *creeping rhizomes*. California mugwort stems are often simple, and leaves are elliptic to oblanceolate, mostly 7–15 cm long, 1–8 cm wide, with *margins entire or coarsely 3–5-lobed near the tip. Lower leaf surfaces are densely covered with white woolly hairs,* and *upper surfaces are gray-green and sparsely hairy.* Flower heads are bell-shaped, 2–4 mm in diameter, and usually nodding. It is not considered a weed in natural areas, but can occasionally be weedy in drainages, ditches, and low places in pastures and elsewhere. California mugwort often inhabits riparian areas throughout California, except deserts and Great Basin, to 2200 m.

Slender aster [*Aster subulatus* Michx. var. *ligulatus* Shinn.][ASTEX]

SYNONYMS: annual aster; slim aster; *Aster divaricatus* (Nutt.) Torrey & Gray; *Aster exilis* Ell.; *Aster neomexicanus* Woot. & Standl.; *Symphyotrichum divaricatus* (Nutt.) Nesom; possibly others

GENERAL INFORMATION: Erect *summer annual* with *small purple and yellow daisylike flower heads*. Most treatments report slender aster to grow to only 0.8 m tall, but plants can commonly reach 1–2 m tall, and on occasion 3 m tall. Slender aster is a ruderal native that is sometimes weedy in agricultural fields and associated irrigation ditches, orchards, and landscaped areas. In natural areas it is usually not considered a weed. The taxonomy of slender aster is unclear. Some ecologists believe there may be more than one variety, subspecies, or closely related species in California.

SEEDLINGS: Cotyledons oval.

MATURE PLANT: Foliage glabrous. Main stem openly branched in the upper portion. Leaves alternate, sessile, *linear to narrowly oblanceolate*, 3–10 cm long, mostly *0.2–1 cm wide, tips acute*, margins smooth or occasionally serrate.

ROOTS AND UNDERGROUND STRUCTURES: *Taproot* often shallow, with fibrous lateral roots.

FLOWERS: July–October. Heads mostly 4–6 mm long, consist of numer-

Slender aster (*Aster subulatus* var. *ligulatus*) plant.   J. M. DiTomaso

Slender aster (*Aster subulatus* var. *ligulatus*) flowering stem.   J. M. DiTomaso

ous pink to purple ray flowers to about 7 mm long and yellow disk flowers. Phyllaries in 2–6 overlapping rows, linear to lanceolate, tips acute to long-tapered. Receptacles flat, lack chaffy bracts.

FRUITS AND SEEDS: Achenes cylindrical-lanceolate, ± angled and ribbed, ± 2 mm long, sparsely short-hairy. Pappus bristles fine, soft, minutely barbed, mostly 3–4 mm long.

POSTSENESCENCE CHARACTERISTICS: Dead plants do not persist beyond winter.

HABITAT: Agronomic and vegetable crops, edges of rice fields, orchards, vineyards, landscaped areas, irrigation ditches, riparian areas, wetlands, alluvial flood plains, and occasionally turf. Usually grows in wet or seasonally wet places, often on saline or alkaline soil.

DISTRIBUTION: Central Valley and Central-western and Southwestern regions of California, to 200 m. Much of the central and southeastern United States.

PROPAGATION/PHENOLOGY: *Reproduces by seed.* Achenes disperse with wind, water, mud, soil movement, and human activities. Seeds germinate early spring through summer.

Slender aster (*Aster subulatus* var. *ligulatus*) flower head.   J. M. DiTomaso

Slender aster (*Aster subulatus* var. *ligulatus*) fruiting head. J. M. DiTomaso

MANAGEMENT FAVORING/DISCOURAGING SURVIVAL: Cultivation or manual removal before seeds develop can control slender aster.

SIMILAR SPECIES: Annual fleabane [*Erigeron annuus* (L.) Pers., synonyms: eastern daisy fleabane; *Aster annuus* L.][ERIAN] is an *annual* to 1.2 m tall and

Slender aster (*Aster subulatus* var. *ligulatus*) immature plant. J. M. DiTomaso

Slender aster (*Aster subulatus* var. *ligulatus*) seedling. J. M. DiTomaso

rough fleabane [*Erigeron strigosus* Muhl. ex Willd., synonyms: prairie fleabane; *Erigeron annuus* (L.) Pers. ssp. *strigosa* (Muhl. ex Willd.) Wagenitz][ERIST] is an annual or sometimes biennial to 0.8 m tall. Both species have *shallow fibrous roots* and white to pale purplish daisylike ray flowers and yellow disk flowers in the center. Unlike slender aster, annual fleabane and rough fleabane have *sparsely hairy foliage, ray flowers that lack pappus bristles*, and *phyllaries arranged in nearly equal rows*. Annual fleabane is distinguished from rough fleabane by having *stems with spreading hairs* and *coarsely toothed leaves*. Annual fleabane inhabits disturbed places in the Klamath Ranges and the north and central Sierra Nevada, to 2000 m. It also occurs in Idaho, Oregon, Utah, and Washington; central, southern, and eastern states; and nearly worldwide. Unlike annual fleabane, rough fleabane typically has *stems with appressed hairs* and *smooth-margined leaves*. Rough fleabane inhabits disturbed sites throughout much of California, except the Great Basin and desert areas, to 1100 m. Rough fleabane also occurs in most contiguous states, except Arizona, New Mexico, Nevada, and Utah. Some botanists recognize 3 varieties of rough fleabane, of which *Erigeron strigosus* Muhl. ex Willd. var. *strigosus* and *Erigeron strigosus* Muhl. ex Willd. var. *septentrionalis* (Fern. & Wieg.) Fern. occur in California. Annual fleabane and rough fleabane generally produce apomictic (asexual) seeds, but some populations appear to intergrade in certain areas. Both species are native to the eastern United States.

Annual fleabane (*Erigeron annuus*) flowering stem. J. M. DiTomaso

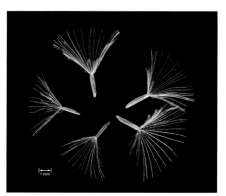

Slender aster (*Aster subulatus* var. *ligulatus*) seeds. J. O'Brien

English daisy [*Bellis perennis* L.][BELPE]

SYNONYMS: bone flower, European daisy, lawndaisy, March daisy, marguerite

GENERAL INFORMATION: Prostrate *perennial* to about 20 cm tall, with *basal rosettes of leaves* and *white daisylike flower heads on leafless stalks*. Under certain conditions, plants form clonal mats from short rhizomes. English daisy has escaped cultivation as an ornamental, especially in coastal regions. It is often a weed of turf and grassy areas. Native to Europe.

SEEDLINGS: Cotyledons oval to nearly round, about 2–3 mm long, glabrous, on stalks that elongate over time. Leaves alternate. First and subsequent few leaves spoon-shaped, about 4–15 mm long, glabrous, margins smooth to finely scalloped or serrate.

MATURE PLANT: Leaves alternate in *basal rosettes, spoon-shaped*, 2–10 cm long, 0.7–2 cm wide, tip rounded, base tapered to a winged stalk, surfaces sparse to moderately covered with soft hairs, especially stalks. Margins smooth to finely scalloped or serrate.

ROOTS AND UNDERGROUND STRUCTURES: Rhizomes short, with shallow fibrous roots.

FLOWERS: March–September. Flower heads mostly 2–3 cm wide, *solitary on leafless stalks* that are covered with stiff to soft hairs, consist of yellow disk flowers and about 30–80 white ray flowers. Phyllaries ovate, 3–6 mm long, in *1–2 equal rows*, hairy. *Receptacles lack chaffy bracts.*

English daisy *(Bellis perennis)* in turf.　　　　　　　　J. M. DiTomaso

English daisy

English daisy *(Bellis perennis)* flower heads.   J. M. DiTomaso

FRUITS AND SEEDS: Achenes lanceolate, 1–1.5 mm long, flattened, yellowish brown, glabrous, longitudinally lined, margins thickened. *Pappus lacking.*

POSTSENESCENCE CHARACTERISTICS: English daisy foliage sometimes dies in cold-winter regions, but regrows from the rhizomes in spring.

HABITAT: Primarily found on turf, but occasionally in moist grassy places,

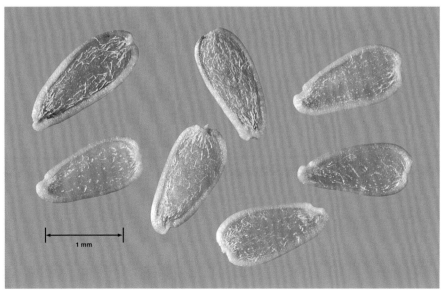

English daisy *(Bellis perennis)* seeds.   J. K. Clark

landscaped areas, gardens, and pastures. Often grows on moist heavy soil.

DISTRIBUTION: Most of California, particularly the Northwestern, Centralwestern, and Southwestern regions, to 200 m. Idaho, Montana, Oregon, Utah, Washington, northeastern states, and some north-central states.

PROPAGATION/PHENOLOGY: *Reproduces vegetatively from rhizomes and by seed.* Under favorable conditions, rhizome fragments can generate new plants. Achenes fall near the parent plant and disperse to greater distances with soil movement, mud, landscape maintenance, animal and human foot traffic, and possibly as a grass seed contaminant. Germination can occur from spring through fall whenever conditions are favorable. Seeds generally appear to be short-lived under field conditions, but have been reported to be long-lived in at least one pasture situation.

MANAGEMENT FAVORING/DISCOURAGING SURVIVAL: Disturbed or trampled moist, grassy areas with bare patches favors English daisy establishment and spread. Closely mowed turf also encourages encroachment. Manually removing plants can help to control English daisy in turf.

SIMILAR SPECIES: English daisy is unlikely to be confused with other species that have white daisylike flower heads.

Hairy beggarticks [*Bidens pilosa* L.][BIDPI]

SYNONYMS: blackjack; cobbler's pegs; common beggarticks; farmer's friend; hairy bur marigold; pitchforks; Spanish needles; *Bidens alausensis* Kunth; *Bidens californica* DC.; *Bidens chilensis* DC.; *Bidens leucantha* (L.) Willd.; *Bidens odorata* Cav.; *Coreopsis leucantha* L.; *Kerneria pilosa* Lowe; others

GENERAL INFORMATION: Erect summer *annual* to 1.8 m high, with *opposite pinnate-compound leaves* and *yellow or yellow and white flower heads*. Hairy beggarticks is a weed of crops, roadsides, and natural areas in tropical and subtropical regions nearly worldwide. It is used medicinally for various ailments in many places and is sometimes consumed as a vegetable. Native to tropical and subtropical America.

Hairy beggarticks (*Bidens pilosa*) plant.   J. M. DiTomaso

Hairy beggarticks

SEEDLINGS: Cotyledons linear-oblong, mostly 2–3 cm long, ± 0.5 cm wide. Leaves opposite. First leaves 1.5–2.5 cm long, ± 1.5 cm wide, pinnate-divided into 3 main lobes. Main lobes coarsely pinnate-lobed. Margins, veins, and leaf stalks short-hairy.

MATURE PLANT: Stems *square* in cross-section. Foliage sparse to moderately hairy. Leaves *opposite, stalked, pinnate-compound*. *Leaflets 3–5*, lanceolate to ovate or ± diamond-shaped, 2–6 cm long, tip acute to long-tapered and slightly pinched-in (acuminate), margins serrate.

ROOTS AND UNDERGROUND STRUCTURES: Taproots shallow, branched, with fibrous lateral roots.

FLOWERS: May–November. *Heads ± 1 cm in diameter*, hemispherical, *consist only of yellow disk flowers or of yellow disk flowers and reduced white ray flowers*. Outer phyllaries 7–9, linear, 4–5 mm long. Inner phyllaries lanceolate, 4–7 mm long. Receptacle chaffy bracts linear, tips pinched in. Insect-pollinated.

Hairy beggarticks (*Bidens pilosa*) flowering stem, head lacking ray flowers.   J. M. DiTomaso

Hairy beggarticks (*Bidens pilosa*) flowering stem, head with white ray flowers.   J. M. DiTomaso

Hairy beggarticks (*Bidens pilosa*) fruit head.   J. M. DiTomaso

Hairy beggarticks

Hairy beggarticks (*Bidens pilosa*) seedlings.
J. M. DiTomaso

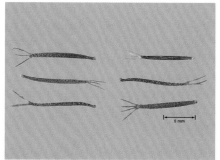

Hairy beggarticks (*Bidens pilosa*) seeds.
J. O'Brien

Devils beggarticks (*Bidens frondosa*) flowering stem.
J. M. DiTomaso

Devils beggarticks
(*Bidens frondosa*) leaf.
J. M. DiTomaso

Hairy beggarticks

Devils beggarticks (*Bidens frondosa*) seedling.
J. M. DiTomaso

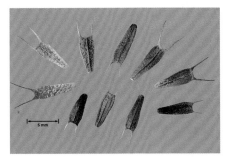

Devils beggarticks (*Bidens frondosa*) fruiting stem.　　J. M. DiTomaso

Devils beggarticks (*Bidens frondosa*) seeds.
J. O'Brien

FRUITS AND SEEDS: Achenes cylindrical, ± 4-sided, mostly 4–16 mm long, 1 mm wide or less, black, ± minutely hairy, with 2–4 barbed pappus awns 2–4 mm long at the apex.

POSTSENESCENCE CHARACTERISTICS: Stems with old heads and a few achenes can persist into winter.

HABITAT: Disturbed open places, waste places, roadsides, crop fields, orchards, and pastures. Designated a facultative wetland indicator species in California.

DISTRIBUTION: South Coast Ranges and southwestern California, to 750 m. Arizona, New Mexico, southern United States, and a few northeastern states. Tropical and subtropical regions nearly worldwide.

PROPAGATION/PHENOLOGY: *Reproduces by seed*. Achenes disperse primarily by clinging to animals, clothing and shoes, and vehicle tires, and with the movement of soil, water, and mud. Seeds germinate in darkness or light. Most seedlings emerge from soil depths to 4 cm.

MANAGEMENT FAVORING/DISCOURAGING SURVIVAL: Manual removal or cultivation before seeds develop can control hairy beggarticks. Some populations in other countries tolerate certain herbicides.

SIMILAR SPECIES: Refer to table 16 (p. 216) for a comparison of beggartick distinguishing features. **Devils beggarticks** [*Bidens frondosa* L., synonyms: sticktight; tickseed sunflower][BIDFR] is a widespread ruderal native *annual* to 1.2 m tall, with compound leaves. It is generally not considered a weed in natural areas but can occasionally be weedy in crop fields and pastures. Under certain conditions, devils beggarticks can accumulate nitrates to levels that are toxic to livestock. Devils beggarticks inhabits disturbed, often moist places in the South Coast Ranges and Southwestern region, to 750 m. It also occurs in Arizona, New Mexico, the southern United States, a few northeastern states, and tropical and subtropical regions nearly worldwide.

Bur beggarticks [*Bidens tripartita* L., synonyms: threelobe beggarticks; *Bidens acuta* (Wieg.) Britt.; *Bidens comosa* (Gray) Wieg.] [BIDTR] is an *annual* to 1.5 m tall, with simple lanceolate leaves to 20 cm long. Bur beggarticks frequently inhabits freshwater wetlands in the central Sierra Nevada and San Joaquin Valley, to 1600 m. It occurs throughout much of the United States. Native to eastern North America.

Tall beggarticks [*Bidens vulgata* E. Greene, synonyms: big devils beggarticks; western sticktight; *Bidens frondosa* L. var. *puberula* Wieg.; *Bidens puberula* Wieg.] [BIDVU] is an uncommon *annual* to 1.5 m tall, with compound leaves. Tall beggarticks typically grows in freshwater wetlands in the North Coast ranges and Sacramento Valley, to 300 m. It is also found throughout much of the United States except some southern states. Native to eastern North America.

Bur beggarticks (*Bidens tripartita*) flowering stem. J. M. DiTomaso

Bur beggarticks (*Bidens tripartita*) flower head. J. M. DiTomaso

Bur beggarticks (*Bidens tripartita*) seedling. J. M. DiTomaso

Table 16. Beggarticks (*Bidens* spp.)

Bidens spp.	Leaves	Achenes	Linear outer phyllaries	Flower head diameter (cm)
B. frondosa devils beggarticks	compound; leaflets 3–5	wedge-shaped, flattened	5–8	± 1
B. pilosa hairy beggarticks	compound; leaflets 3–5	cylindrical, ~ 1 mm wide, ± 4-sided	7–9	± 1
B. tripartita bur beggarticks	simple; stalked	wedge-shaped, flattened	4–5	1–2
B. vulgata tall beggarticks	compound; leaflets 3–5	wedge-shaped, flattened	10–16	1.5–2.5

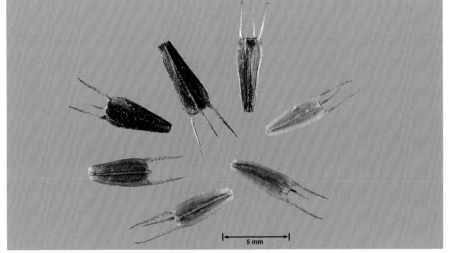

Bur beggarticks (*Bidens tripartita*) seeds. J. O'Brien

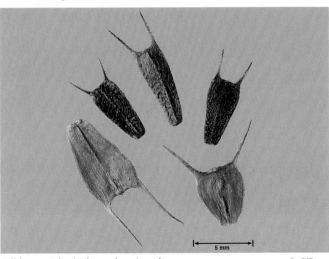

Tall beggarticks (*Bidens vulgata*) seeds. J. O'Brien

Plumeless thistle [*Carduus acanthoides* L.] [CRUAC][Cal-IPC: Limited][CDFA: A]

Musk thistle [*Carduus nutans* L.][CRUNU] [Cal-IPC: Moderate][CDFA: A]

Italian thistle [*Carduus pycnocephalus* L.][CRUPY] [Cal-IPC: Moderate] [CDFA: C]

Slenderflower thistle [*Carduus tenuiflorus* Curtis] [CRUTE] [Cal-IPC: Limited] [CDFA: C]

SYNONYMS: Plumeless thistle: bristly thistle; giant plumeless thistle; spiny thistle; *Carduus fortior* Klok.

Musk thistle: giant plumeless thistle; nodding plumeless thistle; nodding thistle; plumeless thistle. Musk thistle is comprised of a complex of closely related varieties, subspecies, and species, leading to much taxonomic confusion. Four subspecies are currently recognized in North America; of which *Carduus nutans* L. ssp. *nutans* and ssp. *leiophyllus* (Petrovic) Stojanov & Stef. are reported to occur in California. Synonyms for ssp. *leiophyllus* include *Carduus leiophyllus* Petrovic; *Carduus nutans* L. var. *leiophyllus* (Petrovic) Arènes; *Carduus nutans* L. var. *vestitus* (Hallier) Boivin; and *Carduus thoermeri* Weinm.

Plumeless thistle *(Carduus acanthoides)* flowering stem.   J. M. DiTomaso

Plumeless thistle *(Carduus acanthoides)* flower head.   J. M. DiTomaso

Italian thistle: compact-headed thistle; Italian plumeless thistle; Plymouth thistle; shore thistle; slender thistle

Slenderflower thistle: Italian thistle; multiheaded thistle; seaside thistle; shore thistle; winged plumeless thistle

GENERAL INFORMATION: These *Carduus* species have *prickly leaves* and stems with *prickly wings*. Plants exist as basal rosettes until flowering shoots develop at maturity. Refer to table 17 (p. 220) and table 18 (p. 227) for a quick comparison of distinguishing characteristics among *Carduus* species and other thistles. The thistle head weevil *(Rhinocyllus conicus)* is an introduced biocontrol agent that attacks *Carduus* species and several other thistles, including some native thistles *(Cirsium* spp.). The weevil was accidentally released in California. It is established in California and much of the northwestern and north-central United States. Control of *Carduus* thistle infestations by the weevil varies by species and regionally from excellent to poor. The fungus musk thistle rust *(Puccinia carduorum)* has recently been found in California and may soon be state-approved as a biocontrol agent to help control musk thistle.

Plumeless thistle, musk thistle: *Biennials* (or winter annuals) to 1.5 m tall. These species readily hybridize with one another, and plants with intermediate characteristics may be found where their ranges overlap. The hybrid is sometimes referred to as *Carduus* × *orthocephalus* Wallr. Musk thistle is a state-listed noxious weed in 23 states besides California, including Colorado, Idaho, Nevada, New Mexico, Oregon, Utah, and Washington. Plumeless thistle is a state-listed noxious weed in Arizona, California, Colorado, Washington, Wyoming, and a few central and eastern states. Both are native to Europe.

Italian thistle, slenderflower thistle: Winter *annuals*, sometimes biennials, to 2 m tall. Italian thistle and slenderflower thistle infrequently hybridize, since slenderflower thistle is typically self-pollinating and chromosome numbers are variable (Italian thistle: 2n = 62–64, slenderflower thistle: 2n = ± 54, 62–64). However, some botanists believe that Italian thistle and slenderflower thistle may not be distinct species. Besides California, both species are state-listed

Plumeless thistle *(Carduus acanthoides)* seedling.   J. M. DiTomaso

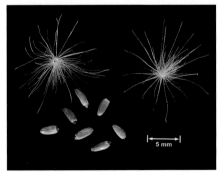

Plumeless thistle *(Carduus acanthoides)* seeds.   J. O'Brien

Plumeless thistle • Musk thistle • Italian thistle • Slenderflower thistle

Musk thistle (*Carduus nutans*) infestation.   J. M.

Musk thistle (*Carduus nutans*) rosette.
J. M. DiTomaso

Musk thistle (*Carduus nutans*) flower head.
J. M. DiTomaso

weeds in Oregon and Washington. Italian thistle and slenderflower thistle are native to the Mediterranean region and central Europe, respectively.

SEEDLINGS: Cotyledons glabrous, dull, lower surface paler than upper surface, base tapered to a short stalk. Leaves alternate, form a basal rosette. First few leaves oblong to elliptic, *margins irregularly prickly-toothed*, base short- to long-tapered, tip broadly acute to rounded, upper surface and lower veins often sparsely hairy. Hairs long, firm, unicellular and multicellular. Later leaves elliptic to (ob)lanceolate, bases long-tapered, margins *irregularly prickly-toothed* and *variably lobed*.

Musk thistle, plumeless thistle: Cotyledons oblong, 8–20 mm long, 3–9 mm wide, tip usually slightly truncate. First 2 leaves often appear opposite.

Italian thistle, slenderflower thistle: Cotyledons oval-oblong, mostly 10–20 mm long, 6–12 mm wide, tip slightly truncate to round. First 2 leaves alternate. Later leaves often oblanceolate, veins often pale.

Plumeless thistle • Musk thistle • Italian thistle • Slenderflower thistle

MATURE PLANT: Stems branched near the top. Basal leaves elliptic to lanceolate, *pinnate lobed*, with *prickly-toothed margins*. Stem leaves alternate, reduced, with *bases that extend down the stem as spiny wings* (decurrent).

Plumeless thistle: Stems strongly winged, glabrous to lightly woolly. Leaves

Table 17. *Carduus* species

Carduus spp.	Flower head shape and arrangement	Flower head width (cm)	Flower stalk length (cm)	Phyllary width and hairs	Other
C. acanthoides plumeless thistle	(hemi-) spherical; solitary or clustered	1–3	< 2	< 2 mm; loosely woolly to glabrous	—
C. nutans musk thistle	(hemi-) spherical; solitary, often nodding	2–7	> 2	> 2 mm; short-hairy to glabrous	—
C. pycnocephalus Italian thistle	cylindrical or elliptical; mostly 2–5-clustered	~ 1–2	< 2	< 2 mm; woolly at bases	lower leaves mostly 4–10-lobed; phyllary tips sandpapery to touch (scabrous)
C. tenuiflorus slenderflower thistle	cylindrical or elliptical; mostly 5–20-clustered	~ 1–2	< 2	< 2 mm; glabrous to sparsely woolly	lower leaves mostly 12–20-lobed; phyllary margins narrowly membranous

Musk thistle (*Carduus nutans*) seedling.
J. M. DiTomaso

Musk thistle (*Carduus nutans*) plant.
J. M. DiTomaso

Musk thistle (*Carduus nutans*) seeds. J. O'Brien

Plumeless thistle • Musk thistle • Italian thistle • Slenderflower thistle

Italian thistle (*Carduus pycnocephalus*) plant.
J. M. DiTomaso

Italian thistle (*Carduus pycnocephalus*) flowering stem.
J. M. DiTomaso

Italian thistle (*Carduus pycnocephalus*) rosette.   J. M. DiTomaso

1-pinnate lobed, typically sparsely hairy. Lower leaves mostly 10–20 cm long.

Musk thistle: Stems narrow winged, glabrous to woolly. Leaves 1–2-pinnate lobed, glabrous to sparsely hairy. Lower leaves mostly 10–40 cm long.

Italian thistle: Stem wings to 5 *mm wide, often interrupted near flower heads*, glabrous to slightly woolly. Leaves mostly covered with woolly hairs. Lower leaves 10–15 cm long, mostly *4–10-lobed*.

Italian thistle (*Carduus pycnocephalus*) seedlings.
J. M. DiTomaso

Italian thistle (*Carduus pycnocephalus*) seeds.
J. O'Brien

Slenderflower thistle: Foliage resembles that of Italian thistle, except stem wings may be up to *10 mm wide*, wings usually *extend continuously to flower heads*, and basal leaves are usually *12–20-lobed*.

ROOTS AND UNDERGROUND STRUCTURES: All have taproots that are generally long, thick, fleshy, occasionally branched, and capable of penetrating the soil to depths of 40 cm or more.

FLOWERS: Heads consist of deeply lobed, purple to pink (rarely white) *disk flowers*. Phyllaries spine-tipped, overlapping in several rows. Receptacles flat, densely covered with cream-colored bristles interspersed among the disk flowers. Insect-pollinated.

Plumeless thistle: May–August. Heads (hemi-)*spherical, 1–3 cm in diameter, solitary or clustered, on stalks less than 2 cm long*. Disk flowers purple, 14–20 mm long, with tubes 7–10 mm long. Phyllaries narrowly lanceolate, *2 mm wide or less*, glabrous to pubescent, tips erect to spreading. Primarily out-crossing, but self-compatible.

Musk thistle: June–September. Heads (hemi-)*spherical, 2–7 cm in diameter, solitary*, often nodding on *stalks usually more than 2 cm long*. Disk flowers purple, 20–25 mm long, with tubes 12–14 mm long. Phyllaries lanceolate to ovate, mostly *3–8 mm wide*, spreading to reflexed at the middle, glabrous to sparsely woolly. Primarily out-crossing, but self-compatible.

Italian thistle: May–July. Heads ± *cylindrical, 1–2 cm in diameter, sessile or nearly sessile, usually 2–5 per cluster*. Disk flowers pink to purple, 10–14 mm long, with tubes 5–8 mm long. Phyllaries linear-lanceolate with erect tips, *2 mm wide or less*, with *persistent patches of woolly hairs at the bases*.

Slenderflower thistle: May–July. Heads, flowers, and phyllaries resemble those of Italian thistle, except that heads usually *5–20 per cluster*, flower tubes mostly *4–6 mm long*, and phyllaries *nearly glabrous to sparsely woolly at the base*, with narrow *membranous margins*.

FRUITS AND SEEDS: Achenes elliptic, curved, slightly compressed, sometimes slightly 4- to 5-sided in cross-section, smooth, glossy, golden to brown. Pappus bristles *numerous*, cream-colored, *fine, minutely barbed* (with magnification), united at the base to form a ring and deciduous as a unit.

Plumeless thistle: Achenes 2–3 mm long, with faint longitudinal stripes. Pappus bristles 11–13 mm long.

Slenderflower thistle (*Carduus tenuiflorus*) plant.   E. Healy

Musk thistle: Achenes 4–5 mm long, with longitudinal dotted stripes. Pappus bristles 13–25 mm long.

Italian thistle: Produces 2 types of achenes 4–6 mm long. Central achenes cream-colored, initially sticky, with about 20 longitudinal stripes and pappus bristles 10–15 mm long. Outer ring of achenes golden to brown, smooth, lack pappus bristles.

Slenderflower thistle: Achenes and pappus bristles resemble those of Italian thistle, except central achenes mostly have 10–13 longitudinal stripes.

POSTSENESCENCE CHARACTERISTICS: Foliage is killed by hard frost, but plants remain intact for an extended period after death. The persistent spiny character of the foliage helps to distinguish plants.

HABITAT: Thistles typically colonize disturbed open sites, roadsides, pastures, annual grasslands, and waste areas.

Plumeless thistle: Often occupies similar but drier, better-drained sites than those inhabited by musk thistle.

Musk thistle: Often associated with sandy fertile soils or soils that are high in calcium. Tolerates a wide range of conditions, from acidic to saline soils and fertile to poor soils. Generally establishes poorly on highly acidic or nutrient deficient soils, or soils with extremes in moisture content.

Italian thistle, slenderflower thistle: Inhabit sandy to clay soils.

DISTRIBUTION: Populations of musk thistle and giant plumeless thistle are currently limited to specific regions; Italian thistle and slenderflower thistle are widely distributed.

Plumeless thistle: Eastern North Coast Ranges (se Humboldt, cw Trinity, w Glenn, ne Lake, and e Colusa Cos.), northern Sierra Nevada (w Nevada Co.), Modoc Plateau (ce Modoc Co.), and San Francisco Bay region (nw Marin Co.), to 1300 m. A previous infestation now eradicated occurred on the Central Coast (cw Monterey Co.). Colorado, Idaho, Montana, Washington, Wyoming, and most central and northeastern states.

Musk thistle: Klamath Ranges, Cascade Range (c and e Siskiyou, n Shasta Cos.) northern Sierra Nevada (s and e Plumas, e Sierra, and c and e Nevada Cos.), and Modoc Plateau (Modoc, n and s Lassen Cos.), to 1200 m. Previous infestations now eradicated occurred in the South Coast (se Los Angeles, ne Riverside Cos.) and Mojave Desert (se San Bernardino Cos.). All contiguous states, except possibly Florida, Maine, and Vermont.

Italian thistle: Southern North Coast, southern North Coast Ranges, Sierra Nevada foothills, Central Coast, San Francisco Bay region, and South Coast Ranges, to 1000 m. Idaho, Oregon, Texas, Alabama, New York, and South Carolina.

Slenderflower thistle: North Coast, North Coast Ranges, Sierra Nevada foothills, Central Coast, San Francisco Bay region, South Coast Ranges, and Southwestern region, to 1000 m. Oregon, Washington, Texas, Pennsylvania, and New Jersey.

Plumeless thistle • Musk thistle • Italian thistle • Slenderflower thistle

Slenderflower thistle *(Carduus tenuiflorus)* seeds.
J. O'Brien

Slenderflower thistle *(Carduus tenuiflorus)* flowering stems.   J. M. DiTomaso

PROPAGATION/PHENOLOGY: *All reproduce by seed.* Seeds fall near the parent plant and disperse to greater distances with wind, water, birds, small mammals, and human activities. Most seeds germinate in fall and spring.

Plumeless thistle, musk thistle: Plants appear to require a period of chilling to induce flowering. First flower heads can produce large numbers of seeds, sometimes 1500 or more seeds per head. Late flower heads produce fewer seeds, to less than 25 seeds per head. Most seeds (about 99%) fall within 50 m of the parent plant. Very few seeds disperse further than 100 m from the parent plant. Seeds generally disperse 1–3 weeks after flowering. Musk thistle seeds appear to have a variable after-ripening period requirement. However, in many areas, most musk thistle seeds generally germinate about 2–4 weeks after dispersal. Plumeless thistle seeds appear to lack an after-ripening period. The optimal temperature for germination of musk thistle seeds is near 20°C, and light is not required. Most plants are biennial, germinating in the winter or early-spring months and existing as a rosette until flower stems develop in the spring or summer of the following year. Seeds that germinate in late summer or fall often behave as winter annuals and flower in the following summer months. Seeds can germinate in constant heavy shade, but seedlings seldom reach maturity under this condition. In one study, buried musk thistle seeds remained viable for 10 years. However, seeds of both species rarely persist in the soil seedbank for more than a few years. A seedbank study of soil in a permanent pasture indicated that seeds on the soil surface and in the top 2 cm of soil did not persist beyond 3 years because of germination. Similarly, a field seedbank study of plumeless thistle suggested that viable seeds of this species rarely persist in the soil seedbank due to decomposition and seed predation by insects, mammals, and birds.

Italian thistle, slenderflower thistle: Plants do not require a period of chilling to induce flowering. Flowering is continuous until soil moisture is depleted. Seeds germinate under a wide range of temperature regimes. High temperatures such as those that often occur in late summer inhibit germination. Seedlings can emerge from soil depths to 8 cm, but depths of 0.5–2 cm appear optimal. Seeds beneath about 1 cm of litter or soil germinate better than those on top of litter or soil. Seed germination on the surface of clay soils is greater than on other soil or litter surfaces. Central seeds are mostly wind- or gravity-dispersed and appear to lack an after-ripening period. Seed coats contain germination inhibitors that can leach out within 2 days in the presence of adequate water. Evidence suggests that the mucilaginous coating can act as an adhesive for seed transport and that it increases water retention of seeds, which improves germination on the surface of clay soils. On average, flower heads produce about 11 central seeds and 2–3 outer seeds. Outer seeds remain attached to the flower head and typically germinate in the head after it falls to the ground. Outer seeds may remain dormant and can persist in the soil seedbank for up to about 7 years.

ADDITIONAL ECOLOGICAL ASPECTS: Musk thistle: A few studies suggest that musk thistle has allelopathic properties. In New Zealand, musk thistle seeds appear to inhibit the germination and radicle growth of other pasture species, but stimulate or have no affect on the germination of other musk thistle seeds. Musk thistle plants also appear to weaken other pasture species by an allelopathic interaction at the early bolting stage, when the larger rosette leaves are decomposing, and at the stage when bolting plants are senescing. However, no specific chemicals have been identified. In addition, the growth of musk thistle seedlings appears to be stimulated when musk thistle tissues are incorporated into the soil. In certain situations, musk thistle may contribute to the decline of soil fertility by inhibiting the survival of nitrogen-fixing species.

MANAGEMENT FAVORING/DISCOURAGING SURVIVAL: Cultivation, manual removal of small plants, or cutting or mowing mature plants with flower stems before seeds mature can help control these species. In general, thistles compete poorly with healthy, established grasses and other vegetation. Disturbances such as fire, overgrazing, or trampling can create prime sites for thistle colonization. Fire can enhance or reduce musk thistle and probably other thistle populations, depending on a variety of factors, such as native plant community type, fire timing and intensity, and seedbank composition.

SIMILAR SPECIES: Canada thistle [*Cirsium arvense* (L.) Scop.][CIRAR], **bull thistle** [*Cirsium vulgare* (Savi) Ten][CIRVU], and **Scotch thistle** [*Onopordum acanthium* L. ssp. *acanthium*][ONRAC] have lobed leaves with prickly margins and may be confused with *Carduus* thistles. Unlike *Carduus* thistles, Canada thistle is a *perennial* with *creeping roots* and *small unisexual flower heads*. Plants are either male or female (dioecious). In addition, Canada thistle has *smooth stems* and *plumose pappus bristles*. Bull thistle is a *coarse biennial* with *plumose pappus bristles* and *upper leaf surfaces covered with stiff bristly hairs* that are rough to touch. Scotch thistle and related *Onopordum* species are distinguished by having *receptacles that are deeply pitted, honeycomb-like*, with *membranous extensions around pits*, and *not densely covered with bristles*. Refer to the entries for **Canada thistle and Bullthistle** (p. 272) and **Scotch thistle** (p. 372), and Table 18 (p. 227), for further information.

Table 18. Spiny-leaved thistles: Flowers pink, purple, or white, except where noted

Plant	Stem wings	Phyllary margins	Receptacle chaff	Pappus	Fruit attachment	Other
Carduus spp. musk, giant plumeless, Italian, and slenderflower thistles	conspicuous (to ~ 1 cm wide)	smooth	dense bristles	soft minutely barbed bristles, fused at base; deciduous	basal, slightly angled	Italian thistle leaves sometimes faintly mottled white
Carthamus spp. whitestem, smooth, and woolly distaff thistles	none	outer leaf-like, spiny pinnate-lobed	narrow scales	0 or many unequal scales in several series; persistent	lateral notch near base	smooth and woolly distaff thistles flowers **yellow**; fruits ± 4-angled
Cnicus benedictus blessed thistle	none or inconspicuous	outer smooth; inner with long, pinnate-branched spines at tips	dense bristles	~ 20 stiff bristles in 2 series, outer long, inner short; persistent	lateral notch near base	flowers **yellow**; heads leafy; fruits cylindrical, with ~ 20 prominent ribs and pale crownlike teeth on apical rim
Cirsium spp. bull, Canada, yellowspine, and wavyleaf thistles	conspicuous (to 1 cm wide; bull thistle), none (Canada thistle), or inconspicuous	smooth	dense bristles	soft feathery bristles, fused at base; deciduous	basal, slightly angled	—
Cynara cardunculus artichoke thistle	none	smooth	dense bristles	long, stiff, feathery bristles, fused at base; deciduous	basal	basal leaves often compound and very large; receptacle fleshy
Onopordum spp. Scotch, Illyrian, and Taurian thistles	conspicuous (mostly 1–4 cm wide)	smooth	membranous extensions around pits	± firm, minutely barbed bristles, fused at base; deciduous	basal	receptacles deeply pitted and honeycomb-like; fruits transversely wrinkled
Scolymus hispanicus golden thistle	conspicuous (mostly to ~ 1 cm wide)	smooth	long scales enclose fruits	2–4 stiff, minutely barbed bristles; ± deciduous	basal	flowers yellow, ligulate; sap milky
Silybum marianum blessed milkthistle	none	spiny-toothed	dense bristles	soft, minutely barbed bristles, fused at base; deciduous	basal, slightly angled	leaves bright green & mottled white

Asteraceae

Smooth distaff thistle [*Carthamus baeticus* (Boiss. & Reuter) Nyman][CDFA list: B]

Woolly distaff thistle [*Carthamus lanatus* L.] [Cal-IPC: Moderate Alert][CDFA list: B]

SYNONYMS: Smooth distaff thistle: *Carthamus lanatus* L. ssp. *baeticus* (Boiss. & Reuter) Nyman; *Carthamus lanatus* L. ssp. *creticus*; *Carthamus nitidus*; California references, not Boiss.; *Kentrophyllum baeticus* Boiss. & Reuter. Some taxonomists consider the correct name to be *Carthamus creticus* L.

Woolly distaff thistle: false starthistle; saffron thistle; woolly safflower; woolly starthistle; *Carthamus lanatus* L. ssp. *lanatus*.

GENERAL INFORMATION: Erect winter *annuals*, with *rigid stems* to 1 m tall and *spiny leaves*. Plants exist as rosettes until flower stems develop in spring or summer. Distaff thistles are highly competitive with cereal crops and desirable rangeland species, and dense populations can develop. In addition, the spiny foliage and flower heads can injure the eyes and mouths of livestock grazing in infested areas. Distaff thistles are closely related to commercial safflower [*Carthamus tinctorius* L.], and these species may hybridize. This precludes the development and release of biocontrol agents in California.

Smooth distaff thistle, woolly distaff thistle: These species are difficult to distinguish from one another, and some botanists classify them as subspecies of *Carthamus lanatus*. However, smooth and woolly distaff thistles have different chromosome numbers (smooth distaff thistle: 2n = 64, woolly distaff thistle: 2n = 44) and are not known to hybridize. Besides California, smooth distaff thistle is a state-listed noxious weed in Oregon (class A). Woolly distaff thistle is a government-listed noxious weed in some southern regions of Australia. Native to the Mediterranean region.

SEEDLINGS: Smooth distaff thistle, woolly distaff thistle: Cotyledons obovate, base tapered. Rosette leaves coarse, 1-pinnate-lobed, lobes spine-tipped.

Smooth distaff thistle (*Carthamus baeticus*) flower head. W. J. FERLATTE

Smooth distaff thistle (*Carthamus baeticus*) seeds. J. O'BRIEN

Woolly distaff thistle (*Carthamus lanatus*) plant.
J. M. DiTomaso

Woolly distaff thistle (*Carthamus lanatus*) immature plant.
J. M. DiTomaso

Woolly distaff thistle (*Carthamus lanatus*) flower head.
J. M. DiTomaso

Woolly distaff thistle (*Carthamus lanatus*) seeds.
J. K. Clark

Woolly distaff thistle (*Carthamus lanatus*) seedling.
J. M. DiTomaso

Cotyledon and first rosette leaf size can vary greatly among populations. Woolly distaff thistle leaves are often less deeply lobed near the tips than those of smooth distaff thistle.

MATURE PLANT: Stems rigid, mostly branched from upper two-thirds of plant. Stem leaves alternate, sessile, *coarse, stiff, spreading or slightly curved downward, 1-pinnate-lobed,* base weakly clasps the stem, conspicuously veined, evenly covered with minute glandular hairs. *Lobes few, narrow, prominently spine-tipped,* mostly opposite. Rosette leaves oblanceolate, mostly pinnate-lobed 1–2 times, lobes spine-tipped, mostly to 20 cm long, often withered at flowering.

Smooth distaff thistle: Stems white to straw-colored, usually sparsely covered with woolly and minute glandular hairs, especially at the bases of flower heads.

Woolly distaff thistle: Stems straw-colored, usually covered with loose *woolly, cobwebby, and glandular hairs,* especially in leaf axils and at bases of flower heads.

ROOTS AND UNDERGROUND STRUCTURES: Taproots slender, long, mostly unbranched, and with numerous fibrous roots.

FLOWERS: Flower heads solitary at stem tips, consist of *numerous disk flowers* and phyllaries in several overlapping rows. *Outer phyllaries spreading, slightly curved, spiny, rigid, leaflike, pinnate-lobed.* Inner phyllaries tipped with spiny appendages. Receptacle dome- to cone-shaped, with *papery, scalelike chaffy bracts.*

Smooth distaff thistle: July–August. Heads mostly 20–25 mm long. *Corollas yellow, 25–35 mm long. Outer phyllaries about twice the length of the inner phyllaries.*

Woolly distaff thistle: July–August. Heads mostly 25–35 mm long. *Corollas yellow, 25–35 mm long. Outer phyllaries mostly less than 1.5 times the length of the inner phyllaries.*

FRUITS AND SEEDS: Achenes buff to brown, sometimes mottled with dark brown, oblong or pyramidal, often weakly 4-sided, with a *lateral notch near the base.* Apex wavy-margined, broader than the base. Outer achenes rough, often

Whitestem distaff thistle (*Carthamus leucocaulos*) flower head.   W. J. Ferlatte

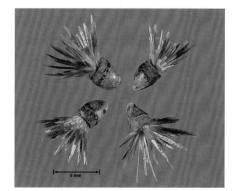

Whitestem distaff thistle (*Carthamus leucocaulos*) seeds.   J. K. Clark

Blessed thistle (*Cnicus benedictus*) flower head. J. M. DiTomaso

Blessed thistle (*Cnicus benedictus*) seedling.
J. M. DiTomaso

Blessed thistle (*Cnicus benedictus*) seeds.
J. K. Clark

dark, *lack a pappus*. Inner achenes mostly smooth, with a *persistent pappus of numerous narrow, unequal, brownish scales*.

Smooth distaff thistle: Achenes 4–6 mm long. *Pappus mostly 8–10 mm long*.

Woolly distaff thistle: Achenes 4–6 mm long. *Pappus mostly 10–13 mm long*.

POSTSENESCENCE CHARACTERISTICS: The rigid stems with old flower heads and leaves often persist through winter and often for a longer period than most other thistles. Phyllaries and seeds remaining in old flower heads can aid species identification.

HABITAT: Disturbed open sites, roadsides, fields, grassland, rangeland, pastures, and agricultural land, especially grain fields. Grows on many soil types. Prolific in areas that receive 40–60 cm annual rainfall.

DISTRIBUTION: Smooth distaff thistle: North and central Sierra Nevada foothills, Central Valley, San Francisco Bay region, southern South Coast Ranges, Southwestern region, and western Mojave Desert, to 500 m. Nevada, Oregon, and South Carolina.

Golden thistle *(Scolymus hispanicus)* plant.
W. J. FERLATTE

Golden thistle *(Scolymus hispanicus)* flower heads.
COURTESY OF ROSS O'CONNELL

Woolly distaff thistle: Northwestern region, Central-western region, central Sierra Nevada, and southern North Coast, to 1100 m. Woolly distaff thistle appears to be rapidly expanding range in the central North Coast area. Texas and Oklahoma.

PROPAGATION/PHENOLOGY: *Reproduce by seed.*

Smooth distaff thistle, woolly distaff thistle: Most achenes fall near the parent plant, but some remain in the persistent seed heads. Achenes and sometimes entire seed heads can disperse to greater distances with animals, human activities and machinery such as tractors and agricultural implements, mud, and water. Newly matured achenes often exhibit a high rate of dormancy, and the proportion of dormant seeds is usually greatest for those retained in the seed heads. Dormant seeds appear to require leaching to remove germination inhibitors and light to break dormancy. Rates of dormancy can vary among populations at different locations. Most seeds germinate after the first significant rain of the cool season, generally 1–3 years following maturation, but some seeds can remain dormant and viable for up to 8 years under field conditions. In most years, few seeds germinate after mid-winter. Germination at locations with low rainfall and temperatures can be slow and drawn-out. Seedlings rarely emerge from soil depths greater than 5 cm. Optimal emergence occurs at or just below the soil surface. Seeds are susceptible to predation by termites.

MANAGEMENT FAVORING/DISCOURAGING SURVIVAL: Mowing just before flower buds develop can prevent most seed production. Mowing earlier can encourage the regrowth of flowering stems. In plants mowed after flower heads have developed, seed can mature in cut flower heads left on the ground. Heavy grazing can aid distaff thistle survival because livestock selectively graze more palatable species, reducing competition with other plants for light and nutrients. Distaff thistles are unlikely to establish in well-managed perennial pastures.

SIMILAR SPECIES: Whitestem distaff thistle [*Carthamus leucocaulos* Sibth. & Smith][CDFA list: A] is an erect winter annual with *shiny white stems* and flower heads mostly 10–13 mm long and *corollas pale purple*, 13–17 mm long. The achenes are 3–5 mm long with a *pappus mostly 5–7 mm long*. No populations are currently documented in California. This species appears to have been eradicated in California, but remains on the state's noxious weed list. It is also a state-listed noxious weed in Oregon (class A). Previous infestations now eradicated occurred in the southern North Coast Ranges (ne Sonoma Co.), to 200 m.

Unlike other pink to purple-flowered thistles (*Carduus, Cirsium, Cynara,* and *Silybum* species), whitestem distaff thistle is the only thistle with *spiny-lobed outer phyllaries* and achenes with short, *narrow, unequal pappus scales*.

Blessed thistle [*Cnicus benedictus* L., synonyms: *Centaurea benedicta* (L.) L.; *Cirsium pugnax* Somm. & Levier] is a spiny-leaved *annual* to 0.6 m tall, with spiny *yellow flower heads that sit just above or are partly concealed by leaflike bracts*. Blessed thistle is further distinguished by having *unlobed outer phyllaries*; inner phyllaries with *long, spiny, pinnate-lobed tips*; and *cylindrical achenes* with ± 20 prominent ribs and pale *crownlike teeth* on the apical rim, and ± 20 pappus bristles in 2 unequal series, the inner bristles short and the outer bristles long. Flower heads appear to consist only of yellow disk flowers, but actually have a

Golden thistle (*Scolymus hispanicus*) rosette.   COURTESY OF ROSS O'CONNELL

Golden thistle *(Scolymus hispanicus)* seeds. J. K. CLARK

few sterile flowers around the margin. Sterile flowers have a very small, linear, 3-lobed corolla. Stems are usually branched throughout, often reddish, slightly succulent, and loosely woolly-hairy. Leaves are 6–25 cm long, thin, conspicuously veined, sparsely hairy, and dotted with glands. Upper leaves are sessile and have a clasping or short-decurrent base. Blessed thistle occurs sporadically on disturbed sites, in fields, and along roadsides in the North Coast Ranges, Central Valley, Central-western region, South Coast, and western Mojave Desert, to 800 m. It also occurs in Arizona, Oregon, Utah, Washington, most southern states, and a few eastern states. Blessed thistle is sometimes cultivated as a medicinal herb. Native to Europe.

Golden thistle [*Scolymus hispanicus* L., synonyms: common golden thistle; scolymus; Spanish oyster; Spanish salsify; *Scolymus grandiflorus* Desf.] [CDFA: A] is an uncommon spiny-leaved *biennial* to 0.8 m tall, with spiny yellow flower heads and *milky sap*. Golden thistle is distinguished by having *ligulate flowers, leaf bases that extend down the stem as spiny wings* (decurrent), *long scalelike chaffy bracts that enclose the achenes on the receptacle*, and achenes with *2–4 stiff, minutely barbed pappus bristles*. In addition, phyllaries are narrow lanceolate, with papery margins and a spiny tip, and are subtended by a few leaflike bracts. Leaves are pinnate spiny-lobed and have *pale veins and markings*. Achenes are 3–5 mm long, flat, winged, and have 2–4 rigid, deciduous pappus bristles. Achenes remain enclosed within a chaffy bract at dispersal. Golden thistle inhabits disturbed, usually grassy sites. Only two sites were found in California (Solano and Alameda Cos.). The Solano County site was eradicated in the 1990s and only a few plants remain in the Alameda County site. Golden thistle also occurs in Alabama, Pennsylvania, and New York. It is a government-listed noxious weed in southeastern Australia. Native to the Mediterranean region.

Purple starthistle [*Centaurea calcitrapa* L.][CENCA] [Cal-IPC: Moderate][CDFA list: B]

Diffuse knapweed [*Centaurea diffusa* Lam.][CENDI] [Cal-IPC: Moderate][CDFA list: A]

Iberian starthistle [*Centaurea iberica* Spreng.][CENIB] [CDFA list: A]

Spotted knapweed [*Centaurea biebersteinii* DC.] [CENMA][Cal-IPC: High][CDFA list: A]

Squarrose knapweed [*Centaurea virgata* Lam. var. *squarrosa* (Boiss.) Gugler][CENSQ][Cal-IPC: Moderate] [CDFA list: A]

SYNONYMS: To date there remains some confusion as to which scientific names are the correct names for some of the following species.

Purple starthistle: red starthistle

Diffuse knapweed: white knapweed; *Acosta diffusa* (Lam.) Soják

Spotted knapweed: *Centaurea maculosa* auct. non Lam. In Europe, *Centaurea maculosa* Lam. and *Centaurea biebersteinii* DC. are treated as distinct species based on chromosome number (2n = 18 and 2n = 36, respectively) even

Purple starthistle (*Centaurea calcitrapa*) infestation.   J. M. DiTomaso

though plants are not reliably distinguishable by morphological characteristics. However, both species occupy different regions in Europe and apparently do not interbreed. In their native ranges, *Centaurea biebersteinii* is typically perennial, while *Centaurea maculosa* is biennial. Currently, the existence of plants with a chromosome count of 2n = 18 (*Centaurea maculosa*) in North America is controversial, and many botanists now apply *Centaurea biebersteinii* to species in North America. *Centaurea micranthos* S.G. Gmel. ex Hayek and *Centaurea stoebe* L. ssp. *micranthos* (Gugler) Hayek are synonyms of *Centaurea biebersteinii* DC. *Centaurea stoebe* L. ssp. *stoebe* and ssp. *maculosa* (Lam.) Hayek are synonyms of *Centaurea maculosa* Lam. Recent evidence indicates that California populations are primarily a different European species, *Centaurea vallesiaca* (DC.) Jord., which can hybridize with *Centaurea biebersteinii*.

Squarrose knapweed: *Centaurea squarrosa* Willd. is an illegally applied name according to the rules of the International Code of Botanical Nomenclature (Saint Louis Code) since the name *Centaurea squarrosa* Roth was previously applied to a different species. *Centaurea virgata* Lam. var. *squarrosa* (Willd.) Boiss. is a synonym of *Centaurea squarrosa* Willd. Some taxonomists are using the name *Centaurea triumfettii* All. for diffuse knapweed. Some floras use *C. virgata* Lam. ssp. *squarrosa* (Willd.) Gugler, but the taxon was given variety ranking first.

Purple starthistle (*Centaurea calcitrapa*) plant.
J. M. DiTomaso

Purple starthistle (*Centaurea calcitrapa*) rosettes.
J. M. DiTomaso

Purple starthistle (*Centaurea calcitrapa*) flowering head.
J. M. DiTomaso

Purple starthistle (*Centaurea calcitrapa*) seeds.
J. O'Brien

Diffuse knapweed (*Centaurea diffusa*) plants in two color forms. J. M. DiTomaso

Diffuse knapweed (*Centaurea diffusa*) buds on rootstalk. J. M. DiTomaso

Diffuse knapweed (*Centaurea diffusa*) flower heads in white and purple form. J. M. DiTomaso

Diffuse knapweed (*Centaurea diffusa*) seedling. J. M. DiTomaso

Diffuse knapweed (*Centaurea diffusa*) leaves and stem. J. M. DiTomaso

Diffuse knapweed (*Centaurea diffusa*) seeds. J. O'Brien

Iberian starthistle *(Centaurea iberica)* flower head.   COURTESY OF ROSS O'CONNELL

Iberian starthistle *(Centaurea iberica)* plant.
COURTESY OF ROSS O'CONNELL

Iberian starthistle *(Centaurea iberica)* seeds.
J. O'BRIEN

GENERAL INFORMATION: Bushy *annuals to perennials*, with flower heads that consist of *spiny or comblike phyllaries* and *white, pink, or purple disk flowers*. Plants exist as basal rosettes until erect, highly branched flowering stems develop at maturity, usually in late spring and summer. These knapweeds and starthistles are highly competitive, noxious weeds, and dense stands can exclude desirable vegetation and wildlife in natural areas. They may also have allelopathic properties. *Centaurea* is a large genus comprised of about 500 species, none of which are native to California and other western states. Thirteen species occur in California as wildland weeds and escaped ornamentals. Numerous biocontrol agents are known to parasitize *Centaurea* species in their native ranges. For the knapweeds, the banded gall fly *(Urophora affinis)*, UV knapweed seed head fly *(Urophora quadrifasciata)*, lesser knapweed flower weevil *(Larinus minutus)*, and broad-nosed seedhead weevil *(Bangasternus fausti)* are established in California to date. The hairy weevil *(Eustenopus villosus)* that primarily attacks yellow starthistle [*C. solstitialis*] has also been reported on spotted knapweed.

Purple starthistle: *Annual to perennial,* to 1 m tall. Besides California, purple starthistle is a state-listed noxious weed in Arizona (prohibited), Nevada, New Mexico (class A), Oregon (class A), and Washington (class A). Native to southern Europe.

Diffuse knapweed: Typically *biennial,* sometimes annual or short-lived perennial, to 0.8 m tall that usually forms large, dense infestations. Besides California,

Spotted knapweed (*Centaurea biebersteinii*) plant. J. M. DiTomaso

Spotted knapweed (*Centaurea biebersteinii*) flower heads from bud to full flower. J. M. DiTomaso

Spotted knapweed (*Centaurea biebersteinii*) white flower head. J. M. DiTomaso

Spotted knapweed (*Centaurea biebersteinii*) rosette. J. M. DiTomaso

Spotted knapweed (*Centaurea biebersteinii*) leaves and stem. J. M. DiTomaso

Spotted knapweed (*Centaurea biebersteinii*) seedling. J. M. DiTomaso

Spotted knapweed (*Centaurea biebersteinii*) seeds. J. O'Brien

Squarrose knapweed (*Centaurea virgata* var. *squarrosa*) plant. J. M. DiTomaso

Squarrose knapweed (*Centaurea virgata* var. *squarrosa*) flower heads. J. M. DiTomaso

Squarrose knapweed (*Centaurea virgata* var. *squarrosa*) leaves and stem. J. M. DiTomaso

Squarrose knapweed (*Centaurea virgata* var. *squarrosa*) seedling. J. M. DiTomaso

Squarrose knapweed (*Centaurea virgata* var. *squarrosa*) seeds. J. O'Brien

diffuse knapweed is a state-listed noxious weed in Arizona (restricted), Colorado, Idaho, Montana (category 1), Nebraska, Nevada, New Mexico (class A), North Dakota, Oregon (class B), South Dakota, Utah, Washington (class B), and Wyoming. Native to southeastern Eurasia.

Iberian starthistle: *Annual, biennial, or short-lived perennial*, to 1 m tall. Closely resembles purple starthistle. Besides California, Iberian starthistle is a state-listed noxious weed in Arizona (prohibited), Nevada, and Oregon (class A). Native to southeastern Eurasia.

Spotted knapweed: *Biennial* to short-lived perennial, to 1 m tall. Besides California, spotted knapweed is a state-listed noxious weed in Arizona (restricted), Colorado, Idaho, Minnesota (secondary), Montana (category 1), Nebraska,

Cornflower (*Centaurea cyanus*) plant.
J. M. DiTomaso

Cornflower (*Centaurea cyanus*) flowering stems.
J. M. DiTomaso

Cornflower (*Centaurea cyanus*) seedling.
J. M. DiTomaso

Cornflower (*Centaurea cyanus*) seeds. J. O'Brien

Nevada, New Mexico (class A), North Dakota, Oregon (class B), South Dakota, Utah, Washington (class B), and Wyoming. Native to Europe.

Squarrose knapweed: *Perennial,* with a woody base, to 0.5 m tall. Besides California, squarrose knapweed is a state-listed noxious weed in Arizona (prohibited), Colorado, Nevada, Oregon (class A), and Utah. Native to Asia.

SEEDLINGS: Cotyledons spatulate to oval. Rosette leaves pinnate-divided. Purple starthistle and Iberian starthistle *develop conspicuous straw-colored spines at the centers of the rosettes.*

MATURE PLANT: *Upper stems lack wings.* Foliage variously covered with short to medium interwoven grayish hairs. Leaves alternate. Lower stem leaves deeply pinnate-lobed 1–2 times, about 10–20 cm long.

Purple starthistle: Leaves resin-dotted. Upper leaves mostly *pinnate-divided.* New leaves densely covered with woolly gray hairs (tomentose).

North African knapweed *(Centaurea diluta)* flower head.
J. M. DiTomaso

North African knapweed *(Centaurea diluta)* seedling.   J. M. DiTomaso

North African knapweed *(Centaurea diluta)* seeds.   J. O'Brien

Black knapweed (*Centaurea nigra*) flower heads. Note all flowers the same length.   J. M. DiTomaso

Black knapweed (*Centaurea nigra*) seedling.
J. M. DiTomaso

Black knapweed (*Centaurea nigra*) seeds.   J. O'Brien

Diffuse knapweed: Upper leaves *entire, linear or bractlike.*

Iberian starthistle: Leaves resin-dotted. Upper leaves mostly *pinnate-divided.* New leaves green and covered with minute bristly hairs.

Spotted knapweed: Leaves resin-dotted. Upper leaves mostly *pinnate-divided.*

Squarrose knapweed: Upper leaves *entire, linear or bractlike, mostly lacking at flowering.*

ROOTS AND UNDERGROUND STRUCTURES: All have a sturdy, ± long taproot.

FLOWERS: Flower heads consist of few to many *fertile disk flowers interspersed with long bristles* on the receptacle. *Phyllaries overlapping in several rows, tips variously spiny or comblike.* Phyllary characteristics are important for species identification. Refer to table 19 (p. 244) for more information.

Table 19. Purple-flowered knapweeds, starthistles, and common crupina (*Acroptilon*, *Centaurea*, and *Crupina* spp.)

Species	Flower head shape	Involucre diameter (mm)	Phyllaries	Pappus	Achene
Acroptilon repens Russian knapweed	hemispheric	10–14	not spiny, margins membranous	deciduous bristles, 6–8 mm long	white, no basal notch
Centaurea biebersteinii spotted knapweed	urn-shaped	~6	not spiny, dark appendages at tip	persist, white bristles, 1–2 mm long	pale brown with fine hairs
Centaurea calcitrapa purple starthistle	hemispheric	6–8	long and spiny, appendages at tip spiny	none	white, often with brown streaks
Centaurea debeauxii meadow knapweed	hemispheric	~12–18	not spiny, phyllaries and appendage brown	none or blackish scales, < 1 mm long	tan with fine hairs
Centaurea diffusa diffuse knapweed	cylindric or slender urn-shaped	~3	short-spiny, appendages green, or straw-colored and short-spiny	none or white scales, < 1 mm long	dark brown
Centaurea diluta North African knapweed	hemispheric	15–18	not spiny, green or straw-colored, appendages green and not spiny	white bristles, ~4.5 mm long	—
Centaurea iberica Iberian starthistle	hemispheric	8–14	similar to purple starthistle except slightly smaller	white bristles, ~1 mm long	white, often with brown streaks
Centaurea nigra black knapweed	hemispheric	~12–20	similar to meadow knapweed	none or blackish bristles, < 1 mm long	—
Centaurea virgata var. *squarrosa* squarrose knapweed	cylindric or slender urn-shaped	3–4	similar to diffuse knapweed except tips curved back or reflexed	none or white bristles, 2–2.5 mm long	pale brown
Crupina vulgaris common crupina	cylindric or slender urn-shaped	~5–12	not spiny, lacking appendages at tip	black-brown, rough, bristles and scales 5–7 mm and < 1 mm long	black-brown with silver-tipped hairs, no basal notch

Purple starthistle: July–October. Flowers 25–40 per head. Corollas purple, 15–24 mm long. Involucre ovoid, 15–20 mm long, body about 6–8(10) mm in diameter. Main phyllaries greenish or straw-colored, spine-tipped. *Central spine stout, spreading, about 10–25 mm long.*

Diffuse knapweed: June–September. Flowers 12–13 average per head, outer flowers sterile. Corollas white, pink, or pale purple, 12–13 mm long. Involucre ovoid-cylindrical, 10–13 mm long, ± 5 mm wide. Main phyllaries pale green, *spine-tipped*, spines straw-colored, *central spine spreading* to 3 mm long. Self-fertile.

Iberian starthistle: July–October. Flowers numerous per head. Corollas rose-pink to whitish, 15–20 mm long. Involucre 15–18 mm long, body mostly 8–14 mm in diameter. Phyllaries and central spines resemble those of purple starthistle, but often slightly smaller.

Spotted knapweed: June–October. Flowers 30–40 per head. Corollas white, pink, or purple, 12–25 mm long. Involucre (unit of phyllaries) ovoid, 10–13 mm long, 8–13 mm wide. Phyllaries pale green or pink-tinged, with parallel veins. *Phyllary tips dark, comblike, not spine-tipped.* Self-fertile.

Squarrose knapweed: June–August. Flowers 4–8(10) per head, outer sterile. Corollas pink to pale purple, 7–9 mm long. Involucre ovoid-cylindrical, 7–8(–10) mm long. Main phyllaries pale green to straw-colored, sometimes purple-tinged, *spine-tipped. Central spine usually reflexed downward,* to 3 mm long.

FRUITS AND SEEDS: Achenes oblong, *2.5–3.5 mm long*, apex flattened, tapered to a rounded and laterally notched base. Pappus (when present) ± whitish, composed of unequal, stiff, minutely barbed bristles or tiny, flat scales.

Meadow knapweed (*Centaurea debeauxii* ssp. *thuillierii*) plant.   J. M. DiTomaso

Meadow knapweed (*Centaurea debeauxii* ssp. *thuillierii*) flower head. Note central flowers shorter than outer flowers. J. M. DiTomaso

Meadow knapweed (*Centaurea debeauxii* ssp. *thuillierii*) seedling. J. M. DiTomaso

Purple starthistle : Achenes white, often brown-streaked. Pappus usually lacking. Achenes white, often brown-streaked. Pappus bristles ± 1 mm long.

Diffuse knapweed: Achenes dark brown, ± 13 per head. Pappus scales less than 1 mm long or lacking.

Spotted knapweed: Achenes pale brown, finely hairy, ± 30 per head. Pappus bristles 1–2 mm long.

Squarrose knapweed: Achenes pale brown, 1–4 per head. Pappus bristles 2–2.5 mm long, or lacking.

POSTSENESCENCE CHARACTERISTICS: Old flower stems can persist for an extended period after senescence (less common for diffuse knapweed). Phyllaries and achenes remaining on old stems can aid species identification when plants are overwintering as rosettes.

HABITAT: Fields, roadsides, disturbed open sites, grassland, rangeland (especially degraded rangeland), and logged areas. Seldom persist in shaded places. Colonize most soil types with a disturbed A horizon.

Purple starthistle : Frequently inhabits heavy, fertile soils, including alluvial soils.

Diffuse knapweed, spotted knapweed: Serious infestations often occur on light, well-drained soils in areas that receive some summer rainfall. Diffuse knapweed requires less moisture than spotted knapweed.

Iberian starthistle: Often colonizes banks of watercourses and other moist sites.

Squarrose knapweed: Often grows on degraded rangeland soils and is more adaptable to drought and cold temperatures than spotted knapweed and diffuse knapweed.

DISTRIBUTION: Purple starthistle: Throughout California, except northern and central Cascade Range, Modoc Plateau, Great Basin, and desert regions, to 1000 m. Arizona, New Mexico, Oregon, Washington, Utah, Alabama, Illinois, Iowa, and some northeastern states.

Diffuse knapweed: North Coast, North Coast Ranges (Del Norte, Humboldt, n Mendocino Cos.), Klamath Ranges (Trinity Co.), Cascade Range (Siskiyou, Shasta Cos.), north and central Sierra Nevada (Plumas, Nevada, sc Placer, e El Dorado, e Amador Cos.), northern Sacramento Valley (c & e Tehama, sc Glenn, s Sutter, n Sacramento Cos.), Modoc Plateau (Modoc, Lassen Cos.), southern San Francisco Bay region (ne Santa Clara Co.), South Coast Ranges (se Monterey Co.), and South Coast (Los Angeles, San Diego Cos.), to 2300 m. A population that has been eradicated occurred in the central area of the border between Mariposa and Madera Counties. All western states, some central and eastern states, especially Illinois and surrounding states.

Iberian starthistle: Southeastern North Coast Ranges (Sonoma, Lake, Napa Cos.), central Sierra Nevada foothills (ne Amador, se Tuolumne, ne Mariposa Cos.), central Central Valley (s Yolo Co.), eastern San Francisco Bay region (s Contra Costa, n & s Alameda, n Santa Clara Cos.), southern South Coast Ranges (sc Santa Barbara Co.), and Peninsular Ranges (c & sw San Diego Co.), to 1000 m. Oregon, Washington, Wyoming, and Kansas.

Spotted knapweed: Populations scattered. North Coast, North Coast Ranges (cw and se Humboldt, Mendocino, nw Sonoma, sc Napa Cos.), Klamath Ranges (n Del Norte, Siskiyou, Trinity Cos.), Cascade Range (Shasta, ne & sc Tehama Cos.), Great Basin (Modoc, Lassen, Mono Cos.), Sierra Nevada (Plumas, Sierra, e Yuba, Nevada, Placer, El Dorado, e Amador, Alpine, e Calaveras, c Tuolomne, ne Fresno Cos.), northern Sacramento Valley (Butte, c Colusa, ne Sutter Cos.), southern San Francisco Bay region (nc Alameda Co.), and southern Peninsular Ranges (c San Diego Co.), to 2000 m. A population that has been eradicated occurred in Glenn County. Most contiguous states except possibly Georgia, Mississippi, Oklahoma, and Texas.

Squarrose knapweed: Klamath Ranges (n Humboldt, c Siskiyou, ce Trinity Cos.), Cascade Range (Shasta Co.), Modoc Plateau (Modoc, ne Lassen Cos.), northern Sierra Nevada (nc Plumas Co.), to 1400 m. Nevada, Oregon, Utah, Michigan.

PROPAGATION/PHENOLOGY: *All reproduce by seeds,* except where noted. Most seeds or seed heads of *Centaurea* species fall near the parent plant, and some can disperse to greater distances with human activities, vehicles and heavy machinery, water, and soil movement, and by clinging to shoes, clothing, and tires and to animals. A few species have other dispersal mechanisms that are described below. Germination can occur over a broad range of environmental conditions. Seedling emergence is typically highest after the first fall rains. Mortality of seedlings that emerge in spring can be high when conditions become dry after emergence. Most seedlings emerge from seeds at or near the soil surface. Plants generally produce fewer viable seeds in dry years. Infestation density generally correlates with the age of the population and degree of disturbance. Spotted knapweed and diffuse knapweed seeds exhibit three germination patterns: nondormant seeds that germinate with or without light exposure, dormant seeds that germinate in response to red light, and dormant seeds that

Meadow knapweed (*Centaurea debeauxii* ssp. *thuillierii*) seeds. J. O'Brien

are not light-sensitive. All germination types occur on each plant. For spotted knapweed and diffuse knapweed, optimal germination is between 10°–28°C.

Purple starthistle: Seeds disperse with the seed head as a unit. Most seed germinates the first year, but buried seed can remain dormant for about 3 years.

Diffuse knapweed: Seeds often disperse when stems break off near the ground and tumble along with the wind. Seedlings can emerge from soil depths to 3 cm. Diffuse knapweed has been shown to occasionally hybridize with spotted knapweed.

Iberian starthistle: The biology of this species is probably similar to purple starthistle.

Spotted knapweed: In addition to seeds, spotted knapweed can reproduce vegetatively from lateral roots just below the soil surface. New rosettes may develop at about 3-cm intervals along lateral roots, expanding populations peripherally. About 2–3 weeks after maturity, drying phyllaries pop seed heads open, ejecting seeds a short distance. Stems typically do not break off and tumble with wind. Some seedlings can emerge from soil depths to 5 cm. Plants may produce up to 40,000 seeds per plant. Spotted knapweed has been shown to occasionally hybridize with diffuse knapweed.

Squarrose knapweed: Seeds disperse with the seed head as a unit from mid- to late summer through winter.

MANAGEMENT FAVORING/DISCOURAGING SURVIVAL: Fertilizer applications and poorly timed mowing can encourage survival. Rosettes are usually too low to be affected by mowing. Mowing at late bud to early bloom growth stage reduces seed production. Mowing after seed-set can disperse seed and can stimulate stem regrowth. Burning removes current growth but may enhance seed germination. Hand-pulling 2–4 times per year or severing plants at least 2 inches below crowns can control small infestations. Maintaining pasture and rangeland health by preventing overgrazing and minimizing disturbance can help limit knapweed establishment and spread. Insect biocontrol agents have been released for all of these knapweed species, but it is too early to determine their effectiveness.

SIMILAR SPECIES: The following 3 species have *fringed or comblike phyllaries that lack spines*. Refer to table 18 (p. 227) for more information.

Cornflower [*Centaurea cyanus* L., synonyms: bachelor's buttons; garden cornflower][CENCY] is an *annual* to 1 m tall, with *showy blue, purple, or white flower heads*. Besides having an annual life cycle, cornflower is distinguished by having *foliage that is covered with short, gray, ± woolly hairs* and *inner corollas mostly 10–15 mm long*. Cornflower sometimes escapes cultivation as an ornamental in most regions of California to 2000 m, except the deserts. Cornflower is naturalized in all contiguous states, except possibly New Mexico. Native to southern Europe.

North African knapweed [*Centaurea diluta* Aiton] is an annual to 1.3 m tall, with pink-purple flowers and a short central spine (0–5 mm long) on the phyllary tips. Unlike diffuse and squarrose knapweed, the corollas of North African knapweed are much longer (20–30 mm). Occasionally found in disturbed areas of southern California and on the central coast. Escaped from cultivation. Native to southwestern Europe.

Black knapweed [*Centaurea nigra* L., synonym: lesser knapweed][CENNI] and meadow knapweed [*Centaurea debeauxii* Gren. & Godr. ssp. *thuillierii* Dostál, synonym: *Centaurea* × *pratensis* Thuill.] [Cal-IPC: Moderate Alert] are less common *perennials* to 1 m tall that are similar to spotted knapweed. Black knapweed is distinguished by having *all flowers fertile and producing fruits*. Like spotted knapweed, meadow knapweed has *sterile outer flowers*, but meadow knapweed is distinguished by having involucres mostly *15–18 mm long*. Black knapweed and meadow knapweed inhabit disturbed places in the Northwestern region and the San Francisco Bay region, to 500 m. Both species also occur in Idaho, Montana, Oregon, and Washington. Black knapweed occurs in most northeastern states and adjacent central states. Meadow knapweed is scattered in some northeastern states. Black knapweed is a state-listed noxious weed in Washington (class B). Meadow knapweed is a state-listed noxious weed in Colorado, Idaho, and Oregon (class B). Native to Europe.

Unlike *Centaurea* species, **Russian knapweed** [*Acroptilon repens* (L.) DC.] and **common crupina** [*Crupina vulgaris* Cass.] have *phyllaries that are ovate* (Russian knapweed) *or narrowly lanceolate* (common crupina) with *papery margins* and *seeds 3–4 mm long that lack a lateral notch near the base*. In addition, only common crupina has *leaf margins with stiff hairs that are tipped with minute barbs* and flower head receptacles with *flattened, scalelike chaffy bracts*. Refer to the entries for **Russian knapweed** (p. 175) and **Common crupina** (p. 299) for more information.

Malta starthistle [*Centaurea melitensis* L.] [CENME][Cal-IPC: Moderate][CDFA list:C]

Yellow starthistle [*Centaurea solstitialis* L.] [CENSO][Cal-IPC: High][CDFA list: C]

Sicilian starthistle [*Centaurea sulphurea* Willd.] [CDFA list: B]

SYNONYMS: Malta starthistle: Napa thistle; tocalote

Yellow starthistle: golden starthistle; St. Barnaby's thistle; yellow cockspur; *Leucantha solstitialis* (L.) A. & D. Löve

Sicilian starthistle: sulphur-colored Sicilian thistle; misidentified as *Centaurea sicula* L. in some older California references

GENERAL INFORMATION: Simple to bushy winter *annuals*, occasionally biennials, with *spiny yellow-flowered heads* and wiry stems. Refer to table 20 (p. 258) for a quick review of important differences among yellow-flowered starthistles.

Malta starthistle: This species is often called tocalote in California. It is most invasive in the Central-western and Southwestern regions of California, but it is generally less prevalent than yellow starthistle statewide. Malta starthistle is not known to cause chewing disease in horses and is used medicinally in Spain. A small seed-head-feeding beetle (*Lasioderma haemorrhoidale*) was unintentionally introduced to California from the Mediterranean region, but it has had little effect in controlling the plant. It is considered a generalist seed head feeder

Malta starthistle (*Centaurea melitensis*) buds and flower heads.   J. M. DiTomaso

Malta starthistle • Yellow starthistle • Sicilian starthistle

Malta starthistle (*Centaurea melitensis*) senesced flower heads.　　J. M. DiTomaso

Malta starthistle (*Centaurea melitensis*) rosette.　　J. M. DiTomaso

Malta starthistle (*Centaurea melitensis*) seedlings.　　J. M. DiTomaso

Malta starthistle (*Centaurea melitensis*) seeds.　　J. O'Brien

and is also known to attack yellow starthistle, Sicilian starthistle, blessed milkthistle, and Italian thistle. The false peacock fly (*Chaetorellia succinea*) and hairy weevil (*Eustenopus villosus*) will also attack Malta starthistle, but to a lesser degree than they attack yellow starthistle. Malta starthistle is thought to have been introduced into California in the late 1700s during the Spanish missionary period. Native to Southern Europe.

Yellow starthistle: Winter annual, sometimes biennial, to 2 m tall. Plants are highly competitive and typically develop dense, impenetrable stands that displace desirable vegetation in natural areas, rangelands, roadsides, and other places. Yellow starthistle is considered one of the most serious rangeland weeds in the western United States. It has spread rapidly since its introduction into California around 1850. In 2002, it was estimated that 5.6 million hectares are infested in California, with the heaviest infestations primarily in the northern and central-western regions. Yellow starthistle is sometimes problematic in grain fields, where the seeds can contaminate the grain harvest and lower its quality and value. Yellow starthistle contains an unidentified compound that causes nigropallidal encephalomalacia, or chewing disease, in horses. The compound affects only horses and permanently damages the area of the brain that controls

Malta starthistle • Yellow starthistle • Sicilian starthistle

Yellow starthistle (*Centaurea solstitialis*) infestation.　　　J. M. DiTomaso

Yellow starthistle (*Centaurea solstitialis*) plant.　　　J. M. DiTomaso

fine motor movements, including mouth and lip movements. Toxic effects are cumulative. Horses must consume 50–150% of an animal's weight in dry-weight plant material over a period of 1 to 3 months to develop symptoms. Because of its bitter taste, horses generally avoid grazing yellow starthistle. However, symptoms of toxicity can occur when horses are allowed to graze infested pastures, especially those that lack adequate amounts of suitable green forage, or are fed contaminated hay over a period of time. Once the toxic threshold has been reached, symptoms occur rapidly. Symptoms include fatigue, lowered head, an uncontrolled rapid twitching of the lower lip, tongue-flicking, involuntary chewing movements, and an unnatural open position of the mouth. Without intervention, affected horses become unable to eat or drink and eventually die from starvation or dehydration. Yellow starthistle has some benefits. Bees foraging on yellow starthistle flowers produce a flavorful, high-quality honey. Several biocontrol agents have been introduced from the Mediterranean region to control yellow starthistle. Currently established in California are the yellow starthistle bud weevil (*Bangasternus orientalis*), hairy weevil (*Eustenopus villosus*), flower weevil (*Larinus curtus*), gall fly (*Urophora sirunaseva*), and peacock fly (*Chaetorellia australis*). A sixth insect, the false peacock fly (*Chaetorellia succinea*), was accidentally introduced in 1991. Like the others, it has a strong affinity for yellow starthistle flower heads. However, these insects have yet to provide significant reduction in yellow starthistle populations in most areas. Native to Southern Europe.

Yellow starthistle (*Centaurea solstitialis*) flower head. J. K. Clark

Yellow starthistle (*Centaurea solstitialis*) seedling. J. M. DiTomaso

Yellow starthistle (*Centaurea solstitialis*) seeds. J. O'Brien

Sicilian starthistle: This species appears to be much less invasive than yellow starthistle or Malta starthistle. Sicilian starthistle was discovered growing in an area near Folsom (Sacramento Co.) in 1923. By 1999, the population had only slightly expanded its range. A few smaller populations have been found elsewhere, but these infestations remain small. Native to Southwestern Europe.

SEEDLINGS: Cotyledons oblong to spatula-shaped, base wedge-shaped, tip truncate to slightly rounded, glabrous. First few leaves typically oblanceolate. Subsequent rosette leaves highly variable, often oblanceolate, entire to pinnate-lobed, terminal lobe larger than lateral lobes. Later rosette leaves to 15 cm long. Hair characteristics are visible with 10–14× magnification.

Malta starthistle: Cotyledons similar to yellow starthistle. Later rosette leaves entire to deeply lobed nearly to the midvein, lobes rounded to slightly rounded. Terminal lobe usually rounded. Upper and lower surfaces evenly covered with stiff thick hairs and resinous dots. Fine cottony hairs sparse on older leaves, sometimes dense on new leaves.

Yellow starthistle: Cotyledons 6–9 mm long, 3–5 mm wide. Later rosette leaves typically deeply lobed nearly to the midvein, often appear ruffled, lobes mostly acute, with toothed to wavy margins. Terminal lobe nearly triangular to lanceolate. Both upper and lower surfaces usually densely covered with fine cottony hairs and stiff thick hairs. Rosette leaves that develop under reduced light levels are typically larger and more erect.

Sicilian starthistle: Cotyledons about 2–4 cm long, 1–2 cm wide, slightly succulent, glabrous. First few leaves oblanceolate, about 4–8 cm long, 1.5–2.5 cm

Starthistle flower heads: Malta *(Centaurea melitensis)* (left), Sicilian *(C. sulphurea)* (middle two), and yellow *(C. solstitialis)* (right). J. K. CLARK

Malta starthistle • Yellow starthistle • Sicilian starthistle

Sicilian starthistle (*Centaurea sulphurea*) immature plant. J. M. DiTomaso

Sicilian starthistle (*Centaurea sulphurea*) stem and leaves. J. M. DiTomaso

wide, sparsely pubescent on both surfaces. Later rosette leaves broader and typically more shallow lobed than those of yellow starthistle. Lateral lobes usually acute. Terminal lobes generally ovate, with a broadly acute tip and finely toothed margins. Surfaces sparsely covered with stiff thick hairs. Fine cottony hairs and resinous dots are lacking.

MATURE PLANT: Stems stiff, wiry, simple in very small plants to openly branched from near or above the base in larger plants. Leaf bases extend down the stems (decurrent) and give stems a winged appearance. Stem leaves alternate, mostly *linear to narrow oblong or oblanceolate*. Margins smooth, toothed, or wavy. Rosette leaves typically wither by flowering time.

Malta starthistle: Largest stem wings usually to about 3 mm wide. Foliage often grayish green. Leaves *evenly covered with thick stiff hairs and minute glandular dots*. Older leaves are usually *sparsely covered with fine, white, cottony hairs that do not hide the stiff hairs and resinous dots*. New leaves sometimes densely covered with fine, cottony hairs.

Yellow starthistle: Largest stem wings usually to about 5 mm wide. Lower stem leaves sometimes deeply pinnate-lobed. Foliage grayish to bluish green, *densely covered with fine white cottony hairs that hide most of the stiff thick hairs and minute glandular dots*.

Sicilian starthistle: Stem leaves toothed. Largest stem wings about 5–6 mm wide. Foliage often yellowish green, *sparsely covered with stiff hairs*.

Malta starthistle • Yellow starthistle • Sicilian starthistle

Sicilian starthistle (*Centaurea sulphurea*) seedling.   J. M. DiTomaso

ROOTS AND UNDERGROUND STRUCTURES: Malta starthistle, Sicilian starthistle: Taproots usually do not penetrate the soil as deeply as those of yellow starthistle. Consequently, plants flower much earlier in the growing season.

Yellow starthistle: Taproots grow vigorously early in the season to soil depths of 2 m or more, allowing plants to access deep soil moisture during the dry summer and early fall months.

FLOWERS: Heads round to ovoid, spiny, solitary on stem tips, consist of numerous yellow disk flowers. Vigorous individuals of Malta and yellow starthistle may develop flower heads in branch axils. Phyllaries palmately spined, with one long central spine and 2 or more pairs of short lateral spines. Receptacles flat, with numerous hairlike chaffy bracts interspersed among the disk flowers. Insect-pollinated.

Malta starthistle: April–July. Heads solitary or in close clusters of 2–3. Corollas typically 10–12 mm long. Involucre mostly 8–15 mm long. Phyllaries often sparsely covered with cobwebby hairs. Central spine of main phyllaries *5–12 mm long, slender, often purple- to brown-tinged.* Lateral spines usually in 3–4 pairs, the *upper pair near the middle of the central spine.* Malta starthistle produces three different types of flower heads, including fully expanded flowers capable of cross-pollination, and two cleistogamous (self-pollinated) types, one with yellow flowers only partially protruding and the other without exerted flowers.

Yellow starthistle: June–December. Grazing and mowing can delay flowering. Corollas mostly 13–20 mm long. Involucre (phyllaries as a unit) mostly 12–18 mm long. Phyllaries dense to sparsely covered with cottony hairs or with patches of hairs at the bases of the spines. Central spine of main phyllaries *10–25 mm long, stout, yellowish to straw-colored throughout*. Lateral spines typically in *2–3 pairs at the base of the central spine*. Mostly self-incompatible.

Sicilian starthistle: May–July. Corollas usually 25–35 mm long, paler than those of yellow starthistle. Involucre 12–30 mm long. Phyllaries glabrous. Central spine of main phyllaries *10–25 mm long, stout, yellowish to straw-colored near the tip, blackish to dark brown near the base*. Lateral spines typically *3–5 pairs at the base of the central spine*.

FRUITS AND SEEDS: Achenes (seeds) generally barrel-shaped, slightly flattened, base laterally notched. Pappus bristles slender, stiff, unequal.

Malta starthistle: Achenes mostly 2–3 mm long, finely *pubescent, grayish to tan*, usually with slightly darker stripes. Base *deeply notched, narrow, hooklike*. Pappus bristles pale tan, 1–3 mm long.

Yellow starthistle: Produces *2 types of achenes*, both *glabrous*, mostly 2–3 mm long, *base broad*. Outer ring of achenes *dull dark brown, often speckled with tan, lack pappus bristles*, often remain in heads. *Inner achenes glossy, gray or tan to mottled cream-colored and tan*, with slender *white pappus bristles* 2–5 mm long.

Sicilian starthistle: Achenes mostly *5–8 mm long*, glossy *dark brown*, often faintly streaked tan. *Base deeply notched, broad, hooklike*. Pappus bristles *dark brown to black*, 6–7 mm long.

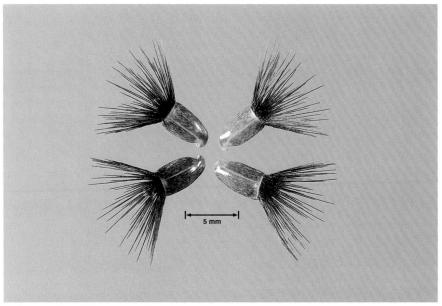

Sicilian starthistle *(Centaurea sulphurea)* seeds. J. O'BRIEN

Table 20. Yellow-flowered starthistles (*Centaurea* spp.)

Characteristic	Malta starthistle *C. melitensis*	Yellow starthistle *C. solstitialis*	Sicilian starthistle *C. sulphurea*
FLOWER HEADS			
length of central spine of phyllaries	~ 5–12 mm	~ 10–25 mm	~ 10–25 mm
color of central spine of phyllaries	purplish to brown-tinged	yellowish to straw-colored throughout	yellowish near tip, dark brown to black near base
length of phyllaries as a unit (involucre)	~ 8–15 mm	~ 12–18 mm	~ 12–30 mm
phyllary hairs	cobwebby, dense to sparse	cottony, dense to sparse	none
senesced flower heads	retain central spines, often shed fuzzy centers	shed central spines, retain fuzzy centers	retain central spines, some shed fuzzy centers
ROSETTE LEAVES			
terminal lobe	± round	lanceolate to ± triangular	± ovate
lateral lobes	± rounded	acute	± acute
margins	entire to deeply lobed ± to midrib	deeply lobed ± to midrib	shallow to lobed ± halfway to midrib
hairs	evenly covered with stiff thick hairs and resinous dots, fine cottony hairs ± sparse	fine white cottony hairs dense, mostly hide thick stiff hairs	stiff thick hairs sparse
SEEDS (ACHENES)			
seed types per head	all have pappus bristles	inner have pappus bristles, outer lack pappus bristles	all have pappus bristles
length without pappus bristles	~ 2–3 mm	~ 2–3 mm	~ 5–8 mm
color	glossy gray to tan, often with slightly darker stripes	inner: glossy gray to tan, or mottled with creamy white; outer: dull dark brown, often speckled with tan	glossy dark brown, often faintly streaked with tan
surface	pubescent	glabrous	pubescent
base and lateral notch	narrow, hooklike	broad, slightly notched	broad, hooklike
pappus bristle color	light tan	white	dark brown to black
pappus bristle length	~ 1–3 mm	~ 2–5 mm	~ 6–7 mm

POSTSENESCENCE CHARACTERISTICS: Stems with old flower heads turn gray-brown and can remain intact for over a year.

Malta starthistle: Plants often senesce earlier than yellow starthistle. Heads *retain the central spines* and often *shed the loose receptacle and dense fuzzy gray hairs,* leaving a shallow bowl of spiny phyllaries.

Yellow starthistle: Plants usually senesce in late summer or fall. Heads *shed the central spines* and *tightly retain a ball of dense fuzzy gray hairs* (chaff) on the receptacle. Often a dense layer of thatch develops on heavily infested sites.

Sicilian starthistle: Plants often senesce earlier than yellow starthistle. Heads *retain the central spines,* and *a large proportion of heads retain the receptacle and dense fuzzy gray hairs.*

HABITAT: Open disturbed sites, open hillsides, grassland, rangeland, open woodlands, fields, pastures, roadsides, and waste places. Yellow and Malta starthistle also inhabit cultivated fields.

DISTRIBUTION: Malta starthistle: Throughout most of California, except Great Basin region, to 2200 m. Common in the Central-western region and Southwestern regions. Uncommon in desert regions. Arizona, Nevada, New Mexico, Oregon, Texas, Utah, Washington, and scattered in a few southern, eastern, and central states.

Yellow starthistle: Throughout most of California, typically to 1800 m, but has been found up to 2600 m. Common in the Sacramento Valley, San Joaquin Valley, Sierra Nevada foothills, Cascade Range, Klamath Ranges, eastern North Coast Ranges, and Central-western region of the state. Less common in southern California and uncommon in the desert regions, moist coastal areas, and east of the Sierra Nevada. Expanding range in mountainous areas and in the dry southern coastal valleys. Most contiguous states, except a few southern and northeastern states.

Sicilian starthistle: Southern Sacramento Valley (Folsom, ne Sacramento Co.), San Francisco Bay region (Hwy 35 in se San Mateo and nw Santa Clara Cos.), to 300 m. Previous small populations located in the adjacent Sierra Nevada foothills (ne El Dorado Co.) and northern San Joaquin Valley (n San Joaquin Co.), may have been eradicated.

PROPAGATION/PHENOLOGY: *Reproduce by seed.* Seeds fall near the parent plant and are dispersed short distances with wind and to greater distances with human activities, animals, water, mud, and soil movement. Most seeds germinate after the first fall rains. Plants exist as basal rosettes through winter and early spring until flower stems develop in late spring or early summer.

Malta starthistle: Seed production is highly variable. Plants can produce 1–60 or more seeds per head and 1–100 heads or more per plant. Wild oat (*Avena* spp.) litter appears to have allelopathic properties that reduce Malta starthistle seed germination. Young seedlings are especially resistant to the effects of fall drought.

Yellow starthistle: Seed heads typically contain about 30–80 seeds. Seed head production is highly variable (1–1000 heads per plant) and depends on a variety of factors, including soil moisture and competition. Large plants can produce nearly 75,000 seeds. Some seed is viable 8 days after flower initiation. Most seed does not appear to require an after-ripening period. Seed germination is closely correlated with rainfall events, and light is required. Large flushes of seeds typically germinate after the first fall rains, but smaller germination flushes can occur in winters and early spring. Non-pappus-bearing seeds appear to require higher temperatures for germination to occur than pappus-bearing seeds. Seeds can survive for up to about 10 years in the field under certain environmental conditions, but it appears that few seeds survive beyond 2–3 years in the Central Valley. Shaded conditions reduce flower production and root growth. In mild-winter areas, a few flowering plants may be found year-round, particularly those that have been grazed or mowed earlier in the season. Photoperiod or a period of cool, moist conditions (vernalization) does not appear to influence the time of flowering. Yellow starthistle litter appears to inhibit germination of its seeds, probably by blocking light penetration.

Sicilian starthistle: The biology of this species is largely undocumented. Where species occur together, seedlings and rosettes of Sicilian starthistle often outcompete those of yellow starthistle.

MANAGEMENT FAVORING/DISCOURAGING SURVIVAL: Monitoring and spot eradication of plants when they are discovered can prevent the spread of starthistles.

Malta starthistle, Sicilian starthistle: Although undocumented, cultural strategies used to control yellow starthistle are likely to control these starthistles as well.

Yellow starthistle: Management techniques such as grazing, mowing, burning, and cultivation can prevent seed production and control infestations when employed over a period of 2–3 years or more, depending on the degree of infestation and other factors. These methods must be properly timed to be effective. High-intensity short-duration grazing by sheep, goats, or cattle should be implemented during the period when plants have bolted to just before they produce spiny heads. Mowing is most effective when plants are cut below the height of the lowest branches and 2–5% of the total population of seed heads is in bloom. Mowing too early can result in higher seed production. Prescribed burns can provide control if implemented after other annual plants have dried but before yellow starthistle seed is produced. Burning at other times may enhance yellow starthistle survival by removing the thatch and encouraging seed germination in the fall. Repeated shallow cultivation throughout the germination period and prior to seed production can control plants in crop fields. To prevent reinfestation, vigilant monitoring and spot eradication may be required indefinitely.

SIMILAR SPECIES: Purple starthistle [*Centaurea calcitrapa* L.] and **Iberian starthistle** [*Centaurea iberica* Spreng.] have *purple flowers, upper stem leaves that are mostly pinnate-divided,* and *immature rosettes have straw-colored spines in the center*. Refer to the entry for these species (p. 235) for more information.

Pineappleweed [*Chamomilla suaveolens* (Pursh) Rydb.] [MATMT]

SYNONYMS: disc mayweed; rayless chamomile; *Matricaria discoidea* DC.; *Matricaria matricarioides* (Less.) Porter; *Matricaria suaveolens* (Pursh) Buch.; non L.; *Santolina suaveolens* Pursh; *Tanacetum suaveolens* (Pursh) Hook.; *Tanacetum suaveolens* (Pursh) Hook.; a few others

GENERAL INFORMATION: Erect summer or winter *annual* to 0.5 m tall, with *finely dissected leaves* and *yellowish green egg-shaped flower heads*. Foliage and flower heads have a pineapple-like fragrance when crushed. Native to northwestern North America and northeastern Asia.

SEEDLINGS: Cotyledons narrowly oblong to linear, 2–12 mm long, ± 1 mm wide, fused at the base, pointed or rounded at the tip, glabrous. First leaf pair opposite, smooth-margined to pinnate-lobed, about 5–8 mm long, glabrous. Subsequent leaves alternate, pinnate-dissected, form a rosette.

MATURE PLANT: Leaves alternate, glabrous, sessile, to 5 cm long, *finely dissected 1–3 times into linear segments*.

ROOTS AND UNDERGROUND STRUCTURES: Taproot shallow, with fibrous lateral roots.

FLOWERS: May–August. Heads ovoid, 0.5–1 cm wide, *densely flowered with yellowish green disk flowers*. Receptacle conical, lacks chaffy bracts. Phyllaries membranous on the margins.

Pineappleweed (*Chamomilla suaveolens*) flower heads. J. M. DiTomaso

Pineappleweed

Pineappleweed (*Chamomilla suaveolens*) plant. J. M. DiTomaso

Pineappleweed (*Chamomilla suaveolens*) seedling. J. K. Clark

Pineappleweed (*Chamomilla suaveolens*) seeds. J. O'Brien

FRUITS AND SEEDS: Heads shatter at maturity. Achenes lanceolate, 1–2 mm long, 3–5 veined, yellow, brown, or gray, become sticky with mucilage from linear brown glands when moistened. *Pappus reduced to a minute crown.*

POSTSENESCENCE CHARACTERISTICS: Dead plants generally do not persist.

HABITAT: Roadsides, waste places, gardens, orchards, vineyards, landscaped areas, nursery and other crops, turf, pastures, sand bars, riverbanks, and other disturbed places. Tolerates compacted soils.

DISTRIBUTION: Throughout California, to 2400 m, except Great Basin and deserts. Most contiguous states, except possibly Texas, Alabama, Georgia, and Florida.

PROPAGATION/PHENOLOGY: *Reproduces by seed.* Fruits fall near the parent plant and disperse to greater distances with human activities such as crop, orchard, or landscape maintenance, and by clinging to shoes, tools, vehicle tires, and to animals, and possibly by other means.

MANAGEMENT FAVORING/DISCOURAGING SURVIVAL: Manual removal or cultivation readily controls pineappleweed. Plants tolerate mowing, but improving turf health with adequate water and fertilizer combined with regular mowing can eventually eliminate pineappleweed in turf.

Pineappleweed

Pineappleweed (*Chamomilla suaveolens*) young plant. J. K. Clark

SIMILAR SPECIES: Pineappleweed in flower is unlikely to be confused with other weedy species although it can appear similar to the nonweedy native *Chamomilla occidentalis* (E. Greene) Rydb. In its vegetative state, pineappleweed can be confused with other weedy species, including **mayweed chamomile** [*Anthemis cotula* L.], **corn chamomile** [*Anthemis arvensis* L.], and **yellow chamomile** [*Anthemis tinctoria* L.]. Unlike pineappleweed, the crushed foliage of mayweed chamomile has an *unpleasant scent*. The crushed foliage of corn chamomile and yellow chamomile *lacks a noticeable fragrance*. Refer to the entry for **Mayweed chamomile** (p. 189) for more information about these species.

Rush skeletonweed [*Chondrilla juncea* L.][CHOJU] [Cal-IPC:Moderate][CDFA List: A]

SYNONYMS: devil's-grass; gum succory; hogbite; naked weed; skeleton weed

GENERAL INFORMATION: Herbaceous *perennial* or biennial, with *rigid, wiry flower stems* to 1 m tall and *milky sap*. Plants exist as basal rosettes until flower stems develop at maturity. Persistent flower stems can hinder harvest machinery. There are several forms or biotypes that differ in leaf width, branching pattern, and flowering time. Characteristics sometimes vary between, but rarely within, populations since all reproduction is clonal by vegetative means and asexual seed production (apomixis). Plants are highly competitive for water and nutrients. Besides California, rush skeletonweed is a state-listed noxious weed in Arizona (prohibited noxious weed), Colorado, Idaho, Montana (category 3), Nevada, Oregon (class B), and Washington (class B). It is also a significant problem in several other countries, particularly Australia. The skeletonweed gall midge (*Cystiphora schmidti*), skeletonweed gall mite (*Eriophyes chondrillae*), and rush skeletonweed rust (*Puccinia chondrillina*) have been introduced as biocontrol agents to help to control some infestations. All are established in California. Of the three agents, the rust fungus appears to be providing the best control in California. Rush skeletonweed is native to southern Europe.

SEEDLINGS: Cotyledons spatulate to oval. First leaves elliptic with backward-pointing teeth. Require a continuous moisture supply for up to 6 weeks to develop a persistent root system.

MATURE PLANT: Rosettes develop 1 or more flowering stems with numer-

Rush skeletonweed (*Chondrilla juncea*) plant. J. M. DiTomaso

Rush skeletonweed (*Chondrilla juncea*) flower head. J. M. DiTomaso

Rush skeletonweed (*Chondrilla juncea*) fruiting head. J. M. DiTomaso

Rush skeletonweed (*Chondrilla juncea*) immature foliage is very similar to dandelion and chicory. J. M. DiTomaso

Rush skeletonweed (*Chondrilla juncea*) seedling. J. M. DiTomaso

ous branches. *Lower stems typically have dense, bristly, downward-pointing hairs.* Upper stems glabrous or nearly glabrous. Rosette leaves oblanceolate, *irregularly shallow-lobed, with lateral lobes opposite one another and usually pointing backward toward the leaf base*, terminal lobe usually acute, 4–12 cm long, 1–5 cm wide, prostrate, usually glabrous or nearly glabrous, typically wither as flower stems develop, margins often purple-tinged. Stem leaves often lacking, bractlike when present, resemble reduced rosette leaves.

ROOTS AND UNDERGROUND STRUCTURES: Taproot slender, deep, persistent, with short lateral branches along the length. Taproots become somewhat woody with age and can penetrate soil to a depth of 2–3 m or more. Most lateral roots are short-lived, nonwoody, and less than 8 cm long, but a few lateral roots near the surface can become *rhizomelike* and grow laterally for 15–20 cm before turning downward. Adventitious buds near the top of the taproot and on major lateral roots can generate new rosettes. Roots are easily fragmented, and pieces as small as 1–2 cm can produce new rosettes from a depth to 1 m.

FLOWERS: May until flowering stems are killed by frost (fall or winter). Flower heads axillary or terminal, sessile or short-stalked, and solitary or in interrupted spikelike clusters of 2–5. Each flower head consists of *7–12 bright yellow ligulate flowers*. Flowers strap-shaped, with *5-lobed corollas 12–18 mm long*. Involucre cylindrical, with *phyllaries in 2 unequal rows*. Outer phyllaries much smaller than the inner phyllaries. Receptacles lack chaffy bracts.

FRUITS AND SEEDS: Achene body oblong, 3–4 mm long, *tapered at both ends*, lack hairs, pale to dark brown, *longitudinally ribbed*, with *acute tubercles near the apex* and *6 small scales surrounding the point of beak attachment at the apex*. Beak slender, 5–6 mm long, excluding the *pappus of many equal, fine, white bristles* about 5 mm long.

POSTSENESCENCE CHARACTERISTICS: *Persistent rigid stems with downward-pointing hairs, reddish tinge to the leaves,* and *clusters of old flower heads*, sometimes with a few seeds, distinguish rush skeletonweed from dandelion (*Taraxacum officinale* Wigg.) and Brassicaceous weeds such as mustards (*Brassica* spp.) and radish (*Raphanus* spp.).

HABITAT: Disturbed soils of roadsides, croplands (especially nonirrigated grain fields), semiarid pastures, rangelands, and residential properties. Grows best on well-drained sandy or gravelly soils in climates with cool winters and hot, relatively dry summers. Tolerates a wide variety of environmental conditions, including rainfall less than 25 cm to more than 120 cm, and cold winter climate. Large or dense populations are less common on heavy clay soils.

DISTRIBUTION: Uncommon. North Coast (ce and cw Mendocino Co.), Cascade Range (ne Shasta Co.), northern Sierra Nevada (s Plumas, Sierra, Nevada, Placer, El Dorado, and Calaveras Cos.), Central Valley (n Sacramento, e Yolo, and Fresno Cos.), San Francisco Bay region (sc Napa, n Santa Clara Cos.), South Coast Ranges (s Monterey, San Luis Obispo Cos.), and South Coast (Los Angeles Co.), to 600 m. Previous infestations now eradicated were in Tehama, Butte, Solano, San Mateo, Madera, Santa Barbara, and San Diego Counties. Idaho, Montana, Oregon, Washington, Georgia, and some northeastern states and a few nearby central states.

PROPAGATION/PHENOLOGY: Triploid. *Reproduces only by clones produced vegetatively from adventitious buds on roots* and *asexually by apomictic seed*. Seeds primarily disperse with wind, but also disperse with water, animals, and human activity. A large proportion of seeds initially exhibit high viablity, with about 90% germination the first year. However, seeds appear to be short-lived since viability often diminishes to about 2% by the third year. Sometimes seeds survive for less than 6 months. Newly matured seeds lack a dormancy period and can germinate within 24 hours under optimal conditions. Fresh seeds germinate without light and at a temperature range of 7°–40°C (optimum 15°–30°C).

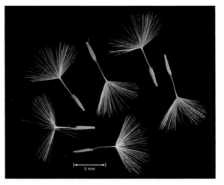

Rush skeletonweed (*Chondrilla juncea*) seeds.
J. O'Brien

Rush skeletonweed (*Chondrilla juncea*) hairs at base of stem point downward.   J. M. DiTomaso

Seed germination and new bud growth typically begin in fall after first rains in mild-winter areas or in early spring in colder climates. Seedling emergence is reduced in water-saturated or heavy clay soils and during drought conditions. First-year plants on deep, sandy soil sometimes produce viable seed earlier in the season than usual. Plants can develop from rosette to mature seed production within about 1 month. A temperature of at least 15°C for a period is necessary to induce flower production. Flower stems typically develop in early summer. One plant can produce 15,000–20,000 seeds per season.

MANAGEMENT FAVORING/DISCOURAGING SURVIVAL: Moderate soil disturbance, such as grain cultivation alternating with grazing on a yearly basis, can increase populations by dispersing rootstocks. Under moist conditions, shallow burial of seed by hooves of grazing livestock appears to promote seed germination. Increasing nutrient levels of poor soils appears to discourage survival by increasing competition from other vegetation.

SIMILAR SPECIES: Rosette leaves of rush skeletonweed and **dandelion** [*Taraxacum officinale* Wigg.] share the following characteristics: *glabrous leaves with lateral lobes in pairs and pointed backward* and *milky sap*. Unlike rush skeletonweed, dandelion has *flower stems that are unbranched, slightly succulent, leafless, hollow, and nonpersistent*, and *seeds without a crown of small scales at the apex*. In addition, dandelion is typically found in turf and gardens. Rosette leaves of **chicory** [*Cichorium intybus* L.] are similar to those of rush skeletonweed and dandelion. However, chicory has *lateral lobes of rosette leaves that point outward or forward*. In addition, lateral lobes are *not always opposite*, and basal leaves typically have *a few rough, coarse hairs*. Rush skeletonweed rosettes can also be separated from both these species by their reddish leaf coloration. Refer to the entries for **Dandelion** (p. 412) and **Chicory** (p. 268) for more information.

Chicory [*Cichorium intybus* L.][CICIN]

SYNONYMS: blue daisy; blue sailors; blueweed; bunk; coffeeweed; succory

GENERAL INFORMATION: *Biennial or perennial* to 1 m tall, with *milky sap*, a *basal rosette of lobed leaves*, and *blue dandelion-like flowers*. The red salad herb known as radicchio is a chicory cultivar. A green cultivar is cooked and eaten as a vegetable. Chicory root is used as a coffee additive and substitute. Chemical compounds extracted from the roots are used commercially to enhance the flavor of foods. Native to Europe.

SEEDLINGS: Cotyledons obovate to oval, 5–15(20) mm long, 3–5 mm wide, glabrous, base tapered and fused, tip truncate or slightly indented (emarginate). Leaves alternate. First few leaves oblanceolate, mostly 3–5 cm long, base long-tapered, margin weakly toothed.

MATURE PLANT: Foliage nearly glabrous to short-bristly-hairy. Basal leaves generally oblanceolate, mostly 8–25 cm long, 1–7 cm wide, margins toothed to deeply lobed. Lobe pairs opposite to alternate on the same plant, point forward, outward, or backward. Flowering stems erect, spreading-branched, wiry, hollow. Upper stem leaves few, sessile, reduced, margins entire or lobed, base lobed and clasping the stem.

ROOTS AND UNDERGROUND STRUCTURES: Taproot sometimes weakly woody, deep, often contorted, exudes milky juice when broken or cut. New leaves and stems grow from the crown.

Chicory (*Cichorium intybus*) plant.
J. M. DiTomaso

Chicory (*Cichorium intybus*) blue and white flower forms.
J. M. DiTomaso

Chicory

Chicory (*Cichorium intybus*) rosettes with previous year's stems. J. M. DiTomaso

Chicory (*Cichorium intybus*) typical blue flower head. J. M. DiTomaso

Chicory (*Cichorium intybus*) seeds. J. O'Brien

Chicory (*Cichorium intybus*) seedlings. R. Uva

Chicory

Blue lettuce (*Lactuca tatarica* ssp. *pulchella*) flower head. J. M. DiTomaso

Blue lettuce (*Lactuca tatarica* ssp. *pulchella*) flowering stem. J. M. DiTomaso

Blue lettuce (*Lactuca tatarica* ssp. *pulchella*) leaf. J. M. DiTomaso

FLOWERS: June–October. Individual plants flower over a long period of time. Heads sessile or nearly sessile, mostly 2.5–5 cm in diameter, consist only of ligulate flowers. *Flowers blue,* occasionally white or purplish. Phyllaries in 2 series, the outer row spreading and smaller than the inner row.

FRUITS AND SEEDS: Achenes wedge-shaped to nearly rectangular, ± 4–5-sided, 1.5–3 mm long. Pappus scales generally crownlike, less than 0.5 mm long.

POSTSENESCENCE CHARACTERISTICS: Dead flowering stems with remnants of flower heads can persist for a few months. A few achenes sometime remain within old heads.

HABITAT: Roadsides, fields, fence lines, pastures, perennial crops, orchards, waste places, urban sites, and low-maintenance turf. Grows best on calcium-rich soils and in poorly drained sites. Prefers moist areas and tolerates wet conditions.

DISTRIBUTION: Throughout California, except the deserts, uncommon in the interior regions of southern California, to 1500 m. All contiguous states.

PROPAGATION/PHENOLOGY: *Reproduces by seed.* Seeds fall near the parent plant or disperse to greater distances with water, mud, human activities, and possibly animals. One plant can produce thousands of seeds. Some seeds are hard-coated and can survive up to ~ 4 years submerged in water.

Chicory

MANAGEMENT FAVORING/DISCOURAGING SURVIVAL: Chicory does not tolerate cultivation.

SIMILAR SPECIES: Blue lettuce [*Lactuca tatarica* (L.) C. Meyer ssp. *pulchella* (Pursh) Stebb., synonym: *Lactuca pulchella* (Pursh) DC.][LACPU] is a *native perennial* to 1 m tall, with *long rhizomes* and blue flowers that resemble those of chicory. Blue lettuce is not usually considered a weed in California, but it is occasionally weedy elsewhere. Unlike chicory, blue lettuce has *flower heads on stalks* and *short-beaked achenes with a pappus of bristles*. In addition, flowering stems have many leaves. Lower leaves are linear to narrowly elliptic or lanceolate, with smooth or lobed margins. Blue lettuce inhabits sagebrush scrub, pinyon-juniper woodands, and dry to moist alluvial valleys in the Great Basin and Cascade Range, from about 1000 to 2000 m. It also occurs in all western states, most central states, and several northeastern states.

Dandelion [*Taraxacum officinale* Wigg.] rosettes resemble those of chicory, except dandelion rosette leaves usually have *leaf lobes or teeth in opposite pairs and that point backward*. Dandelion leaves sometimes have a few hairs on the midvein and lower surfaces, but *lack bristly hairs* that are rough to touch. Mature plants are easy to distinguish because dandelion does not have leafy stems. Refer to the entry for **Dandelion** (p. 412) for more information.

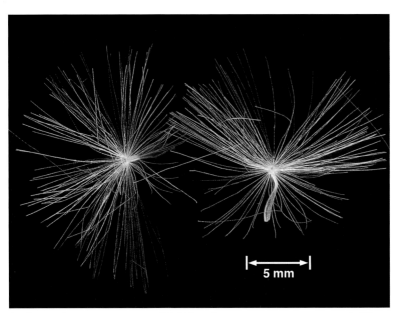

Blue lettuce (*Lactuca tatarica* ssp. *pulchella*) seeds.   J. O'BRIEN

Canada thistle [*Cirsium arvense* (L.) Scop.] [CIRAR] [Cal-IPC: Moderate][CDFA list: B]

Yellowspine thistle [*Cirsium ochrocentrum* A. Gray] [CIROH][CDFA list: A]

Wavyleaf thistle [*Cirsium undulatum* (Nutt.) Spreng.] [CIRUN][CDFA list: A]

Bull thistle [*Cirsium vulgare* (Savi) Ten.] [CIRVU] [Cal-IPC: Moderate] [CDFA list: C]

SYNONYMS: Canada thistle: corn thistle; creeping thistle; *Breea arvensis* (L.) Less.; *Carduus arvensis*(L.) Robson; *Cirsium incanum* (Gmel.) Fish; *Cirsium ochrolepideum* Juz.; *Cirsium setosum* (Willd.) Bess. ex Bieb.; *Cnicus arvensis* (L.) Hoffm.; *Serratula arvensis* L.; others

Yellowspine thistle: *Cnicus ochrocentrus* (Gray) Gray

Wavyleaf thistle: gray thistle; wavy-leaved thistle; *Carduus helleri* Small; *Carduus undulatus* Nutt.; *Cirsium helleri* (Small) Cory; *Cirsium megacephalum* (Gray) Cockerell; *Cirsium ochrocentrum* Gray var. *helleri* (Small) Petrak; *Cirsium undulatum* (Nutt.) Spreng. var. *undulatum* and var. *megacephalum* (Gray) Fern.; *Cnicus undulatus* (Nutt.) Gray; *Cnicus undulatus* Gray var. *megacephalus* Gray

Canada thistle (*Cirsium arvense*) population, male (left) and female (right). J. M. DiTomaso

Canada thistle • Yellowspine thistle • Wavyleaf thistle • Bull thistle

Bull thistle: bank thistle; bird thistle; black thistle; blue thistle; bur thistle; button thistle; common thistle; Fuller's thistle; plume thistle; roadside thistle; spear thistle; *Ascalea lanceolata* (L.) Hill; *Carduus lanceolatus* L.; *Carduus vulgaris* Savi; *Cirsium abyssinicum* Sch.Bip.ex A.Rich; *Cirsium lanceolatum* (L.) Scop.; *Cirsium lanceolatum* (L.) Scop. var. *hypoleucum* DC.; *Cnicus lanceolatus* (L.) Willd.; others

GENERAL INFORMATION: Erect plants with *prickly foliage* and flower heads of *purple, pink, or white disk flowers*. Refer to table 21 (p. 283) for a comparison of some important characteristics of *Cirsium* species.

Canada thistle: Clump- or patch-forming *perennial* to 1 m tall, with *extensive creeping roots* and *small unisexual flower head that lack prickles*. Plants are male or female (dioecious), and dense patches of a single sex often occur. Canada thistle is widespread in cool, temperate regions of Eurasia and North America. In addition to California, it is a state-listed noxious weed in 27 states, including Arizona (prohibited noxious weed), Colorado, Idaho, Montana (category 1), Nevada, New Mexico (class A), Oregon (class B), Utah, Washington (class C), and Wyoming. Some taxonomists recognize 4 varieties or biotypes that differ in growth habit, phenology, seed germination, and leaf characteristics. The Canada thistle stem weevil (*Ceutorhynchus litura*), bud weevil (*Larinus planus*), and thistle stem gall fly (*Urophora cardui*) have been released as biocontrol agents, but have not had significant detectable impacts. Currently, only the thistle

Canada thistle (*Cirsium arvense*) male flowers.
J. M. DiTomaso

Fruiting heads of Canada thistle (*Cirsium arvense*).
J. K. Clark

Canada thistle (*Cirsium arvense*) female flowers.
J. M. DiTomaso

stem fly has achieved limited establishment in California. The thistle is native to Europe.

Yellowspine thistle, wavyleaf thistle: Clump- or patch-forming *perennials* to 1 m tall. Yellowspine thistle often spreads vigorously from *deep, creeping roots* to form clonal patches, while wavyleaf thistle typically forms compact clumps from *short, thick roots* that spread only a short distance. These species hybridize, and plants with intermediate characteristics may occur where their ranges overlap. Native to the central states.

Bull thistle: Coarse *biennial*, sometimes annual or short-lived perennial, to about 2 m tall, with *stiff-hairy foliage* and conspicuously *prickly-winged stems and prickly flower heads*. Bull thistle is common throughout temperate and Mediterranean climate regions of the world. Regional biotypes vary primarily in life cycle patterns and seed dormancy and longevity. The bull thistle gall fly (*Urophora stylata*) was released in the Pacific Northwest as a biocontrol agent but has had little impact. It is not yet established in California. The thistle is native to Eurasia.

SEEDLINGS: Cotyledons oval to oblong, fused at the base, thick, dull, glabrous or slightly granular. First leaves alternate, elliptic to oblanceolate, tapered at the base into a winged stalk, about 2–4 times longer than cotyledons. Margin teeth terminate with a weak prickle. Subsequent few to several leaves typically resemble first leaves except increasingly larger.

Canada thistle: Cotyledons mostly 5–14 mm long, 3–6 mm wide, lower surface midvein shiny. First leaf margins slightly wavy to unevenly toothed. Surfaces covered with stiff hairs. Lower surfaces often sparsely covered with soft cobwebby hairs. Seedlings initially develop a deep taproot. Creeping roots develop in about 2–4 months. Seedlings sometimes initiate stems early and have poorly developed rosettes.

Bull thistle: Cotyledons mostly 7–20 mm long, 3–7 mm wide, paler on lower surface. First leaf margins unevenly toothed. Leaf surfaces, especially upper, *cov-*

Canada thistle (*Cirsium arvense*) young sprout from rhizome.　　　　　　　　　　　J. M. DiTomaso

Canada thistle (*Cirsium arvense*) rhizomes.
　　　　　　　　　　　J. K. Clark

Canada thistle • Yellowspine thistle • Wavyleaf thistle • Bull thistle

Canada thistle (*Cirsium arvense*) seeds.   J. O'BRIEN

ered with *long, stiff, papillae-based hairs*. Lower surfaces sometimes granular, initially cobwebby. Rosettes often reach a diameter of about 60 cm the first spring.

MATURE PLANT: Leaves variable, sessile or nearly sessile, toothed to lobed, sometimes with lobes toothed.

Canada thistle: Stems slender, *glabrous or nearly glabrous*, leafy, several to numerous from creeping roots. Leaf bases sometimes extend briefly down stem internodes as inconspicuous prickly wings to about 1 cm long. Rosette leaves few or lacking. All leaves oblong to lanceolate, mostly 5–20 cm long, upper surface *glabrous or nearly glabrous, green*, lower surface sometimes sparsely woolly, margin *nearly entire to shallow-lobed* and toothed, main prickles mostly 3–6 mm long.

Yellowspine thistle: *Stems and lower leaf surfaces densely covered with white woolly hairs*. Leaf bases extend down stems as prickly wings to about 2 cm long. Upper surfaces of all leaves *grayish, loosely covered with white woolly hairs*. Rosette leaves elliptic to oblanceolate, about 10–25 cm long, mostly *less than 3 cm wide*, margin *deeply coarse-lobed* and toothed, with yellowish main prickles (spinelike) mostly *3–12 mm long along the leaf margin*. Lobes often narrow, ruffled, crowded, and stiffly spreading. Stem leaves smaller and spinier, with prickles 5–15 mm long.

Wavyleaf thistle: Leaf bases extend down stems as inconspicuous prickly wings to about 1 cm long. Foliage resembles that of yellowspine thistle, except rosette leaves are often *more than 3 cm wide* and slightly longer (mostly 15–30 cm long), with margins are often conspicuously wavy or undulating. Rosette leaf margins moderate to *shallowly coarse-lobed* and toothed, with main prickles mostly *2–5 mm long*. Stem leaf prickles 2–10 mm long.

Canada thistle • Yellowspine thistle • Wavyleaf thistle • Bull thistle

Yellowspine thistle *(Cirsium ochrocentrum)* rosette. J. M. DiTomaso

Yellowspine thistle *(Cirsium ochrocentrum)* plant. J. M. DiTomaso

Yellowspine thistle *(Cirsium ochrocentrum)* flowering stem. J. M. DiTomaso

Yellowspine thistle *(Cirsium ochrocentrum)* seeds. J. O'Brien

Bull thistle: Stems coarse, usually single, branched, loosely covered with *white cobwebby hairs*, sometimes glandular. *Leaf bases extend nearly all the way down stem internodes as conspicuous prickly wings.* Rosette leaves elliptic to oblanceolate, mostly 10–40 cm long, margin moderate to deeply coarse-lobed and toothed, with main prickles mostly 5–15 mm long. Stem leaves smaller, more deeply lobed and spinier than rosette leaves. All leaves have upper surfaces *green, evenly covered with stiff, sharp-pointed, papillae-based hairs* (± 1 mm long), giving the leaf a sandpaper feel, sometimes sparsely cobwebby, lower surfaces variably covered with cobwebby hairs.

ROOTS AND UNDERGROUND STRUCTURES: Canada thistle: Root systems consist of an extensive network of vertical and creeping horizontal roots. Most roots occur in the top 45 cm of soil, but vertical roots 2–3 m deep are common. Horizontal roots may spread several meters in all directions. Individual roots can survive for up to 2 years. Roots are brittle and fragment easily. Some roots survive in frozen soil.

Yellowspine thistle: Taproot thick, deep, typically with many vigorously *creeping lateral roots* that produce new shoots from soil depths to 1.2 m. Small rosettes from creeping roots are often conspicuous in late summer.

Wavyleaf thistle: Taproot thick, deep, usually branched into *short-creeping*, often *tuberlike lateral roots* that can generate new shoots. Plants often develop into compact clonal clumps but are typically not widely spreading.

Bull thistle: Taproots thick, fleshy, to about 70 cm deep, often branched into several arms.

FLOWERS: Flower heads consist of several overlapping rows of spine-tipped phyllaries (Canada thistle phyllaries have tiny, relatively soft spines) and numerous disk flowers interspersed with bristles on the receptacle. Insect- and self-pollinated.

Canada thistle: June–October. Heads *unisexual*, numerous, often clustered, cylindrical or narrow ovoid to bell-shaped. Involucre 1–2(2.5) cm long, *0.5–2 cm in diameter*, often purplish, glabrous or with white woolly hair. Outer phyl-

Wavyleaf thistle (*Cirsium undulatum*) plant.
J. M. DiTomaso

Wavyleaf thistle (*Cirsium undulatum*) flowering stem.
J. M. DiTomaso

Wavyleaf thistle (*Cirsium undulatum*) rosette.
J. M. DiTomaso

Canada thistle • Yellowspine thistle • Wavyleaf thistle • Bull thistle

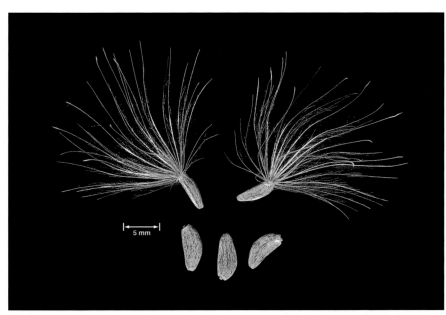

Wavyleaf thistle (*Cirsium undulatum*) seeds. J. O'Brien

laries ovate, appressed, often with spreading tips. Spines ± 1 mm long, fine, often dark. Corollas pink, purple, or white, male mostly 10–14 mm long with a shorter pappus, female mostly 14–20 mm long with a longer pappus. Corolla lobes 2–3 mm long.

Yellowspine thistle: April–August. Heads 1–few, ovoid to bell-shaped, initially loosely covered with white woolly hairs, later becoming nearly glabrous. Involucre mostly 2.5–4 cm long, 2–4 cm in diameter. Outer phyllaries ovate to lanceolate, *appressed* to stiffly erect, each with a conspicuous *white, often glandular midvein* and a *stout yellowish spine 5–12 mm long* that is *abruptly spreading to bent downward* (reflexed). Corollas white, pink, or pale purple, mostly 25–40 mm long, lobes mostly *9–14 mm long*.

Wavyleaf thistle: June–October. Heads resemble those of yellowspine thistle, except involucres are sometimes larger (to 5 cm long, 5 cm in diameter), and outer phyllary spines are *slender* and mostly *2–5 mm long*. Corollas white, pink, or pale to bright purple, mostly 25–50 mm long, with lobes about *7–10 mm long*.

Bull thistle: June–October. Heads 1–many, hemispheric to bell-shaped, usually loosely covered with cobwebby hairs. Each head has *at least 1 bractlike leaf just below*. Involucre (phyllaries as a unit) 2.5–4 cm long, 2–4 cm in diameter. Phyllaries lanceolate to linear, spreading to reflexed. Spines 1–5 mm long, sometimes yellowish. Corollas purple (rarely white), mostly 25–35 mm long, with lobes mostly 5–8 mm long.

Canada thistle • Yellowspine thistle • Wavyleaf thistle • Bull thistle

Bull thistle *(Cirsium vulgare)* infestation.
J. M. DiTomaso

Bull thistle *(Cirsium vulgare)* plant.
J. M. DiTomaso

Bull thistle *(Cirsium vulgare)* flowering stem.
J. M. DiTomaso

FRUITS AND SEEDS: Achenes ovate to elliptic, slightly compressed, often slightly curved or oblique at the apex, glabrous, glossy, with a slightly angled basal attachment scar and a short beak (± 0.5 mm long) surrounded by a collar at the apex. Pappus bristles *feathery* (plumose), nearly equal, *fused at the base into a ring*, readily *deciduous as a unit*.

Canada thistle: Achenes 2–4 mm long, ± 1 mm wide, tan. Pappus bristles mostly 12–20 mm long, tan.

Yellowspine thistle: Achenes *5–8 mm long*, ± 3 mm wide, pale orangish brown, sometimes with fine red-brown striations, slightly sticky with mucilage when moistened. Pappus bristles mostly 25–30 mm long, tan.

Wavyleaf thistle: Achenes *4–7 mm long*, ± 2.5 mm wide, pale straw-colored to tan, slightly sticky with mucilage when moistened. Pappus bristles mostly 25–40 mm long, white to pale tan.

Bull thistle: Achenes 3–5 mm long, ± 1 mm wide, gray or tan, sometimes with darker longitudinal striations. Pappus bristles mostly 15–30 mm long, tan to whitish.

POSTSENESCENCE CHARACTERISTICS: Flower stems of all four species typically senesce in fall, often with the onset of frosty nights. Old flower stems with flower head remnants usually persist for an extended period. Dead bull thistle stems can remain standing for 1–2 years.

Canada thistle • Yellowspine thistle • Wavyleaf thistle • Bull thistle

Bull thistle (*Cirsium vulgare*) rosette.   J. M. DiTomaso

HABITAT: Open disturbed sites, roadsides, crop fields, pastures, hillsides, rangeland, and forest openings. Thistles typically do not tolerate deep shade or constantly wet soils.

Canada thistle: Also agronomic and vegetable crops, stream banks, moist depressions, and gardens. Tolerates a wide range of soil types but grows best in moist soils.

Yellowspine thistle, wavyleaf thistle: Often inhabit shallow, sandy soils.

Bull thistle: Also agronomic crops, orchards, and recently logged and newly planted forestry sites. Grows best on heavy fertile soils.

DISTRIBUTION: Canada thistle: Scattered throughout California, except southern

Bull thistle (*Cirsium vulgare*) seedling.   Courtesy of Jim McHenry

Canada thistle • Yellowspine thistle • Wavyleaf thistle • Bull thistle

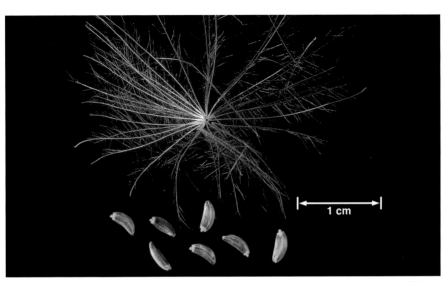

Bull thistle (*Cirsium vulgare*) seeds.   J. O'Brien

Sierra Nevada, Sonoran and Mojave deserts, and Channel Islands, to 1800 m. Most contiguous states, except many southeastern states and a few south-central states.

Yellowspine thistle: Uncommon, populations scattered. Cascade Range (ne and ce Siskiyou Co.), Modoc Plateau (Modoc, Lassen Cos.), eastern Sierra Nevada (se Plumas Co.), south-central Transverse Ranges (sc Ventura Co.), and Santa Rosa Island, to 1700 m. Arizona, Colorado, New Mexico, Utah, Wyoming, and South Dakota to Texas.

Peregrine thistle (*Cirsium cymosum*), a native nonweedy thistle similar to wavyleaf thistle (*Cirsium undulatum*).   J. M. DiTomaso

Peregrine thistle (*Cirsium cymosum*) flower head.   J. M. DiTomaso

Wavyleaf thistle: Uncommon, populations scattered. Cascade Range (ne Siskiyou, se Shasta Cos.), Modoc Plateau (e Modoc, c and ne Lassen, se Plumas, and e Sierra Cos.), San Francisco Bay region (sc Contra Costa Co.), southern South Coast Ranges (cw San Luis Obispo Co.), South Coast (se Los Angeles Co.), Santa Catalina Island, and Peninsular Ranges (sw San Diego Co.), to 1600 m. All western and central states.

Bull thistle: Common throughout California, except deserts; less common in arid hillsides of southern California, to 2300 m. All contiguous states.

PROPAGATION/PHENOLOGY: Most seeds fall near the parent plant or disperse short distances with wind, although a small proportion may move longer distances. Some seeds disperse to greater distances with human activities, water, soil movement, and as seed or hay contaminants. Birds and small mammals can consume and disperse some seeds. Seed dormancy at maturity is variable, depending on environmental conditions and biotype. Soil disturbance facilitates seed germination and seedling establishment. Seedlings typically emerge from soil depths to about 5 cm. Rosette foliage may be killed by a hard freeze and can regrow from roots in spring.

Canada thistle: *Reproduces vegetatively from creeping roots* and *by seed*. Root reserves are lowest when flowering begins in early summer. New roots and shoot buds develop in winter, and shoots emerge in spring. Subterranean shoots can develop roots and buds at any node. Root and subterranean shoot fragments about 1 cm long or more can develop into a new plant. Female plants only produce viable seed if male plants are within pollinator range, within 400 m in one study. Abundant seed typically develops only when the sexes are less than about 50 m apart. If pollinated, viable seeds develop in about 8–10 days after flowering. Male flowers occasionally have functional ovaries and a few seeds may develop. Most seeds germinate within 3 years of maturing, but deeply buried seeds can survive 10 years or more. Seeds generally germinate in mid- to late spring. Rosettes typically do not flower the first season. Plants flower only when day length is 14 hours or more, depending on biotype.

Yellowspine thistle, wavyleaf thistle: *Reproduce by seed* and *vegetatively from creeping roots*. The biology of these species is poorly understood.

Bull thistle: *Reproduces by seed*. Plants exist as rosettes until flowering stems develop at maturity. Seeds germinate in fall after the first rains or in spring. Fluctuating temperatures and moisture stimulate germination. Germination occurs under a wide temperature range (5°–40°C) and with or without light. Compared to other thistles, bull thistle seeds can germinate under low moisture conditions (to −0.75 MPa). First year rosettes usually persist through summer, but may die back during a dry summer and regrow in fall. Rosettes (typically second-year) require a cold period (vernalization) and the presence of sufficient soil nitrogen to initiate growth of flower stems. Thus, plants on very poor soil or in the shade of other plants may take more than 2 seasons to mature. Seed and flower head production is highly variable, depending on environmental conditions. Number of flower heads per plant ranges from 1 to over 400, and the number of seeds per flower head ranges from less than 100 to more than 400, with an average of about 200. Plants in grazed pastures often produce more seed than plants in adjacent ungrazed areas due to reduced competition from grazed plants. Most seeds either germinate within the first year or die, but seeds buried to about 15 cm or deeper may survive for up to 3 years or more.

MANAGEMENT FAVORING/DISCOURAGING SURVIVAL: Heavy grazing and disturbances that create bare soil patches facilitate seedling establishment and survival.

Perennial thistles: Repeated cultivation, mowing, or hand-cutting reduces and can eventually eliminate infestations. However, occasional cultivation may increase infestations by dispersing root fragments. Plantings that create dense shade may reduce infestations. Planting agricultural fields to alfalfa and mowing at least twice a year can significantly reduce and eventually eliminate Canada thistle.

Bull thistle: Cultivation, mowing, or hand-pulling just before flowering can control infestations. Cut flower heads can still develop viable seed.

SIMILAR SPECIES: There are numerous native thistle species in California, some of which are rare or endangered. Bull thistle is easily distinguished by having *stiff, sharp-pointed, papillae-based hairs on the upper surface of the leaves*, and Canada thistle is the only thistle with *unisexual flower heads*. However, several native species are difficult to distinguish from *yellowspine and wavyleaf thistles*, and plants should be positively identified before a control plan is implemented. Only yellowspine and wavyleaf thistles have *all* of the following characteristics: *perennial life cycle, short to long creeping roots, appressed to erect phyllaries* with *abruptly spreading to reflexed yellow spines* and *smooth margins*, outer phyllaries with a *conspicuous white midvein* that often has a *sticky glandular area*, and *corollas 29–50 mm long*.

Table 21. *Cirsium* species

Cirsium spp.	Roots	Stems	Upper leaf surfaces	Head diameter (cm)	Phyllary spine length (mm)	Other
C. arvense Canada thistle	creeping network	not winged; ± glabrous	± glabrous	0.5–2	± 1, very slender	unisexual flower heads
C. ochrocentrum yellowspine thistle	taproot and long creeping lateral roots	not winged; densely woolly	loosely woolly	2–4	5–12, stout	phyllary midvein white, often glandular; corolla lobes mostly 9–14 mm long
C. undulatum wavyleaf thistle	taproot and short creeping or tuber-like lateral roots	not winged; densely woolly	loosely woolly	2–5	2–5, slender	phyllary midvein white, often glandular; corolla lobes mostly 7–10 mm long
C. vulgare bull thistle	taproot	prickly-winged; cobwebby	stiff-bristly	2–4	1–5, moderate to slender	flower heads with 1 or more bractlike leaves immediately below

Hairy fleabane [*Conyza bonariensis* (L.) Cronq.][ERIBO]
Horseweed [*Conyza canadensis* (L.) Cronq.][ERICA]

SYNONYMS: Hairy fleabane: asthmaweed; flaxleaved fleabane; South American conyza; *Conyza ambigua* DC.; *Conyza hispida* Kunth; *Conyza linearis* DC.; *Conyza linifolia* Willd.; *Conyzella linifolia* (Willd.) Greene; *Erigeron bonariensis* L.; *Erigeron linifolium* Willd.; *Erigeron undulatus* Moench.; *Leptilon bonariense* (L.) Small; *Leptilon linifolium* (Willd.) Small; *Marsea bonariensis* (L.) V.M. Badillo; others

Horseweed: butterweed; Canada fleabane; Canadian horsetail; coltstail; mare's tail; pride weed; *Erigeron canadensis* L.; *Erigeron pusillus* Nutt.; *Leptilon canadense* (L.) Britton & A. Br.; *Marsea canadensis* (L.) V.M. Badillo

GENERAL INFORMATION: Hairy fleabane: Gray-hairy summer *annual* or *biennial* to 1.2 m tall, with erect stems that are usually branched near the base of the plant and *whitish flower heads that consist of disk flowers surrounded by female flowers that appear disklike*. Native to South America.

Horseweed: Native summer *annual* or *biennial to 2(–3) m tall*, typically with a *single erect main stem that branches only in the upper half* and *whitish flower heads that consist of disk flowers surrounded by ray flowers with small corollas*. Horseweed is a ruderal species that occurs nearly worldwide. Some botanists

Hairy fleabane (*Conyza bonariensis*) plant.
J. M. DiTomaso

Hairy fleabane (*Conyza bonariensis*) flowering and fruiting stem.
J. M. DiTomaso

recognized 3 varieties in the United States. Varieties *canadensis* and *glabrata* (Gray) Cronq. occur in California and most other western states.

SEEDLINGS: Seedlings of both species are similar. Cotyledons oval to ovate, about 2–3 mm long, ± 2 mm wide, glabrous or nearly glabrous, short-stalked. Leaves alternate. First leaf ovate to elliptic, about 5–12 mm long, 4–8 mm wide, surfaces sparsely or densely covered with hairs, on a hairy stalk shorter than the blade. Subsequent leaves ± elliptic, margins usually few-toothed.

MATURE PLANT: Leaves alternate, margins smooth to few-toothed or lobed. Lower leaves generally oblanceolate, taper to a short stalk. Upper leaves linear to lanceolate, sessile.

Hairy fleabane: *Main stems usually branch near the base.* Foliage gray-green, covered with short stiff and long soft hairs. Leaves mostly 1–8 cm long, to 0.8 cm wide.

Horseweed: *Main stem usually branches only in the upper half.* However, cutting of the main stem can lead to numerous basal branches. Foliage glabrous to densely covered with short stiff hairs. Leaves 1–10 cm long, to about 1 cm wide.

ROOTS AND UNDERGROUND STRUCTURES: Taprooted, with fibrous lateral roots.

Hairy fleabane (*Conyza bonariensis*) seedling.
J. M. DiTomaso

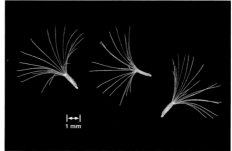

Hairy fleabane (*Conyza bonariensis*) flower heads. J. M. DiTomaso

Hairy fleabane (*Conyza bonariensis*) seeds.
J. O'Brien

Horseweed (*Conyza canadensis*) population near an agricultural field. J. M. DiTomaso

Horseweed (*Conyza canadensis*) flowering stem.
J. M. DiTomaso

Horseweed (*Conyza canadensis*) flowering and fruiting heads. J. M. DiTomaso

FLOWERS: June–September. Flower heads initially urn-shaped. Corollas cream-colored. Phyllaries in 2–3 slightly unequal rows, linear to narrowly lanceolate, margins ± membranous. Receptacles lack chaffy bracts.

Hairy fleabane: *Lateral stems often taller than the central stem. Heads 5–7 mm*

wide near the base, mostly 4–6 mm long, consist of about 10–20 *central disk flowers surrounded by numerous disklike female flowers* (disciform heads). Phyllaries mostly green or purplish, midveins sometimes brownish, *not conspicuously filled with resin*, densely hairy.

Horseweed: *Lateral stems shorter than the central stem. Heads 2.5–4 mm in diameter near the base*, usually 3–5 mm long, consists of about 7–13 central disk flowers surrounded by 20–40 ray flowers with ligules (petal-like part of corolla) ± 1 mm long. Phyllary midveins brown, conspicuously *filled with resin*, glabrous to hairy.

FRUITS AND SEEDS: Phyllaries reflex downward in fruit, inner surfaces whitish to dull brown. Achenes narrow, elliptic or oblong, ± 1.5 mm long, slightly flattened, minutely hairy.

Hairy fleabane: Pappus bristles mostly 3–4 mm long, brownish white, often reddish with age.

Horseweed: Pappus bristles mostly 2.5–3 mm long, dirty white.

POSTSENESCENCE CHARACTERISTICS: Dead stems retain flower heads with reflexed phyllaries and bare receptacles until winter.

HABITAT: Roadsides, agronomic crops, landscaped areas, orchards, vineyards,

Horseweed (*Conyza canadensis*) seedling.
J. K. Clark

Horseweed (*Conyza canadensis*) rosettes.
J. M. DiTomaso

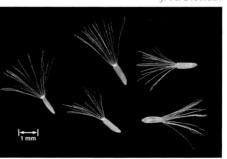

Horseweed (*Conyza canadensis*) seeds.   J. O'Brien

Coulter's conyza (*Conyza coulteri*) plant.
J. M. DiTomaso

Coulter's conyza (*Conyza coulteri*) immature plant.
J. M. DiTomaso

waste places, ditch and canal banks, and urban sites. Hairy fleabane typically inhabits dry places and is mostly an urban and agricultural weed of highly disturbed and often compacted soils (i.e., roadsides) in southern California. Horseweed has been designated a facultative wetland indicator species in California. In wet years it can be found in coastal sage scrub on hillsides.

DISTRIBUTION: Hairy fleabane: Common. Central Valley, Central-western region, and Southwestern region, to 1000 m. Arizona, Nevada, New Mexico, Oregon, Utah, and most southern states.

Horseweed: Common throughout California, to 2000 m. All contiguous states. Nearly worldwide.

PROPAGATION/PHENOLOGY: *Reproduce by seed.* Achenes disperse with wind, soil movement, water, and human activities. Plants exist as rosettes until flower stems develop at maturity. Seeds can germinate year-round under favorable conditions. Spring-germinating plants are annual. Late-summer and fall-germinating plants are usually biennial.

MANAGEMENT FAVORING/DISCOURAGING SURVIVAL: Manual removal or cultivation before seed develops can control both species.

SIMILAR SPECIES: Coulter's conyza [*Conyza coulteri* Gray, synonyms: conyza; *Conyza coulteri* Gray var. *virgata* (Benth.) Gray; *Laennecia coulteri* (Gray) Nesom; *Eschenbachia coulteri* (Gray) Rydb.] is a ruderal native *annual* to 2 m tall that usually resembles horseweed in habit, but has flower heads *without marginal ray flowers*. Unlike horseweed and hairy fleabane, Coulter's conyza has

Coulter's conyza (*Conyza coulteri*) foliage.
J. M. DiTomaso

Coulter's conyza (*Conyza coulteri*) seedling.
J. M. DiTomaso

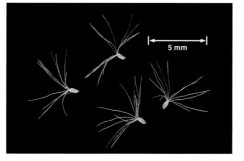

Coulter's conyza (*Conyza coulteri*) seeds.
J O'Brien

densely glandular-hairy foliage and phyllaries and toothed upper leaves that are *sessile* and *weakly clasping the stem*. Coulter's conyza is often associated with springs, stream margins, and seasonal pools that dry very late in the season. It can also inhabit disturbed places in the southern Sierra Nevada foothills, San Joaquin Valley, Central-western region, Southwestern region, White and Inyo mountains, and deserts, to 1000 m. It also occurs in Arizona, Colorado, Nevada, New Mexico, Texas, and Utah.

Tall fleabane [*Conyza floribunda* Kunth, synonyms: asthmaweed; *Conyza bonariensis* (L.) Cronq. var. *leiotheca* (Blake) Cuatrec; *Erigeron floribundus* (Kunth) Schultz-Bip.; *Erigeron bonariensis* L. var. *floribundum* (Kunth) Cuatrec] [ERIFL] is a summer *annual* or *biennial* to 2 m tall that closely resembles hairy fleabane. Like hairy fleabane, tall fleabane has *marginal disklike female flowers that lack a petal-like corolla*. Unlike hairy fleabane, tall fleabane has *smaller flower heads, lateral stems that are often shorter than the main stem*, and dried phyllaries that are usually reddish brown on the inner surface. Tall fleabane inhabits disturbed places in the Northwestern region, Sacramento Valley, Southwestern region, and possibly elsewhere, to 100 m. It appears to be rapidly expanding its range on the coastal slopes of southern California. Tall fleabane also occurs in Alabama, Georgia, and Mississippi. Some taxonomists believe hairy fleabane and tall fleabane are different biotypes of the same species and refer to tall fleabane as

Conyza bonariensis (L.) Cronq. var. *leiotheca* (Blake) Cuatrec. Native to tropical America.

Tall fleabane (*Conyza floribunda*) plant.
J. M. DiTomaso

Tall fleabane (*Conyza floribunda*) flowering stem.
J. M. DiTomaso

Tall fleabane (*Conyza floribunda*) seedling.
J. M. DiTomaso

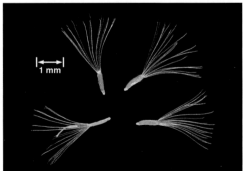

Tall fleabane (*Conyza floribunda*) seeds.
J. O'Brien

Southern brassbuttons [*Cotula australis* (Sieber) Hook. f.][CULAU]

Brassbuttons [*Cotula coronopifolia* L.][CULCO] [Cal-IPC: Limited]

SYNONYMS: Southern brassbuttons: Australian waterbuttons; carrot weed; small brassbuttons; soldier's button; *Anacyclus australis* Sieber ex Spreng.; *Lancisia australis* (Sieber ex Spreng.) Rydb.; *Soliva tenella* A. Cunn.; *Strongylosperma australe* (Sieber ex Spreng.) Less.

Brassbuttons: common brassbuttons; *Lancisia coronopifolia* (L.) Rydb.

GENERAL INFORMATION: Both species have *aromatic foliage* and *buttonlike flower heads*.

Southern brassbuttons: Low, spreading *annual* to 0.2 m tall. Native to Australia.

Brassbuttons: Emersed to terrestrial *perennial* to about 0.5 m tall. Native to South Africa.

SEEDLINGS: Southern brassbuttons: Cotyledons linear. First leaves appear opposite, pinnate-dissected into 3 linear lobes.

Brassbuttons: Cotyledons oblong, base tapered and ± sheathing, tip rounded, about 1–2 mm long, glabrous. First leaves appear opposite, ± sessile, glabrous, narrowly oblong, base tapered, tip rounded, about 4–7 mm long.

MATURE PLANT: Leaves alternate.

Southern brassbuttons *(Cotula australis)* plant.   J. M. DiTomaso

Southern brassbuttons (*Cotula australis*) flowering stem.   J. M. DiTomaso

Southern brassbuttons (*Cotula australis*) seeds.   J. O'Brien

Southern brassbuttons: Foliage *sparsely covered with soft spreading hairs*. Stems slender, spreading. Leaves *pinnate-dissected 2–3 times*, 2–6 cm long, stalked or sessile above.

Brassbuttons: Foliage *glabrous, slightly fleshy*, light or yellowish green. Stems thick, ± spreading. Leaves sessile, *linear to lanceolate or oblong*, sheathing stem at the base, mostly 2–7 cm long. *Margins smooth to coarsely toothed or lobed.*

ROOTS AND UNDERGROUND STRUCTURES: Southern brassbuttons: Plants often have a weak taproot and numerous fibrous roots, but under certain conditions, especially in turf, lower stems sometimes root at the nodes.

Brassbuttons: Roots generally fibrous. Stems usually creep by rooting at the lower nodes.

FLOWERS: Heads *solitary on long stalks*, appear to consist only of disk flowers, but actually *consist of bisexual disk flowers at the center and pistillate flowers without corollas* (disciform) *around the margin*. Flowers lack a pappus. *Pistillate flowers have a stalk below the ovary* (immature achene) *that is as long as the ovary*. Disk flower corollas 4-lobed. *Receptacle flat, lacks chaff or hair*. Stamens 4.

Southern brassbuttons: January–May. *Heads pale yellow to white, 3–6 mm in diameter*. Phyllaries in 2 series.

Brassbuttons: March–December. *Heads bright yellow, 5–15 mm in diameter*. Phyllaries in 2–3 series.

FRUITS AND SEEDS: Achenes flattened, of 2 types. Pistillate achenes ovate, *broadly winged, stalked*. Disk achenes barely winged, nearly sessile.

Southern brassbuttons: Achenes minutely glandular hairy on both sides. Pistillate achenes about 1 mm long, on a stalk slightly longer than the achene. Disk achenes oblong, less than 1 mm long.

Brassbuttons: Achenes smooth on the outer face, covered with minute papillae on the inner face. Pistillate achenes 1.5–2 mm long, *on a stalk about as long as the achene*. Disk achenes oblanceolate, 1–1.5 mm long.

HABITAT: Southern brassbuttons: Turf, landscaped areas, gardens, yards, orchards, vineyards, annual crops, and waste places, especially in coastal areas.

Brassbuttons: Freshwater and salt marshes, wetlands, vernal pools, ditches, and seasonally wet places in many plant communities, occasionally turf. Does not tolerate significant frost.

DISTRIBUTION: Southern brassbuttons: North, Central, and South Coast, to 250 m. Oregon, Arizona, Texas, Florida, and Maine.

Brassbuttons: North, Central, and South Coast, San Francisco Bay region, Central Valley, and South Coast Ranges, to 300 m. Oregon, Washington, Arizona, and Nevada.

Brassbuttons (*Cotula coronopifolia*) along stream side. J. M. DiTomaso

Brassbuttons (*Cotula coronopifolia*) stem and flower heads. J. M. DiTomaso

Brassbuttons (*Cotula coronopifolia*) flower head. J. M. DiTomaso

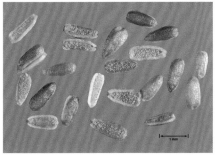

Brassbuttons (*Cotula coronopifolia*) achenes from disk flowers. J. O'Brien

Mexican brassbuttons (*Cotula mexicana*) infestation in turf.   Robin Breckenridge

Mexican brassbuttons (*Cotula mexicana*) plants.   J. M. DiTomaso

PROPAGATION/PHENOLOGY: *Reproduce by seed.* Seeds fall near the parent plant or disperse with water and mud. Most seeds germinate after the first fall rains through winter. Seeds survive about 1–2 years under field conditions. Brassbuttons also *reproduces vegetatively from stems that root at the nodes.*

SIMILAR SPECIES: Mexican brassbuttons [*Cotula mexicana* (DC.) Cabrera, synonyms: *Cotula pygmaea* (H.B.K.) Benth. & Hook. f. ex Hemsley; *Cotula minuta* (L.f.) Schinz; *Soliva mexicana* DC.; *Hippia minuta* L.f.] is a low-growing *perennial* with sparsely hairy foliage and *pinnate-dissected leaves*. Currently, it is not included in California floras. Unlike brassbuttons and southern brassbuttons, Mexican brassbuttons has pale flower heads about *2 mm in diameter,* hairy *receptacles, 3-lobed disk corollas, 3 stamens per disk flower*, and *achenes of 1 type*. Mexican brassbuttons primarily inhabits turf in the San Francisco Bay region, Sacramento Valley, and Cascade Range (Mt. Shasta area) and is expanding its range. It is difficult to control in turf and has the potential to become a serious weed. Mexican brassbuttons typically roots more deeply than turf grasses, thrives with low clipping regimes, and can exclude turf on golf courses, but it competes poorly with taller vegetation. Native to southern Mexico, Central America, and northern South America, where it grows in wet subalpine meadows and riparian areas at 3000 to 4000 m.

Mexican brassbuttons (*Cotula mexicana*) seeds.   J. O'Brien

Smooth hawksbeard [*Crepis capillaris* (L.) Wallr.] [CVPCA]

SYNONYMS: *Crepis virens* L.; *Lapsana capillaris* L.

GENERAL INFORMATION: Erect winter *annual*, sometimes summer annual or *biennial*, to about 1 m tall, with *milky sap* and *small, yellow, dandelion-like flower heads*. Plants exist as rosettes until flower stems develop at maturity. Smooth hawksbeard typically colonizes open, dry disturbed sites. Native to Europe.

SEEDLINGS: Cotyledons ovate, ± 3 mm long, 2–3 mm wide, glabrous. Leaves alternate. First leaf elliptic, about 8–15 mm long, 4–5 mm wide, margins sparsely toothed, lower surface often purplish, surfaces ± pubescent, especially lower midvein. Leaf stalk ± glandular-hairy.

MATURE PLANT: Stems branched in the upper portion. Foliage nearly glabrous to minutely hairy, especially on lower leaf veins. Stem leaves alternate. Rosette and lower leaves ± oblong, 3–30 cm long, margins sparsely *toothed to pinnate-lobed*. Upper leaves few, ± bractlike.

ROOTS AND UNDERGROUND STRUCTURES: Taproot short.

FLOWERS: June–September. Flower heads consist of 20–60 *yellow ligulate flowers*. Involucre (phyllaries as a unit) cylindrical, mostly 5–8(10) mm long, ± minutely glandular-hairy. Outer phyllaries 8, linear, much shorter than the inner phyllaries. Inner phyllaries 8–16, lanceolate. All phyllary margins membranous. Receptacles lack chaffy bracts.

FRUITS AND SEEDS: Achenes *cylindrical*, 1.5–2.5 mm long, taper at both ends,

Smooth hawksbeard (*Crepis capillaris*) plant. J. M. DiTomaso

Smooth hawksbeard (*Crepis capillaris*) seedling.
J. M. DiTomaso

Smooth hawksbeard (*Crepis capillaris*) flower heads.
J. M. DiTomaso

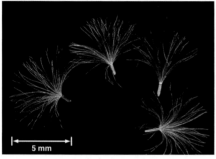

Smooth hawksbeard (*Crepis capillaris*) seeds.
J. O'Brien

Smooth hawksbeard (*Crepis capillaris*) leaves.
J. M. DiTomaso

lack a distinct beak, longitudinally 10-ribbed, brown. Pappus bristles fine, soft, about 3–4 mm long, white.

POSTSENESCENCE CHARACTERISTICS: Old flower stems with reflexed phyllaries and empty receptacles can persist into winter.

HABITAT: Open dry disturbed sites, fields, orchards, pastures, roadsides, turf, and waste places.

DISTRIBUTION: Northwestern region, Cascade Range, northern and central Sierra Nevada, San Francisco Bay region, and South Coast, to 1300 m. Northern half of the United States, except a few central states.

PROPAGATION/PHENOLOGY: *Reproduces by seed.* Seeds disperse primarily with wind, but also with water, soil movement, mud, and human activities. Newly dispersed seeds can germinate immediately if conditions are favorable.

MANAGEMENT FAVORING/DISCOURAGING SURVIVAL: Manual removal, mowing, or cultivation before seed develops can control smooth hawksbeard.

SIMILAR SPECIES: Italian hawksbeard [*Crepis bursifolia* L.] is an uncommon, short-lived, taprooted *perennial* to 0.4 m tall, with lower leaves that are pinnate-lobed and have toothed margins. Lateral lobes are lanceolate, and the terminal lobe is ovate. Similar to bristly hawksbeard, the phyllaries of Italian hawksbeard have a few yellowish bristly hairs. Italian hawksbeard is distinguished primarily by having *achenes with a very slender beak that is about 2 times the length of the body*. Italian hawksbeard generally inhabits turf in the San Francisco Bay region and possibly elsewhere, to 100 m. Native to southern Europe.

Bristly hawksbeard [*Crepis setosa* Haller f., synonym: rough hawksbeard] [CVPSE] is an uncommon *annual* to 0.8 m tall, with lower leaves that are pinnate-lobed near the base and sparsely toothed in the upper portion. Foliage is generally covered with fine, spreading hairs, but unlike smooth hawksbeard, bristly hawksbeard *stems and phyllaries are also evenly covered with yellowish bristly hairs*. In addition, achenes have a *slender beak that is slightly less than or equal to the length of the body*. Flower heads are mostly 8–12 mm long. Bristly

Italian hawksbeard (*Crepis bursifolia*) flowering and fruiting stem. J. M. DiTomaso

Italian hawksbeard (*Crepis bursifolia*) leaves.
J. M. DiTomaso

Italian hawksbeard (*Crepis bursifolia*) seeds.
J. O'Brien

Bristly hawksbeard (*Crepis setosa*) plant.
J. M. DiTomaso

Bristly hawksbeard (*Crepis setosa*) flowering stem.
J. M. DiTomaso

Bristly hawksbeard (*Crepis setosa*) seedling.
J. M. DiTomaso

Bristly hawksbeard (*Crepis setosa*) seeds. Note that pappus is not present.
J. O'Brien

hawksbeard inhabits disturbed sites in the western North Coast Ranges, to 300 m. It also occurs in Oregon, Texas, and a few scattered eastern states. Native to Europe.

European hawksbeard [*Crepis vesicaria* L. ssp. *taraxacifolia* (Thuill.) Thell.] [CVPVT] is an uncommon *annual or biennial* to 0.8 m tall with achenes that have *a slender beak that is slightly less than or equal to the length of the body*, similar to those of bristly hawksbeard. Lower leaves are pinnate-lobed with a toothed terminal lobe. European hawkbeard is distinguished by having stems and phyllaries that are *sparsely covered with woolly hairs* and have a *few blackish bristles*. European hawksbeard occurs in the western North Coast Ranges and San Francisco Bay region, to 300 m. It also occurs in Oregon and a few eastern states. Native to Europe.

Common crupina [*Crupina vulgaris* Cass.][CJNVU]
[Cal-IPC: Limited][CDFA List: A][Federal Noxious Weed]

SYNONYMS: bearded creeper; *Centaurea crupina* L.; *Serratula crupina* (L.) Vill.

GENERAL INFORMATION: Erect cool-season *annual* to 0.6(1) m tall at maturity, with purple flower heads that appear to consist only of disk flowers. Plants exist as basal rosettes until flower stems develop in spring. Rosette leaves whither as flowering commences in late spring or early summer. Common crupina is primarily a noxious rangeland weed. Plants adapt to many environmental conditions and are highly competitive for water and nutrients. Dense populations that displace desirable forage species can develop on large tracts of poor rangeland. Currently, common crupina is most problematic in Idaho. Besides California, it is a state-listed noxious weed in Oregon (class A), Washington (class A), Idaho, Montana (category 3), South Dakota, North Carolina, and Florida. Native to southern Europe.

SEEDLINGS: Cotyledons oblong, 10–30 mm long, 6–16 mm wide, fleshy, glabrous, upper surface slightly glossy, 3-veined from base. Midvein slightly raised, often purplish red. Lower surfaces paler than upper surfaces. First and second leaves alternate, but initiated close together so as to appear opposite, lanceolate with serrate margins, about 25–40 mm long, 6–15 mm wide. Margins and lower midvein covered with short ± papilla-based glandular hairs. Third and subsequent leaves deeply pinnate-lobed with serrate, glandular-hairy margins.

MATURE PLANT: Flower stems openly branched, longitudinally ridged. Rosette leaves sessile or petioled, to 8 cm long, oblong to obovate in outline, *deeply pinnate-divided, divisions narrow and opposite*, covered with short, stiff hairs.

Common crupina (*Crupina vulgaris*) flowering stem. J. M. DiTomaso

Common crupina

Common crupina (*Crupina vulgaris*) habitat.
J. M. DiTomaso

Common crupina (*Crupina vulgaris*) flower head. J. M. DiTomaso

Common crupina (*Crupina vulgaris*) leaves and stem. J. M. DiTomaso

Stem leaves *alternate, deeply pinnate-lobed once or twice, lobes narrow*, reduced near stem tops. Rosette and stem *leaf margins appear spiny-toothed with stiff hairs that are barbed at the tip* (glochidiate).

ROOTS AND UNDERGROUND STRUCTURES: Dense, fibrous roots.

FLOWERS: May, until soil moisture is depleted. Flower heads *cylindrical to ovoid or slender urn-shaped*, on stalks 1–3 cm long, consist of *1–2 fertile disk flowers in the center* and *2–4 sterile disklike flowers around the margin*, both types with a *pappus of bristles*. Corollas slender, purple, ± 14 mm long, tubular with linear lobes. Phyllaries lanceolate, in *several unequal overlapping series*, as a unit, 8–20 mm long. Receptacle with *flat, scalelike chaffy bracts*. Insect-pollinated. Primarily outcrossing, but self-compatible. Self-pollinates when pollinators are scarce.

FRUITS AND SEEDS: Fertile achene *nearly cylindrical*, 3–6 mm long, 1.5–3.5 mm wide, *base rounded*, apex ± flat-topped, black-brown, ± covered with minute, lustrous golden-brown scales that rub off easily. *Pappus bristles black-brown, glossy, stiff, radiate outward at a 90° angle from the achene axis in 2 very unequal series*. Outer bristles inconspicuous, scalelike, flattened, to 1 mm long. Inner bristles stiff, minutely barbed, 5–10 mm long.

POSTSENESCENCE CHARACTERISTICS: Old flower heads that remain on dead stems can aid in species identification. Refer to table 19 (p. 244) for a comparison of purple-flowered knapweeds, starthistles, and common crupina.

HABITAT: Disturbed grassland, pastures, rangeland, forested areas, canyons, riparian areas, roadsides, and waste places. Adapts to many moisture conditions, temperature regimes, and soil types.

DISTRIBUTION: Uncommon. Modoc Plateau (sc Modoc Co.) and southern North Coast Ranges (ce Sonoma Co.), currently to 250 m in California, to 1000 m in other western states. Idaho, Oregon, Washington, and Massachusetts.

PROPAGATION/PHENOLOGY: *Reproduces by seed.* Flower heads often disperse as units. Achenes fall near the parent plant or disperse to greater distances with rodents, livestock, game birds, human activity, and by floating on water. Seeds typically do not disperse beyond 1.5 m from the parent plant with wind or 90 m on the hooves of livestock. Plants produce 1(2) achenes per flower head and from about 2 to 850 achenes per plant. About 85% of seeds produced germinate the following fall. Seeds germinate under a wide range of temperatures. Most germination occurs after the first significant rains of fall or early winter, but germination can continue throughout the rainy season. Seeds can survive ingestion by most animals, except sheep, and can remain viable for up to about 3 years under field conditions. Plants are typically grazed only when more palatable forage is unavailable.

MANAGEMENT FAVORING/DISCOURAGING SURVIVAL: Quarantining cattle and horses for at least 6 days after foraging on rangeland infested with flowering plants can help prevent the introduction of common crupina to noninfested sites. Grazing by deer or livestock, or cutting common crupina, can dramatically increase flower and seed production by stimulating lateral branching.

SIMILAR SPECIES: Several knapweeds in the genus *Centaurea* and **Russian knapweed** [*Acroptilon repens* (L.) DC.] superficially resemble common crupina. Unlike common crupina, the knapweeds have flower head receptacles with *bristles* and leaf margins that *lack* bristly, barb-tipped hairs. Phyllary and achene characteristics are important in distinguishing these species. Refer to table 19 (p. 244) for a comparison of these species.

Common crupina (*Crupina vulgaris*) seedling.
J. M. DiTomaso

Common crupina (*Crupina vulgaris*) seeds.
J. M. DiTomaso

Artichoke thistle [*Cynara cardunculus* L.]
[Cal-IPC: Moderate][CDFA list: B]

SYNONYMS: cardoon; desert artichoke; wild artichoke

GENERAL INFORMATION: Large *perennial* to 2.5 m tall, with coarse, spiny, deeply pinnate-lobed leaves and large showy flower heads of purple disk flowers. Artichoke thistle usually invades disturbed grassland primarily in coastal regions. Dense colonies displace desirable vegetation and wildlife and can exclude livestock. Artichoke thistle is a progenitor of the commercially cultivated spineless **globe artichoke** [*Cynara scolymus* L.]. Some taxonomists consider globe artichoke and artichoke thistle to be the same species, *C. cardunculus* L. The two species readily hybridize. Also, a few spiny wild types often develop among globe artichoke seedlings. Artichoke thistle is sometimes cultivated as an ornamental and for its leaf stalks, which are consumed as a vegetable. It is available commercially. The artichoke fly *(Terellia fuscicornis)* was accidentally introduced into California, but it is not a CDFA approved biocontrol agent. Preliminary studies suggest that some native thistles (*Cirsium* spp.) may be vulnerable to attack by the fly. The fly's impact on artichoke thistle populations is unknown. Because larvae feed only on mature flower heads, commercial artichokes are not significantly affected since they are harvested while immature. The thistle is native to the Mediterranean region.

SEEDLINGS: Cotyledons obovate, 3–5 cm long, bases gradually long-tapered, tips rounded, glabrous or with scattered short woolly hairs. First and subsequent few leaves elliptic, mostly 3–20 cm long, tapered to a long stalk, margins

Artichoke thistle (*Cynara cardunculus*) flower head. J. M. DiTomaso

often weakly toothed, teeth tipped with a fine yellowish spine mostly 0.5–3 mm long, sparse to moderately covered with short white woolly hairs. Seedlings develop a deep taproot during the first year. Rosette leaves often die during the first summer and regrow when rains commence in fall.

MATURE PLANT: Stems erect, thick, branched near the top, ribbed. Basal leaves *pinnate-lobed or pinnate-divided 1–2 times, often appear ± compound, up to 1.5 m long, lobes tipped with stiff, yellowish to pale orange spines 0.5–2 cm long, similar spines 0–3 at each side of the lobe base,* upper surface loosely covered with white to gray woolly hairs, lower surface densely covered with white to gray woolly hairs. Stem leaves alternate, resemble basal leaves except smaller, bases extend down the stem about 1–3 cm as a spiny wing (decurrent). Cultivated types may lack or have weak spines.

ROOTS AND UNDERGROUND STRUCTURES: Tapoot thick, fleshy, can penetrate soil to depths of 2 m. Root fragments can generate new shoots.

FLOWERS: April–July. Heads solitary at stem tips, ovoid to hemispheric, *3–15 cm in diameter,* consist of numerous blue-violet to purple or rarely white disk flowers ± 5–8 cm long. Phyllaries *ovate, overlapping in several series of 12–16, tapered to a stout point.* Receptacle *fleshy, covered with bristles 20–45 mm long.* Insect-pollinated.

FRUITS AND SEEDS: Achenes conical to cylindrical, slightly compressed to

Artichoke thistle *(Cynara cardunculus)* plant.
J. M. DiTomaso

Artichoke thistle *(Cynara cardunculus)* seedling.
J. M. DiTomaso

Artichoke thistle *(Cynara cardunculus)* seeds.
J. O'Brien

± 4-angled, 6–8 mm long, with an attachment scar at the base, glabrous, dark brown to tan, sometimes with black, brown, or dark green longitudinal striations. Pappus bristles *feathery* (plumose), *25–40 mm long*, fused into a ring at the base, tan, attached slightly off-center, deciduous.

POSTSENESCENCE CHARACTERISTICS: Stems typically die after flowering and can remain standing for several months to more than a year. Old flower heads with a few achenes sometimes persist on the stems.

HABITAT: Disturbed, open sites in grassland, pasture, chaparral, coastal sage scrub, riparian areas, abandoned agricultural fields. Often associated with areas impacted by historic or recent overgrazing. Grows best on deep clay soils. Does not tolerate heavy shade.

DISTRIBUTION: Throughout low elevation areas of California, except Great Basin and desert regions, primarily near the coast in southern California, to 500 m.

PROPAGATION/PHENOLOGY: *Reproduces primarily by seed* and less frequently by root fragments usually resulting from mechanical disturbance. Most achenes fall near the parent plant or disperse up to about 20 m with wind. Some achenes disperse to greater distances with water, mud, soil movement, animals, and human activities. Most seeds germinate after the first rain in fall, but some germination can occur year-round under favorable conditions. Field observations suggest that most seeds survive about 5 years under field conditions. One-year-old plants sometimes flower, but most plants do not flower until their second year. Individual plants often live for many years.

MANAGEMENT FAVORING/DISCOURAGING SURVIVAL: For manual removal, a large portion of the taproot must be removed, otherwise the remaining root will generate new shoots. Burning does not kill the taproots of plants. Cutting flower stems before maturity or browsing by goats can reduce seed production. On agricultural land, repeated cultivation can eventually eliminate troublesome populations.

SIMILAR SPECIES: Other weedy thistles have smaller, less-deeply-lobed leaves and generally lack a fleshy receptacle. In addition, all other thistles except *Cirsium* species have a pappus of scales or minutely barbed bristles.

Cape-ivy [*Delairea odorata* Lem.]
[SENMI][Cal-IPC: High][CDFA list: C]

SYNONYMS: German-ivy; Italian ivy; ivy groundsel; parlor ivy; water ivy; *Senecio mikanioides* Otto ex Walp.

GENERAL INFORMATION: Vigorous *perennial vine*, with stems to about 9 m long and *palmate-lobed leaves*. Cape-ivy can invade various plant communities, but it is especially noxious in coastal riparian areas. Vines grow over trees and shrubs and can form dense mats that smother underlying vegetation. Such problematic infestations also reduce native species richness and seedling recruitment in the community. Cape-ivy contains pyrrolizidine alkaloids and can be *toxic* to animals when ingested. However, toxicity problems due to Cape-ivy ingestion are rare. Plant material in contact with water or dissolved extract above a threshold concentration may cause fish kill. Cape-ivy was introduced to the United States in the late 1800s as a house plant. It is also considered an invasive weed problem in Australia. Native to the moist mountain forests of South Africa.

SEEDLINGS: Cotyledons oblong, 3–10 mm long, 1–3 mm wide, glabrous, lower surface and stalk below (hypocotyl) often reddish purple. First leaf ovate, about 4–8 mm long, on a stalk about as long, margins usually smooth, sometimes weakly 3-lobed, ± sparsely ciliate. Subsequent leaves broadly ovate, 8–12 mm long, 7–10 mm wide, margin usually weakly 5-lobed, on stalks ± 5 mm long.

MATURE PLANT: Foliage *glabrous*, evergreen in mild climates, leaves and stems deciduous elsewhere. Stems sometimes purplish. Leaves alternate,

Cape-ivy (*Delairea odorata*) flower clusters.   J. M. DiTomaso

nearly triangular to kidney-shaped, palmately 5–9-lobed, mostly 2–8 cm long, often slightly broader than long, thin, spongy, light green, sometimes purplish or glossy. Lobes *shallow, pointed*. Stalks about equal to or slightly longer than leaf length. Some or all plants of certain populations have a pair of kidney-shaped stipules at the base of the leaf stalk.

ROOTS AND UNDERGROUND STRUCTURES: Stolons ± lustrous, covered by a thick cuticle. Rhizomes often purple-mottled, thick, contain much carbohydrate. Stolons and rhizomes root at the nodes. Small fragments that include a node can regenerate into a new plant under favorable conditions.

FLOWERS: September–March, depending on local and environmental conditions. Flower head clusters in corymbs, often panicle-like on short branches in upper leaf axils. Heads consist of about 20–40 yellow disk flowers. Phyllaries ± 8, mostly 3–4 mm long, tips green. Self-incompatible. Insect-pollinated.

FRUITS AND SEEDS: Achenes glabrous to slightly hairy, cylindrical, with a pappus of minutely barbed, deciduous bristles.

HABITAT: Disturbed riparian sites, seasonal wetlands, coastal bluffs and scrub, moist canyons, oak woodlands, and coastal grassland, as well as Monterey or Bishop pine, eucalyptus, and redwood forests. Most infestations are associated with urban areas or former human habitations. Grows in deep shade or under

Cape-ivy (*Delairea odorata*) habitat.
J. M. DiTomaso

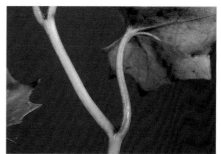

Cape-ivy (*Delairea odorata*) biotype lacking stipules.
J. M. DiTomaso

Cape-ivy (*Delairea odorata*) biotype with stipules.
J. M. DiTomaso

Cape-ivy *(Delairea odorata)* flower heads and leaves.   J. M. DiTomaso

cloudy conditions, but does not tolerate full sunlight. Tolerates serpentine soils. Established plants can tolerate drought.

DISTRIBUTION: Can be found in cultivation in the Central Valley and other areas where shade and summer water are provided. Invasive in the North, Central, and South Coast, San Francisco Bay region, and San Gabriel Mountains, especially the south slope, to about 500 m. Southwestern Oregon and Hawaii.

PROPAGATION/PHENOLOGY: *Reproduces vegetatively from rhizomes, stolons, and fragments of rhizomes, stolons, and stems,* and in some locations, by seed. Stem fragments as small as 2.5 cm that include a node can generate a new plant. Even small fragments of dying brown stems can resprout, although the regeneration rate is reduced by about one-third. Under experimental laboratory conditions, stem fragments left to dry on a table for 10 weeks resprouted when moistened. Seeds disperse with wind, water, soil movement, and probably human activities. Most seeds germinate within a temperature range of 13–34°C, with the optimal temperature between 21–26°C. While most seeds produced are not viable, some viable seeds develop in most sites throughout California and Oregon.

MANAGEMENT FAVORING/DISCOURAGING SURVIVAL: Manual removal of plants, including roots and rhizomes, before viable seed develops can help control infestations in areas where plants are accessible. Removing all plant material from the site will help reduce the incidence of regrowth from rhizome, stolon, or stem fragments. Follow-up removal of resprouts is essential.

SIMILAR SPECIES: **Wild cucumbers** (*Marah* spp.) and **wild grapes** (*Vitis* spp.) are *native vines* with palmate-lobed leaves that grow in some of the same

Cape-ivy

Cape-ivy (*Delairea odorata*) seedling. J. M. DiTomaso

habitats as Cape-ivy, but unlike Cape-ivy, they are desirable components of the natural vegetation. In the vegetative state, wild cucumbers and wild grapes are readily distinguishable by having *stems with tendrils*. Cape-ivy can also look somewhat like **English** [*Hedera helix*] and **Algerian ivy** [*H. canariensis*]. The two **Hedera** species are woody at the base and do not have yellow flowers.

Cape-ivy (*Delairea odorata*) filled viable seeds (3 lower) and non-viable unfilled seed (above, with pappus). J. K. Clark

Cutleaf burnweed [*Erechtites glomerata* (Desf. ex Poir.) DC.][Cal-IPC: Moderate]

Australian burnweed [*Erechtites minima* (Poir.) DC.] [EREPR][Cal-IPC: Moderate]

SYNONYMS: Cutleaf burnweed: Australian fireweed; bushman's burnweed; cutleaf fireweed; cut-leaved coast fireweed; New Zealand fireweed; *Erechtites arguta* (A. Rich.) DC.; *Senecio argutus* A. Rich.; *Senecio glomeratus* Desf. ex Poir.

Australian burnweed: coastal burnweed; little fireweed; toothed coast fireweed; *Erechtites prenanthoides* (A. Rich.) DC.; *Senecio minimus* Poir.

GENERAL INFORMATION: Erect *annuals to short-lived perennials* to 2 m tall, with *unpleasant-scented foliage* and *small, pale yellow, cylindrical flower heads*. Cutleaf burnweed and burnweed can become locally dominant on disturbed sites, especially those in coastal regions that are newly burned or logged. However, dense populations typically do not persist beyond 10 years. Native to Australasia.

SEEDLINGS: Leaves alternate.

Cutleaf burnweed: Cotyledons narrowly oblong, mostly 4–9 mm long, ± 2 mm wide. First leaves narrowly elliptic, long soft-hairy. First few leaves initiated close together, appear opposite or whorled.

Australian burnweed: Cotyledons ovate, about 1–2 mm long, glabrous. First

Cutleaf burnweed (*Erechtites glomerata*) flowering stem. J. M. DiTomaso

Cutleaf burnweed • Australian burnweed

Cutleaf burnweed (*Erechtites glomerata*) plant. J. M. DiTomaso

Cutleaf burnweed (*Erechtites glomerata*) stem and leaves. J. M. DiTomaso

Cutleaf burnweed (*Erechtites glomerata*) immature plants. J. M. DiTomaso

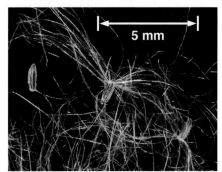

Cutleaf burnweed (*Erechtites glomerata*) seeds. J. O'Brien

leaves elliptic, slightly larger than cotyledons, short-stalked, margins smooth or few-toothed.

MATURE PLANT: Leaves alternate, fairly evenly distributed on stems, sessile, base narrowly long-tapered and weakly clasping stem. Stems and lower surfaces of leaves sparse to densely covered with short woolly grayish hairs.

Cutleaf burnweed: Leaves (2)5–15 cm long, to 4 cm wide, margins weakly sharp-toothed. *Lower leaves pinnate-lobed.* Upper leaves mostly narrow-elliptic to lanceolate.

Australian burnweed: *All leaves narrow-elliptic to lanceolate,* 7–20 cm long, mostly to 2 cm wide, sometimes to 4 cm wide, margins finely sharp-toothed.

Cutleaf burnweed · Australian burnweed

Australian burnweed (*Erechtites minima*) immature plants. J. M. DiTomaso

Australian burnweed (*Erechtites minima*) plant.
J. M. DiTomaso

Australian burnweed (*Erechtites minima*) seedling. J. M. DiTomaso

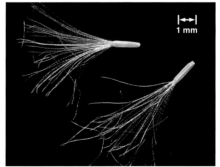

Australian burnweed (*Erechtites minima*) seeds.
J. O'Brien

American burnweed (*Erechtites hieracifolia*) flowering and fruiting stem. J. M. DiTomaso

American burnweed (*Erechtites hieracifolia*) plant. J. M. DiTomaso

American burnweed (*Erechtites hieracifolia*) leaves. J. M. DiTomaso

ROOTS AND UNDERGROUND STRUCTURES: Both species usually have a taproot.

FLOWERS: April–October. Heads numerous, *cylindrical, mostly 5 mm wide or less, appear to consist only of pale yellow disk flowers*, but actually consist of bisexual or male disk flowers surrounded by a ring of disklike female ray flowers (disciform heads). *Main phyllaries green, 4–8 mm long, equal in a single row,* with a few highly reduced phyllaries or bracts at the base.

Cutleaf burnweed: Main phyllaries ± 13. Heads mostly 2–3(4) mm wide at the base.

Australian burnweed: Main phyllaries ± 8. Heads mostly 1–2(3) mm wide at the base.

FRUITS AND SEEDS: Achenes *cylindrical, longitudinally ribbed,* dark, ~ 1–2 mm long excluding pappus bristles. Pappus bristles numerous, fine, white, persistent.

Cutleaf burnweed: Pappus bristles ± 5 mm long.

Australian burnweed: Pappus bristles ± 6–7 mm long.

POSTSENESCENCE CHARACTERISTICS: Dead flower stems have head remnants that consist of a small, bare, nearly flat receptacle with the main phyllaries reflexed downward.

HABITAT: Disturbed places, burned sites, roadsides, fields, coastal forest and woodland, grassland, coastal scrub.

DISTRIBUTION: Both species occur in the North Coast, western North Coast Ranges, San Francisco Bay region, Central Coast, and western South Coast Ranges, to 500 m. Both species also occur in Oregon. Cutleaf burnweed also occurs in Washington.

PROPAGATION/PHENOLOGY: *Reproduce by seed.* The biology of these species is poorly understood. Seeds disperse primarily with wind. Soil disturbance, including fire, appears to enhance germination. Both species appear to have fairly long-lived seeds and develop persistent seedbanks.

MANAGEMENT FAVORING/DISCOURAGING SURVIVAL: Periodic manual removal of plants before seeds mature may help control troublesome populations.

SIMILAR SPECIES: American burnweed or eastern fireweed [*Erechtites hieracifolia* (L.) DC., synonyms: *Erechtites hieracifolius* (L.) Raf. ex DC.; *Senecio hieracifolius* L.][EREHI] is a less-common *annual to short-lived perennial*, with or without lobed leaves. American burnweed is distinguished by having whitish to pale yellow cylindrical heads *4–8 mm wide* at the base with *phyllaries mostly 10–18 mm long*. In addition, seeds are mostly 2–3 mm long, excluding the pappus bristles, and longitudinally ribbed with a few short, coarse, appressed hairs between ribs. American burnweed occurs in disturbed habitats along the North and northern Central Coast, to 500 m. It also occurs in Oregon, Washington, and most central and eastern states. Native to the eastern United States.

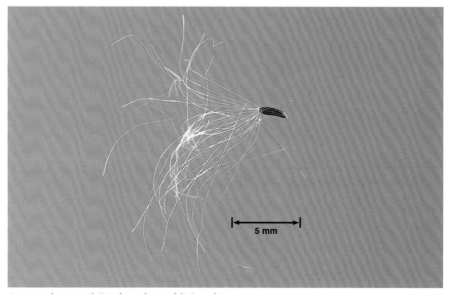

American burnweed (*Erechtites hieracifolia*) seeds.   J. O'BRIEN

Smallflower galinsoga [*Galinsoga parviflora* Cav. var. *parviflora*][GASPA]
Hairy galinsoga [*Galinsoga quadriradiata* Ruiz Lopez & Pavon][GASCI]

SYNONYMS: Smallflower galinsoga: gallant-soldier; quickweed; *Adventina parviflora* Raf.; *Galinsoga quiqueradiata* Ruiz & Pav.; *Galinsoga semicalva* (Gray) St. John & White; *Stemmatella sodiroi* Hieron.; *Wiborgia acmella* Roth; *Wiborgia parviflora* (Cav.) Kunth

Hairy galinsoga: ciliate quickweed; fringed quickweed; shaggy soldier; *Galinsoga aristulata* Bickn.; *Galinsoga bicolorata* St. John & White; *Galinsoga caracasana* (DC.) Schultaz-Bip; *Galinsoga ciliata* (Raf.) Blake; *Galinsoga parviflora* Cav. var. *hispida* DC.

GENERAL INFORMATION: Erect summer *annuals* to 0.6 m tall, with *opposite ovate leaves* that are *3-veined from the base* and small flower heads with a *few short, white* (occasionally pink or purplish red) *ray flowers* and several yellow disk flowers. Both species occur nearly worldwide as weeds. Smallflower galinsoga is consumed as a vegetable in southeast Asia. Smallflower galinsoga is native to South America. Hairy galinsoga is native to Mexico.

SEEDLINGS: Smallflower galinsoga: Cotyledons spoon-shaped to nearly square, broader than long (about 2–6 mm long, 3–7 mm wide), tip often shal-

Smallflower galinsoga (*Galinsoga parviflora*) leaf, flower head, and fruiting head. J. M. DiTomaso

Smallflower galinsoga (*Galinsoga parviflora*) flowering stem with persistent chaff on receptacle. J. M. DiTomaso

Smallflower galinsoga (*Galinsoga parviflora*) flowering stem. J. M. DiTomaso

Smallflower galinsoga (*Galinsoga parviflora*) seeds. J. O'Brien

lowly indented (emarginate) with an abrupt minute point in the center, glabrous. Leaves opposite. First 2 leaves ovate, about 8–18 mm long, 6–12 mm wide, 3-veined from the base, margin ciliate and smooth or weakly few-toothed. Upper surface sparsely covered with minute hairs. Subsequent leaves resemble first leaves, except margins ± regularly fine-toothed.

Hairy galinsoga: Seedlings resemble those of smallflower galinsoga.

MATURE PLANT: Stems simple or branched. Leaves *opposite, ovate, 3-veined from the base*, margins fine to coarsely serrate.

Smallflower galinsoga: Upper stems glabrous to sparsely soft-hairy, sometimes also glandular. Leaf stalks to 2.5 cm long. *Blades longer than wide*, 1–11 cm long.

Hairy galinsoga: Upper stems usually moderate to densely soft-hairy, sometimes also glandular. Leaf stalks 2–6 cm long. *Blades nearly as wide as long*, 1.5–9.5 cm long.

ROOTS AND UNDERGROUND STRUCTURES: Roots mostly fibrous, shallow. Taproot small or lacking.

FLOWERS: May–December. Head ± bell-shaped, in oppositely branched cymes. *Ray flowers ± 5, corolla usually white* (sometimes pink or purplish red). Disk flowers several to numerous, yellow. Phyllaries ovate, separate, in 2 rows. Receptacles conical, with chaffy bracts of 2 types. Outer chaffy bracts fused in

Hairy galinsoga (*Galinsoga quadriradiata*) leaves and flower heads.   J. M. DiTomaso

Hairy galinsoga (*Galinsoga quadriradiata*) seedling.   J. M. DiTomaso

Hairy galinsoga (*Galinsoga quadriradiata*) seeds.   J. O'Brien

clusters of 2–3 along with a phyllary around each ray flower. Inner chaffy bracts narrow, flat, membranous.

Smallflower galinsoga: Heads 2–6 mm in diameter. *Outer phyllaries 2–4, margins membranous. Inner chaffy bracts deeply 3-lobed.* Ray flower corollas 1–1.5 mm long, rarely pink.

Hairy galinsoga: Heads 2–10 mm in diameter. *Outer phyllaries 1–2, margins ± herbaceous. Inner chaffy bracts unlobed or shallowly 2–3-lobed.* Ray flower corollas 1–3 mm long, rarely dark purplish red.

FRUITS AND SEEDS: Achenes obconical, angled or flattened, 1–2.5 mm long, glabrous or hairy, dark.

Smallflower galinsoga: Ray achene pappus scales 5–8, unequal, ± 1 mm long. Disk achene pappus scales 15–20, 1–2 mm long, tips acute to slightly rounded. *Phyllaries and chaffy bracts persist after the achenes fall off.*

Hairy galinsoga: Ray achene pappus scales lacking or 8–20, unequal, 0.2–1.5 mm long. Disk achene pappus scales lacking or few-20, 1–1.5 mm long, acute to bristle-tipped. *Phyllaries and chaffy bracts deciduous with the achenes.*

POSTSENESCENCE CHARACTERISTICS: Senesced plants generally do not persist.

HABITAT: Gardens, nurseries, irrigated orchards, fields, vegetable crops, irrigation ditches, and other disturbed places. Thrive on moist fertile soil.

DISTRIBUTION: Both species are more common in the eastern United States than in the west.

Smallflower galinsoga: San Francisco Bay region, Southwestern region, and area east of the central Sierra Nevada, to 1000 m. Arizona, Colorado, Oregon, New Mexico, Washington, and most central and eastern states.

Hairy galinsoga: San Francisco Bay region, to 200 m. Colorado, Montana, Oregon, Washington, Wyoming, and most central and eastern states.

PROPAGATION/PHENOLOGY: *Reproduce by seed.* Achenes fall near the parent plant and disperse to greater distances with water, mud, human activities, and in nursery containers. Seeds mature about 11–14 days after flower heads first appear. Most seeds are nondormant at maturity. Germination requirements are highly variable. Germination occurs throughout spring and summer when conditions are favorable. Multiple generations can mature and produce seed within one season. One plant can produce up to 7500 seeds. Seedlings establish from the soil surface to a depth of about 3 cm.

MANAGEMENT FAVORING/DISCOURAGING SURVIVAL: Manually removing plants before seeds develop can control both species.

SIMILAR SPECIES: Hairy galinsoga leaves look similar to those of **purple deadnettle** [*Lamium purpureum* L.][LAMPU]. Unlike hairy galinsoga, purple deadnettle has square stems. Refer to the entry for **Henbit and Purple deadnettle** (p. 868, vol. 2) for more information.

Everlasting cudweed [*Gnaphalium luteo-album* L.]
Purple cudweed [*Gnaphalium purpureum* L.][GNAPU]

SYNONYMS: Everlasting cudweed: common cudweed; Jersey cudweed; white cudweed; *Pseudognaphalium luteo-album* (L.) Hilliard & Burtt

Purple cudweed: catfoot; chafe-weed; everlasting cudweed; featherweed; spoon-leaf purple everlasting; rabbit tobacco; *Gamochaeta purpurea* (L.) Cabrera; *Gamochaeta ustulata* (Nutt.) Nesom; *Gnaphalium perigrinum* Fern.; *Gnaphalium purpureum* L. var. *ustulatum* (Nutt.) Boivin; *Gnaphalium spathulatum* auth.; *Gnaphalium ustulatum* Nutt.

GENERAL INFORMATION: Both species have *densely woolly foliage, small flower heads with translucent papery phyllaries,* and *lack an aromatic scent.* Plants exist as rosettes until flowering stems develop at maturity.

Everlasting cudweed: Annual to 0.6 m tall. Native to Eurasia.

Purple cudweed: Summer/winter annual and/or biennial to 0.6 m tall. Purple cudweed is a widespread ruderal native that is sometimes weedy in agricultural fields, pastures, orchards, and other non-natural habitats. Under certain conditions, plants can accumulate nitrates to levels that are lethally *toxic* to cattle.

SEEDLINGS: Leaves develop alternate, but appear opposite.

Everlasting cudweed: Cotyledons oval to oblong, 1–2 mm long, glabrous. First few leaves elliptical to oblanceolate, mostly 4–6 mm long, tips broadly acute, covered with woolly hairs.

Everlasting cudweed (*Gnaphalium luteo-album*) infestation. J. M. DiTomaso

Everlasting cudweed (*Gnaphalium luteo-album*) plant. J. M. DiTomaso

Everlasting cudweed (*Gnaphalium luteo-album*) flowering stem. J. M. DiTomaso

Purple cudweed: Cotyledons oval to oblong, 1.5–2.5 mm long, ± 1 mm wide, gray-green, usually hairy. First few leaves obovate to oblanceolate, tips rounded or with an abrupt nipplelike point (mucronate), ± covered with woolly hairs, especially lower surfaces.

MATURE PLANT: Stems erect, simple or branched, densely covered with white woolly hairs. Leaves alternate, sessile.

Everlasting cudweed: Stems densely leafy in the lower portions, often horizontal and bent upward at the base (decumbent). Leaves linear to narrowly oblanceolate or spoon-shaped, tips slightly rounded to acute, 1–6 cm long, *upper and lower surfaces densely covered with white to grayish woolly hairs.*

Purple cudweed: Leaves oblanceolate to spoon-shaped, tips rounded to broadly acute. Lower leaves 1.5–12.5 cm long, sometimes purplish. Hairs of stems and leaves woolly, *appressed,* white to grayish. *Lower leaf surfaces densely woolly. Upper leaf surfaces moderate to sparsely covered with woolly hairs,* hairiness decreases over time.

ROOTS AND UNDERGROUND STRUCTURES: Both species are taprooted.

FLOWERS: Heads small, numerous, in dense sessile clusters, consist of a few disk flowers and many ray flowers with minute corollas that superficially resemble disk flowers (disciform heads). Phyllaries numerous, overlapping, papery, ± shiny when dry. Receptacles flat, lack chaffy bracts.

Everlasting cudweed: Year-round. Flower head clusters *spherical.* Heads 3–5 mm long, usually consist of *50–120 flowers* per head. *Phyllaries white, yellowish, or golden,* nearly transparent.

Flower clusters of everlasting cudweed (*Gnaphalium luteo-album*). J. K. Clark

Everlasting cudweed (*Gnaphalium luteo-album*) seedling.   J. M. DiTomaso

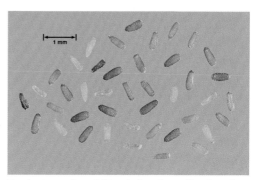

Everlasting cudweed (*Gnaphalium luteo-album*) seeds.   J. O'Brien

Purple cudweed: April–October. Flower head clusters *spikelike*. Heads 4–6 mm long. *Phyllaries brown or purplish in the upper portion,* translucent.

FRUITS AND SEEDS: Achenes oblong, 0.5–0.75 mm long. Pappus bristles fine, whitish, numerous.

Everlasting cudweed: Pappus bristles *weakly fused at the base, detach as a unit or in clusters.*

Purple cudweed: Pappus bristles *fused at the base to form a ring, detach as a unit.*

POSTSENESCENCE CHARACTERISTICS: Reflexed phyllaries remain attached to the heads after seeds disperse, but dead stems generally do not persist.

HABITAT: Vacant fields, agricultural fields, orchards, vineyards, waste places, and disturbed areas. Both species inhabit dry or seasonally wet sites.

Purple cudweed (*Gnaphalium purpureum*) plant.
J. M. DiTomaso

Purple cudweed (*Gnaphalium purpureum*) flowering stem.
J. M. DiTomaso

Purple cudweed (*Gnaphalium purpureum*) seedlings.
J. Neal

Purple cudweed (*Gnaphalium purpureum*) seeds.
J. O'Brien

DISTRIBUTION: Everlasting cudweed: Throughout California, except northern North Coast, Klamath Ranges, northern Cascade Range, and Modoc Plateau, to 2100 m. Oregon, Washington, Nevada, Utah, Arizona, Florida, and New York.

Purple cudweed: North Coast, western North Coast Ranges, central Sierra Nevada foothills, San Joaquin Valley, western Central-western region, and Channel Islands, to 1200 m. Oregon, Washington, Montana, Arizona, New Mexico, and southern and eastern United States.

PROPAGATION/PHENOLOGY: *Reproduce by seed.* Under favorable conditions, seeds germinate readily upon dispersal.

Everlasting cudweed: The biology of this species is poorly understood.

Lowland cudweed (*Gnaphalium palustre*) plant.   J. M. DiTomaso

Lowland cudweed (*Gnaphalium palustre*) flowering stems.   J. M. DiTomaso

Purple cudweed: In California, seeds mostly germinate November through April.

MANAGEMENT FAVORING/DISCOURAGING SURVIVAL: Cultivation or close mowing before seeds mature can usually control white and purple cudweed.

SIMILAR SPECIES: Refer to table 22 (p. 324) for a comparison of distinguishing features among cudweeds and licorice plant.

Lowland cudweed [*Gnaphalium palustre* Nutt., synonyms: western marsh cudweed; *Filaginella palustris* (Nutt.) Holub.] is a *nonaromatic native annual* to 0.3 m tall. Lowland cudweed occurs in dry or moist places throughout California, except the Modoc Plateau, to 2700 m. Although native plants are desirable in natural areas, lowland cudweed can be weedy in pastures,

Licoriceplant (*Helichrysum petiolare*) habitat along coast.   J. M. DiTomaso

Licoriceplant (*Helichrysum petiolare*) patch.   J. M. DiTomaso

Licoriceplant (*Helichrysum petiolare*) stem and leaves. J. M. DiTomaso

Licoriceplant (*Helichrysum petiolare*) inflorescence. J. M. DiTomaso

orchards, agricultural fields, and other disturbed non-natural habitats.

Japanese cudweed [*Gnaphalium japonicum* Thunb., synonyms: father-and-child plant; *Euchiton japonicus* (Thunb.) A. Anderb.] is an *annual* to 0.8 m tall that has escaped cultivation as an ornamental. It occurs in disturbed places in the North Coast, eastern North Coast Ranges, and San Joaquin Valley, to 700 m. Native to Australasia.

Creeping cudweed [*Gnaphalium collinum* Labill., synonym: *Euchiton gymno-*

Licoriceplant (*Helichrysum petiolare*) seeds. J. O'Brien

cephalus (DC.) A. Anderb.] is a *perennial* to 0.4 m tall, with *creeping rhizomes and leafy stolons*. Creeping cudweed has also escaped cultivation as an ornamental. It inhabits moist grasslands and open woodlands in the North Coast, western North Coast Ranges, and northern San Joaquin Valley, to 800 m. Native to Australia.

Licoriceplant [*Helichrysum petiolare* Hillard & Burtt, synonyms: *Helichrysum petiolatum* (L.) DC.; *Gnaphalium petiolatum* L.] is a spreading *shrub* to 0.6 m tall that has escaped cultivation as an ornamental in California and is not included in the current California flora. It is locally naturalized in certain forested areas and shrubland on the Central coast (south side of Mt. Tamalpais and the Monterey Peninsula). Native to South Africa.

Table 22. Cudweeds (*Gnaphalium* spp.) and licorice plant (*Helichrysum petiolare*)

Species	Life cycle	Upper leaf surface	Flower head clusters	Outer phyllaries	Other
Gnaphalium collinum creeping cudweed	perennial with creeping rhizomes and stolons	green, nearly glabrous, ± glandular	spherical	brown to purplish; translucent	basal leaves more numerous than stem leaves; leafy bracts below flower head clusters short
Gnaphalium japonicum Japanese cudweed	annual	green, nearly glabrous, ± glandular	spherical	brown to purplish; translucent	basal leaves lacking or fewer than stem leaves; leafy bracts below flower head clusters long
Gnaphalium luteo-album everlasting cudweed	annual	gray, densely loose woolly-hairy	spherical	white to pale yellow or pale golden; ± transparent	50–150 flowers per head; heads mostly 4–5 mm long
Gnaphalium palustre lowland cudweed	annual	gray, densely woolly-hairy	spherical	tips white, translucent; lower part brownish	40–60 flowers per head; heads mostly 3–4 mm long
Gnaphalium purpureum purple cudweed	annual/biennial	gray to gray-green, matted woolly-hairy	spikelike or oblong	brown to purplish; translucent	pappus bristles strongly fused at the base, detach as a unit
Helichrysum petiolare licoriceplant	spreading shrub	gray, densely woolly-hairy	± spherical clusters form a flat-topped to slightly rounded inflorescence	white, opaque	leaves ovate; corollas yellow

Curlycup gumweed [*Grindelia squarrosa* (Pursh) Dunal var. *serrulata* (Rydb.) Steyerm.]

SYNONYMS: curlytop gumweed; gumweed; resin-weed; rosinweed; tarweed; *Donia squarrosa* Pursh; *Grindelia serrulata* Rydb.

GENERAL INFORMATION: Erect *biennial*, sometimes annual or short-lived perennial, to about 1 m tall, with *resinous yellow flower heads* and *outwardly coiled phyllaries*. Plants growing on high-selenium soils can accumulate selenium, sometimes to levels that are *toxic* to livestock when ingested. However, curlycup gumweed contains bitter alkaloids, tannins, resins, and glucosides, and livestock typically avoid consuming the plant when more palatable forage is available. Curlycup gumweed tends to increase in density under grazing pressure. It is a state-listed secondary noxious weed in Minnesota. Native Americans have used the plant medicinally. Native to the Great Plains region of North America.

SEEDLINGS: Cotyledons oblong, about 3–7 mm long, 1–2 mm wide, fused at the base, glabrous. Leaves alternate. First and subsequent few leaves ± elliptic, at least twice the size of the cotyledons, base tapered to a stalk as long as the blade, covered with short glandular hairs. Later leaves elongated, ± narrowly oblanceolate.

MATURE PLANT: Foliage glabrous, aromatic, *sticky with a resinous coating*. Stems branched, white to yellowish, weakly woody, especially at the base. Leaves alternate, ± oblong, 1.5–7 cm long, sessile, base ± lobed and clasping the stem. Margins serrate, *teeth tips thickened, rounded, with a sessile resin gland*

Curlycup gumweed (*Grindelia squarrosa* var. *serrulata*) flower heads.   J. M. DiTomaso

Curlycup gumweed

Curlycup gumweed (*Grindelia squarrosa* var. *serrulata*) leaves and stem. J. M. DiTomaso

Curlycup gumweed (*Grindelia squarrosa* var. *serrulata*) plants. J. M. DiTomaso

Curlycup gumweed (*Grindelia squarrosa* var. *serrulata*) immature plant. J. M. DiTomaso

at the apex (visible with 10× magnification). Leaf surfaces dotted with sessile resin glands.

ROOTS AND UNDERGROUND STRUCTURES: Taproots can grow to about 2 m deep. Lateral roots typically shallow, sometimes extensive.

FLOWERS: June–October. Flower heads hemispheric, mostly 1.2–2 cm in diameter across the phyllaries, consist of 0 or 24–36 *yellow ray flowers* 8–10 mm long and numerous *yellow disk flowers*. Heads occasionally have only disk flowers. *Phyllaries overlapping in several rows, coated with sticky resin. Outermost phyllaries outwardly coiled about 360°*. Receptacles flat, ± pitted, *lack chaffy bracts. Pappus awns 2–6 per flower*, slender, white, shorter than disk corollas, flat, rolled and ± U-shaped in cross-section, *deciduous*.

FRUITS AND SEEDS: Achenes ± oblong, mostly 2–4 mm long, slightly flattened, ± 4-angled in cross-section, often curved, smooth, pale gray or tan. Apex truncate with a minute beak in the center.

POSTSENESCENCE CHARACTERISTICS: Dead stems with old flower heads can persist into the following year.

HABITAT: Disturbed areas, fields, roadsides, streamsides, desert scrub, and Joshua tree woodland.

DISTRIBUTION: Eastern Klamath Ranges, east side of the Sierra Nevada, central Sacramento Valley, South Coast, and Mojave Desert, to 1000 m. Probably elsewhere. Throughout much of the United States, except for the southeastern states.

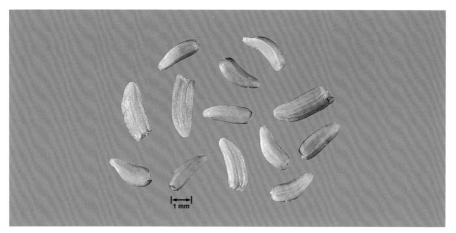

Curlycup gumweed (*Grindelia squarrosa* var. *serrulata*) seeds. J. O'Brien

Common gumplant (*Grindelia camporum* var. *camporum*) plant. J. M. DiTomaso

Curlycup gumweed

Common gumplant (*Grindelia camporum* var. *camporum*) flower head.   J. M. DiTomaso

Common gumplant (*Grindelia camporum* var. *camporum*) seeds.   J. O'Brien

PROPAGATION/PHENOLOGY: *Reproduces by seed. Deeply buried seeds can survive up to about 5 years.*

MANAGEMENT FAVORING/DISCOURAGING SURVIVAL: Manually removing plants before seeds develop can help control populations. Reducing grazing pressure may help to limit the increase of curlycup gumweed. Healthy, competitive plant communities generally suppress curlycup gumweed. Pasture fertilization can stimulate grasses and reduce gumweed populations.

SIMILAR SPECIES: Curlycup gumweed is the only non-native *Grindelia* species among the six recognized in California. Distinguishing among these species can be difficult. In most cases, native *Grindelia* are very important components of natural ecosystems. **Common gumplant** [*Grindelia camporum* Greene var. *camporum*], synonyms: Great Valley gumplant; *Grindelia procera* Greene; *Grindelia robusta* Nutt. var. *robusta* in part; others] is occasionally weedy in orchards. It is a variable *native perennial* from a woody root crown. Unlike curlycup gumweed, common gumplant has *phyllaries that are mostly spreading to curved downward, but not coiled to 360°*, and *leaf teeth tips are acutely pointed and lack a resin gland at the apex*. Common gumplant inhabits fields, roadsides, plains, and saline areas in Central Valley, eastern North Coast Ranges, San Francisco Bay region, western South Coast Ranges, and Southwestern region, to 1400 m. It also occurs in Nevada, Massachusetts, and Pennsylvania. Another variety of common gumplant, *Grindelia camporum* Greene var. *bracteosum* (J. Howell) M.A. Lane, has phyllaries that are outwardly coiled to nearly 360° and acutely pointed leaf teeth that lack a resin gland at the apex. This variety may be derived from the hybridization of curlytop gumweed and another native species.

Common sunflower [*Helianthus annuus* L.][HELAN]
Texas blueweed [*Helianthus ciliaris* DC.][HELCI]
[CDFA list: A]

SYNONYMS: Common sunflower: annual sunflower; wild sunflower; *Helianthus annuus* ssp. *jaegeri* (Heiser) Heiser and ssp. *lenticularis* (Douglas) Cockerell; *Helianthus annuus* var. *macrocarpus* (DC.) Cockerell; *Helianthus aridus* Rydb.; *Helianthus lenticularis* Dougl. ex Lindl.; others

Texas blueweed: blueweed; yerba parda

GENERAL INFORMATION: Common sunflower: Erect summer *annual* to 3 m tall, with coarse ovate to heart-shaped leaves and showy yellow flower heads. Common sunflower is the highly variable wild type of the cultivated sunflower. Common sunflower is a secondary state-listed noxious weed in Iowa and Minnesota. Native to the United States.

Texas blueweed: Erect herbaceous *perennial* to 0.7 m tall, with bluish green oblong to lanceolate leaves and an extensive system of *creeping roots*. Dense clonal patches often develop from the creeping root system. Besides California, Texas blueweed is a state-listed noxious weed in Arizona (restricted), Oregon (class A), and Washington (class A). Native to the south-central states and northern Mexico.

Common sunflower (*Helianthus annuus*) flower head and leaf. J. M. DiTomaso

Common sunflower (*Helianthus annuus*) seedling. J. K. CLARK

Common sunflower (*Helianthus annuus*) plant.
J. M. DiTomaso

SEEDLINGS: Common sunflower: Cotyledons oblong, 15–40 mm long, joined at the bases, smooth. First few true leaves opposite, ± elliptic, margins weakly round-toothed, dull green, covered with short bristly hairs that are rough to touch. Subsequent leaves alternate, elliptic to ovate or lanceolate.

MATURE PLANT: Boths species have a strong pungent odor, especially when crushed.

Common sunflower: Foliage covered with *stiff hairs that are rough to touch, giving them the feel of sandpaper.* Stems usually highly branched. Lower leaves often opposite, upper leaves mostly alternate. Leaves *heart-shaped to ovate, base often*

Common sunflower (*Helianthus annuus*) seeds. J. O'BRIEN

Common sunflower • Texas blueweed

Texas blueweed (*Helianthus ciliaris*) flower head.
J. L. Johnson

Texas blueweed (*Helianthus ciliaris*) flowering stems.
Courtesy of Robin Breckenridge

slightly lobed, 10–40 cm long, margin coarsely toothed, usually 3-veined from the base, on a long stalk.

Texas blueweed: Stems often sparsely covered with short stiff hairs. Leaves mostly opposite, sessile, typically *glabrous or hairy only on the margins* (ciliate), oblong to lanceolate, 3–8 cm long, margins wavy to shallow-lobed or smooth or sometimes crinkled, *bluish green, covered with a whitish film* (glaucous).

ROOTS AND UNDERGROUND STRUCTURES: Common sunflower: Seedlings initially have a taproot. Mature root system mostly fibrous.

Texas blueweed: Plants can develop an extensive system of woody horizontal creeping roots with buds every few inches. Root fragments with a bud can generate a new shoot. Generally shallow-rooted in uncultivated soils, but can

Texas blueweed (*Helianthus ciliaris*) infestation.
Courtesy of Ray Smith

Texas blueweed (*Helianthus ciliaris*) seeds.
J. O'Brien

Maximilian sunflower (*Helianthus maximilianii*) flower heads and leaves.   J. M. DiTomaso

develop a deep root system in cultivated fields.

FLOWERS: Flower heads showy, solitary on long peduncles. Ray flowers yellow.

Common sunflower: February–October. *Flower head receptacle 1.5–3.5(20) cm across. Central disk nearly flat on top, usually dark reddish or purplish brown.* Ray flower corollas mostly more than 2.5 cm long. Disk flower corollas 5–8 mm long with red to purple or yellow lobes. Receptacle chaffy bracts scalelike, tip deeply 3-lobed, mostly glabrous. Phyllaries typically *more than 4 mm wide, ciliate.*

Texas blueweed: June–November. *Flower head receptacle 1.2–2.5 cm across. Central disk rounded on top, yellowish.* Ray flower corollas about 1 cm long. Disk flower corollas 4–5 mm long with reddish lobes. Receptacle chaffy bracts scalelike, hairy at the tips, tip rounded or 3-lobed. Phyllaries shorter than the disk.

FRUITS AND SEEDS: Achenes resemble those of commercial sunflower. Pappus scales 2(4), deciduous.

Common sunflower: Achenes 3–15 mm long. Pappus scales 2–3.5 mm long.

Texas blueweed: Achenes about 3 mm long. Pappus scales about 2 mm long.

HABITAT: Common sunflower: Disturbed sites, roadsides, fields (especially cereal and safflower fields), and shrubland. Often infests fertile soils.

Texas blueweed: Roadsides, irrigated fields, stream and ditch banks, and low drainage areas. Grows best on cultivated soil. Often inhabits alkaline or saline soils.

DISTRIBUTION: Common sunflower: Throughout California, to 1900 m. All

Maximilian sunflower (*Helianthus maximilianii*) seeds. J. O'BRIEN

Prairie sunflower (*Helianthus petiolaris*) seeds. J. O'BRIEN

contiguous states.

Texas blueweed: Sacramento Valley (c Tehama Co.), San Francisco Bay region (w Alameda Co.), South Coast (cw San Luis Obispo, c Santa Barbara, n and e Ventura, Los Angeles, and sw San Diego Cos.), and western Mojave Desert (w San Bernardino Co.), to 300 m. Arizona, New Mexico, Nevada, Utah, Nebraska to Texas, and Illinois.

PROPAGATION/PHENOLOGY: Common sunflower: *Reproduces by seed.* Seedlings can emerge from depths of at least 10 cm. Mature plants can tolerate temperatures to −2°C, and seedlings are less sensitive to freezing temperatures than mature plants.

Texas blueweed: *Reproduces vegetatively from the creeping roots* and *by seed.* Seed viability is often low.

MANAGEMENT FAVORING/DISCOURAGING SURVIVAL: Cultivation before flowers develop viable seed can help to control common sunflower. For Texas blueweed, repeated cultivation to a depth of about 20 cm at least every 2 months, or more frequently, can help to control troublesome populations. However, occasional cultivation may facilitate the spread of Texas blueweed by dispersing root fragments.

SIMILAR SPECIES: Maximilian sunflower [*Helianthus maximilianii* Schrader, synonym: *Helianthus dalyi* Britt.][HELMA] is a *perennial* to 0.3 m tall, with *short rhizomelike roots.* Unlike Texas blueweed, Maximilian sunflower has leaves *10–30 cm long* with *smooth margins* and that are *folded along the midrib,* and foliage *is not* blue-green and glaucous. In addition, most leaves are alternate, *disk flower corolla lobes are yellow,* and flower heads are often in *racemelike clusters.* Flower heads usually have more than 25 narrow-lanceolate phyllaries 15–25 mm long. Maximilian sunflower has escaped cultivation as an ornamental and inhabits disturbed sites in southern San Joaquin Valley (Fresno Co.), to 100 m. It also occurs in Colorado, Idaho, Montana, New Mexico, Washington, Wyoming, central states, most eastern states, and a few southern states. Native to central and eastern North America.

Prairie sunflower [*Helianthus petiolaris* Nutt. ssp. *petiolaris*] [HELPE] is a summer *annual* to 2 m tall, with lanceolate to triangular leaves on long stalks. Unlike common sunflower, prairie sunflower has phyllaries *less than 4 mm wide* that are not conspicuously ciliate on the margins and *truncate to wedge-shaped leaf bases*. Foliage is green and stiff-hairy to nearly glabrous. *Central chaffy bracts are tipped with stiff white hairs*. Prairie sunflower inhabits disturbed sites in the San Francisco Bay region and the Southwestern region, to 450 m. Subspecies *petiolaris* also occurs in most contiguous states, except Arizona, Utah, and a few southern states. Subspecies *fallax* Heiser occurs in Arizona, Colorado, New Mexico, Nevada, and Utah. Native to the central states.

Spikeweed [*Hemizonia pungens* (Hook. & Arn.) Torrey & A. Gray ssp. *pungens* and ssp. *septentrionalis* Keck] [HEZPU]

SYNONYMS: common spikeweed; common tarweed; piny tarweed; *Hartmannia pungens* Hook. & Arn.

GENERAL INFORMATION: Native winter or summer *annual* to 1.2 m tall, with *spine-tipped upper leaves* and small *yellow flower heads*. Plants exist as rosettes until flowering stems develop in late spring and early summer. In natural areas in its native range, spikeweed is a desirable component of the ecosystem. Small animals forage on the seeds. It can become a pest in grazed pastures and rangeland since most animals avoid consuming the prickly foliage. Thus, grazing pressure can increase the abundance of spikeweed to undesirable levels. Spikeweed is native to northern and central California, but has been introduced into southern California and other states. It is a state-listed noxious weed in Oregon (class B) and Washington (class C), where it is problematic in the agricultural areas of the Columbia River basin. Another subspecies, **smooth tarplant** [*Hemizonia pungens* (Hook. & Arn.) Torrey & A. Gray ssp. *laevis* Keck], is a rare native grassland species in the South Coast region and Peninsular Ranges and is threatened by development and flood control efforts. Identification to subspecies is important in southern California, where both the common and rare subspecies occur. Refer to table 23 (p. 339) for a comparison of distinguishing characteristics.

Spikeweed (*Hemizonia pungens* ssp. *pungens*) plant.   J. M. DiTomaso

Spikeweed

Spikeweed (*Hemizonia pungens* ssp. *pungens*) flower heads   J. M. DiTomaso

Spikeweed (*Hemizonia pungens* ssp. *pungens*) seedling.   J. M. DiTomaso

Spikeweed (*Hemizonia pungens* ssp. *pungens*) seeds.   Jim O'Brien

SEEDLINGS: Cotyledons narrowly oblanceolate to elliptic, about 5–10 mm long, 1–2 mm wide, glabrous. First 2 leaves opposite, similar to cotyledons in shape and size, margins few-toothed, surfaces finely hairy.

MATURE PLANT: Stem branches rigidly ascending. Foliage ± unscented, *scabrous* in subspecies *pungens*, or *glabrous* in subspecies *septentrionalis*. Rosette and lower leaves linear-lanceolate, mostly 5–15 cm long, deeply pinnate-divided two times into narrow segments, yellowish green. Upper stem leaves alternate, linear to narrowly awl-like, *stiff, spine-tipped*, mostly 1–2 cm long. Often there are clusters or fascicles of leaves in the axils.

ROOTS AND UNDERGROUND STRUCTURES: Taprooted.

FLOWERS: April–October. Flower heads yellow, nearly sessile, shorter than subtending leaves, consist of numerous ray and disk flowers. Ray corollas mostly 3–5 mm long, 2-lobed. Disk flowers sometimes sterile (staminate only), anthers yellow. Phyllaries in 1 row, keeled, spine-tipped, each partially encloses a ray achene. *Chaffy bracts conspicuously spine-tipped,* scattered on recep-

Spikeweed

Fitch's tarweed (*Hemizonia fitchii*) flowering stems. J. M. DiTomaso

Fitch's tarweed (*Hemizonia fitchii*) plant.
J. M. DiTomaso

Fitch's tarweed (*Hemizonia fitchii*) rosette.
J. M. DiTomaso

tacle. Subspecies *pungens* heads 4–10 mm in diameter, clustered or scattered. Subspecies *septentrionalis* heads mostly 4–6 mm in diameter, scattered. Insect-pollinated.

FRUITS AND SEEDS: Ray achenes oblanceolate, 3-angled, ± 2 mm long, surface ± wrinkled, with a minute oblique projection at the apex. Disk achenes ellipsoid or football-shaped, ± 2 mm long. *Ray and disk achenes lack a pappus.* Sterile disk flowers do not develop achenes.

Fitch's tarweed (*Hemizonia fitchii*) seedling.
J. M. DiTomaso

Fitch's tarweed (*Hemizonia fitchii*) seeds.
J. O'Brien

Spikeweed

Parry's tarweed (*Hemizonia parryi* ssp. *rudis*) plant. J. M. DiTomaso

Parry's tarweed (*Hemizonia parryi* ssp. *rudis*) flower head. J. M. DiTomaso

Parry's tarweed (*Hemizonia parryi* ssp. *rudis*) seedling. J. M. DiTomaso

Parry's tarweed (*Hemizonia parryi* ssp. *rudis*) seeds. J. O'Brien

POSTSENESCENCE CHARACTERISTICS: Dead stems with spine-tipped leaves and old flower heads can persist through winter.

HABITAT: Grassland, pastures, rangeland, and seasonally moist depressions, sometimes on saline or alkaline soil.

DISTRIBUTION: To 500 m in California.

Subspecies *pungens*: Native to San Joaquin Valley and South Coast Ranges. Introduced to Southwestern region and Peninsular Ranges. Arizona, Nevada, Oregon, and New York.

Subspecies *septentrionalis*: Native to Cascade Range foothills and Sacramento Valley. Idaho, Nevada, Oregon, and Washington.

PROPAGATION/PHENOLOGY: *Reproduces by seed.* Seeds fall near the parent plant, disperse short distances with wind and long distances with animals and sometimes water.

MANAGEMENT FAVORING/DISCOURAGING SURVIVAL: High grazing pressure can increase the density of spikeweed populations. Manual removal early in the growing season can control small patches. Sheep have been reported to consume spikeweed in enclosed pastures. Burning can reduce spikeweed the following season, but the effect generally does not persist in subsequent years.

SIMILAR SPECIES: A few other native yellow-flowered annual *Hemizonia* species are occasionally weedy in pastures and rangeland.

Fitch's tarweed [*Hemizonia fitchii* A. Gray] and **Parry's tarweed** [*Hemizonia parryi* Greene ssp. *parryi* and ssp. *rudis* Keck, synonym: pappose tarweed] have *spine-tipped leaves*. Refer to table 23 (below) for a comparison of distinguishing features of *Hemizonia* species with spine-tipped leaves. Fitch's tarweed occurs in the Northwestern region, Cascade Range foothills, northern and central Sierra Nevada foothills, Sacramento Valley, northern San Joaquin Valley, South Coast Ranges, northern San Bernardino Mountains, and Santa Cruz Island, to 1000 m. Fitch's tarweed also occurs in Oregon.

Parry's tarweed subspecies *parryi* inhabits grassland and alkaline or salt marshes in the southern North Coast Ranges, southern Sacramento Valley, San Francisco Bay region, Central Coast except southern portion, and northern and central South Coast Ranges. Parry's tarweed subspecies *rudis* inhabits grassland in the central-western Central Valley. Both subspecies occur up to about 100 m elevation.

Kellogg's tarweed [*Hemizonia kelloggii* Greene] and **threeray tarweed** [*Hemizonia lobbii* Greene] *lack spine-tipped leaves* and more closely resemble **virgate tarweed** [*Holocarpha virgata* (Gray) Keck ssp. *virgata*]. Refer to the entry for **Virgate tarweed** (p. 347) for pictures and more information about how to distinguish these species. Kellogg's tarweed is native to the southern Sierra Nevada foothills, San Joaquin Valley, east San Francisco Bay, and Southwestern region, to 700 m, and is generally considered alien in the southeastern North Coast Ranges. Kellogg's tarweed also occurs in Arizona. Threeray tarweed occurs in the Cascade Range foothills, eastern San Francisco Bay region, and eastern South Coast Ranges, to 700 m.

Table 23. Common *Hemizonia* species with spine-tipped leaves and yellow anthers

Hemizonia spp.	Height (m)	Disk flower pappus	Chaffy bract tips	Foliage	Distinguishing features of rare subspecies
H. fitchii Fitch's tarweed	to 0.5	mostly 8–12 scales	rounded with numerous long fine interwoven hairs	upper foliage with stalked brown glands, unpleasant scent	no subspecies in CA
H. parryi Parry's tarweed	to 0.7	mostly 3–5 scales	thickened, fleshy or resinous, acute to rounded, with a few to several coarse long hairs	foliage sessile-glandular and long-hairy or scabrous, mild scent, phyllaries ± with sessile glands, especially margins	ssp. *congdonii*: foliage long-hairy, nonglandular, phyllaries nonglandular (c and s Central-western region); ssp. *australis*: anthers dark (South Coast)
H. pungens spikeweed	to 1.2	lacking	spine-tipped, mostly glabrous	leaves and bracts glabrous or scabrous, ± unscented	ssp. *laevis*: chaffy bracts rounded or acute, not spine-tipped

Telegraphplant [*Heterotheca grandiflora* Nutt.] [HTTGR]

SYNONYMS: telegraphweed; *Heterotheca floribunda* Benth.

GENERAL INFORMATION: Erect summer *annual, biennial, or short-lived perennial* to 2 m tall, with yellow flower heads. Telegraphplant is a ruderal native that is common on roadsides and other disturbed places. It is usually considered a weed only when it inhabits agricultural land and landscaped areas.

SEEDLINGS: Cotyledons spoon-shaped, about 5–12 mm long, 4–5 mm wide, glabrous. Leaves alternate. First and subsequent few leaves elliptic, mostly 15–20 mm long, 5–10 mm wide, covered with short hairs, tip usually abruptly minute-pointed (mucronulate).

MATURE PLANT: Foliage aromatic, covered with *glandular hairs and long stiff hairs*. Main stem simple or branched at the base, lowest section densely long-gray-hairy and weakly woody. Branches long, erect, usually only branched again near the top. Leaves alternate, 2–7 cm long, margins serrate. Lower leaves elliptic to lanceolate, on a stalk that is usually lobed and clasping the stem at the base. Upper leaves, ascending, ± oblanceolate, *sessile, base not clasping the stem*.

ROOTS AND UNDERGROUND STRUCTURES: Taproot deep, upper portion weakly woody.

Telegraphplant (*Heterotheca grandiflora*) plant. J. M. DiTomaso

Telegraphplant (*Heterotheca grandiflora*) flowering stem. J. M. DiTomaso

Telegraphplant (*Heterotheca grandiflora*) flower heads.   J. M. DiTomaso

FLOWERS: Primarily summer, but can flower at nearly anytime depending on location. Panicles ± flat-topped. Flower heads *yellow*, consist of about 25–40 *ray* and 30–75 *disk flowers*. Phyllaries *separate, overlapping in 3–5 rows*, glandular. Involucre (phyllaries as a unit) mostly 5–10 mm in diameter. *Receptacles lack chaffy bracts*, ± pitted. Ray corollas 5–8 mm long, often coiled downward.

FRUITS AND SEEDS: Achenes of 2 types, both lanceolate, 2–5 mm long. Ray achenes glabrous, slightly 3-angled, *lack pappus bristles*. Disk achenes hairy, flattened, with tawny to orange-brown *inner pappus bristles* 3–5 mm long and *outer pappus scales* 0.2–0.7 mm long.

POSTSENESCENCE CHARACTERISTICS: Dead flowering stems can persist for a few months. Old flower heads have reflexed phyllaries and ± pitted receptacles.

HABITAT: Disturbed areas, roadsides, waste places, fields, dry streambeds, sand dunes, coastal sage scrub, oak woodland, agricultural fields, orchards, and vineyards.

DISTRIBUTION: Sierra Nevada foothills, southern Northwestern region, Central Valley, Central-western region, southern Tehachapi Mountains, Southwestern region, and possibly adjacent desert regions, mostly to 300 m. Nevada and northwestern Mexico. Introduced to Arizona and Utah.

PROPAGATION/PHENOLOGY: *Reproduces by seed.* Ray achenes generally fall near the parent plant. Disk achenes disperse with wind. Both types also disperse with water, soil movement, and human activities. Germination occurs fall through spring. Disk achenes are not dormant at maturity and can germinate with or without light as soon as moisture conditions become favorable, typi-

Telegraphplant

Telegraphplant (*Heterotheca grandiflora*) seedling. J. M. DiTomaso

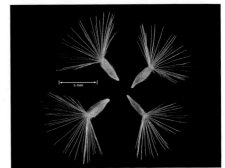

Telegraphplant (*Heterotheca grandiflora*) immature stem. J. M. DiTomaso

Telegraphplant (*Heterotheca grandiflora*) seeds. J. O'Brien

cally in fall. Ray achenes have a harder seed coat and remain dormant until the seed coat degrades or is scarified and conditions become favorable. Soil disturbance enhances ray achene germination.

MANAGEMENT FAVORING/DISCOURAGING SURVIVAL: Manual removal or cultivation before plants develop seed can control telegraphplant. Cultivation can deplete the soil seedbank of hard-coated achenes by stimulating germination.

SIMILAR SPECIES: Horseweed [*Conyza canadensis* (L.) Cronq.] and **Coulter's conyza** [*Conyza coulteri* Gray] have a flowering stem growth habit that is similar to that of telegraphplant. Unlike telegraphplant, horseweed and Coulter's conyza have *white to cream-colored flower heads*. Horseweed flowering stems have linear to narrowly lanceolate leaves and foliage that lacks glandular hairs. Coulter's conyza has glandular foliage and toothed stem leaves that are *broadly sessile and nearly clasping the stem*. Refer to the entry for **Hairy fleabane and Horseweed** (p. 284) for more information.

Orange hawkweed [*Hieracium aurantiacum* L.][HIEAU]

SYNONYMS: devil's paintbrush; grim-the-collier; orange paintbrush; red daisy; *Pilosella aurantiaca* (L.) F. Schultz & Schultz-Bip.

GENERAL INFORMATION: *Long-hairy perennial rosette*, with flowering stems to 0.7 cm tall, *orange dandelion-like flowers, milky sap*, and *slender stolons*. One plant can quickly spread vegetatively from long stolons or rhizomes to form a dense clonal mat of rosettes. Orange hawkweed often thrives on disturbed rangeland, pastures, meadows and open forests, especially those that are degraded, and once established, populations are difficult to control because of its ability to colonize an area by vegetative reproduction. Since its introduction to the region about 30 years ago, orange hawkweed has rapidly invaded large areas of British Columbia, Oregon, northern Idaho, northern Washington, and northwestern Montana. It is a state-listed noxious weed in Colorado, Idaho, Minnesota (secondary), and Washington (class B). Native to Europe, where it

Orange hawkweed (*Hieracium aurantiacum*) flowering stem.   J. M. DiTomaso

Orange hawkweed (*Hieracium aurantiacum*) foliage. J. M. DiTomaso

Orange hawkweed (*Hieracium aurantiacum*) seedling. J. M. DiTomaso

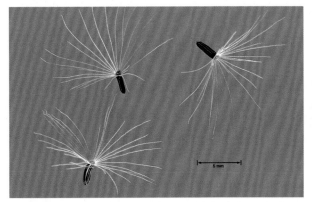
Orange hawkweed (*Hieracium aurantiacum*) seeds. J. O'Brien

is a ruderal species that inhabits pastures, roadsides, old fields, and other disturbed sites.

SEEDLINGS: Cotyledons oval to elliptic, about 2–4 mm long, 1–2 mm wide, glabrous. Leaves alternate. First and subsequent few leaves elliptic to ± obovate, base tapered to a stalk nearly equal to the length of the blade, upper surface sparsely hairy, lower surface glabrous or nearly glabrous, become increasingly larger than cotyledons. Later leaves develop hair characteristics described for mature plants.

MATURE PLANT: Leaves alternate, in rosettes, oblong-elliptic to oblanceolate, 2–20 cm long, usually more than 1 cm wide, margins smooth or barely toothed. Upper surface *densely covered with long, soft, spreading, papillae-based hairs*. Lower surface covered with simple hairs and minute star-shaped (stellate) hairs.

ROOTS AND UNDERGROUND STRUCTURES: Roots fibrous, shallow, develop adventitiously on the long, slender stolons.

FLOWERS: Mostly June–September. Flower stem covered with black hairs, *leafless or with a few bractlike leaves*, with a dense to open *terminal cluster of few to several flower heads*. Heads consist of *orange ligulate flowers* that turn purplish when dry. Phyllaries in 2–4 unequal rows, 5–10 mm long as a unit, covered with glandular hairs and black hairs. Receptacles lack chaffy bracts.

FRUITS AND SEEDS: Achenes cylindrical, 1.5–2 mm long, longitudinally ribbed, dark. Pappus bristles white, brittle, persistent.

POSTSENESCENCE CHARACTERISTICS: Senesced individuals do not persist.

HABITAT: Turf, fields, pastures, roadsides, rangeland, gardens, meadows, open forests, and other disturbed places. Often grows on poor, sandy soils but also grows in fine-textured, fertile soils.

DISTRIBUTION: Northern Sierra Nevada foothills (especially Grass Valley, Nevada Co.), to about 300 m. Expected to expand its range in California. Colorado, Idaho, Montana, Oregon, Washington, Wyoming, most north-central and northeastern states, a few southern states, and British Columbia. Expanding rapidly in Oregon and Colorado.

PROPAGATION/PHENOLOGY: *Reproduces vegetatively from stolons, rhizomes, and by seeds.* Stolon fragments can generate new plants. Seed reproduction is generally less important than vegetative reproduction in a given locality. Most seeds fall near the parent plant but can disperse long distances with wind. Populations vary in chromosome number. Polyploid populations generally produce asexual seed (apomixis). Seeds can germinate upon dispersal if conditions are favorable.

MANAGEMENT FAVORING/DISCOURAGING SURVIVAL: Manual removal and cultivation can disperse stolon fragments and facilitate the spread of orange hawkweed. Grazing and mowing at the appropriate time can limit seed production, but also promotes vegetative reproduction. However, careful manual removal of the entire plant, including as many roots and stolons as possible,

Meadow hawkweed (*Hieracium caespitosum*) flower heads.   J. M. DiTomaso

before seeds develop can help to control small populations. Cultivation before seeds develop and planting a smother crop can help to control large populations in pastures and on other sites where cultivation is feasible. Planting perennial grasses and then proper fertilizing after grasses have established can reduced reinvasion by hawkweeds.

SIMILAR SPECIES: There about 11 species of invasive hawkweeds in western North America estimated to infest over 200,000 hectares. All are of European origin and can be distinguished from native species because they are stoloniferous and reproduce apomictically (all belong to the subgenus *Pilosella*). In California, the native *Hieracium* species have leaves along the flowering stem, *lack stolons,* and are further distinguished by having *white or yellow flower heads.*

Meadow hawkweed [*Hieracium caespitosum* Dumort., synonyms: yellow hawkweed; *Hieracium pratense* Tausch] [HIECA] is an invasive *perennial rosette with long stolons, yellow flower heads,* and foliage that closely resembles that of orange hawkweed. Meadow hawkweed is an expanding problem in many eastern and northern states and is expected to eventually arrive in California. It is a state-listed noxious weed in Idaho and Washington (class B, noxious quarantine plant). Like orange hawkweed, yellow hawkweed often forms dense clonal mats that displace other forage species on rangeland and pastures. Seeds appear to survive up to about 7 years under field conditions. Native to Europe.

Virgate tarweed [*Holocarpha virgata* (Gray) Keck ssp. *virgata*][HOQVI]

SYNONYMS: sticky tarweed; yellowflower tarweed; *Hemizonia virgata* Gray

GENERAL INFORMATION: *Aromatic native annual*, with erect stems to 1.2 m tall, *resinous foliage*, and yellow flower heads. Plants exist as basal rosettes until flower stems develop in spring. Virgate tarweed is usually not considered a weed in natural areas. Birds and mammals consume the seeds, and the pollen is an important food source for bees. However, mature virgate tarweed is unpalatable to livestock, and populations can increase to an undesirable density in grazed pastures and rangeland. Late-spring rainfall after annual grasses have matured generally increases population density.

SEEDLINGS: Cotyledons oval, glabrous. First and subsequent few leaves opposite, linear to narrowly oblanceolate, hairy, glandular, tips rounded.

MATURE PLANT: Foliage *sweetly aromatic*, covered with a *sticky resin*. Rosette leaves linear, mostly 6–15 cm long, 0.3–1 cm wide, ± bristly-hairy, often withered by flowering time. Main stem branched well above the base. *Branches and branchlets ascending, rigidly straight*. Lower stem leaves opposite or alternate. Upper stem leaves linear, ± whorled, overlapping, ± 5 mm long, lay close to the stem (appressed), *mostly on short branchlets. Stem leaves tipped with a sessile open-pit resin gland.*

Virgate tarweed (*Holocarpha virgata*) population in a rangeland.   J. M. DiTomaso

ROOTS AND UNDERGROUND STRUCTURES: Taproot usually deeper than the roots of winter annual grasses.

FLOWERS: June–December. Flower heads sessile and short-stalked, mostly 5–6 mm long, subtended by many leaflike bracts, with 9–25 *yellow disk flowers with black anthers* surrounded by 3–7 *yellow ray flowers*. Ray corollas mostly 4–6 mm long, tip 3-lobed. Disk flower corollas 3.5–4.5 mm long, style glabrous below branches. Ray and disk flowers *lack a pappus*. Phyllaries in 1 row, *margins folded around ray flower achenes*, each with about 5–20 *open-pit resin gland projections*. Receptacle 1–2 mm in diameter, nearly flat, with chaffy bracts. Self-incompatible. Insect-pollinated.

FRUITS AND SEEDS: Ray achenes obovoid, mostly 2.4–3.5 mm long, ± 3-sided, slightly compressed front to back, with a short thick off-center beak at the apex, surface black and minutely roughened. Ray achenes detach with the enfolding phyllary. Disk achenes oblanceolate, ± 4 mm long, a large proportion typically sterile. *Both achene types lack a pappus.*

POSTSENESCENCE CHARACTERISTICS: Brown flower stems with persistent gland-tipped leaves and a few remnant seed heads can persist into winter. Dead foliage remains aromatic and sticky.

HABITAT: Grassland, woodland, fields, pastures, rangeland, and roadsides.

DISTRIBUTION: Central Valley, eastern North Coast Ranges, northern and central Sierra Nevada foothills, and northeastern Central-western region, to 800 m. Subspecies *elongata* Keck is uncommon and occurs only along the central and southern South Coast.

Virgate tarweed (*Holocarpha virgata*) plant.
J. M. DiTomaso

Virgate tarweed (*Holocarpha virgata*) flowering stem.
J. M. DiTomaso

Virgate tarweed (*Holocarpha virgata*) seedling.
J. M. DiTomaso

Virgate tarweed (*Holocarpha virgata*) seeds.
J. O'Brien

PROPAGATION/PHENOLOGY: *Reproduces by seed.* Most achenes fall near the parent plant or disperse short distances with wind, rain, and animals. Seeds germinate in fall after the rainy season begins, through mid-spring. Ray achenes are hard-coated and may require scarification or degradation of the seed coat to germinate. Fertile disk achenes lack a hard seed coat and readily germinate when conditions become favorable.

Kellogg's tarweed (*Hemizonia kelloggii*) plant.
J. M. DiTomaso

Kellogg's tarweed (*Hemizonia kelloggii*) flowering stem.
J. M. DiTomaso

Kellogg's tarweed (*Hemizonia kelloggii*) seeds.
J. O'Brien

Threeray tarweed *(Hemizonia lobbii)* flower head. J. K. CLARK

Threeway tarweed *(Hemizonia lobbii)* plant.
J. K. CLARK

Threeray tarweed *(Hemizonia lobbii)* seedling.
J. M. DiTomaso

MANAGEMENT FAVORING/DISCOURAGING SURVIVAL: Mowing pasture to a height of 4 inches in July or August can greatly reduce virgate tarweed density the following year.

SIMILAR SPECIES: Kellogg's tarweed [*Hemizonia kelloggii* Greene] and **threeray tarweed** [*Hemizonia lobbii* Greene] are yellow-flowered native *annuals* to 1 m and 0.6 m tall, respectively, with glandular, somewhat aromatic foliage and ray achenes that have a short off-center beak near the apex. Both species are a desirable component of the vegetation in natural areas under most circumstances, but like virgate tarweed, they can increase to undesirable levels on rangeland and pastures that are intensively grazed. Kellogg's tarweed and threeray tarweed are distinguished by their *lack* of large open-pit resin glands on the phyllaries and leaf tips. Refer to table 24 (p. 352) for a comparison of other distinguishing features. Kellogg's tarweed is native to the southern Sierra Nevada foothills, San Joaquin Valley, east San Francisco Bay, and Southwestern region, to 700 m and is considered alien in the southeastern North Coast Ranges. It also occurs in Arizona. Threeray tarweed occurs in the Cascade Range foothills, eastern San Francisco Bay region, and eastern South Coast Ranges, to 700 m.

Stinkwort [*Dittrichia graveolens* (L.) Greuter] [Cal-IPC: Moderate Alert] is an erect, fall-flowering, aromatic annual to about 0.7 m tall, with sticky glandular-hairy foliage and flower heads about 5–7 mm in diameter that consist of short

Virgate tarweed

Stinkwort (*Dittrichia graveolens*) habitat.
J. M. DiTomaso

Stinkwort (*Dittrichia graveolens*) plant.
J. M. DiTomaso

yellow ray flowers and yellow to reddish disk flowers. Stinkwort is a member of the Inuleae tribe and is related to the cudweeds (*Gnaphalium* spp.), but more closely resembles plants in the tarweed group (*Holocarpha, Hemizonia*). Unlike virgate tarweed, stinkwort lacks open-pit resin glands and has flower stems with sparse alternate linear leaves 1–4 cm long and 0.1–0.8 cm wide throughout; phyllaries in 2–3 unequal rows that are flat, narrow lanceolate, and have a membranous margin; and ± 3-sided achenes that are pubescent, pale, ± 2 mm long,

Stinkwort (*Dittrichia graveolens*) seedling.
J. M. DiTomaso

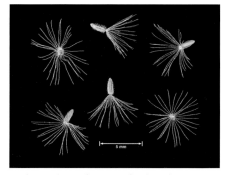

Stinkwort (*Dittrichia graveolens*) flowering and fruiting stem.
J. M. DiTomaso

Stinkwort (*Dittrichia graveolens*) seeds.
J. O'Brien

Virgate tarweed

and abruptly narrowed at the apex to a very short stalk with an expanded disk to which a single row of ± deciduous, nearly equal pappus bristles are attached. Stinkwort is not included in most California floras. It inhabits disturbed places, roadsides, pastures, fields, riparian woodlands, levees, washes, and margins of tidal marshes, primarily in the San Francisco Bay region, especially the southern portion. Populations have also been located in San Diego and near Sacramento (southern Sacramento Valley). Stinkwort appears to be expanding range rapidly. Native to Europe.

Table 24. Tarweeds (*Hemizonia* and *Holocarpha* spp.) with yellow flower heads and soft-tipped leaves

Species	Glands	Disk pappus	Ray flowers	Disk flowers	Disk flower anther color
Hemizonia kelloggii Kellogg's tarweed	sessile and slender-stalked	scales	5	6	yellow
Hemizonia lobbii threeray tarweed	mostly sessile	scales	3	3	black
Holocarpha virgata virgate tarweed	sessile and thick-stalked open pit resin glands	lacking	3–7	9–25	black

Virgate tarweed leaf with sessile open pit resin gland at tip (left); phyllary with stalked open pit resin glands (right).

Smooth catsear [*Hypochaeris glabra* L.][HRYGL] [Cal-IPC: Limited]

Common catsear [*Hypochaeris radicata* L.][HRYRA] [Cal-IPC: Moderate]

SYNONYMS: Smooth catsear: false dandelion; flatweed; glabrous catsear; *Hypochoeris glabra* L.

Common catsear: coast dandelion; false dandelion; flatweed; frogbit; gosmore; hairy catsear; long-rooted catsear; rough catsear; spotted catsear

GENERAL INFORMATION: Both species have *milky juice, a basal rosette of leaves,* and mostly *branched stems* with *yellow dandelion-like flower heads.* Smooth and common catsear often thrive on overgrazed pastures and rangeland. Both are palatable to livestock. Native to Europe.

Smooth catsear: *Annual* with flower stems to 0.4 m tall.

Common catsear: *Perennial* with flower stems to 0.8 m tall.

SEEDLINGS: Cotyledons oblanceolate, about 8–12 mm long, glabrous. Leaves alternate, form a rosette.

Smooth catsear: First few leaves glabrous, oblanceolate, about 10–20 mm long. Margins often have a few irregular shallow-rounded teeth.

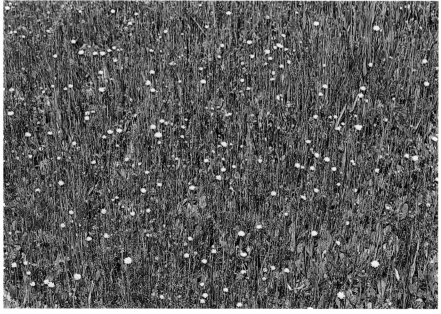

Smooth catsear (*Hypochaeris glabra*) in an annual grassland.   J. M. DiTomaso

Smooth catsear (*Hypochaeris glabra*) seedling.
J. M. DiTomaso

Smooth catsear (*Hypochaeris glabra*) plant.
J. M. DiTomaso

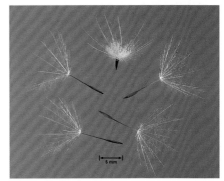
Smooth catsear (*Hypochaeris glabra*) seeds. Note two achene types.
J. O'Brien

Common catsear: First few leaves usually hairy, oblanceolate, about 8–15 mm long. Margins smooth or with a few small teeth.

MATURE PLANT: Leaves ± oblong to oblanceolate. Rosettes usually prostrate on open ground.

Smooth catsear: Foliage *glabrous*. Leaves 2–10 cm long, tips rounded. *Margins smooth to shallow-lobed*.

Common catsear: Foliage covered with *coarse yellowish hairs that are rough to touch*. Leaves 6–14 cm long. *Margins coarsely sharp-toothed to deeply lobed*.

ROOTS AND UNDERGROUND STRUCTURES: Smooth catsear: Taproot slender.

Common catsear: Roots fibrous, but plants often have several deep, thick, fleshy roots and appear taprooted.

FLOWERS: Flower stems mostly *branched*, occasionally simple, leafless, typically have a few *small bracts*. Flower heads *yellow, dandelion-like*, consist only of ligulate flowers. Phyllaries unequal, in *more than 2 overlapping rows*. Receptacle flat, with *long, thin chaffy bracts*.

Smooth catsear: March–June. Ligulate flowers barely extend beyond the phyllaries. Involucre ± bell-shaped, mostly 12–16 mm long.

Common catsear (*Hypochaeris radicata*) rosettes.
J. M. DiTomaso

Common catsear (*Hypochaeris radicata*) plant.
J. M. DiTomaso

Common catsear (*Hypochaeris radicata*) seedling.
J. M. DiTomaso

Common catsear: May–November. Ligulate flowers conspicuously longer than the phyllaries. Involucre ± oblong-cylindrical, mostly 16–25 mm long.

FRUITS AND SEEDS: Achenes brown, narrowly oblong or ellipsoid, longitudinally 10-ribbed, with minute sharp projections in the upper part, often with a slender beak at the apex. Pappus bristles stiff, *featherlike* (plumose), about 8–10 mm long, dull white to tan. A few outer bristles are sometimes short and simple.

Smooth catsear: Achenes of *2 types*. Outer achenes ± 4 mm long, lack a beak at the apex, pappus bristles short. Inner achenes more than 4 mm long, with a slender beak nearly equal to the achene body.

Common catsear: Achenes of *1 type*. Achene bodies 3–4 mm long, with a beak equal to or shorter than the body.

HABITAT: Disturbed places, fields, grassland, pastures, roadsides, orchards, vineyards, landscaped areas, and gardens.

Smooth catsear: Also agricultural fields. Can grow on serpentine soil.

Common catsear: Also turf. Tolerates drought.

DISTRIBUTION: Common.

Smooth catsear: Throughout California, except deserts and Great Basin, to 1200 m. Oregon, Washington, many southern states, and a few eastern states.

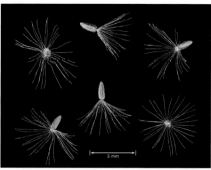

Common catsear (*Hypochaeris radicata*) seeds.
J. O'Brien

Common catsear (*Hypochaeris radicata*) flower heads.   J. M. DiTomaso

Common catsear: Northwestern region, Cascade Range foothills, northern Sierra Nevada foothills, Sacramento Valley, Central-western region, and South Coast, to 500 m. Western states, except possibly Arizona and Wyoming, and most southern and eastern states.

PROPAGATION/PHENOLOGY: *Reproduce by seed.* Seeds disperse with wind, by clinging to animals, through human activities, and as seed contaminants. Seeds do not require an after-ripening period. Germination can occur fall through spring, depending on the region and environmental conditions. Smooth and common catsear hybridize with one another, but most hybrids produce few viable seed.

Common catsear: In heavily grazed areas or mowed turf, plants can *reproduce vegetatively* by offsets from the crown, and diffuse clonal patches can develop. However, root fragments do not regenerate when detached from the crown. Seeds generally do not persist in the soil seedbank. Established seedlings can mature and produce seed within 2 months.

MANAGEMENT FAVORING/DISCOURAGING SURVIVAL: Mowing and light to moderate grazing usually facilitate survival. Burning can stimulate germination of common catsear and probably smooth catsear seed. Undisturbed vegetation cover discourages catsear establishment and reproduction. Cultivation can control both species.

SIMILAR SPECIES: Dandelion [*Taraxacum officinale* Wigg.] [TAROF] and **lesser hawkbit** [*Leontodon taraxacoides* (Villars) Mérat] are *perennial rosettes* that may be confused with catsear species. Unlike smooth catsear and common catsear, dandelion and lesser hawkbit have *unbranched flower stems that lack small bracts*, flower head receptacles that *lack chaffy bracts*, and phyllaries that are in *2 unequal rows*. In addition, dandelion has glabrous foliage, acute leaf lobes that point toward the base, *simple* (not feathery) *pappus bristles*, and *outer phyllaries spreading or curved downward* (reflexed). Lesser hawkbit has stiff-hairy foliage and outer phyllaries that are much smaller than the inner phyllaries. Refer to the entry for **Dandelion** (p. 412) for more information about these species.

Poverty sumpweed [*Iva axillaris* Pursh]
[IVAAX][CDFA list: C]

SYNONYMS: bozzleweed; death-weed; devil's-weed; povertyweed; salt sage; small-flowered marsh elder; *Iva axillaris* Pursh ssp. *robustior* (Hook.) Bassett; *Iva axillaris* Pursh var. *pubescens* A. Gray

GENERAL INFORMATION: Erect *perennial* to 0.6 m tall, with *nodding green flower heads* and *vigorous creeping roots*. The foliage has an *unpleasant aromatic scent*. Poverty sumpweed is a widespread native of western North America and is generally a desirable component of salt marsh and alkali plains communities. However, poverty sumpweed can form large *clonal colonies* on sites subjected to cultivation, grazing, or other types of disturbance. Once established, colonies are difficult to eliminate. In cultivated fields, colonies can significantly reduce crop yields. Studies suggest that poverty sumpweed has allelopathic properties. Handling the foliage may cause contact dermatitis, and inhaling pollen can trigger allergic responses in sensitive individuals. Poverty sumpweed is generally more problematic in western states other than California. Poverty sumpweed is a noxious weed in the state of Victoria, Australia.

MATURE PLANT: Foliage nearly glabrous to moderately covered with short stiff hairs. Stems mostly branched at the base. Leaves nearly sessile, opposite on lower portions of stems, often alternate near tips, narrowly elliptic to

Poverty sumpweed (*Iva axillaris*) population.　　　　　　　　J. M. DiTomaso

Poverty sumpweed

Poverty sumpweed (*Iva axillaris*) plant.
J. M. DiTomaso

Poverty sumpweed (*Iva axillaris*) flowering stem. J. M. DiTomaso

Poverty sumpweed (*Iva axillaris*) young sprouts from a rhizome. J. M. DiTomaso

obovate, *tip rounded, margins smooth*, 1–4 cm long, thick, leathery, typically *dotted with reddish, resinous glands*.

ROOTS AND UNDERGROUND STRUCTURES: Roots extensively *creeping*, woody, highly branched, slender, to 2.5 m deep, with numerous shoot buds. *Roots store abundant reserves*. Fragments can generate new plants.

FLOWERS: May–September. Flower heads solitary in upper leaf axils, *curved downward on short stalks*, greenish, 5–7 mm wide, consist of 5–20 male disk flowers surrounded by 1–5 disklike female flowers with tiny tubular corollas. Phyllaries *fused into a shallow cup* with 5–8 lobes. Both flower types lack a pappus. Wind-pollinated.

FRUITS AND SEEDS: Only the peripheral female flowers develop achenes. Achenes obovate, slightly flattened, 2–3 mm long, brown to near black, covered with minute glistening bumps, lack a pappus.

POSTSENESCENCE CHARACTERISTICS: Foliage dies back in cold winter climates, but roots survive.

HABITAT: Alkaline plains, edges of salt marshes, cultivated fields (especially agronomic crop fields), pastures, rangeland, roadsides, and waste places. Often grows on poorly drained, heavy alkaline or saline soils, but it is not limited to these soil types.

DISTRIBUTION: Throughout California, to 2500 m. All western states and many central states.

PROPAGATION/PHENOLOGY: Reproduces vegetatively from creeping roots and by seed. Roots seldom exhaust their energy reserves. Deep roots can remain dormant for long periods under intense competition and produce new shoots under drought conditions. Most seeds fall near the parent plant, but some seeds probably disperse to greater distances with water, soil movement, mud, animals, and agricultural equipment and other human activities. Newly matured seed can be dormant. Seeds submerged in water can survive about 8 months. Individual plants are generally long-lived.

MANAGEMENT FAVORING/DISCOURAGING SURVIVAL: Cultivation can disperse root fragments. Deep roots can survive repeated cultivation for several years. Alternating wheat with summer fallow or growing perennial hay crops such as alfalfa can help suppress infestations. Managing pastures to prevent overgrazing helps to suppress the development of large colonies.

SIMILAR SPECIES: San Diego marsh-elder [*Iva hayesiana* A. Gray] is a *rare native perennial* in southwestern San Diego County that closely resembles poverty sumpweed, but is not a weed. In this region, *Iva* species should be correctly identified before any control strategies are implemented. Unlike poverty sumpweed, San Diego marsh-elder has flower heads with *separate phyllaries*.

The ragweeds (*Ambrosia* spp.) have flower heads that are similar to those of *Iva* species. Ragweeds are most easily distinguished by having *leaves with margins that are toothed to lobed*. Refer to the entry for **Annual bursage and Giant ragweed** (p. 182) for more information.

Poverty sumpweed (*Iva axillaris*) seeds.   J. O'Brien

Willowleaf lettuce [*Lactuca saligna* L.][LACSL]
Prickly lettuce [*Lactuca serriola* L.][LACSE]
Bitter lettuce [*Lactuca virosa* L.]

SYNONYMS: Willowleaf lettuce: willow lettuce; *Lactuca saligna* L. var. *runcinata* Gren. & Godr.

Prickly lettuce: china lettuce; common wild lettuce; compass plant; English thistle; horse thistle; whip thistle; *Lactuca scariola* L. Lance-leaved prickly lettuce refers to a variety with leaves that lack lobes.

Bitter lettuce: wild lettuce

GENERAL INFORMATION: Erect winter or summer annuals, sometimes biennials, to about 2 m tall, with *milky sap* and panicles of *pale yellow flower heads*. These lettuces have *leaves with a row of prickly bristles on the lower midvein*, except sometimes the lower midvein of willowleaf lettuce is smooth. The bristles in willowleaf lettuce are much softer than those of prickly lettuce. Plants exist as basal rosettes until flowering stems develop at maturity. Refer to table 25 (p. 368) for a quick comparison of distinguishing characteristics of lettuces.

Willowleaf lettuce (*Lactuca saligna*) plant.
J. M. DiTomaso

Willowleaf lettuce (*Lactuca saligna*) flower heads.
J. M. DiTomaso

Willowleaf lettuce (*Lactuca saligna*) stem and leaves.
J. M. DiTomaso

Willowleaf lettuce • Prickly lettuce • Bitter lettuce

Prickly lettuce is a prolific colonizer of disturbed habitats in California. Native to Europe.

SEEDLINGS: Cotyledons elliptic-oblong to ovate, often with tips slightly indented, bases abruptly taper into a short stalk, usually with a few fine gland-tipped hairs, especially on the margins. Leaves alternate. First and subsequent few leaves oblanceolate to elliptic, ± glandular-hairy, tips rounded, bases gradually taper into a short stalk, margins weakly few-toothed to smooth. First blade and stalk 8–20 mm long. Later rosette leaves have stiff bristles on the lower midvein (sometimes lacking in willowleaf lettuce).

Willowleaf lettuce: Cotyledons 5–15 mm long, 3–9 mm wide. Leaves mostly 2–2.5 times longer than wide.

Prickly lettuce: Cotyledons mostly 5–10 mm long, 3–8 mm wide. Leaves mostly 2–2.5 times longer than wide. Midvein hairs much stiffer than in other species of *Lactuca*.

Bitter lettuce: Cotyledons 8–15 mm long. Leaves mostly 1.5–2 times longer than wide.

MATURE PLANT: Stems glabrous or bristly-hairy in the lower portion, branched in the panicle. Rosette leaves sometimes withered and lacking at flowering. Stem leaves alternate, base lobed and clasping the stem. *Lower midveins typically have a row of prickly bristles.*

Willowleaf lettuce: Stems often few per rosette, spreading and arched upward at the base. Leaves *linear to narrowly lanceolate* in outline, pinnate-lobed with rounded indentations or entire, *margins smooth or with a few teeth*. Basal lobes of stem leaves *very narrow*.

Prickly lettuce: Stems often 1 per rosette. Leaves *oblanceolate to obovate or*

Willowleaf lettuce (*Lactuca saligna*) seedling.
J. M. DiTomaso

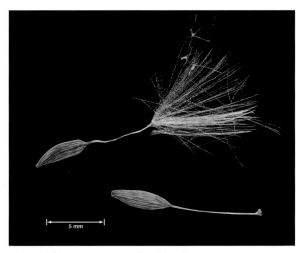

Willowleaf lettuce (*Lactuca saligna*) seeds. J. O'Brien

elliptic in outline, pinnate-lobed with nearly rounded indentations or entire, *margins prickly-toothed*. Stem leaves often twist at the base to lie in a vertical plane, basal lobes narrow to moderately broad. Plants with entire leaves have been referred to as *Lactuca serriola* L. var. *integrata* Gren. & Godr. (synonym: *L. serriola* L. forma *integrifolia* (Gray) S.D. Prince & R.N. Carter). However, this taxon is no longer recognized. The pinnate-lobed leaf form and entire leaf form are both common and often occur together in the same population. The

Prickly lettuce (*Lactuca serriola*) habit. J. K. Clark

Prickly lettuce (*Lactuca serriola*) flower heads.
J. K. Clark

Prickly lettuce (*Lactuca serriola*) type with deeply lobed leaves. J. M. DiTomaso

Prickly lettuce (*Lactuca serriola*) type without deeply lobed leaves.   J. M. DiTomaso

Prickly lettuce (*Lactuca serriola*) leaf type without deeply lobed leaves.   J. M. DiTomaso

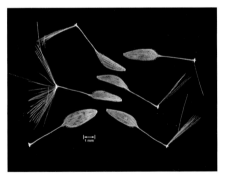

Prickly lettuce (*Lactuca serriola*) seedling.
J. M. DiTomaso

Prickly lettuce (*Lactuca serriola*) seeds.
J. O'Brien

different leaf shapes are probably an expression of genetic variability within a population. Midvein hairs of prickly lettuce are much stiffer than other species of *Lactuca*, even in the seedling stage.

Bitter lettuce: Stems often 1 per rosette. Rosette leaves usually several at flowering. Leaves *broadly obovate* in outline, pinnate-lobed with rounded indentations or entire, *margins prickly-toothed*. Stem leaves mostly 2–4 cm wide, *basal lobes very broad*.

ROOTS AND UNDERGROUND STRUCTURES: Taproot often deep (> 1 m), with fibrous lateral roots. Deep roots allow plants to flower and survive late into the growing season. Bitter lettuce roots are reported to have an unpleasant, fetid scent.

Willowleaf lettuce • Prickly lettuce • Bitter lettuce

FLOWERS: Flower heads consist only of *pale yellow ligulate flowers*. Phyllaries in 2 or more overlapping rows, cylindrical as a unit. Receptacles lack chaffy bracts.

Willowleaf lettuce: July–December. Rarely flowers in the spring. Panicles narrow, often *spikelike*. Flower heads *sessile or nearly sessile, lay close to stem*.

Bitter lettuce (*Lactuca virosa*) habitat.   J. M. DiTomaso

Bitter lettuce (*Lactuca virosa*) flowering stem.
J. M. DiTomaso

Bitter lettuce (*Lactuca virosa*) flower heads.
J. M. DiTomaso

Bitter lettuce (*Lactuca virosa*) rosette.
J. M. DiTomaso

Bitter lettuce (*Lactuca virosa*) seedling.
J. M. DiTomaso

Bitter lettuce (*Lactuca virosa*) leaf.
J. M. DiTomaso

Bitter lettuce (*Lactuca virosa*) seeds.   J. O'Brien

Prickly lettuce: April–October. Panicles open, branches spreading. Flower heads stalked and sessile.

Bitter lettuce: June–November. Panicles open, branches spreading. Flower heads stalked and sessile.

FRUITS AND SEEDS: Achenes lanceolate, flattened, *5 or more ribs per face*, body 3–4 mm long, with a *long slender beak* at the apex. Pappus bristles white, detach separately. Achenes and pappus characteristics are useful for species identification.

Willowleaf lettuce: Achenes brown and mottled with darker brown, ribs minutely barbed near the apex, beak equal to or longer than body. Pappus bristles ± 6 mm long.

Prickly lettuce: Achenes light to dark brown, ribs minutely barbed and hairy near the apex, beak equal to or longer than body. Pappus bristles 4–5 mm long.

Bitter lettuce: Achenes dark brown to nearly black, surface rough, *margin thick-*

Willowleaf lettuce • Prickly lettuce • Bitter lettuce

Tall lettuce (*Lactuca canadensis*) bolted stem with foliage. J. M. DiTomaso

Tall lettuce (*Lactuca canadensis*) seedling. J. M. DiTomaso

winged, beak nearly equal to body. Pappus bristles ± 8 mm long.

HABITAT: Prickly lettuce, willowleaf lettuce: Disturbed sites, annual grasslands, roadsides, seasonal wetlands, waste places, ditchbanks, fields, agronomic and vegetable crops, orchards, vineyards, landscaped areas, and urban places.

Bitter lettuce: Disturbed sites, roadsides, waste places, fields, moist ditchbanks, shrubby slopes, woodlands, and coastal prairie.

DISTRIBUTION: Willowleaf lettuce: North Coast Ranges, Sacramento Valley, San Francisco Bay region, Central Coast, and western South Coast Ranges, to 800 m. Arizona, Nevada, New Mexico, Oregon, Washington, and most central and eastern states.

Prickly lettuce: Throughout California, to 2000 m. All contiguous states.

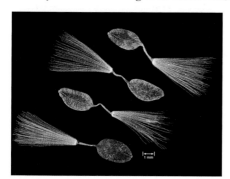

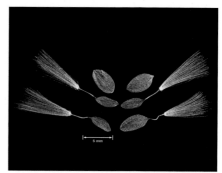

Tall lettuce (*Lactuca canadensis*) seeds. J. O'Brien

Western lettuce (*Lactuca ludoviciana*) seeds. J. O'Brien

Bitter lettuce: San Francisco Bay region, Cascade Range, and western South Coast Ranges, to 300 m. Alabama.

PROPAGATION/PHENOLOGY: *Reproduce by seed.* Seeds disperse with wind, water, mud, soil movement, human activities including agricultural and landscape maintenance equipment, and possibly animals. Newly matured prickly lettuce seeds generally remain dormant for several months, and germination does not require light.

MANAGEMENT FAVORING/DISCOURAGING SURVIVAL: Manual removal, cultivation, and mowing flower stems before seed matures can control these species.

SIMILAR SPECIES: The following species are more uncommon than the preceding species. Refer to table 25 (p. 368) for a quick comparison of distinguishing characteristics.

Biennial lettuce [*Lactuca biennis* (Moench.) Fern., synonyms: tall blue lettuce; woodland lettuce; *Lactuca spicata* Hitchc. ex Britt. & Br.][LACBI] is usually a *biennial* to 2.5 m tall, with a *short taproot* and large, dense panicles of *pale blue to cream-colored flower heads 10–15 mm wide*. It is distinguished by having achenes 5–7 mm long that *lack a beak or taper into a short thick beak*. Pappus bristles are brownish. Leaves are ovate to elliptic in outline and pinnate-lobed with coarsely toothed margins. Biennial lettuce occurs sporadically in disturbed places in the western North Coast Ranges, North Coast, and San Francisco Bay region, to 500 m. It also occurs in Colorado, Idaho, Montana, Oregon, Utah, Washington, Wyoming, and north-central and northeastern states. Native to northeastern North America.

Tall lettuce [*Lactuca canadensis* L., synonyms: Canada lettuce; *Lactuca steelei* Britt.; *Lactuca sagittifolia* Ell.][LACCA] is usually a *biennial* to 2.5 m tall, with a taproot and panicles of *pale yellow flower heads 10–15 mm long*. Tall lettuce is further distinguished by having *long-beaked achenes* with *1 rib per face* and flower heads mostly with *13–25 flowers per head*. Open flower heads are mostly 5–10 mm wide. Leaves are elliptic in outline and pinnate-lobed with curved, linear lobes and smooth margins. Tall lettuce inhabits disturbed places and forests in the Klamath Ranges and northern Sierra Nevada, to 1050 m. It occurs in all contiguous states, except possibly Arizona and Nevada. Native to eastern North America.

Western lettuce [*Lactuca ludoviciana* (Nutt.)DC., synonyms: biannual lettuce; *Lactuca campestris* Greene] is a *biennial* to 2 m tall that has *long-beaked achenes* with *1 rib per face* like tall lettuce. Unlike tall lettuce, western lettuce has flower heads *15–22 mm long* and mostly with *20–56 flowers per head*. Open flower heads are 4–5 mm wide. Leaves are lanceolate in outline and pinnate-lobed with broadly lanceolate lobes and finely toothed margins. Western lettuce inhabits disturbed sites and waste places in the Cascade Range foothills and South Coast, to 300 m. It also occurs in all western states, except possibly Nevada, and most central states, including Texas, Louisiana, and Mississippi. Western lettuce is under state protection in Indiana, where it is listed as extirpated on the rare, threatened, and endangered species plant list. Native to the central states.

Table 25. Lettuces (*Lactuca* spp.)

Lactuca spp.	Flower color	Achene beak length	Ribs per face of achene	Achene body margin wings	Life cycle	Other important characteristics
L. biennis biennial lettuce	pale blue to cream	lacking or half the body length; thick	3	none or inconspicuous	biennial	open flower heads 10–15 mm diameter; panicles large, dense, to 2.5 m tall; pappus bristles tawny
L. canadensis tall lettuce	pale yellow	nearly equal to body	1	none or inconspicuous	biennial (annual, short-lived perennial)	flower heads mostly 10–15 mm long; leaves usually lack prickles on lower midvein; achenes finely transverse-wrinkled
L. ludoviciana western lettuce	blue to ± pale yellow	nearly equal to body	1	none or inconspicuous	biennial (annual, short-lived perennial)	flower heads mostly 15–22 mm long
L. saligna willowleaf lettuce	pale yellow	equal to or longer than body	> 3	none or inconspicuous	annual or biennial	stem leaves narrow (pinnate-lobed or entire) with smooth to sparsely toothed margins; leaf base lobes narrow; panicles narrow, spikelike, with flower heads ± sessile, lay close to stem
L. serriola prickly lettuce	pale yellow	equal to or longer than body	> 3	none or inconspicuous	annual or biennial	stem leaves oblong-elliptic (pinnate-lobed or entire) with prickly-toothed margins; leaf base lobes broad; panicles open, with flower heads stalked and sessile
L. tartarica ssp. *pulchella* blue lettuce*	bright blue	half the body length or less; thick	> 3	none or inconspicuous	perennial with deep creeping roots	open flower heads 2–3 cm diameter
Lactuca virosa bitter lettuce	pale yellow	nearly equal to body	> 3	thick, conspicuous	biennial (annual)	stem leaves broadly obovate, with prickly-toothed margins; leaf base lobes very broad

Note: *Blue lettuce [*Lactuca tartarica* (L.) C. Meyer ssp. *pulchella* (Pursh) Stebb.] is a native species that is included in the table for identification purposes only.

Oxeye daisy [*Leucanthemum vulgare* Lam.] [CHYLE][Cal-IPC: Moderate]

SYNONYMS: butter daisy; dog daisy; dun daisy; Dutch curse; field daisy; golden flower; goldens; great oxeye; horse daisy; kellup-weed; maudlin daisy; maudlinwort; marguerite; moon daisy; poorland flower; thunderflower; white daisy; whiteweed; *Chrysanthemum leucanthemum* L.; *Leucanthemum leucanthemum* (L.) Rydb.; a few others

GENERAL INFORMATION: *Clumping perennial* to 1 m tall, with *white daisy flower heads* and *creeping roots*. Dense colonies can develop on favorable sites. Oxeye daisy is cultivated as an ornamental, but has escaped cultivation in all contiguous states. In California, it is most invasive in moist grassland and coastal scrub. Livestock generally avoid grazing the foliage, and milk from dairy cattle that have consumed the plant can have an unpleasant flavor. Oxeye daisy is a state-listed noxious weed in Colorado, Minnesota (secondary), Montana (category 1), Ohio (prohibited), Washington (class B, plant quarantine), and Wyoming. It is also a noxious weed in southeastern Australia. Oxeye daisy can host the yellow dwarf potato virus. Native to Europe.

SEEDLINGS: Cotyledons ovate to oblong, about 5–7 mm long, tip slightly rounded, glabrous, on stalks ± 1 mm long, ± fused and sheathing at the base. Leaves alternate, but appear opposite. First few leaves oblong-lanceolate, about 7–15 mm long, tips weakly rounded, tapered to a stalk about 5–10 mm long,

Oxeye daisy (*Leucanthemum vulgare*) plant. J. M. DiTomaso

glabrous. Later leaves similar in size, oblong-obovate, coarsely round-toothed or lobed along the distal margins.

MATURE PLANT: Foliage glabrous or nearly glabrous. Leaves *alternate*, ± oblanceolate, typically *irregular-pinnate-lobed about half-way to the midvein, margin coarsely toothed*, teeth often rounded. Rosette and lower leaves taper to a stalk about as long as the blade, blade and stalk to 15 cm long. Upper stem leaves sessile, ± oblong, shorter than lower leaves. Middle stem leaves usually less than 5 cm long.

ROOTS AND UNDERGROUND STRUCTURES: Main roots creeping, horizontal, shallow, extensive on favorable sites, seasonally develop new shoots. Root fragments can regenerate into new plants.

FLOWERS: May–August. Flower heads mostly *3–7 cm in diameter, solitary on a long stalk*. Ray flowers about 20–30 per head, corollas white, mostly 1–2 cm long. Disk flowers yellow. Receptacle convex to slightly conical, *lacks chaffy bracts*. Phyllaries *unequal, overlapping in 2–3 rows, margins membranous*.

FRUITS AND SEEDS: Achenes lanceolate, ± round in cross-section, often curved, mostly 1.5–2.5 mm long, dark brown to black with 10 pale, longitudinal ribs. *Pappus lacking*.

Oxeye daisy (*Leucanthemum vulgare*) flower heads. J. M. DiTomaso

Oxeye daisy (*Leucanthemum vulgare*) rosette. J. M. DiTomaso

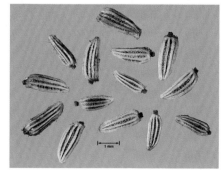

Oxeye daisy (*Leucanthemum vulgare*) seeds. J. O'Brien

POSTSENESCENCE CHARACTERISTICS: Foliage typically dies in late summer or fall after seed matures.

HABITAT: Disturbed places, roadsides, pastures, grassland, and coastal scrub. Often grows on poor soil. Thrives on moist clay soils.

DISTRIBUTION: Klamath Ranges, western North Coast Ranges, higher elevations of the northern and central Sierra Nevada, San Francisco Bay region, western Transverse Ranges, and Peninsular Ranges, to 2000 m. All contiguous states.

PROPAGATION/PHENOLOGY: *Reproduces by seed* and *vegetatively from creeping roots and root fragments.* Seed production is often prolific, especially when abundant moisture is available. Seeds disperse with water, mud, soil movement, animals, and vehicle tires and other human activities. Seed is often a component of commercial "wildflower" packets. Seeds germinate whenever sufficient moisture is available, generally from fall through late spring in California. Seeds can survive ingestion by animals. Some seeds can remain viable for up to about 20 years or more under field conditions. Cultivation and earth-moving and agricultural equipment can disperse root fragments.

MANAGEMENT FAVORING/DISCOURAGING SURVIVAL: Occasional shallow cultivation can spread root fragments and enlarge population. However, cultivation at a depth of about 15 cm in summer followed by repeated shallow cultivations can help control problematic patches in pastures and other places where cultivation is appropriate by desiccating the root systems and killing the seedlings. Mowing during or after flowering will disperse seeds. Removal of smaller patches with hand tools is feasible, and mulching is effective in controlling seedlings and reducing regrowth of roots.

SIMILAR SPECIES: Max chrysanthemum [*Leucanthemum maximum* (Ramond) DC., synonym: *Chrysanthemum maximum* Ramond] is an uncommon *perennial* to 0.7 m tall, with white daisylike flower heads *7–9 cm in diameter* and *rhizomes*. Max chrysanthemum is *further distinguished by having more regularly toothed leaves* and *middle stem leaves more than 5 cm long.* In addition, lower leaves are up to 25 cm long and have slightly winged stalks. Max chrysanthemum has escaped cultivation as an ornamental in some disturbed areas of the North Coast, Central Coast, western South Coast Ranges, and western Transverse Ranges, to 200 m. It also occurs in Oregon, Washington, Louisiana, New York, and Ohio. Native to Europe.

Scotch thistle [*Onopordum acanthium* L. ssp. *acanthium*][ONRAC][Cal-IPC: High)[CDFA list: A]

Illyrian thistle [*Onopordum illyricum* L.][CDFA list: A]

Taurian thistle [*Onopordum tauricum* Willd.] [CDFA list: A]

SYNONYMS: Scotch thistle: asses' thistle; cotton thistle; downy thistle; heraldic thistle; jackass thistle; Queen Mary's thistle; Scotch cottonthistle; silver thistle; winged thistle; woolly thistle

Illyrian thistle: Illyrian cottonthistle

Scotch thistle (*Onopordum acanthium*) plant.   J. M. DiTomaso

Scotch thistle • Illyrian thistle • Taurian thistle

Scotch thistle (*Onopordum acanthium*) rosette.
J. M. DiTomaso

Scotch thistle (*Onopordum acanthium*) flower head and winged spiny stem.   J. M. DiTomaso

Scotch thistle (*Onopordum acanthium*) seeds.
J. O'Brien

Taurian thistle: bull cottonthistle

GENERAL INFORMATION: Coarse *biennials*, occasionally annuals or short-lived perennials, with *spiny leaves, conspicuously spiny-winged stems*, and *spiny flower heads*. These species exist as rosettes throughout the first year until flower stems develop during the second spring or summer season. Although generally uncommon compared to other genera of thistles, *Onopordum* thistles can become locally abundant and develop tall, dense, impenetrable stands, especially on fertile soil. To date, biological control agents have been unsuccessful in controlling these species in the United States.

Scotch thistle: Scotch thistle is the most frequently encountered species of the three in the United States. Besides California, it is a state-listed noxious weed in Arizona (restricted), Colorado, Idaho, Missouri, Nevada, New Mexico (class A), Oklahoma, Oregon (class B), Utah, Washington (class B, plant quarantine), and Wyoming. Native to Europe.

Illyrian thistle: Currently, Illyrian thistle occurs only in California. Native to southeastern Europe.

Taurian thistle: Besides California, Taurian thistle is a state-listed noxious weed in Colorado. Native to the Mediterranean region.

SEEDLINGS: Cotyledons oval to oblong, gradually tapered at the base, fleshy, about 1.5–2 cm long. Leaves alternate, form a rosette, elliptic to oblanceolate, margin irregularly spiny-toothed.

Scotch thistle, Illyrian thistle: Leaves covered with *white woolly hairs*, with the lower surface more densely covered than the upper.

Taurian thistle: Leaves usually covered with *glandular hairs*. Seedlings do not compete well with established perennial grasses.

MATURE PLANT: Stems erect, branched, with *conspicuous spiny wings that are continuous along the entire length of the stems*. Rosette and stem leaves alternate, very spiny.

Scotch thistle: Generally 1.5–3 m tall. Foliage covered with *woolly, pale gray hairs*. Stem wings broad, mostly 2–3 cm wide, but can be up to 5 cm wide. Leaves 10–50 cm long, generally broadly elliptic, *margin spiny-toothed to shallow spiny-lobed*.

Illyrian thistle: To 2.5 m tall. Foliage covered with *woolly pale gray hairs*. Stem wings 0.5–2 cm wide. Leaves 10–50 cm long, often narrow elliptic, *margin deeply pinnate-lobed 1 or 2 times*, lobes spiny-toothed.

Taurian thistle: To 2 m tall. Foliage typically covered with *short, sticky glan-*

Illyrian thistle (*Onopordum illyricum*) flower head. G. D. Barbe

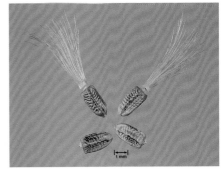

Illyrian thistle (*Onopordum illyricum*) flowering stems. G. D. Barbe

Illyrian thistle (*Onopordum illyricum*) seeds. Jim O'Brien

Scotch thistle • Illyrian thistle • Taurian thistle

Taurian thistle *(Onopordum tauricum)* flower head. W. J. FERLATTE

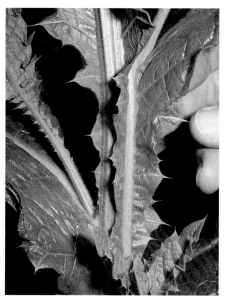

Taurian thistle *(Onopordum tauricum)* leaves with spiny margiins. COURTESY OF ROSS O'CONNELL

Taurian thistle *(Onopordum tauricum)* rosette. D.O. CLARK

Taurian thistle *(Onopordum tauricum)* seedling. J. M. DITOMASO

dular hairs. Stem wings usually 0.5–2 cm wide. Leaves 10–25 cm long, acutely triangular, *margins deeply pinnate-lobed* with 6–8 pairs of spiny-toothed lobes.

ROOTS AND UNDERGROUND STRUCTURES: All have a thick taproot.

FLOWERS: Heads spheric to hemispheric, mostly solitary but often in clusters (cymes) of 2–7. Heads consist of numerous spine-tipped phyllaries in many overlapping rows and *numerous disk flowers.* Phyllary spines mostly less than 5 mm long. Receptacle slightly fleshy, *lack chaffy bracts or bristles, deeply pitted,* with the *depressions surrounded by membranous extension.*

Scotch thistle: July–September. Phyllaries linear to narrow-lanceolate, covered with short hairs and a few cobwebby hairs. Disk flowers white or purple, *glabrous,* 20–25 mm long.

Taurian thistle (*Onopordum tauricum*) seeds. J. O'Brien

Illyrian thistle: Summer. Phyllaries lanceolate to ovate, glabrous or with a few cobwebby hairs. Disk flowers purple, *short glandular-hairy on the lobes*, 25–35 mm long.

Taurian thistle: Summer. Phyllaries lanceolate, glabrous. Disk flowers purplish pink, *glabrous or nearly glabrous*, 25–30 mm long.

FRUITS AND SEEDS: Achenes narrowly obovate, usually 4–5-angled, glabrous, mottled brown to blackish, surface roughened with wavy transverse ridges. Pappus bristles equal, numerous, minutely barbed, fused into a ring at the base, detach from the achene as a unit.

Scotch thistle: Achenes 4–5 mm long. Pappus bristles *pink to reddish*, 7–9 mm long.

Illyrian thistle: Achenes 4–5 mm long. Pappus bristles *cream-colored*, 10–12 mm long.

Taurian thistle: Achenes 5–6 mm long. Pappus bristles *pink to reddish*, especially near the tips, 8–10 mm long.

POSTSENESCENCE CHARACTERISTICS: Spiny-winged stems can persist into the next season with spiny phyllaries and receptacles attached.

HABITAT: Disturbed sites, roadsides, fields, annual grassland, pasture, rangeland, and riparian areas. Thrive on fertile soil.

Scotch thistle: Often inhabits sites with high soil moisture. Can also be found along canals and ditch banks.

DISTRIBUTION: Scotch thistle: Klamath and Cascade Ranges (c, e, and n Siskiyou, ce Trinity, w and ce Shasta Cos.), Modoc Plateau (Modoc, Lassen, e Plumas, and e Sierra Cos.), North Coast Ranges (s Mendocino, cw and ne

Sonoma, se Lake Cos.), Sierra Nevada (c and se Nevada, n Placer, se Calaveras, Tulare Cos.; intersection of El Dorado, Alpine, and Amador Cos.), Central Valley (sc Sacramento Co.), San Francisco Bay region (nw Santa Cruz, sw San Mateo Cos.), South Coast Ranges (e Monterey, cw and s San Benito Cos.), Mojave Desert (se and c Inyo Co.), and northwestern Transverse Ranges (ne Ventura, sw Kern Cos.), to 1600 m.

Previous infestations now eradicated in Butte, Alameda, Santa Barbara, San Bernardino, and San Diego Counties. All western states, most central and north eastern states, a few southern states.

Illyrian thistle: San Francisco Bay region (c and cw Santa Clara Co.), to 500 m.

Taurian thistle: Modoc Plateau (ne Siskiyou Co.), northern North Coast Ranges (ce Monterey Co.), Sierra Nevada foothills (sc Yuba, ne Nevada, and c Madera Cos.), to 1400 m. Colorado.

PROPAGATION/PHENOLOGY: *Reproduce by seed*. Achenes fall near the parent plant, disperse short distances with wind and animals, and to greater distances with water, animals, including livestock, and human activities. In California, most seeds germinate in the fall after the first rains, but seeds can germinate year-round under favorable moisture and temperature conditions. Buried seed of Scotch and Illyrian thistle can remain viable in the soil seedbank for at least 7 years and possibly up to 20 years or more. Yearly seed production and dormancy vary greatly, depending on environmental conditions.

Scotch thistle: A low percentage of newly matured seeds (about 8–14%) can germinate immediately under favorable conditions. Newly matured seeds are generally sensitive to light and contain a water-soluble germination inhibitor that is removed by leaching with water or can be negated with gibberellic acid. Nitrogen, cold stratification, and fluctuating temperatures stimulate germination of newly matured seeds. Chilling sensitizes recently matured seeds to photoperiod, with 8 hours of light being optimal for germination. Seed burial induces dormancy. Seeds recovered from soil are sensitive to light quality, but not photoperiod, and are less responsive to the germination stimulators listed above. Low-intensity burning stimulates germination of dormant seeds in the upper soil layer, but imbibed seeds are less tolerant of heating by fire. One plant can produce an average of 20,000 to 40,000 seeds. Seedlings emerge from soil depths to 4.5 cm, with 0.5 cm being optimal. Fall-germinating seedlings overwinter as rosettes and bolt the following season to produce large plants. Early-spring-germinating seedlings bolt the same season to produce smaller plants. Late-spring-germinating seedlings that fail to receive adequate chilling during that season will overwinter and bolt the next season. This germination pattern leads to the largest plants.

Illyrian thistle: Germination responses are probably similar to those of Scotch thistle. Seedlings and flowering stems are highly palatable to goats. In one study, less than 1% of viable seed germinated after passage through the digestive tracts of goats and sheep, although it is unclear if the lack of germination was due to dormancy or seed death.

Scotch thistle • Illyrian thistle • Taurian thistle

Snowy thistle (*Cirsium occidentale* var. *candidissimum*), a native nonweedy species that can resemble *Onopordum* species.   J. M. DiTomaso

Snowy thistle (*Cirsium occidentale* var. *candidissimum*) rosettes also resemble *Onopordum* species.   J. M. DiTomaso

MANAGEMENT FAVORING/DISCOURAGING SURVIVAL: Managing rangeland and perennial grass pastures to minimize open gaps from fall through spring helps to discourage invasion by these species. Unlike sheep and cattle, goats will consume Illyrian thistle rosettes and flowering stems. Utilizing goats to browse Illyrian thistle before flower heads develop can help control populations of Illyrian thistle. Manually removing *Onopordum* species before seeds mature can control small populations. Establishing or encouraging perennial grasses can increase Scotch thistle seedling mortality due to increased competition for moisture.

SIMILAR SPECIES: Unlike other genera of thistles with spiny foliage, *Onopordum* thistles have flower head *receptacles that lack bristles among the flower* and that have *deep pits surrounded by membranous extensions*.

Bristly oxtongue [*Picris echioides* L.][PICEC] [Cal-IPC: Limited]

SYNONYMS: bugloss; bugloss-picris; *Helminthia echioides* (L.) Gaertn.; *Helminthotheca echioides* (L.) Holub

GENERAL INFORMATION: Erect winter and/or summer *annual or biennial* to nearly 1 m tall, with *milky sap, stiff-bristly foliage,* and *yellow dandelion-like flower heads.* Plants exist as basal rosettes until branched flower stems develop at maturity. Native to Europe.

SEEDLINGS: Cotyledons oval to obovate, about 8–12 mm long, glabrous, on stalks ± 2 mm long. Leaves alternate. First leaf ± oblong, base tapered, about 8–20 mm long, evenly covered with stiff papilla-based hairs. Hair tips branched. Subsequent leaves resemble first leaf except increasingly larger.

MATURE PLANT: Variable. *Foliage evenly covered with hairs that are minutely branched at the tip* (magnification required). Stems coarse, *branched.* Leaves alternate, oblong, 5–20 cm long, covered with *stiff, coarse, papilla-based hairs.* Margins smooth, coarse-toothed, or shallow-lobed. Rosette and lower leaves taper to a winged stalk. Upper leaves sessile, sometimes lobed at the base and clasping the stem.

ROOTS AND UNDERGROUND STRUCTURES: Taprooted, with spreading lateral roots.

Bristly oxtongue (*Picris echioides*) infestation in a meadow. J. M. DiTomaso

Bristly oxtongue

Bristly oxtongue *(Picris echioides)* plant.
J. M. DiTomaso

Flower head of bristly oxtongue *(Picris echioides)*.
J. K. Clark

FLOWERS: May–September (October). *Stems branched.* Flower heads *terminal and axillary*, mostly 2–4 cm in diameter, on short to long stalks, consist only of yellow ligulate flowers. Phyllaries separate to the base, in 2–3 overlapping rows. Outer phyllaries broadly ovate, *loosely spreading*, bristly. Inner phyllaries narrow. Involucre (phyllaries as a unit) 1.5–2 cm long. Receptacles lack chaffy bracts. Insect-pollinated.

FRUITS AND SEEDS: Achenes brown to pale orange-brown, cylindrical-lanceolate with a slender beak at the apex, surface minutely horizontally ridged, body 2.5–4 mm long, beak about as long as the body. Pappus bristles white, feathery (plumose), equal, mostly 4–7 mm long.

POSTSENESCENCE CHARACTERISTICS: Stems with hairs that are minutely branched at the tip remain intact until winter. Papillae-based hairs are visible on dead leaves and phyllaries.

HABITAT: Roadsides, waste places, fields, pastures, crop fields, orchards, vineyards, landscaped areas, gardens, and other disturbed open places. Most common in seasonally wet places. Thrives on clay soils, especially those high in calcium.

DISTRIBUTION: Common throughout most of California, except deserts and Great Basin, to 450 m. More common near the coast than in interior regions of southern California. Oregon, Washington, North Dakota, a few central states, and northeastern states.

PROPAGATION/PHENOLOGY: *Reproduces by seed.* Seeds probably disperse short distances with wind. Some seeds disperse greater distances with water, soil movement, and human activities, including clinging to tools, vehicle tires, and landscape and agricultural machinery. Seeds germinate in fall after the rains begin and/or in spring depending on the location and climate.

Bristly oxtongue (*Picris echioides*) leaves.
J. M. DiTomaso

Bristly oxtongue (*Picris echioides*) seedlings.
J. M. DiTomaso

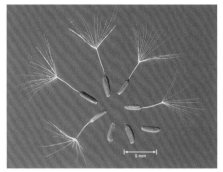

Bristly oxtongue (*Picris echioides*) seeds.
J. O'Brien

MANAGEMENT FAVORING/DISCOURAGING SURVIVAL: Cultivation, manual removal, or repeated mowing before seeds develop can control bristly oxtongue.

SIMILAR SPECIES: Prickly goldenfleece [*Urospermum picroides* (L.) Scop. ex F.W. Schmidt] is an uncommon *annual to biennial* to about 0.5 m tall, with *milky sap* and yellow flower heads that closely resemble those of bristly oxtongue. Unlike bristly oxtongue, prickly goldenfleece has hairs that *lack* minutely branched tips and flower heads with only a *single row of broad phyllaries*. In addition, foliage is usually sparsely covered with simple, coarse, bristly hairs that are *not papilla-based*. In addition, achenes have a *beak that is longer than the body*, and the body is covered with minute tubercles or projections. Prickly goldenfleece inhabits disturbed places in the San Francisco Bay region and South Coast, to about 100 m. Native to Europe.

Tansy ragwort [*Senecio jacobaea* L.][SENJA] [Cal-IPC: Limited][CDFA list: B]

Common groundsel [*Senecio vulgaris* L.][SENVU]

SYNONYMS: Tansy ragwort: stinking willie

Common groundsel: birdseed; chickenweed; grimsel; old-man-of-the-spring; ragwort; simson

GENERAL INFORMATION: Erect herbs with *alternate, pinnate-lobed leaves* and *yellow flower heads*. Many *Senecio* species, including tansy ragwort and common groundsel, contain pyrrolizidine alkaloids and are *toxic* to humans and livestock when ingested in a single large quantity or in small amounts over several weeks or months. Ingested alkaloids damage the liver, and liver failure can continue to develop long after alkaloid ingestion has ceased. Cattle, horses, goats, and young animals are more susceptible to the toxic effects than sheep. Native to Eurasia.

Tansy ragwort: Noxious cool-season *annual, biennial, or perennial* to 1.2 m tall. Tansy ragwort has caused heavy livestock losses on pastures and rangelands in the Pacific Northwest. Animals generally avoid ingesting mature plants, but poisoning can occur when seedlings are grazed accidentally along with other forage or when contaminated hay is ingested. Tansy ragwort populations have

Tansy ragwort (*Senecio jacobaea*) infestation.　　　　　　　　J. M. DiTomaso

Tansy ragwort (*Senecio jacobaea*) rosette.
J. M. DiTomaso

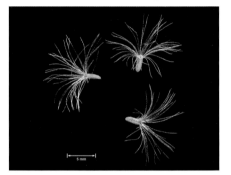

Tansy ragwort (*Senecio jacobaea*) plant.
J. M. DiTomaso

Tansy ragwort (*Senecio jacobaea*) seeds.
J. O'Brien

been reduced dramatically in California and Oregon since the introduction of the cinnabar moth (*Tyria jacobaeae*) in 1959, and later, the ragwort flea beetle (*Longitarsus jacobaeae*). Unfortunately, these insects do not tolerate colder inland climates. Bees foraging on tansy ragwort flowers produce bitter honey tainted with alkaloids. Besides California, tansy ragwort is a state-listed noxious weed in Arizona (prohibited), Colorado, Idaho, Oregon (class B), and Washington (class B).

Common groundsel: Nearly ubiquitous winter or summer *annual* to 0.6 m tall. Infestations are most troublesome during cool, moist periods. Livestock losses due to the ingestion of common groundsel are generally uncommon. Most poisonings are due to ingestion of contaminated hay or hay cubes over a period of time. Common groundsel is primarily a problem in alfalfa fields, especially the first cutting. Common groundsel is a state-listed noxious weed in Colorado.

SEEDLINGS: Remain as rosettes until maturity. First leaves alternate, subsequent leaves variable, margins toothed to deeply pinnate-lobed.

Tansy ragwort: Cotyledons oval, ± 3 mm long, tip truncate or slightly indented, base rounded-wedge-shaped. First leaves oval with wavy margins, 6–8 mm long, sometimes with a few glandular hairs.

Common groundsel (*Senecio vulgaris*) plant. J. K. CLARK

Common groundsel (*Senecio vulgaris*) flowering stem. J. M. DiTomaso

Common groundsel (*Senecio vulgaris*) seedling. J. K. CLARK

Common groundsel (*Senecio vulgaris*) fruiting head. J. M. DiTomaso

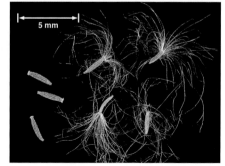
Common groundsel (*Senecio vulgaris*) seeds. J. O'Brien

Woodland groundsel (*Senecio sylvaticus*) in a logged opening.   J. M. DiTomaso

Common groundsel: Cotyledons narrowly elliptic to oblong, 3–11 mm long, tip rounded-acute, base tapered and sheathing, glabrous or scurfy, often purplish below. First leaves ovate, 8–12 mm long, tip broadly acute, margin coarsely shallow-toothed, glabrous or with a few hairs.

MATURE PLANT: Foliage glabrous to lightly covered with long wavy to cottony hairs, especially on the midveins, lower leaf surfaces, and new growth. Leaves highly variable, generally *evenly spaced on stems*. Lower leaves taper into indistinct petioles. Upper leaves reduced, sessile, ± clasping stem.

Tansy ragwort: Stems erect, single or branched from the crown, branched near the top, to 1.2 m tall. Leaves *deeply pinnate-dissected 1–2 times* on a 3-dimensional plane so the leaves do not appear flat, mostly 5–20 cm long. Lower leaves deciduous.

Common groundsel: Stems ± erect, single or branched from the crown, sometimes rooting at the lower nodes. Leaves *deeply pinnate-lobed 1 time*, 2–10 cm long, margin toothed.

ROOTS AND UNDERGROUND STRUCTURES: Tansy ragwort: Crown or short taproot usually develops many spreading fleshy lateral roots about 15 cm long, with numerous, deeper secondary fibrous roots. Roots and crowns, especially of rosettes, readily develop new shoot and root buds when injured or disturbed.

Common groundsel: Taproot small, sometimes not evident. Secondary fibrous roots shallow.

FLOWERS: Flower heads clustered at stem tips, with *yellow flowers*. Main phyllaries in *1 equal row, often black-tipped*, typically with a few *highly reduced phyllaries or bracts at the base*. Insect-pollinated.

Woodland groundsel (*Senecio sylvaticus*) foliage.
J. M. DiTomaso

Woodland groundsel (*Senecio sylvaticus*) plant.
J. M. DiTomaso

Woodland groundsel (*Senecio sylvaticus*) flowering and fruiting heads.
J. M. DiTomaso

Tansy ragwort: July–September. Flower heads *showy*, in *dense, flat-topped to slightly rounded clusters* of about 20–60 heads. Clusters woolly-hairy at the base. Heads consist of numerous disk flowers surrounded by *12–15 well-spaced ray flowers*. Ray corollas 8–12 mm long, ± 2 mm wide. Main phyllaries *± 13 per head*, mostly *3–5 mm long*, hemispheric to cylindrical as a unit.

Common groundsel: Nearly year-round. Flower heads *not showy*, often nodding, *few to several initially in dense clusters, later in loose clusters* as the stalks elongate. Heads consist of numerous disk flowers, usually *lack ray flowers*. Main phyllaries *± 21 per head*, mostly *4–6 mm long*, as a unit cylindrical to bell- or urn-shaped. *Reduced outer phyllaries conspicuously black-tipped*, usually 2–6 per head. Often self-pollinated.

FRUITS AND SEEDS: Achenes cylindrical, 1.5–3 mm long, shallow-ribbed, light brown, often pubescent. Pappus bristles numerous, soft, white, about twice the length of the achene.

Tansy ragwort: Ray achenes glabrous. Disk achenes pubescent on ribs. Pappus ± persistent.

Common groundsel: Achenes sparsely pubescent between ribs. Pappus generally deciduous.

POSTSENESCENCE CHARACTERISTICS: Dead brown stems can persist for several months.

HABITAT: Tansy ragwort: Disturbed sites, waste places, roadsides, pastures, fields, rangeland, near riparian areas, and in forested areas. Grows best on light, well-drained soils in cool, moist climates. Seldom tolerates high water tables or acidic soils. Inhabits grassland, woodland, and dune communities in its native range.

Common groundsel: Disturbed sites, waste places, roadsides, fields, vegetable and agronomic crop fields, gardens, nurseries, orchards, vineyards, landscaped areas, and yards. Grows best on moist, fertile soil. Plants die during extended dry, hot periods. Adapts to many environmental conditions.

DISTRIBUTION: Tansy ragwort: North Coast, western Klamath Ranges, southwestern Cascade Range, northern Sierra Nevada, northern Sacramento Valley, and San Francisco Bay region, to 1500 m. Idaho, Montana, Oregon, Washington to British Columbia, Illinois, Michigan, and some northeastern states.

Common groundsel: Throughout California, except deserts, to 1500 m. All contiguous states.

PROPAGATION/PHENOLOGY: Achenes disperse with wind, water, and human activities, including agricultural equipment, and by clinging to vehicle tires, to animals, and to the shoes and clothing of humans.

Tansy ragwort: *Reproduces by seed* and *vegetatively from the fleshy roots*. Achenes that disperse with wind usually travel only a few meters. Some seeds survive ingestion by birds. Seeds do not require an after-ripening period and are often highly viable. Germination occurs soon after seeds are shed from summer through fall, but germination can also occur year-round under favorable conditions. Ray flower seeds have thicker coats (pericarp), disperse later, and germinate slower than disk flower seeds. Frost, drought, and burial often induce seed domancy. Seeds typically remain viable for at least 6 years under field conditions, but have been reported to survive for up to 20 years. Most seedlings emerge from soil depths of 1–2 cm. Open sites with little competing vegeta-

Woodland groundsel (*Senecio sylvaticus*) seedling.   J. M. DiTomaso

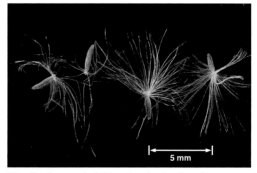

Woodland groundsel (*Senecio sylvaticus*) seeds.   J. O'Brien

Purple ragwort *(Senecio elegans)* plant. J. M. DiTomaso

tion favor seedling survival. Plants are typically biennial, but some regenerate after flowering to become short-lived perennials. Crowns and fleshy roots can develop new root and shoot buds, especially in response to disturbance or injury. Fragments of the fleshy roots can generate new shoots.

Common groundsel: *Reproduces by seed.* Seed dormancy varies among populations. Seeds typically germinate in early spring through late fall, but can germinate year-round in mild climate areas. Fluctuating temperatures, light, cold stratification, leaching with water, or scarification stimulate germination. Seed longevity is poorly documented. In one case, seeds buried in a meadow were reported to survive for more than 50 years. Seedlings mature rapidly, and 3 or more generations may complete a life cycle in one season.

MANAGEMENT FAVORING/DISCOURAGING SURVIVAL: Tansy ragwort: Mowing or mechanical control often enhances survival by stimulating vegetative reproduction. Managing pastures to maintain continuous vegetative cover decreases survival of seedlings.

Common groundsel: Manual removal and other methods that prevent plants from producing seed can control common groundsel.

SIMILAR SPECIES: Woodland groundsel [*Senecio sylvaticus* L.][SENSI] is an *annual* to 0.8 m tall that closely resembles common groundsel. Unlike common groundsel, woodland groundsel *lacks the outer reduced phyllaries* or has only a few reduced phyllaries with *green tips*. In addition, woodland groundsel has less than 8 ray flowers with corollas that barely extend beyond the main phyllaries. Woodland groundsel inhabits open disturbed forests, woodland, and rocky areas in the Central Coast, North Coast, and Klamath Ranges, to 200 m. It also occurs in Oregon, Washington, Louisiana, and some northeastern states. Woodland groundsel is aromatic and can cause headaches with constant exposure to its odor. Seeds disperse with wind and can germinate immediately if conditions are favorable, but most seeds typically germinate in fall and spring. Native to Europe.

Purple ragwort [*Senecio elegans* L., synonym: redpurple ragwort] is an uncommon *annual* to 0.6 m tall that has escaped cultivation in a few areas. It is easily distinguished from other ragworts by having showy flower heads with *purple*

Purple ragwort *(Senecio elegans)* seedling.
J. M. DiTomaso

Purple ragwort *(Senecio elegans)* flower heads.
J. M. DiTomaso

Purple ragwort *(Senecio elegans)* seeds. J. O'Brien

ray flowers and *sticky glandular foliage*. Phyllaries are black-tipped. Purple ragwort inhabits disturbed sites in the San Francisco Bay region, Central Coast, and South Coast, to 100 m. Native to South Africa.

Oxford ragwort [*Senecio squalidus* L.] [CDFA list: B] is a winter *annual, biennial or short-lived perennial* to 0.3(0.5) m tall, with *showy yellow flower heads* and foliage that resembles that of common groundsel. Oxford ragwort was discovered growing at the Berkeley Botanic Garden a number of years ago. It has since been eradicated, but remains on the California state noxious weed list. In England, Oxford ragwort was introduced as an ornamental to the Oxford Botanic Garden in the late 1700s. It spread rapidly throughout the British Isles, where it hybridizes with common groundsel. Most hybrids are infertile, but in a few places the cross has produced a larger, fertile triploid species [*Senecio cambrensis* Rosser]. Oxford ragwort is distinguished by having *few to several showy flower heads* arranged in *loose, flat-topped to slightly rounded clusters*. Heads consist of numerous disk flowers surrounded by *10–14 ray flowers* with corollas 8–15 mm long, 2–4 mm wide. There are ± *21 main phyllaries 5–7 mm long*. In addition, involucres are bell-shaped to cylindric, fruiting heads are often nodding, and achenes are ± pubescent. Native to the mountainous regions of central and southeastern Europe, where it grows on scree and is not weedy.

Sweet resinbush [*Euryops subcarnosus* DC. ssp. *vulgaris* B. Nord.] is a *shrub* to about 1.2 m tall with yellow flower heads, that is related to the *Senecio* group. Sweet resinbush was previously referred to as *Euryops multifidus* (Thunb.) DC. in North America. However, *Euryops multifidus* is a different South African species that is often confused with sweet resinbush. To date, sweet resinbush is not known to be naturalized in California and is not included in California floras. Sweet resinbush is becoming increasingly problematic in some Sonoran Desert areas of Arizona, where large stands of the shrub displace the native vegetation. It is expected to eventually expand its range into southern California. One specimen has been collected from the entrance of Trancas Canyon in Los Angeles County. Sweet resinbush foliage is glabrous, except for woolly tufts of hairs in the leaf axils. Leaves are *linear, 3–6-lobed at the tip*, and up to 1.5 cm long. The yellow flower heads are ± 1 cm wide on stalks (peduncles) a few centimeters long. Leaves and flower heads are mostly clustered at the ends of short branchlets. Unlike members of the genus *Senecio*, sweet resinbush has fruits with deciduous, soft pappus bristles that have *conspicuously long barbs*, and flower heads with *phyllaries that are slightly fused at the base* and that *lack a row of very short phyllaries* below. Sweet resinbush was introduced to Arizona as an experimental shrub for erosion control. Native to South Africa.

Blessed milkthistle [*Silybum marianum* (L.) Gaertn.] [SLYMA][Cal-IPC: Limited]

SYNONYMS: cabbage thistle; holy thistle; lady's thistle; milkthistle; spotted thistle; St. Mary's thistle; variegated thistle; white thistle; *Carduus marianus* L.; *Carduus mariae* Crantz; *Carthamus maculatum* (Scop.) Lam.; *Cirsium maculatum* Scop.; *Mariana lactea* Hill; *Silybum maculatum* (Scop.) Moench; *Silybum mariae* (Crantz) Gray

GENERAL INFORMATION: Erect winter or summer *annual or biennial* generally to 2 m tall, with *white-variegated prickly leaves*. Blessed milkthistle often occurs in dense, competitive stands. Depending on the amount of soil moisture, plants range from extremely small to very tall. Blessed milkthistle is a state-listed noxious weed in Oregon (class B) and Washington (class A) and is a government-listed noxious weed in much of southern Australia. Under stressful conditions, such as drought, herbicide treatment, or mowing, the foliage can occasionally accumulate levels of nitrates that are *toxic* to cattle. The seedhead weevil (*Rhinocyllus conicus*) was released in 1971 in southern California as a biocontrol agent for blessed milkthistle, but control has been poor to date. In addition, the weevil attacks several native thistle species. Seeds contain silybin, a compound that stimulates liver tissue growth and is used as an antidote for poisoning by the death cap mushroom [*Amanita phalloides*]. Blessed milkthistle has been used medicinally for at least 2000 years. The young foliage, with prickles

Blessed milkthistle (*Silybum marianum*) patch.   J. M. DiTomaso

Blessed milkthistle

Blessed milkthistle (*Silybum marianum*) flower head.　　　　　　　　J. M. DiTomaso

Blessed milkthistle (*Silybum marianum*) fruiting head with dispersing seed.　　J. M. DiTomaso

Blessed milkthistle (*Silybum marianum*) immature plant.　　　　　　　J. M. DiTomaso

removed, is sometimes consumed as a salad green or cooked vegetable. Native to the Mediterranean region.

SEEDLINGS: Cotyledons broadly obovate, about 1–1.5 cm long, thick, glabrous. First leaf pair alternate, elliptic-oblong, mostly 1–2 cm long, margin prickly-toothed, nearly glabrous.

MATURE PLANT: Stems branched, thick, hollow, ribbed, *lack wings or spines*, sparsely hairy. Leaves coarsely pinnate-lobed, prickly-toothed, ruffled, nearly glabrous. Upper surfaces shiny, *green and conspicuously variegated with white*.

Basal leaves 15–70 cm long. Stem leaves reduced, sessile, ± clasping the stem at the base, often curved downward.

ROOTS AND UNDERGROUND STRUCTURES: Taproot thick, usually with numerous lateral roots.

FLOWERS: April–July. *Flower heads consist of numerous pink to purple disk flowers*, base 2–6 cm in diameter, on long stalks. Phyllaries spine-tipped, pinnate prickly-toothed. Outer phyllaries typically bent downward (reflexed). Receptacles have *bristly chaffy bracts*.

FRUITS AND SEEDS: Achenes ± lanceolate, 6–8 mm long, slightly flattened, mottled black and brown, with a yellowish ring at the apex. Basal attachment point slightly angled. Pappus bristles numerous, *minutely barbed,* flat, mostly 15–20 mm long, *fused at the base to form a ring, detach as a unit.*

POSTSENESCENCE CHARACTERISTICS: Dead stems with bristly flower head receptacles can persist throughout the summer and fall and often into the winter.

HABITAT: Disturbed sites, roadsides, pastures, fields, agronomic crops, waste places, orchards, and trail margins in chaparral and woodlands. Grows best on fertile soils.

DISTRIBUTION: North Coast, North Coast Ranges, Central Valley, San Francisco Bay region, South Coast Ranges, South Coast, and Channel Islands, to 500 m. Oregon, Washington, Nevada, Arizona, some southern states, and many northeastern states.

Blessed milkthistle (*Silybum marianum*) seedling.   J. M. DiTomaso

Blessed milkthistle (*Silybum marianum*) seeds. J. O'Brien

PROPAGATION/PHENOLOGY: *Reproduces by seed.* Seeds probably disperse only short distances with wind, but they disperse to longer distances with human activities, water, soil movement, animals, and as a crop seed or feed contaminant. Most seeds germinate after the first fall rain, but some can germinate throughout winter and early spring. Seeds can survive at least 9 years under field conditions.

MANAGEMENT FAVORING/DISCOURAGING SURVIVAL: Cultivation can control seedlings. Mowing mature plants before flowers open can help control stands. Burning can encourage seed germination and establishment.

SIMILAR SPECIES: Italian thistle [*Carduus pycnocephalus* L.] is sometimes confused with blessed milkthistle because of its weakly variegated leaves. Unlike blessed milkthistle, Italian thistle has *spiny-winged stems* and *flower heads 1–2 cm wide*. Refer to the entry for **Plumeless thistle, Musk thistle, Italian thistle, and Slenderflower thistle** (p. 217) for more information.

Lawn burweed [*Soliva sessilis* Ruiz Lopez & Pavon] [SOVPT]

SYNONYMS: carpet burweed; common soliva; field burweed; jo-jo; onehunga; soliva; spurweed; *Gymnostyles chilensis* Spreng.; *Soliva daucifolia* Nutt.; *Soliva microloma* Phil.; *Soliva pterosperma* (Juss.) Less.

GENERAL INFORMATION: Low-growing winter *annual* to about 25 cm in diameter and to 7 cm tall, with *dissected compound leaves* and *inconspicuous flower heads*. Lawn burweed has bur-like fruits that can puncture the skin of humans and other animals, making it a nuisance weed in turf. It is a state-listed noxious weed in Washington (class A) and is a noxious weed in western Australia. Native to South America.

SEEDLINGS: Cotyledons oblong to narrowly ovate, fused at the base. First leaf pair lanceolate, bases long-tapered, often sparsely hairy. Subsequent leaves pinnate-dissected into 4–5 elliptic-lanceolate segments, often sparsely hairy.

MATURE PLANT: Stems *prostrate to ascending*, often branched at the base, often dark or purple-spotted. Foliage covered with soft, short hairs. Leaves alternate, carrotlike, *pinnate-compound*, about 1–3 cm long, usually with *palmate-dissected leaflets*. Leaflet lobes usually 2–8, narrowly elliptic-lanceolate.

ROOTS AND UNDERGROUND STRUCTURES: Roots fibrous.

FLOWERS: (February) April–June. Heads *sessile, axillary*, mostly 5–10 mm in

Lawn burweed (*Soliva sessilis*) plants. J. M. DiTomaso

diameter, 2.5–3 mm long, greenish, consist of *inconspicuous male disk flowers with yellow corollas surrounded by fertile disklike female flowers that lack corollas* (disciform heads). Disk flowers 7–9 per head, with a *long, thick style at the apex*. Phyllaries 5–12 per head, ovate, tip acute, in 1 or 2 nearly equal rows. Peripheral female flowers 9–12 per head. Receptacles flat, lack chaffy bracts.

FRUITS AND SEEDS: Achenes ovate, flattened, back slightly convex, body minutely hairy, with a *thick spinelike style at the apex*, about 3.5–5.5 mm long including style. Achene body margin *winged, with a hornlike, slightly incurved tooth on each side of the style*. Wing bases sometimes notched.

POSTSENESCENCE CHARACTERISTICS: Dead plants generally do not persist.

HABITAT: Turf, compacted paths, roadsides, pastures, waste places, and other disturbed places.

DISTRIBUTION: Northwestern region, Central-western region, Southwestern region, Central Valley, Sierra Nevada foothills, and possibly elsewhere. Elevation limit undocumented. Oregon, Washington, and southern states.

PROPAGATION/PHENOLOGY: *Reproduces by seed.* Achenes disperse by clinging to animals, the clothing of humans, tires, machinery, and in lawn clippings. In California, seeds generally germinate during the cool season after the fall or winter rains begin.

MANAGEMENT FAVORING/DISCOURAGING SURVIVAL: Manually removing plants or covering plants with a thick layer of mulch before fruits develop can help control lawn burweed. Removing the fruits from shoes, clothing, pets, and

Lawn burweed (*Soliva sessilis*) fruiting head. J. M. DiTomaso

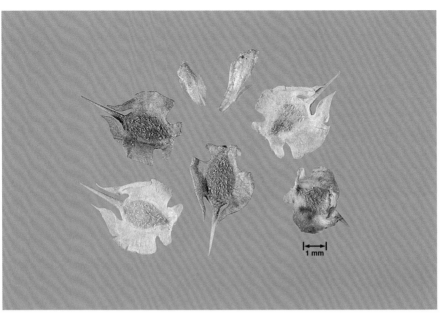

Lawn burweed (*Soliva sessilis*) seeds. J. O'BRIEN

landscape machinery can help prevent the spread of lawn burweed. Promoting growth of cool-season turf species in fall and maintaining higher mowing heights can increase turf competition and suppress lawn burweed.

SIMILAR SPECIES: Pineapple-weed [*Chamomilla suaveolens* (Pursh) Rydb.] [MATMT], **southern brassbuttons** [*Cotula australis* (Sieber) Hook. f.][CULAU], and **Mexican brassbuttons** [*Cotula mexicana* (DC.) Cabrera] have dissected leaves that can be confused with those of lawn burweed. Only lawn burweed has *achenes with a spinelike style at the apex*. In addition, pineapple-weed has a more erect habit and conspicuous greenish yellow egg-shaped flower heads on stalks. Southern brassbuttons has conspicuous yellow buttonlike flower heads on long stalks. Mexican brassbuttons has green flower heads about 2 mm wide on stalks to 10 mm long. Refer to the entries for **Pineappleweed** (p. 261) and **Brassbuttons** (p. 291) for more information.

Spiny sowthistle [*Sonchus asper* (L.) Hill ssp. *asper*] [SONAS]

Annual sowthistle [*Sonchus oleraceus* L.][SONOL]

SYNONYMS: Spiny sowthistle: prickly sowthistle; spiny-leaf sowthistle; *Sonchus oleraceus* var. *asper* L.

Annual sowthistle: colewort; common sowthistle; hares lettuce; pualele; smooth sowthistle; *Sonchus ciliatus* Lam.

GENERAL INFORMATION: Coarse, erect, winter or summer *annuals* to 1.4 m tall, with *milky juice* and *yellow dandelion-like flower heads*. Plants exist as rosettes

Spiny sowthistle (*Sonchus asper*) plant.   J. M. DiTomaso

Spiny sowthistle (*Sonchus asper*) flowering stem. J. M. DiTomaso

Spiny sowthistle (*Sonchus asper*) seeds. J. O'Brien

until flower stems develop in spring or summer. Both species can harbor economically important pests that affect vegetables and fruit trees, such as nematodes, lettuce aphid (*Nasonovia ribis-nigri*), lettuce-root aphid (*Pemphigus bursarius*), and green-peach aphid (*Myzus persicae*), which are potential vectors of certain viral diseases. Both species were introduced with the Spanish settlement or perhaps even earlier. Native to Europe.

SEEDLINGS: Species are not easy to distinguish. Cotyledons ovate to elliptic-oblong, about 4–8 mm long, 1.5–7 mm wide, glabrous, short-stalked. Leaves alternate. First leaf ± ovate, about 4–15 mm long, sparsely hairy, margin lined with backward pointing teeth, base abruptly narrowed or gradually tapered to a stalk 2–10 mm long.

MATURE PLANT: Foliage glabrous, ± glaucous. Stems *thick*, hollow between nodes, smooth, leafy. Leaves alternate, margins toothed and ± ruffled. Lower leaves mostly 10–20 cm long, base tapered or winged, deeply pinnate-lobed, lateral lobes lanceolate, terminal lobe deltoid. Upper leaves alternate, sessile, smaller than lower leaves, *base lobed and clasping the stem*.

Spiny sowthistle: Stems angled. Leaf margin teeth *prickly* to touch. *Basal lobes of upper leaves rounded, strongly curved downward or coiled.*

Annual sowthistle: Terminal lobe of lower leaves much larger than lateral lobes. Leaf margin teeth *soft* to touch. *Basal lobes of upper leaves acute*, straight to slightly curved downward.

ROOTS AND UNDERGROUND STRUCTURES: Both species have a short, thick, usually unbranched taproot, with numerous fine lateral roots.

FLOWERS: Primarily spring and summer, sometimes year-round under favorable conditions. Flower heads clustered at stem tips, mostly 1.5–2.5 cm in diameter, consist only of yellow ligulate flowers. Closed flower heads urn-shaped. Stalks and phyllaries glabrous or with bristly-glandular hairs. Phyllaries in 3 unequal rows, the outer row short and ± triangular, the inner rows linear. Receptacle nearly flat, lacks chaffy bracts. Heads typically open morning through midday and track the sun. Insect-pollinated. Self-compatible.

Annual sowthistle (*Sonchus oleraceus*) in an agricultural field. J. K. CLARK

Annual sowthistle (*Sonchus oleraceus*) flower head and achenes. J. K. CLARK

Annual sowthistle (*Sonchus oleraceus*) flower heads. J. M. DiTomaso

Spiny sowthistle: *Corolla tube longer than straplike segment* above. Glandular hairs often dark, tack-shaped.

Annual sowthistle: Heads sometimes cottony-hairy at base. *Corolla tube about equal to the straplike segment* (ligule) above.

FRUITS AND SEEDS: Phyllaries reflex downward in fruiting heads. Achenes *strongly flattened*, elliptic to oblanceolate, lack a beak. Pappus bristles numerous, soft, fine, white.

Spiny sowthistle: Achenes 2–3 mm long, ± 1.5 mm wide, *faces 3-ribbed lengthwise, otherwise smooth, margin narrow-winged*. Pappus bristles ± 3 times the length of the achene.

Annual sowthistle: Achenes 2.5–4 mm long, ± 1 mm wide, *faces 2–4-ribbed lengthwise, finely transverse-wrinkled*. Pappus bristles ± 2 times the length of the achene.

POSTSENESCENCE CHARACTERISTICS: Plants turn dark after the first frost and decay within a short period.

HABITAT: Roadsides, fields, pastures, riparian areas, ditches, yards, gardens, vegetable and agronomic crops, orchards, vineyards, urban waste areas, logged areas in forests, and other disturbed sites. Tolerate some soil salinity.

DISTRIBUTION: Both species occur in all contiguous states.

Spiny sowthistle: Common throughout California, to 1900 m.

Annual sowthistle: Abundant throughout California, to 1500 m.

PROPAGATION/PHENOLOGY: *Reproduce by seed.* Achene production can be high, especially for spiny sowthistle. Achenes disperse primarily with wind, but also with animals, water, mud, soil movement, human activities, and as a commercial grass seed contaminant. Achenes float on water, and some can survive ingestion by birds and mammals. Seeds primarily germinate in fall and spring. A period of cool, moist conditions and light stimulates germination. Seeds are reported to survive up to 2–8 years under field conditions.

Clasping leaf of annual sowthistle (*Sonchus oleraceus*). J. K. CLARK

Annual sowthistle (*Sonchus oleraceus*) immature plant. J. M. DiTomaso

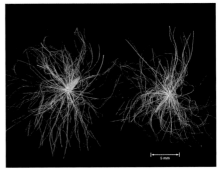

Annual sowthistle (*Sonchus oleraceus*) seeds. J. O'Brien

Rosettes of spiny sowthistle (*Sonchus asper*) (right) and annual sowthistle (*Sonchus oleraceus*) (left). J. M. DiTomaso

Seedlings of spiny sowthistle (*Sonchus asper*) (right) and annual sowthistle (*Sonchus oleraceus*) (left). J. M. DiTomaso

MANAGEMENT FAVORING/DISCOURAGING SURVIVAL: Manual removal or cultivation as needed to prevent fruits from maturing can control the sowthistles.

SIMILAR SPECIES: Slender sowthistle [*Sonchus tenerrimus* L.] is an uncommon *annual to perennial*, with *slender flower stems* to about 1 m tall and yellow dandelion-like flower heads. Unlike annual sowthistle and spiny sowthistle, slender sowthistle has only *slightly flattened achenes* (that otherwise resemble those of annual sowthistle), corolla tubes that are *shorter than the straplike segment* above, deeply pinnate-lobed leaves with *lateral lobes that are narrow or constricted at the base and sometimes lobed again*, and *terminal leaf lobes that are about the same size as the lateral lobes*. Clasping leaf bases are acute, and blade margins are weakly soft-toothed. Slender sowthistle occurs on disturbed sites in the Central Valley, southern San Francisco Bay region, South Coast, and Peninsular Ranges, to 500 m. It appears to be expanding range in California. Slender sowthistle also occurs in Alabama and New York. Native to southern Europe.

Perennial sowthistle [*Sonchus arvensis* L., synonyms: field sowthistle; *Sonchus arvensis* L. ssp. *arvensis* and ssp. *uglinosus* (Bieb.) Nyman][SONAR][CDFA list: A] is a noxious *perennial* to 1.8 m tall, with vigorous *creeping roots* to about 2 m deep. In addition to having *creeping roots*, perennial sowthistle is distinguished primarily by having *slightly flattened achenes that are 3–4-angled, with 2 lengthwise ribs between each angle, and strongly transverse-wrinkled*. Flower heads and stalks are glabrous or have stiff glandular hairs. Lower leaves are entire to deeply pinnate-lobed, with prickly margins. Upper leaves have basal clasping lobes that are curled. Plants are self-incompatible. Mature seeds disperse about 10 days after flower heads open. Most seed germinates in spring, and light is not required. Seeds survive up to about 3 years, sometimes longer, under field conditions. Root fragments 1 cm long or more can develop into new plants under favorable conditions. Perennial sowthistle is uncommon in moist, disturbed places in the northwestern Modoc Plateau and northeastern Cascade Range, to 1800 m. Known infestations that have been eradicated occurred in the Sacramento Valley, northern South Coast Ranges, and South Coast, but widespread in eastern states. Besides California, perennial sowthistle is a state-listed noxious weed in Arizona (prohibited), Nevada, Idaho, Wyoming, North Dakota, South Dakota, Illinois, Iowa (primary), Minnesota (primary), and Hawaii. Native to Europe.

Perennial sowthistle *(Sonchus arvensis)* plant in landscape garden.   J. Neal

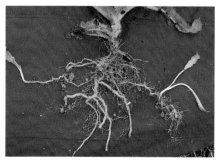

Perennial sowthistle *(Sonchus arvensis)* rhizome.
J. Neal

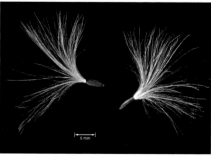

Perennial sowthistle *(Sonchus arvensis)* seeds.
J. O'Brien

Small wirelettuce [*Stephanomeria exigua* Nutt.]

SYNONYMS: plumed ptilory; stephanomeria; whiteplume wirelettuce; *Ptiloria exigua* (Nutt.) Greene; others specific to subspecies names

GENERAL INFORMATION: Native summer *annual*, occasionally biennial, with erect flower stems to about 0.6 to 1.5 m tall, *milky juice*, and *white to pink ligulate flower heads*. Plants exist as rosettes until flower stems develop in late spring to summer. In natural areas, small wirelettuce is a desirable component of the ecosystem. However, it is generally a ruderal native that inhabits disturbed places. Small wirelettuce consists of a highly variable complex of closely related taxa that are currently recognized as 5 subspecies. *Stephanomeria exigua*

Small wirelettuce (*Stephanomeria exigua* ssp. *coronaria*) plant.   J. M. DiTomaso

Small wirelettuce

Small wirelettuce *(Stephanomeria exigua* ssp. *coronaria)* flowering stem. J. M. DiTomaso

Small wirelettuce *(Stephanomeria exigua* ssp. *coronaria)* foliage. J. M. DiTomaso

Nutt. ssp. *coronaria* (E. Greene) Gottlieb is the subspecies that is most often an occasional pest in orchards, vineyards, and other cultivated crops.

MATURE PLANT: Plants ± cone-shaped in outline, with a *single main stem* and many ascending branches. Foliage glabrous to minutely glandular. Stem leaves alternate. Rosette and lower stem leaves narrowly (ob)*lanceolate, elliptic, or oblong*, mostly 2–5 cm long, *margins toothed to deeply pinnate-lobed, lobes narrow and pointing toward the base.* Rosette leaves usually wither by the time flowers mature. Bases of lower stem leaves sessile, slightly lobed and clasping the stem. Upper leaves sessile, *small and bractlike, linear, few or lacking at flowering time.*

ROOTS AND UNDERGROUND STRUCTURES: *Taprooted.*

FLOWERS: May–October. Panicles open or flower heads clustered or solitary on the branch tips. Flower heads mostly 6–7 mm long, consist of 5–6 *white to pale pink or lavender ligulate flowers* 6–8 mm long. Involucre cylindrical, with phyllaries in *2 very unequal rows*, outer phyllaries erect or reflexed. Receptacle lacks chaffy bracts.

FRUITS AND SEEDS: Achenes narrowly *cylindrical, 2–7 mm long, ± 5-angled*, with a *narrow, lengthwise groove on each face.* Sometimes there is a row of minute tubercles on each side of the groove. Pappus bristles several to numerous, *nearly equal, plumose, at least in the upper half*, stiff, white, thickened and sometimes fused at the base, often deciduous above the base, leaving a ± scalelike crown.

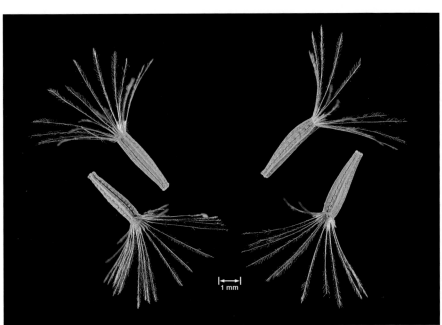

Small wirelettuce (*Stephanomeria exigua* ssp. *coronaria*) seeds.   J. O'Brien

POSTSENESCENCE CHARACTERISTICS: Dead plants do not persist beyond winter.

HABITAT: Disturbed dry areas in many plant communities, open slopes and shrubland, desert scrub, and occasionally orchards, vineyards, and crop fields.

DISTRIBUTION: Throughout California, to 2000 m. All western states, except possibly Montana and Washington, and Texas. New York.

PROPAGATION/PHENOLOGY: *Reproduces by seed.* Achenes disperse primarily with wind, but also with water, mud, and possibly animals and human activities, including vehicle tires. Seeds generally germinate in spring.

MANAGEMENT FAVORING/DISCOURAGING SURVIVAL: Manual removal or cultivation before plants produce seeds can control small wirelettuce.

SIMILAR SPECIES: There are several native *Stephanomeria* species in California, most of which are not weedy in on agricultural lands and other human-disturbed places. The majority of species are perennials with a woody root crown. Of the few annuals, small wirelettuce is the species that is most often encountered as a weed in orchards, vineyards, and crop fields.

Feverfew [*Tanacetum parthenium* (L.) Schultz-Bip.] [CHYPA]

Common tansy [*Tanacetum vulgare* L.][CHYVU] [Cal-IPC: Moderate]

SYNONYMS: Feverfew: *Chrysanthemum parthenium* (L.) Bernh.; *Matricaria odorata* Lam.; *Matricaria parthenium* L.; *Matricaria vulgaris* S. F. Gray; *Pyrethrum parthenium* J. E. Smith

Common tansy: bitter buttons; garden tansy; hindhead; parsley fern; tansy; *Chrysanthemum uliginosum* Pers.; *Chrysanthemum vulgare* (L.) Bernh.; *Tanacetum vulgare* L. var. *crispum* DC.

GENERAL INFORMATION: Erect, *aromatic perennials* to 1 m tall. Both species are cultivated as garden ornamentals and medicinal herbs. Native to Europe.

Feverfew: Plants have a taproot, leaves that are *pinnate-dissected 2–3 times,* and *white and yellow daisylike flower heads.* Plants are generally short-lived perennials.

Common tansy: Erect, *aromatic perennial* to 1.5 m tall, with *creeping roots, deeply pinnate-lobed leaves,* and *yellow buttonlike flower heads.* Plants can form dense colonies from the creeping roots, especially in riparian areas along rivers. Common tansy can be fatally *toxic* to humans and animals when large quantities of fresh plant material or medicinal teas or oils are ingested. However, quantities

Feverfew (*Tanacetum parthenium*) plant.   J. M. DiTomaso

Feverfew (*Tanacetum parthenium*) flower head.
J. M. DiTomaso

Feverfew (*Tanacetum parthenium*) flowering stems.   J. M. DiTomaso

Feverfew (*Tanacetum parthenium*) seedling.
J. M. DiTomaso

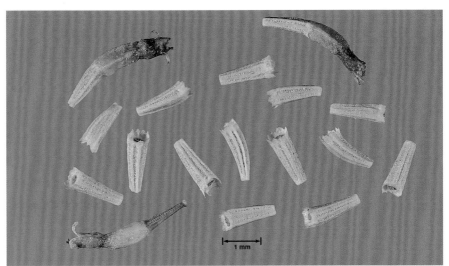

Feverfew (*Tanacetum parthenium*) seeds.   J. O'Brien

large enough to cause problems are rarely ingested. The toxic principle affects the nervous system and/or the heart, but it has not yet been identified. Sensitive individuals can develop contact dermatitis from handling the foliage, especially when exposed for extended periods of time. Common tansy is a state-listed noxious weed in Colorado, Minnesota (secondary), and Washington (class B).

SEEDLINGS: Leaves alternate.

Feverfew: Cotyledons oval to nearly round, mostly 2–3 mm long, 1–2 mm wide, glabrous. First leaf ± ovate, 5–10 mm long, 3–5 mm wide, margins smooth or few-toothed, sparsely glandular. Subsequent leaves lobed.

Common tansy: Cotyledons ovate to elliptic, ± 3 mm long, ± 2 mm wide, glabrous. First leaf ± elliptic, mostly 6–12 mm long, ± 5 mm wide, margins lobed or toothed, sparsely glandular.

MATURE PLANT: Foliage ± glandular, glabrous to sparsely hairy. Stem bases sometimes weakly woody. Leaves alternate, mostly 4–10 cm long.

Feverfew: Short-lived. Leaves stalked. Primary divisions ± broad, usually 3–5, tips rounded or acute. Secondary divisions ± irregular-toothed or lobed.

Common tansy: Leaves sessile or short-stalked, evenly dotted with flat or sunken glands. Primary divisions narrow, mostly 4–10 pairs, tips acute. Secondary divisions narrow, toothed.

ROOTS AND UNDERGROUND STRUCTURES: Feverfew: Taproot deep, with numerous lateral roots and highly branched feeder rootlets.

Common tansy: Creeping roots thick, extensive, with numerous lateral roots.

FLOWERS: Phyllaries unequal, overlapping in 2–3 rows, margins translucent. Receptacles rounded or conelike, minutely tubercled, *lack chaffy bracts*.

Common tansy (*Tanacetum vulgare*) plant.
J. M. DiTomaso

Common tansy (*Tanacetum vulgare*) flower heads.
J. M. DiTomaso

Common tansy (*Tanacetum vulgare*) seedling.
J. M. DiTomaso

Common tansy (*Tanacetum vulgare*) foliage. J. M. DiTomaso

Feverfew: May–August. Flower clusters ± open, rounded or flat-topped. Flower heads *daisylike* (radiate), consist of *yellow disk flowers* surrounded by *white ray flowers*. Ray corollas 4–8 mm long.

Common tansy: August–October. Flower clusters *dense, flat-topped or slightly rounded*. Flower heads *buttonlike, yellow, appear to consist only of disk flowers*, but actually consist of disk flowers surrounded by a ring of ray flowers that lack corollas (disciform heads).

FRUITS AND SEEDS: Achenes cylindrical, 1–3 mm long, usually with a *minute crownlike pappus of toothed scales* to 0.5 mm long.

Feverfew: Achenes 5–10-ribbed. Pappus scales sometimes lacking.

Common tansy: Achenes ± 5-angled, minutely glandular.

HABITAT: Disturbed places (especially in urban areas), gardens, yards, fields, roadsides, and ditch banks.

DISTRIBUTION: Feverfew: North Coast Ranges, Klamath Ranges, northern Sierra Nevada foothills, Central Valley, San Francisco Bay region, western South Coast Ranges, and South Coast, to at least 1400 m. All western states, except Arizona and New Mexico; most eastern states.

Common tansy: North Coast, western North Coast Ranges, Cascade Range, and western South Coast Ranges, to at least 1900 m. Most contiguous states, except a few southern states. Particularly a problem in Idaho, where it forms dense monotypic stands. Also increasing in Wyoming, including the Grand Teton National Park.

PROPAGATION/PHENOLOGY: *Reproduce by seed.* Common tansy also *reproduces vegetatively from creeping roots.*

MANAGEMENT FAVORING/DISCOURAGING SURVIVAL: Cultivation, repeated mowing, and manual removal (including the roots of common tansy) can control both species.

SIMILAR SPECIES: Mayweed chamomile [*Anthemis cotula* L.][ANTCO] is an *annual* with an *unpleasant scent* and white and yellow daisylike flowers. Unlike feverfew, mayweed has leaves that are *dissected 2–3 times into linear segments*, flower heads that are *solitary on long stalks* from the leaf axils, and receptacles with *chaffy bracts*. Refer to the entry for **Mayweed chamomile** (p. 189) for more information.

Common tansy (*Tanacetum vulgare*) senesced stems. J. M. DiTomaso

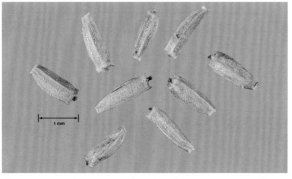

Common tansy (*Tanacetum vulgare*) seeds. J. O'Brien

Dandelion [*Taraxacum officinale* Weber ex Wigg.][TAROF]

SYNONYMS: blowball; cankerwort; common dandelion; face clock; lions-tooth; monk's head; *Leontodon taraxacum* L.; *Taraxacum laevigatum* (Willd.) DC. in part; *Taraxacum retroflexum* Lindb. f.; *Taraxacum vulgare* (Lam.) Schrank.

GENERAL INFORMATION: Variable *perennial rosette*, with *milky juice* and *yellow ligulate flower heads on leafless stalks* usually to about 0.3 m, but can grow to 0.5 m in height. Dandelion consists of a complex of numerous biotypes that also vary with different environmental conditions. North American plants exhibiting certain characteristics due to environment were previously considered a distinct species *Taraxacum laevigatum* (Willd.) DC. Dandelion occurs as a weed nearly worldwide and is especially common in turf. It is a host of aster yellows disease, which also affects a number of vegetable crops. All plant parts are edible. The greens are cultivated and consumed as a vegetable, and roots are dried, roasted, and used as a coffee substitute and for medicinal purposes. Dandelion contains high amounts of certain minerals and is a good minor component of pasture forage for livestock. Native to Europe.

SEEDLINGS: Cotyledons oval, mostly 4–10 mm long, 2–4 mm wide, glabrous, midvein terminates with a gland at the apex. Leaves alternate. First and subsequent few leaves elliptic to oblong, base tapered, margins weakly few-toothed. Later leaves oblanceolate, bases gradually taper, margins sparsely toothed with backward-pointing teeth.

MATURE PLANT: Leaves oblong, mostly 7–25 cm long, glabrous to sparsely

Dandelion (*Taraxacum officinale*) fruiting plants.   J. M. DiTomaso

hairy, especially lower midveins. Margins variably pinnate-lobed and/or toothed, but usually pinnate-lobed about halfway or more to the midvein, *lateral lobes point toward the leaf base*, terminal lobe often larger than lateral lobes.

ROOTS AND UNDERGROUND STRUCTURES: Taproots simple or branched, thick, can grow to more than 2 m deep, but are often up to about 0.5 m deep. Roots often break near the crown when plants are hand-pulled.

FLOWERS: Nearly year-round in mild climates. Flower heads develop *singly on a hollow leafless stalk* (peduncle), mostly 2–3.5 cm in diameter, *consist only of yellow ligulate flowers*. Peduncles elongate as the flower heads mature. Phyllaries in 2 rows, *outer phyllaries reflexed downward, inner phyllaries erect*. Receptacle convex, lacks chaffy bracts.

FRUITS AND SEEDS: Fruiting heads resemble fuzzy gray-white spheres. Achenes flattened, lanceolate, body mostly 3–5 mm long, with a beak 8–10 mm long at the apex, minutely toothed or tubercled in the upper portion, gray to olive-brown. *Pappus bristles fine, minutely barbed,* white to pale gray, mostly 6–8 mm long, persistent.

POSTSENESCENCE CHARACTERISTICS: Rosettes remain green year-round, but the leafless flower stalks with a solitary naked receptacle and reflexed phyllaries remain for a short period after fruits disperse.

HABITAT: Turf, orchards, vineyards, perennial crops such as alfalfa, nursery

Dandelion (*Taraxacum officinale*) plant.
J. M. DiTomaso

Dandelion (*Taraxacum officinale*) seedlings.
J. M. DiTomaso

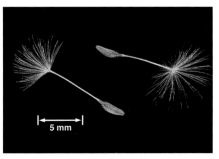

Dandelion (*Taraxacum officinale*) seeds.
J. O'Brien

Lesser hawkbit (*Leontodon taraxacoides*) plant. J. M. DiTomaso

Lesser hawkbit (*Leontodon taraxacoides*) flowering and fruiting heads. J. M. DiTomaso

crops, pastures, fields, and roadsides. A common weed of mountain meadows, particularly in southern California. Often grows in moist places, especially those that receive some water year-round.

DISTRIBUTION: Throughout California, except deserts, to 3300 m. All contiguous states.

PROPAGATION/PHENOLOGY: *Reproduces by seed* and occasionally vegetatively from root fragments. Seeds typically develop asexually (apomixis) and rarely sexually (with fertilization). Seeds disperse primarily with wind, but also with mud, water, animals, hay, and landscape maintenance equipment and other human activities. Plants often produce thousands of seeds per plant. In mild climates, seeds can germinate year-round when conditions are favorable. Light enhances germination. Seeds generally survive up to 3–4 years under field conditions. Root fragments of only a few millimeters long can develop new shoots and roots under favorable conditions, especially in summer.

Lesser hawkbit (*Leontodon taraxacoides*) fruiting head. J. M. DiTomaso

MANAGEMENT FAVORING/DISCOURAGING SURVIVAL: Dandelion readily tolerates mowing. Manual removal is most effective when the entire root is removed, since plants will regrow from remaining root fragments. Despite its ability to regenerate from root fragments, regular cultivation can control dandelion. Fabric mulch or a layer of organic mulch at least 8 cm thick can control seedlings.

SIMILAR SPECIES: Lesser hawkbit [*Leontodon taraxacoides* (Vill.) Mérat, synonym: *Leontodon lesseryi* (Wallr.) G. Beck] is a less-common *annual, biennial, or perennial rosette* with a taproot and leaves and yellow flower heads on leafless stalks that resemble those of dandelion. Unlike dandelion, lesser hawkbit has foliage that is usually *evenly covered with short bristly hairs* and central achenes that have featherlike *pappus bristles* (pumose) and a *beak 1–3 mm long*. In addition, leaves are typically shallow lobed with lobes usually pointing outward or slightly forward and without a distinct terminal lobe. Outer phyllaries are very small and ± erect. Two subspecies are recognized in California. Subspecies *longirostris* Finch & Sell is typically annual, rarely biennial and has achene beaks 2–3 mm long. It inhabits roadsides, fields, pastures, vernal pools, hillsides, and other disturbed places in the Northwestern region, northern Sierra Nevada, San Joaquin Valley, San Francisco Bay region, and Central Coast, to 1000 m. Subspecies *taraxacoides* is a perennial, occasionally biennial and has achene beaks ± 1 mm long. It occurs mostly along the North Coast. Subspecies *taraxacoides* also occurs in Arizona, Nevada, Oregon, Texas, Washington, and many eastern states. Native to Europe.

Rosettes of the following species more or less resemble dandelion rosettes: **chicory** (*Cichorium intybus*, p. 268), **rush skeletonweed** (*Chondrilla juncea*, p. 264), **annual and perennial sowthistles** (*Sonchus* spp., p. 398), and **smooth and common catsear** (*Hypochaeris* spp., p. 353). Refer to the entries for these species for more information.

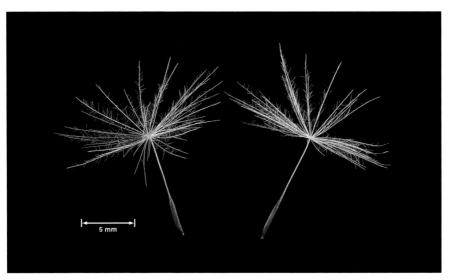

Lesser hawkbit (*Leontodon taraxacoides*) seeds.   J. O'Brien

Western salsify [*Tragopogon dubius* Scop.][TRODM]
Common salsify [*Tragopogon porrifolius* L.][TROPS]

SYNONYMS: Western salsify: western goat's beard; wild oyster plant; yellow goat's beard; yellow salsify; *Tragopogon dubius* Scop. ssp. *major* (Jacq.) Voll.; *Tragopogon major* Jacq.

Common salsify: goat's-beard; Jerusalem star; noon-plant; oyster plant; salsify; vegetable oyster

GENERAL INFORMATION: Erect, slightly fleshy herbs to 1 m tall, with *grass-like leaves, milky juice,* and *large dandelion-like flower heads on stalks that are swollen just below the heads.* Plants exist as grasslike rosettes until flower stems develop at maturity. Salsify species are edible. Animals consume the foliage and fruits. Native to Europe.

Western salsify: *Annual or biennial,* with *yellow flower heads.*

Western salsify (*Tragopogon dubius*) plant. J. M. DiTomaso

Western salsify • Common salsify

Western salsify (*Tragopogon dubius*) fruiting head. J. M. DiTomaso

Western salsify (*Tragopogon dubius*) seedling with seed coat attached. J. M. DiTomaso

Common salsify: Biennial with *purple flower heads*. Garden cultivars are grown for their roots, which are consumed as a cooked vegetable and are reported to taste mildly of oyster.

SEEDLINGS: Cotyledons linear, about 11–13 cm long, 2–3 mm wide, sessile and ± fused at the base, glabrous. Leaves alternate. First and subsequent leaves mostly 9–12 cm long, 4–5 mm wide, resemble the cotyledons. Western salsify leaves often have long cobwebby hairs at the base.

MATURE PLANT: Stem branches few, erect to ascending. Rosette and stem leaves *grasslike with parallel veins*, mostly to 15–50 cm long, slightly fleshy, exude *milky juice* when torn, mostly glabrous, sometimes with long, wavy hairs at the base in western salsify. Stem leaves alternate, sessile, base broad and clasping the stem.

ROOTS AND UNDERGROUND STRUCTURES: Taproots fleshy, thick, usually branched, contain milky juice.

FLOWERS: April–July. Flower heads *solitary on long fleshy stalks that are nearly as wide as the base of the heads where they join*, consist only of ligulate flowers. Phyllaries in 1 row, *much longer than the flowers*. Involucre (phyllaries as a unit) narrowly conical in closed heads. Receptacle lacks chaffy bracts. Flower heads typically open and track the sun morning through midday. Insect-pollinated.

Western salsify: Flower heads *yellow*.

Common salsify: Flower heads *purple*.

FRUITS AND SEEDS: Mature fruiting heads 4–7 cm in diameter, resemble large fuzzy balls. Achenes oblong-elliptic, ± 5-angled in cross-section, minutely

Western salsify (*Tragopogon dubius*) flower heads. J. M. DiTomaso

Western salsify (*Tragopogon dubius*) seeds. J. O'Brien

toothed and ribbed lengthwise, gradually taper into a *long, thick, persistent beak*. Pappus bristles *featherlike* (plumose), stiff, unequal, deciduous.

Western salsify: Achenes including beak mostly 2.5–3.5 cm long. Pappus bristles nearly white.

Common salsify: Achenes including beak mostly 2.5–4 cm long. Pappus bristles tan.

POSTSENESCENCE CHARACTERISTICS: Senesced stems with grasslike leaves and old flower heads can persist through fall. Old flower heads have reflexed phyllaries and bare receptacles, but sometimes a few achenes remain attached.

HABITAT: Disturbed sites, waste places, vacant urban lots, fields, orchards, vineyards, perennial crops, roadsides, trail sides in woodlands and grasslands, and open areas in coniferous forests.

DISTRIBUTION: Western salsify: Northern Sierra Nevada, San Francisco Bay region, Central Valley, western South Coast Ranges, and San Bernardino Mountains, to 2700 m. Most contiguous states, except a few southern states.

Common salsify: Throughout California, except Great Basin and deserts, to 1700 m. Most contiguous states, except a few southern and north-central states.

PROPAGATION/PHENOLOGY: *Reproduce by seed.* Achenes disperse with wind, water, mud, human activities, and animals. Seeds germinate in fall and/or spring, and light is not required. Seeds generally survive about 1–2 years under field conditions.

MANAGEMENT FAVORING/DISCOURAGING SURVIVAL: Mowing flower stems to prevent seeds from developing or cultivation can control salsify species.

SIMILAR SPECIES: Salsify rosette leaves can be confused with grasses and sedges (*Carex* spp.). Unlike grasses and sedges, salsify leaves have *milky juice* and *lack ligules and/or sheaths*.

Meadow salsify [*Tragopogon pratensis* L.] [TROPR] is a less common *biennial* to about 0.8 m tall, with *bright yellow flower heads*. Unlike western salsify, meadow salsify has flower head stalks that are *only slightly broadened at the top* and *phyllaries that are shorter than the flowers*. In addition, fruiting heads are 1.8–4.5 cm in diameter, and the achenes, including the beak, are mostly 1.5–2.5 cm long and have nearly white pappus bristles. Meadow salsify inhabits disturbed, often moist places in the Klamath Ranges, high regions of the Cascade Range and Sierra Nevada, and the Modoc Plateau, usually from 900 to 1600 m. Native to Europe.

Common salsify (*Tragopogon porrifolius*) plant.
J. M. DiTomaso

Meadow salsify (*Tragopogon pratensis*) flower head. R. Uva

Common salsify (*Tragopogon porrifolius*) flower head. J. M. DiTomaso

Common salsify (*Tragopogon porrifolius*) fruiting heads. J. M. DiTomaso

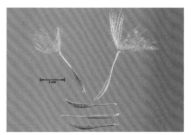

Common salsify (*Tragopogon porrifolius*) seeds. J. O'Brien

Meadow salsify (*Tragopogon pratensis*) seeds. J. O'Brien

Spiny cocklebur [*Xanthium spinosum* L.][XANSP]
Common cocklebur [*Xanthium strumarium* L.][XANST]

SYNONYMS: Spiny cocklebur: Bathurst burr; cocklebur; clotweed; dagger cocklebur; Spanish thistle; spiny clotbur; *Acanthoxanthium spinosum* (L.) Fourr.; *Xanthium spinosum* L. var. *inerme* Bel.

Common cocklebur: Bathurst burr; buttonbur; clotbur; ditchbur; hedgehog bur-weed; noogoora bur; rough cocklebur; sea burdock; sheep bur; *Xanthium californicum* Greene; *Xanthium calvum* Millsp. & Sherff.; *Xanthium campestre* Greene; *Xanthium canadense* Mill.; *Xanthium chinense* Mill.; *Xanthium macrocarpum* var. *glabratum* DC.; *Xanthium palustre* Greene; *Xanthium pennsylvanicum* Wallr.; *Xanthium strumarium* L. var. *canadense* (Mill.) Torrey & A. Gray and var. *glabratum* (DC.) Cronquist; many others

Spiny cocklebur (*Xanthium spinosum*) plant. J. M. DiTomaso

Spiny cocklebur • Common cocklebur

Spiny cocklebur (*Xanthium spinosum*) flowering and fruiting stem.   J. K. CLARK

Spiny cocklebur (*Xanthium spinosum*) seedling.
J. M. DiTomaso

GENERAL INFORMATION: Coarse, native plants that produce *large burs covered with hook-tipped prickles*. Both species occur as weeds in temperate to subtropical regions nearly worldwide. Ingestion of seedlings and seeds at 1% or more of body weight can be fatally *toxic* to livestock, especially pigs and calves. Seeds and sprouts contain a high concentration of a diterpene glycoside that causes an acute metabolic disorder characterized by a sudden drop in blood glucose and an increase of certain liver enzymes. Symptoms include weakness, loss of appetite, uncoordination, muscular spasms, coma, and death. Death can occur before other symptoms become apparent. Seedling toxicity diminishes with growth. In humans, handling either species can cause contact dermatitis in sensitive individuals.

Spiny cocklebur: Summer *annual* to 1 m tall, with leaves that are *densely white-hairy below* and *long, yellowish three-branched spines in the axils of the stems*. Plants are typically less variable than common cocklebur.

Common cocklebur: Summer *annual* to 1.5 m tall, with *green leaves* and *stems without spines*. Plants are highly variable within and between populations. The most recent taxonomic treatment describes 2 subspecies, one with a center of distribution in North and South America, the other with a center of distribution in Europe, Asia, and the Mediterranean. The Eurasian subspecies is composed of a complex of intergrading biotypes. The North and South American subspecies is composed of several complexes. Previously, numerous biotypes within the complexes have been named as separate species, subspecies, and varieties, creating much taxonomic confusion.

Spiny cocklebur • Common cocklebur

Common cocklebur *(Xanthium strumarium)* immature plant. J. M. DiTomaso

Common cocklebur *(Xanthium strumarium)* plant. J. M. DiTomaso

Common cocklebur *(Xanthium strumarium)* flowering and fruiting stem. J. M. DiTomaso

SEEDLINGS: Stalk below cotyledons (hypocotyl) thick, fleshy. Cotyledon stalks short, fused at base. Leaves alternate, but first pair is often opposite. Prickly burs typically remain attached underground.

Spiny cocklebur: Cotyledons oblong-elliptic, mostly 1–4 cm long, 0.2–1 cm wide, glabrous. First and subsequent few leaves lanceolate, 1–4 cm long, 1–1.2 cm wide, margins few-toothed in the lower half, surfaces sparsely covered with short, appressed hairs, sometimes weakly 3-veined from the base.

Common cocklebur: Cotyledons oblong-elliptic to lanceolate, mostly 3–7 cm long, 0.6–1 cm wide, thick, ± fleshy, glabrous, usually 3-veined from the base, upper surface ± glossy green, lower surface paler. First and subsequent few leaves ovate, margins ± toothed, surfaces ± covered with minute hairlike projections that are rough to touch (scabrous) and often a few glandular hairs, 3-veined from the base.

MATURE PLANT: Leaves alternate.

Spiny cocklebur: Stems tough, branched, with *coarse, 3-branched, yellowish spines 1–3 cm long in the leaf axils*. *Leaves lanceolate or elliptic to linear*, margin entire to toothed or pinnate-lobed, 3–10 cm long, 0.6–3 cm wide, with *1 main vein from the base*. Upper surface gray-green, nearly glabrous to minutely stiff-hairy. *Lower surface densely covered with short, white woolly hairs*. Leaf stalks short.

Common cocklebur: Foliage has a distinctive scent. Stems thick, branched, often slightly fleshy, often reddish or black-spotted, or tinged dull red. Leaves broadly triangular, often weakly 3-lobed, with *3 main veins from the base* 3–15 cm long and wide, base sometimes slightly lobed, margin variably toothed. *Upper and lower surfaces green,* scaberous, and rough to touch, ± glandular. Leaf stalks mostly 3–15 cm long.

ROOTS AND UNDERGROUND STRUCTURES: Spiny cocklebur: Taproot often branched.

Common cocklebur: Taproot thick, sometimes weakly woody, with numerous highly branched lateral roots to 60 cm deep or more.

FLOWERS: July–October. *Male and female flowers develop in separate heads on the same plant* (monoecious). Male flower heads small, green to rusty red, *develop in clusters in the upper leaf axils.* Male flowers have tubular 5-lobed corollas, free anthers, and lack phyllaries or have minute phyllaries in 1–3 rows. Female flower heads develop in the leaf axils below the male flower heads. Each female head usually consists of 2 flowers that lack corollas. Female involucres become *hardened prickly burs* that enclose the 2 seeds at maturity. Receptacles have numerous chaffy bracts with spiny hooked tips. Wind-pollinated, self-compatible.

Spiny cocklebur: Female flower heads mostly develop singly in the leaf axils below the male flower heads and sometimes develop in small clusters in the upper axils.

Common cocklebur (*Xanthium strumarium*) stem with purple blotches.   J. K. CLARK

Common cocklebur (*Xanthium strumarium*) fruiting stem.   J. M. DiTOMASO

Common cocklebur (*Xanthium strumarium*) seedling.   J. K. Clark

Common cocklebur: Female flower heads develop in clusters in the leaf axils below the male flower heads.

FRUITS AND SEEDS: *Bur prickles numerous, hook-tipped.*

Spiny cocklebur: Burs ± cylindrical, mostly 1–1.5 cm long, 0.5–1 cm wide including prickles, body minutely glandular, yellowish green to light tan, with 1–2 slender inconspicuous straight beaks at the apex. Prickles slender, glabrous.

Common cocklebur: Burs ellipsoid, mostly 1.5–3.5 cm long, green to yellowish brown, with 2 conspicuous thick straight or curved beaks at the apex. Prickles stout, lower portion often stiff glandular-hairy or sometimes minutely glandular to glabrous.

POSTSENESCENCE CHARACTERISTICS: Old stems with burs, and in spiny cocklebur, spines, can persist into winter.

HABITAT: Open, often moist disturbed places, roadsides, ditches, valley bottomlands, fields, pastures, cultivated crops, orchards, riparian areas, seasonal wetlands, and waste places.

Spiny cocklebur: Typically associated with open grasslands and in dried pools, but not vernal pools. Often found in areas where sheep or cattle graze.

Common cocklebur: Particularly competitive in cotton fields. Although an agricultural weed, it is very often found in riparian areas, where it is probably native.

DISTRIBUTION: Both species occur in all contiguous states.

Spiny cocklebur: Throughout California, except Great Basin and deserts, to 900 m.

Common cocklebur: Throughout California, except Great Basin, to 500 m.

PROPAGATION/PHENOLOGY: *Reproduce by seed.* Burs disperse primarily with water and by clinging to animals, vehicle tires, and the shoes, clothing and equipment of humans. Burs are buoyant and can float for up to about 1 month.

Spiny cocklebur: Burs contain 2 seeds, but appear to lack a difference in germination requirements. Some seeds survive up to about 3 years under field conditions.

Common cocklebur: Each bur contains 2 seeds that usually have different germination requirements. The larger seed typically germinates in the spring following the season it was produced. The smaller seed usually germinates a year after the larger seed, or sometimes germinates at the same time or later in the season than the larger seed, but it can remain dormant for several years. Germination does not require light. Scarification can hasten germination.

MANAGEMENT FAVORING/DISCOURAGING SURVIVAL: Manual removal, mowing, or cultivation before burs develop can control the cockleburs. Cut plants with immature burs can still develop viable seed.

SIMILAR SPECIES: Cockleburs are unlikely to be confused with other weedy species.

Coast fiddleneck [*Amsinckia menziesii* (Lehm.) Nelson & J.F. Macbr. var. *intermedia* (Fisch. & C.A. Mey.) Ganders][AMSIN]

Menzies fiddleneck [*Amsinckia menziesii* (Lehm.) Nelson & J.F. Macbr. var. *menziesii*]

SYNONYMS: A highly variable species consisting of numerous biotypes, many of which have been previously named as distinct species.

Coast fiddleneck: coast buckthorn; common fiddleneck; fingerweed; fireweed fiddleneck; rancher's fireweed; tarweed; yellow burweed; yellow forget-me-not; yellow tarweed; zaccoto gordo; *Amsinckia arizonica* Suksd.; *Amsinckia arvensis* Suksd.; *Amsinckia attenuata* Eastw.; *Amsinckia californica* Suksd.; *Amsinckia campestris* Greene; *Amsinckia demissa* Suksd.; *Amsinckia echinata* Gray; *Amsinckia hanseni* Brand.; *Amsinckia intactilis* Macbr.; *Amsinckia intermedia* Fisch. & C.A. Mey.; *Amsinckia irritans* Brand.; *Amsinckia longituba* Brand.; *Amsinckia obovallata* Greene; *Amsinckia valens* Macbr.; many others

Menzies fiddleneck: small-flowered fiddleneck; *Amsinckia copelandii* Suksd.; *Amsinckia eatonii* Suksd.; *Amsinckia helleri* Brand.; *Amsinckia hirticaulis* Suksd.; *Amsinckia micrantha* Suksd.; *Amsinckia parviflora* Heller; *Amsinckia retrorsa* Suksd.; *Echium menziesii* Lehm.; others

GENERAL INFORMATION: Native winter *annual*, with erect stems to 1.2 m tall, *bristly-hairy foliage*, and *coiled infloresences of small tubular yellow-orange to*

Coast fiddleneck (*Amsinckia menziesii* var. *intermedia*) plant. J. M. DiTomaso

Coast fiddleneck (*Amsinckia menziesii* var. *intermedia*) flowering stem. J. M. DiTomaso

Coast fiddleneck • Menzies fiddleneck

Menzies fiddleneck (*Amsinckia menziesii* var. *menziesii*) plant.　　J. M. DiTomaso

Menzies fiddleneck (*Amsinckia menziesii* var. *menziesii*) flowering stem.　　J. M. DiTomaso

pale yellow flowers. Plants exist as rosettes until flower stems develop at maturity. Fiddlenecks are a desirable component of wildlands, but coast and Menzies fiddleneck can be problematic in agronomic crops, orchards, and pastures. Although they can coexist, they are not known to hybridize. Fruits of coast and Menzies fiddleneck, as well as **tarweed fiddleneck** [*Amsinckia lycopsoides* Lehm.] [AMSLY] and **western fiddleneck** [*Amsinckia tessellata* A. Gray] [AMSTE] contain pyrrolizidine alkaloids and can be *toxic* to livestock, especially horses and cows, when ingested in quantity. Sheep, goats, and chickens appear less susceptible to the toxic effects. Affected animals develop neurological symptoms and/or liver disease, the manifestations of which are commonly known as wandering disease, hard liver disease, protein poisoning, and winter-wheat poisoning. Poisonings most often occur when contaminated grain feed or hay is consumed over a period of days to months. In addition, the foliage of *Amsinckia* species can occasionally accumulate *toxic* levels of nitrates under certain conditions, particularly during periods of cold cloudy weather or drought stress. Coast fiddleneck and Menzies fiddleneck are government-listed noxious weeds in some regions of Australia. Menzies fiddleneck is an introduced weed in South America. Of the *Amsinckia* species and varieties, coast fiddleneck is most frequently reported to be a weed in California.

SEEDLINGS: Cotyledons deeply lobed in a Y shape, lobes oblanceolate with rounded tips, sparsely hairy. Leaves alternate. First and subsequent leaves linear-oblanceolate, margins usually smooth, stiff-hairy. Cotyledon and leaf hairs sometimes have minute papillate bases.

MATURE PLANT: Foliage sparse to moderately covered with stiff bristly hairs,

Coast fiddleneck • Menzies fiddleneck

Coast fiddleneck (*Amsinckia menziesii* var. *intermedia*) young rosette.   J. M. DiTomaso

Coast fiddleneck (*Amsinckia menziesii* var. *intermedia*) seedling.   J. M. DiTomaso

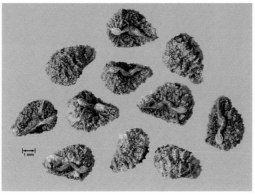

Coast fiddleneck (*Amsinckia menziesii* var. *intermedia*) seeds.   J. O'Brien

Eastwood fiddleneck (*Amsinckia eastwoodiae*), a nonweedy native. Note length of tube compared to *A. menziesii*.
J. M. DiTomaso

often with papillate bases. Leaves alternate, sessile or lowest stem leaves ± short-stalked, linear to narrowly oblong or lanceolate, about 2–15 cm long, *margins smooth*.

ROOTS AND UNDERGROUND STRUCTURES: Taprooted.

FLOWERS: Mostly March–June. Inflorescences are *coiled, one-sided, spikelike cymes* at the stem tips. *Sepal lobes 5, bristly-hairy. Corolla tubes straight, radial, 5-lobed, 10-veined near base, throat open.*

Coast fiddleneck: Corollas usually yellow-orange with interior orange-red spots, 4–10 mm wide at the top, 7–11 mm long.

Menzies fiddleneck: Corollas ± pale yellow, usually lack interior orange-red spots, mostly 2–3 mm wide at the top, 4–7 mm long.

FRUITS AND SEEDS: Fruits consist of 4 erect 1-seeded nutlets that eventually separate after dispersal. Sometimes only 2–3 nutlets develop to maturity. Nutlets

± triangular-ovoid, 2–3.5 *mm long*, ± 1 mm wide, *surface sharply tubercled or ridged*, ± gray, with a small to large lateral attachment scar.

POSTSENESCENCE CHARACTERISTICS: Senesced plants turn gray-brown, often retain stiff hairs and many fruits, and can persist into winter.

HABITAT: Disturbed open and often dry places, grassland, fields, pastures, roadsides, agronomic crops, orchards, vineyards, and waste places. Menzies fiddleneck is generally found on deeper soils, in old fields, and in crops; coast fiddleneck is more common on shallower soils and on hillsides.

DISTRIBUTION: Both varieties occur throughout California, to 1700 m.

Coast fiddleneck: Western states, except possibly Colorado and New Mexico. Scattered in a few central states and some northeastern states.

Menzies fiddleneck: Colorado, Idaho, Nevada, Oregon, Utah, Washington, Connecticut, North Carolina, and South Carolina.

PROPAGATION/PHENOLOGY: *Reproduces by seed.* Fruits at first remain attached to the bristly calyx and disperse with water, mud, soil movement, and by clinging to animals, human shoes and clothing, and agricultural equipment. Seeds germinate in fall through early spring.

MANAGEMENT FAVORING/DISCOURAGING SURVIVAL: In agronomic fields, preplanting cultivation of fiddleneck seedlings or plants before seeds develop can help control fiddleneck. Manual removal of fiddleneck plants that emerge after the crop is planted can improve control. Mowing before seeds develop can also help control fiddleneck.

SIMILAR SPECIES: There are several native *Amsinckia* species and varieties in California. All are annuals, and most are not problematic. However, tarweed fiddleneck and western fiddleneck are occasionally weedy.

Like coast fiddleneck and Menzies fiddleneck, **tarweed fiddleneck** has 5-lobed corolla tubes and sharply tubercled or ridged nutlets. Tarweed fiddleneck differs by having *corolla tube throats that are closed by hairy bulges at the top of the tube.* Tarweed fiddleneck inhabits open disturbed sites throughout California, except Great Basin and deserts, to 400 m. It occurs throughout the northwestern states, is scattered in the central and northeastern states, and is a government-listed noxious weed in parts of Australia. In addition, tarweed fiddleneck hybridizes with coast fiddleneck.

Western fiddleneck is distinguished from the previously described species by having *2–4 sepal lobes, corolla tubes 20-veined near the base,* and *smooth to smoothly round-tubercled nutlets.* Western fiddleneck occurs in the eastern North Coast Ranges, Central Valley, San Francisco Bay region, South Coast Ranges, western Transverse Ranges, Great Basin, and deserts, to 2200 m. It also occurs in most western states.

Houndstongue [*Cynoglossum officinale* L.][CYWOF] [Cal-IPC:Moderate]

SYNONYMS: beggar's-lice; common houndstongue; dog bur; dog's tongue; glovewort; gypsyflower; sheeplice; sticktight; woolmat

GENERAL INFORMATION: Coarse *biennial*, sometimes annual or short-lived perennial, with erect flower stems to 1.2 m tall and panicles of *reddish purple flowers*. Plants exist as a rosette until flower stems develop at maturity. Dense stands can develop. Foliage, especially young leaves, and fruits contain pyrrolizidine alkaloids and are *toxic* to livestock, especially horses, when ingested in small amounts over time or in a single large quantity. Most poisonings occur when animals consume contaminated hay over time. Chronic toxicity symptoms include weight loss, poor body condition, and sometimes liver disease and/or neurological effects. Live plants have a distinctive scent that appears to deter animals from consuming foliage. Prickly nutlets can matt fur of animals, causing mechanical irritation and reducing the value of wool in sheep. Houndstongue is most problematic in some northwestern states and the western Great Plains region. It is a state-listed noxious weed in Oregon (class B), Washington (class C), Colorado, Wyoming, and Montana (category 1). In Europe, houndstongue roots have been used medicinally for centuries. Native to Eurasia and probably introduced as a grain contaminant.

SEEDLINGS: Cotyledons ± oval, about 1–1.5 cm long, tip rounded, base tapered to a stalk to 1 cm long, surfaces short-hairy. Leaves alternate. First and subsequent few leaves elliptic-oblong, ~ 1–2 cm long, tip acute, base tapered to a stalk to ± 1 cm long, surfaces short-hairy.

Houndstongue (*Cynoglossum officinale*) plant.
J. M. DiTomaso

Houndstongue (*Cynoglossum officinale*) flowers.
J. M. DiTomaso

Houndstongue (*Cynoglossum officinale*) fruiting stem. J. M. DiTomaso

MATURE PLANT: Foliage moderate to densely covered with long, soft, white hairs. Leaves alternate, oblanceolate to narrow elliptic, mostly 8–20 cm long, 2–5 cm wide. Lower leaf bases tapered to a stalk mostly 4–10 cm long. Upper leaves sessile, smaller than lower leaves.

ROOTS AND UNDERGROUND STRUCTURES: Taproot thick, ± woody, often branched, outer layer dark, can grow to 1 m deep or more.

FLOWERS: May–July. Flowers are in panicles of racemes, with leaves usually *slightly overlapping the base of the panicle* and *leaflike bracts*. Flowers 4–9 mm long, ± 5 mm wide, often horizontal to slightly drooping. Sepals 5, fused at the base, densely long soft-hairy. Corolla *dark reddish purple*, 5-lobed, lobes bowl-shaped and *slightly longer than sepals*, tube hidden by sepals, throat with 5 hairy, scalelike appendages. Stamens 5, attached in the upper tube. Flower stalks 2–12 mm long. Typically self-pollinated.

FRUITS AND SEEDS: Fruits consist of *4 spreading-ascending nutlets on a pyramid-shaped receptacle* with a *persistent style attached to the center of the receptacle*. Nutlets ± ovoid, *upper surface flat with raised margins*, mostly 5–8 mm long, light brown to gray-brown, covered with *short prickles tipped with downward-pointing barbs*.

POSTSENESCENCE CHARACTERISTICS: Old stems with fruits can persist through winter or longer.

HABITAT: Open disturbed, often moist places, roadsides, fields, pastures, rangeland, open woodland and forests, sand dunes, waste places, abandoned crop-

Houndstongue

Houndstongue (*Cynoglossum officinale*) young rosette.   J. M. DiTomaso

Houndstongue (*Cynoglossum officinale*) seedling.   J. M. DiTomaso

land, and ditch and canal banks. Typically colonizes bare soil. Often grows on sandy or gravelly soil.

DISTRIBUTION: Cascade Range, mostly 800–1525 m, possibly higher. Most contiguous states, except some southern states.

PROPAGATION/PHENOLOGY: *Reproduces by seed.* Nutlets remain attached to or fall near the parent plant and disperse to greater distances primarily by clinging to animals and to the shoes, clothing, and equipment of humans. Seeds require a cool, moist period (vernalization) in soil before they can germinate. Seeds persisting on stems through winter or longer require vernalization in soil the following winter to germinate. Seeds germinate in spring or survive about 1–3 years under field conditions. Most seedlings emerge from a soil depth of

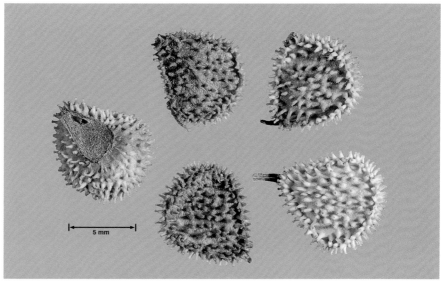

Houndstongue (*Cynoglossum officinale*) seeds.   J. O'Brien

about 1 cm. Seedlings are generally robust and rapidly develop a strong taproot. Most plants flower once and die, but a few may flower every season for up to 4 years.

MANAGEMENT FAVORING/DISCOURAGING SURVIVAL: First-year rosettes tolerate mowing, grazing, and drought stress. Mowing flowering stems before nutlets develop greatly reduces seed production. Maintaining pasture and rangeland health by preventing overgrazing and minimizing disturbance can help limit houndstongue spread and establishment.

SIMILAR SPECIES: Two common *native perennial* houndstongue species [*Cynoglossum grande* Dougl. ex Lehm. and *Cynoglossum occidentale* A. Gray] are primarily distinguished by having *leaves that do not overlap the base of the inflorescence, inflorescence bracts that are highly reduced and scalelike or lacking*, flowers with *bluish corollas that are usually conspicuously longer than the sepal lobes*, and *nutlets with a rounded upper surface*. The native perennials are not weedy and are included here only for identification purposes.

Native houndstongue (*Cynoglossum grande*) flowering stem, a nonweedy species that does not have leafy stems.   J. M. DiTomaso

Pride of Madeira [*Echium candicans* L.f.] [Cal-IPC: Limited]

Vipers bugloss [*Echium plantagineum* L.]

SYNONYMS: Pride of Madeira: *Echium branchyanthum* Hornem.; *Echium cynoglossoides* Desf.; *Echium densiflorum* DC.; *Echium fastuosum* auct. non Dryander ex Aiton; *Echium macrophyllum* Lehm.; *Echium pallidum* Salisb.; a few others

Vipers bugloss: blue weed; Lady Campbell weed; Paterson's curse; purple bugloss; purple echium; purple viper's bugloss; Riverina bluebell; salvation jane; viper's bugloss; *Echium lycopsis* L.

GENERAL INFORMATION: Pride of Madeira: Coarse-branched *shrub* to 3 m tall, with *dense, conical panicles to 0.4 m long that consist of coiled spike- or racemelike cymes of blue to purple flowers*. Pride of Madeira is cultivated as an ornamental, but it has escaped cultivation in some coastal areas. Native to Madeira and the Canary Islands.

Vipers bugloss: Erect winter *annual or biennial* to about 1 m tall, with *spikelike cymes that are coiled at the tips. Flowers are purplish blue, pink, or white*. Plants exist as basal rosettes until flowering stems develop at maturity. Vipers bugloss is considered a noxious weed in southern and eastern Australia, where several biocontrol agents are being investigated for release. It appears to be less

Pride of Madiera (*Echium candicans*) plants along a hillside. J. M. DiTomaso

Pride of Madiera (*Echium candicans*) inflorescences. J. M. DiTomaso

Pride of Madiera (*Echium candicans*) plant.
J. M. DiTomaso

Pride of Madiera (*Echium candicans*) rosette.
J. M. DiTomaso

problematic elsewhere, including California. Vipers bugloss contains variable levels of *toxic* pyrrolizidine alkaloids and can cause fatal liver damage to livestock, especially horses and pigs, when consumed in quantity over a period of

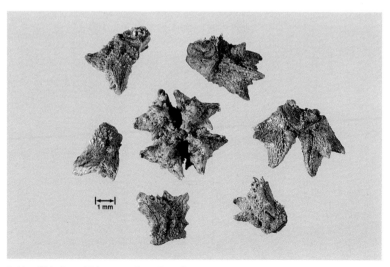

Pride of Madiera (*Echium candicans*) seeds. J. O'Brien

Vipers bugloss *(Echium plantagineum)* plant.
J. M. DiTomaso

Vipers bugloss *(Echium plantagineum)* flowering stem.
J. M. DiTomaso

weeks to years. Vipers bugloss is a useful honey bee plant because it produces abundant pollen and nectar early in the season. Native to southern Europe.

SEEDLINGS: Pride of Madeira: Seldom encountered.

Vipers bugloss: Cotyledons oblong, ~ 5–12 mm long, 4–8 mm wide, tip ± truncate to slightly indented. Margins ciliate. Upper surface glabrous. Lower surface sparsely hairy to ± glabrous. Leaves alternate. First leaf elliptic, with a rounded-acute tip, slightly larger than cotyledons, covered with papillae-based hairs.

MATURE PLANT: Pride of Madeira: Bark whitish, papery. Branches many, thick. Foliage densely covered with short, silky, flattened (appressed), grayish white hairs. Some hairs often papillae-based. Leaves alternate, densely clustered near branch tips, lanceolate to narrowly elliptic, 6–25 cm long, mostly 2–4 cm wide.

Vipers bugloss *(Echium plantagineum)* flowers.
J. M. DiTomaso

Vipers bugloss *(Echium plantagineum)* seedling.
J. M. DiTomaso

Vipers bugloss: Foliage evenly covered with long, erect, soft to bristly, gray to white, papillae-based hairs. Basal rosette leaves stalked, oblong-oval to elliptic, mostly 5–16 cm long, 1–2 cm wide. Stem leaves alternate, narrowly oblong, 2–10 cm long, sessile, bases sometimes weakly lobed and clasping the stem.

ROOTS AND UNDERGROUND STRUCTURES: Pride of Madeira: Roots woody. Vipers bugloss: Taproot tough, with many lateral roots.

FLOWERS: Panicles terminal, consist of *coiled raceme-like or spikelike cymes* (helicoid cymes). Flowers ± trumpet-shaped with an expanded, *slightly oblique throat* and *5 slightly unequal lobes*. Calyx 5-lobed nearly to base. Stamens 5, unequal. Style 1, with a 2-lobed or forked stigma. Insect-pollinated.

Pride of Madeira: March–July. *Panicles dense, ± conical, about 10–40 cm long.* Corollas blue to purple, *barely bilateral*, 5–11 mm long, tube about equal in length to the calyx. Calyx 3–5 mm long. All stamens and the style protrude beyond the corolla.

Vipers bugloss: March–August. *Panicles openly branched.* Corollas purplish blue, pink, or white, *noticeably bilateral*, 15–30 mm long, with the tube about twice as long as the calyx. Calyx 5–12 mm long. *Only the lower 2–3 stamens protrude beyond the corolla.* Anthers mature before the stigma. Day length less than 12 hours long and high temperatures inhibit flowering.

FRUITS AND SEEDS: Fruits consist of *4 erect nutlets on a flat receptacle* surrounded by the calyx (sepals as a unit). Nutlets blackish, ovoid, about 2–3 mm long, 3-sided, with a *flat basal attachment scar*. Surfaces wrinkled and/or covered with minute tubercles.

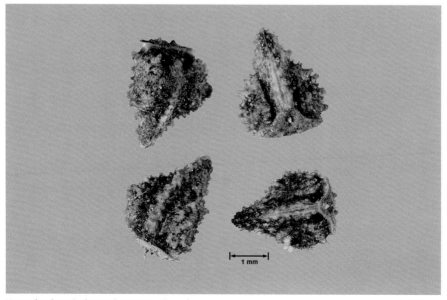

Vipers bugloss (*Echium plantagineum*) seeds.   J. O'BRIEN

Pine echium (*Echium pininana*) flowering stalk.
J. M. DiTomaso

Pine echium (*Echium pininana*) flowers.
J. M. DiTomaso

HABITAT: Both species tolerate poor soil.

Pride of Madeira: Open coastal hillsides and bluffs. Requires a source of summer moisture in inland areas. Grows on many soil types.

Vipers bugloss: Disturbed sites, roadsides, fields, and grassland. Widespread in Mediterranean climate areas. Thrives in high-latitude regions where winter temperatures are low and summer days are long.

DISTRIBUTION: Pride of Madeira: San Francisco Bay region, Central Coast, and South Coast, to 300 m.

Vipers bugloss: Central Coast and South Coast, to 300 m. Pennsylvania and Massachussetts.

PROPAGATION/PHENOLOGY: *Reproduce by seed.*

Pride of Madeira: The biology of this species is poorly documented.

Vipers bugloss: Seeds disperse with water, soil movement, human activities, as a feed contaminant, and via animals, including birds and livestock. Seeds germinate under a broad temperature range (optimal 20°–30°C), and darkness enhances germination. Seeds are hard-coated and survive ingestion by livestock. Few newly matured seeds will germinate under favorable conditions. Most seeds germinate after the first fall rains, but some germinate in late winter, spring, and

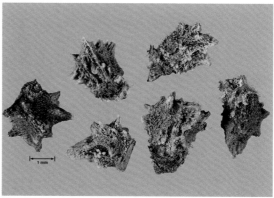

Pine echium (*Echium pininana*) seedling. J. M. DiTomaso

Pine echium (*Echium pininana*) seeds. J. O'Brien

early summer. A small proportion of seeds remain dormant in the soil seedbank for up to 5 years or more. Seedlings emerge from soil depths to about 7 cm. Seedlings establish deep roots and can tolerate severe moisture stress for at least 1 month. Seedlings germinated in late spring and early summer may persist through summer as a basal rosette under favorable conditions and flower the following spring.

MANAGEMENT FAVORING/DISCOURAGING SURVIVAL: Vipers bugloss: Hand-pulling or mowing can control small patches. However, cut or pulled plants with immature flowers can continue to mature seed. Repeated cultivation can kill flushes of seedlings. Burning destroys some seeds but may stimulate others to germinate. Vipers bugloss is not yet a major problem in the United States, but it has been demonstrated in Australia that grazing cattle on pastures and rangeland with vipers bugloss can increase populations of the species.

SIMILAR SPECIES: Pine echium [*Echium pininana* Webb & Berth.] is a *shrub to 5.5 m tall* that has escaped cultivation as an ornamental in some coastal areas of California. Unlike pride of Madeira, pine echium typically exists as a vegetative rosette on a short woody stem for several years, flowers once, and then dies (monocarpic). Pine echium is further distinguished by having *few branches or a single unbranched stem* and *1–few dense, pyramidal panicles 1–3.5 m long that are interspersed with leaves*. Panicles sometimes droop at the tip. Flowers are blue, purplish, or white with blue streaks. In addition, leaves are mostly broadly lanceolate, up to 50 cm long, 6–10 cm wide, and covered with flattened, soft to bristly, sometimes papillae-based hairs. The distribution of pine echium is poorly documented. It occurs sporadically along the northern Central Coast (especially Bodega Bay) and in the San Francisco Bay region, and possibly in other coastal areas. Native to the Canary Islands, where it inhabits only laurel forests and tall heath [*Erica arborea* L.] shrublands at 600–1000 m elevation.

Seaside heliotrope [*Heliotropium curassavicum* L.] [HEOCU]

European heliotrope [*Heliotropium europaeum* L.] [HEOEU]

SYNONYMS: Seaside heliotrope: alkali heliotrope; Chinese pusley; cola de mico; devil-weed; quailplant; salt heliotrope; white-weed; wild heliotrope; *Heliotropium curassavicum* L. var. *oculatum* (Heller) I.M. Johnst. and var. *parviflorum* Ball; *Heliotropium lehmannianum* Bruns

European heliotrope: barooga weed; caterpillar weed; common heliotrope; European turnsole; potato weed; wanderie curse; *Heliotropium stevenianum* Andrz.

GENERAL INFORMATION: Both species have foliage and fruits that contain pyrrolizidine alkaloids and are *toxic* to humans and livestock, especially horses, when ingested in small amounts over time or in a single large quantity. The toxic chemicals found in the plant primarily damage the liver. Chronic toxicity symptoms for horses include weight loss and lethargy. Acute symptoms can include aimless pacing, off-balance gait, chewing, head pressing, yawning, and digestive tract disturbance. Young foliage and nutlets are more poisonous than mature foliage. Many livestock poisonings occur when animals ingest contami-

Seaside heliotrope *(Heliotropium curassavicum)* flowering stem. J. M. DiTomaso

Seaside heliotrope • European heliotrope

Seaside heliotrope (*Heliotropium curassavicum*) plants. J. M. DiTomaso

Seaside heliotrope (*Heliotropium curassavicum*) flowers. J. K. Clark

Seaside heliotrope (*Heliotropium curassavicum*) seedling. J. M. DiTomaso

Seaside heliotrope (*Heliotropium curassavicum*) seeds. J. O'Brien

nated hay or grain. Sheep grazing in *Heliotropium*-infested pastures can accumulate copper to toxic levels in the liver.

Seaside heliotrope: Widespread native *perennial*, occasionally summer annual, with prostrate to ascending stems to 0.6 m long, *glabrous foliage*, and *coils of flowers that are white with greenish white, purplish, or yellow centers*. Plants sometimes have creeping roots that develop new shoots. Seaside heliotrope is usually considered a desirable component of the vegetation in natural areas, but troublesome colonies can develop in agronomic fields and associated ditches, canal banks, managed forest clearings, and elsewhere. Some botanists recognize 3 varieties, of which only variety *oculatum* (Heller) I.M. Johnston occurs in California. Some sources suggest this species may be naturalized from Central America.

European heliotrope: Summer *annual*, occasionally biennial, with ascending to erect stems to about 0.5 m tall, short-hairy foliage, and *coils of cream-colored or white flowers with yellow centers*. European heliotrope is more toxic than seaside heliotrope when ingested. It is a government-listed noxious weed in southern Australia. Native to North Africa and southern and eastern Europe.

European heliotrope (*Heliotropium europaeum*) plant. J. M. DiTomaso

European heliotrope (*Heliotropium europaeum*) seedlings. J. M. DiTomaso

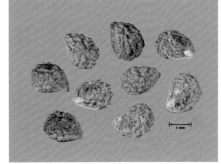

European heliotrope (*Heliotropium europaeum*) flowering stem. J. M. DiTomaso

European heliotrope (*Heliotropium europaeum*) seeds. J. O'Brien

SEEDLINGS: Leaves opposite or alternate.

Seaside heliotrope: Cotyledons ovate, mostly 4–6 mm long, ± 2 mm wide, glabrous, with a stalk of ± equal length. First and subsequent few leaves oblanceolate, 15–35 mm long, 3–6 mm wide, glabrous except sometimes a few long hairs on the midvein near the base.

European heliotrope: Cotyledons ovate, with a stalk of equal length or longer, ± sparsely covered with minute fine hairs. First and subsequent few leaves ovate, finely hairy, upper surface with conspicuous depressed veins.

MATURE PLANT: Leaves alternate, sometimes opposite on lower stems.

Seaside heliotrope: *Foliage glabrous*, slightly succulent, gray-green to bluish green with a whitish waxy bloom (glaucous). Stems mostly prostrate to ascending. Leaves usually oblanceolate, sometimes linear or obovate, 1–6 cm long, tips acute to rounded, nearly sessile to short-stalked.

European heliotrope: *Foliage short-hairy*, especially leaf veins and stems. Hairs often minutely papilla-based. Stems ascending to erect. Leaves ovate to elliptic, 1.5–5 cm long, tip ± rounded, short-stalked, light green to gray-green.

ROOTS AND UNDERGROUND STRUCTURES: Seaside heliotrope: Roots sometimes creeping, can develop new shoots.

European heliotrope: Taproots slender, to 1 m deep or more, with numerous highly branched lateral roots.

FLOWERS: Spikes one-sided, tip coiled. Corollas radial, 5-lobed, 3–5 mm

Clasping heliotrope (*Heliotropium amplexicaule*) plant. J. M. DiTomaso

Clasping heliotrope (*Heliotropium amplexicaule*) seedling. J. M. DiTomaso

Clasping heliotrope (*Heliotropium amplexicaule*) flowering stem. J. M. DiTomaso

Clasping heliotrope (*Heliotropium amplexicaule*) seeds. J. O'Brien

wide. *Style attached at top of ovary*. Style appears forked with a linear stigma and a sterile appendage. Ovary nearly round, 4-chambered. Stamens 5, attached to and included within the corolla tube. Calyx 5-lobed.

Seaside heliotrope: March–October. Corollas ± bell-shaped, *white with a yellow, yellow-green, or bluish purple throat to pale lavender with a darker throat*. Calyx glabrous.

European heliotrope: March–June. Corollas ± trumpet-shaped, *cream-colored or white, often with a yellow throat*. Calyx densely hairy.

FRUITS AND SEEDS: Fruits nearly round, 4-lobed, eventually separate into 4 erect, ovoid nutlets ± 2 mm long. Style deciduous. Nutlet attachment scar lateral.

Seaside heliotrope: Fruits ± 2 mm in diameter. *Nutlets smooth*, glabrous.

European heliotrope: Fruits 2–3 mm in diameter. *Nutlets irregularly tubercled*, sometimes minutely hairy.

POSTSENESCENCE CHARACTERISTICS: Old stems with fruits can persist into winter. Seaside heliotrope plants turn a dark purple-brown within a short time of being cut or pulled.

HABITAT: Seaside heliotrope: Many plant communities in open disturbed places, roadsides, fields, ditches, pastures, margins of agronomic and vegetable crops, riparian areas, seasonal wetlands (including those on serpentine soil), pond margins, forest clearings, and saline or alkaline sites. Often grows on moist or seasonally moist sandy soils.

European heliotrope: Open disturbed areas, fields, pastures, roadsides, waste places, and urban sites. Typically grows in dry places.

DISTRIBUTION: Seaside heliotrope: Throughout California, to 2100 m. One or more varieties of this species occur in most contiguous states, except for some north-central and eastern states.

European heliotrope: Central Valley, northern and central Sierra Nevada foothills, Modoc Plateau, San Francisco Bay region, and Central Coast, to 1400 m. Many eastern states and most southern states.

PROPAGATION/PHENOLOGY: Nutlets fall near the parent plant and disperse to greater distances with water, mud, soil movement, animals, human activities including agricultural equipment, and as feed or seed contaminants.

Seaside heliotrope: *Reproduces by seed* and *sometimes vegetatively from creeping roots*. Seeds germinate and new shoots grow from roots in spring.

European heliotrope: *Reproduces by seed.* Seeds germinate in spring and summer. Plants often begin to flower within 3–4 weeks after emergence. Flower and fruit production continues until plants are killed by frost. When summer moisture is available, more than 1 generation of plants can complete their life cycle in a season. Some seeds survive ingestion by animals.

MANAGEMENT FAVORING/DISCOURAGING SURVIVAL: Cultivation as needed to prevent plants from developing fruits can control these species in areas where cultivation is appropriate. European heliotrope competes poorly with established vegetation. Maintaining healthy pastures without open bare spots can limit the spread of European heliotrope.

SIMILAR SPECIES: Clasping heliotrope [*Heliotropium amplexicaule* M. Vahl, synonyms: blue heliotrope; violet heliotrope; *Helotropium anchusifolium* Poir.][HEOAM] is a *perennial* with ascending or decumbent stems to about 0.5 m tall, *short-hairy foliage*, and *coiled spikes of purple flowers*. The combination of *perennial life cycle, short-hairy foliage,* and *coiled spikes of purple flowers* distinguish it from other *Heliotropium* species. In addition, leaves are elliptic to oblanceolate, about 4–9 cm long, and nearly sessile to short-stalked. Clasping heliotrope inhabits open disturbed places and fields in the northern and central areas of the South Coast, to 300 m. It also occurs in most southern states and a few northeastern states. Clasping heliotrope is generally less toxic to humans and livestock than seaside and European heliotrope. Native to Argentina.

Broadleaf forget-me-not [*Myosotis latifolia* Poir.] [Cal-IPC: Limited]

SYNONYMS: common forget-me-not; wood forget-me-not; *Myosotis sylvatica* Hoffm. misapplied in older California references

GENERAL INFORMATION: Erect to nearly prostrate *perennial* to 0.7 m tall, with **creeping roots** and *coiled racemes of pale blue to pink funnel-shaped flowers*. Although not documented as the cause of livestock poisonings, broadleaf forget-me-not and **true forget-me-not** [*Myosotis scorpioides* L.] contain toxic pyrrolizidine alkaloids similar to those found in **houndstongue** [*Cynoglossum offinale* L.]. Refer to the **Houndstongue** entry (p. 430) for more information about the toxic properties of the alkaloids. Native to northwestern Africa and introduced as a garden ornamental.

SEEDLINGS: Cotyledons nearly round to oval, about 4–6 mm long, with a hairy, short stalk. Leaves alternate. First and subsequent few leaves elliptic to oblanceolate, ~ 5–15 mm long, base tapered to a ± narrow-winged stalk of nearly equal length, blade surfaces and stalk sparsely covered with short, flattened (appressed) hairs.

MATURE PLANT: Stems woody at the base, erect to prostrate with upturned tips (decumbent), upper portions densely covered with short spreading hairs. Leaves broadly elliptic to narrowly oblanceolate, 2–15 cm long, 1–2.5 cm wide, tip usually abruptly tapered to a minute nipplelike point (mucronate), sparsely covered with short, appressed hairs. Lower leaves tapered to a narrowly winged stalk. Upper leaves sessile.

Broadleaf forget-me-not (*Myosotis latifolia*) plant.   J. M. DiTomaso

Broadleaf forget-me-not

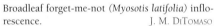

Broadleaf forget-me-not (*Myosotis latifolia*) seedling. J. M. DiTomaso

Broadleaf forget-me-not (*Myosotis latifolia*) inflorescence. J. M. DiTomaso

Broadleaf forget-me-not (*Myosotis latifolia*) seeds. J. O'Brien

ROOTS AND UNDERGROUND STRUCTURES: Horizontal creeping stems woody, with numerous fibrous lateral roots.

FLOWERS: February–July. Racemes initially coiled, later elongate and open, with spaced flowers and reduced leaves at the base. Corollas *pale blue or pink*, ± funnel-shaped, mostly *5–10 mm wide*, with 5 spreading lobes and 5 pale crests that nearly close the top of the tube. Calyx 4–5 mm long, densely covered with straight hairs and *hook-tipped hairs*, deeply 5-lobed, *lobes equal* and linear-lanceolate. Stamens 5, included within corolla tube.

FRUITS AND SEEDS: Fruits consist of 4 erect nutlets surrounded by the calyx. *Fruit stalks 5–10 mm long*, ascending to spreading. Nutlets ovoid, ± 2 mm long, ± 1.5 mm wide, slightly flattened, margin narrow-winged and curved to 1 side, glossy, dark greenish brown to greenish black, with an attachment scar at the base. Receptacle nearly flat.

POSTSENESCENCE CHARACTERISTICS: Old stems with fruits can persist through winter.

HABITAT: Shady moist disturbed places, riparian areas, coastal forests and woodlands, moist meadows and fields, roadsides, and old gardens and other areas near human habitation.

DISTRIBUTION: Scattered throughout California, except Great Basin and

Broadleaf forget-me-not

True forget-me-not *(Myosotis scorpioides)* plant. J. M. DiTomaso

deserts, to 300 m. More common in the San Francisco Bay region, North Coast, and North Coast Ranges.

PROPAGATION/PHENOLOGY: *Reproduces by seed* and *vegetatively from creeping roots*.

SIMILAR SPECIES: True forget-me-not [*Myosotis scorpioides* L., synonym: *Myosotis palustris* (L.) Hill] [MYOPA] is a *perennial*, typically with simple stems to 0.6 m tall or long, creeping roots, and *pale blue flowers with yellow centers*. True forget-me-not is further distinguished by having *sepals with only straight hairs*. True forget-me-not is designated a facultative wetland indicator species in California. It often inhabits places with wet or seasonally wet soil, standing water, drainage ditches, and other moist, disturbed places in the Northwestern region and Sierra Nevada, especially the northern portion, to 2000 m. Native to Europe and introduced as a garden ornamental.

A few other non-native *Myosotis* species occur locally in the Klamath Ranges or sporadically throughout much of California, but all are generally not problematic. All are *annual to biennial* and have some *hook-tipped hairs on the sepals* like broadleaf forget-me-not, but unlike broadleaf forget-me-not, the annual/biennial species have *corollas 1–5 mm wide*. Corolla color is white, blue, or yellow turning blue.

European sticktight [*Lappula squarrosa* (Retz.) Dumort., synonyms: *Lappula echinata* Gilib.; *Lappula myosotis* Moench; *Myosotis lappula* L.; others] [LPLSQ] is an ascending to erect *annual* to 0.8 m tall that resembles *Myosotis* species. Unlike *Myosotis* species, European sticktight has *flower receptacles that are narrowly elongate* and *nutlets that have 2–3 rows of barbed prickles on the margin*

Broadleaf forget-me-not

True forget-me-not *(Myosotis scorpioides)* flowering stem.   J. M. DiTomaso

and a *lateral attachment scar*. In addition, European sticktight lacks hook-tipped hairs on the calyxes and has blue corollas 2–4 mm wide and erect flower stalks 1–3 mm long. European sticktight inhabits moist, disturbed places, cultivated ground, gardens, and urban places in the South Coast (Santa Monica, Upland), to 200 m. It occurs in most contiguous states, except some southern states. Native to Eurasia and probably introduced as a garden ornamental.

True forget-me-not *(Myosotis scorpioides)* seeds.   J. O'Brien

Prickly comfrey [*Symphytum asperum* Lepechin] [SYMAS][CDFA list: B]

Common comfrey [*Symphytum officinale* L.][SYMOF]

SYNONYMS: Prickly comfrey: rough comfrey; *Symphytum asperrimum* Donn ex Sims

Common comfrey: asses-ears; backwort; boneset; bruisewort; consolida; consound; gum plant; healing herb; knitback; knit-bone; slippery-root; *Symphytum officinale* L. ssp. *uliginosum* (Kern.) Nyman; *Symphytum uliginosum* Kern.

GENERAL INFORMATION: Clumped *perennial herbs* to about 1 m tall, with a deep taproot, coarse foliage, and coiled inflorescences. Once established, these species are difficult to remove because small pieces of broken roots generate new plants. Both comfrey species contain *toxic* pyrrolizidine alkaloids that can cause liver damage when consumed in sufficient quantity over time or in a single large dose. Toxicity problems most often arise in humans who ingest exces-

Prickly comfrey (*Symphytum asperum*) plant.   COURTESY OF ROSS O'CONNELL

sive quantities of herbal remedies. The toxic compounds are most concentrated in the roots.

Prickly comfrey: This species is a noxious weed in northwestern California. A hybrid between prickly and common comfrey [*S. peregrinum,* synonym: *S.* × *uplandicum*] is sometimes cultivated for silage in Russia. Native to Southwestern Asia.

Common comfrey: Common comfrey is cultivated as an ornamental and as a medicinal herb. In Europe, it is also cultivated for use as green manure, and the young leaves are sometimes eaten as a vegetable. Native to Europe.

MATURE PLANT: Stems ascending to erect, thick, angular, hollow, usually branched, covered with rough, bristly hairs. Leaves alternate, ovate to lanceolate, mostly stalked. Upper stem leaves sometimes sessile.

Prickly comfrey: Petioles usually winged. *Stem wings lacking on internodes.* Leaves 5–30 cm long, sparsely covered with short bristly hairs.

Common comfrey: Stems and petioles winged. *Stem wings continuous on internodes.* Leaves 5–15 cm long, moderately covered with short bristly hairs.

ROOTS AND UNDERGROUND STRUCTURES: Taproot usually branched, *deep, thick, brittle,* internally white, covered with a thin blackish bark, contains much mucilage. *A small piece of the taproot can generate a new plant.*

FLOWERS: May–Summer. Cymes loosely coiled, racemelike, axillary or terminal. Flowers *bell-shaped,* stalked, in pairs on the upper sides of stems. Corolla 10–15 mm long, with 5 *lanceolate appendages* in the throat that alternate with 5 stamens. Throat appendages and anthers are at about the same level. Sepals fused at the base, deeply 5-lobed, covered with bristly hairs. Ovary superior, deeply 4-lobed. *Style 1, with 1 stigma, attached to the ovary base between the ovary lobes, persistent in fruit.* Stamens mature before the ovary. Insect-pollinated, especially by bees.

Prickly comfrey: Corolla initially pink, turns deep blue or purple, tube appendages *linear-lanceolate.* Calyx 2–4 mm long.

Common comfrey: Corolla typically reddish to purple, sometimes blue, pale yellow, or cream-colored, tube appendages lanceolate. Calyx 3–6 mm long.

FRUITS AND SEEDS: Fruits consist of 4 nutlets (sometimes only 1–3 mature) that separate at maturity. Nutlets ovoid, attached at the base, contain 1 seed, do not open to release the seed.

Prickly comfrey: Nutlets 3–4 mm long, 2–2.5 mm wide, glossy brown to blackish brown, surface finely granular.

Common comfrey: Nutlets 4–5 mm long, ± 3 mm wide, glossy black.

HABITAT: Typically grows on moist soils of open sites, waste places, fields, ditches, and gardens. Plants thrive on fertile garden soils. Tolerates cold winter climates.

Prickly comfrey • Common comfrey

DISTRIBUTION: Prickly comfrey: North Coast and low-lying regions in the northern North Coast Ranges (Humboldt Co.) and Sacramento Valley (Yolo Co.), to 100 m. Oregon, Washington, and some northeastern and north-central states.

Common comfrey: San Francisco Bay region, to 100 m. Expected to expand its range in California. Colorado, Idaho, Montana, Oregon, Utah, and Washington. Most eastern states.

Common comfrey (*Symphytum officinale*) plant. J. M. DiTomaso

Common comfrey (*Symphytum officinale*) flowering stem.   J. M. DiTomaso

Common comfrey (*Symphytum officinale*) stem and leaf base.   J. M. DiTomaso

PROPAGATION/PHENOLOGY: *Reproduces by seed* and vegetatively from root fragments. In cold winter climates, foliage dies during the winter and regenerates from the roots in spring.

Seeds of common comfrey germinate rapidly in moist soil, especially peaty or loam soil, and in water.

MANAGEMENT FAVORING/DISCOURAGING SURVIVAL: Although fragmented roots can generate new plants, repeated cultivation can eventually eliminate weedy populations, especially when timed to prevent seed production. Mowing before seed is produced can also help to control populations.

SIMILAR SPECIES: Common borage [*Borago officinalis* L.] is a bristly to stiff-hairy *annual* to about 0.7 m tall, with loose cymes of bright blue flowers. Besides its *annual life* cycle, borage is distinguished by having *bowl-shaped flowers* with *petals that are fused at the base* and have *widely spreading lobes. Stamens form a conspicuous cone in outline.* Leaves are ovate to oblanceolate, and the lower leaves are mostly 8–20 cm long and 3–8 cm wide. Borage sometimes escapes cultivation as a garden ornamental in North Coast, San Francisco Bay region, Central Coast, South Coast, and northern and central Sierra Nevada foothills, to 300 m. It is also naturalized in Montana, Oregon,

Common comfrey (*Symphytum officinale*) seed. Similar in appearance to common comfrey (*Symphytum asperum*) seeds. J. O'Brien

Common borage (*Borago officinalis*) plants. J. M. DiTomaso

Common borage (*Borago officinalis*) flowering stems. J. M. DiTomaso

Common borage (*Borago officinalis*) seeds. J. O'Brien

Utah, Washington, a few north-central states, and most northeastern states. The foliage and roots of common borage also contains small quantities of *toxic* pyrrolizidine alkaloids, but the risk of developing toxicity problems due to the ingestion of this species appears low. The flowers of common borage are attractive to bees. Native to southern Europe.

Yellow rocket [*Barbarea vulgaris* W.T. Aiton.][BARVU]

SYNONYMS: bitter cress; common winter cress; garden yellow rocket; potherb; rocket cress; St. Barbaras cress; upland cress; water mustard; yellow weed; *Barbarea arcuata* (Opiz ex J. Presl & K. Presl) Reichenb.; *Barbarea stricta* auct. non Andrz.; *Barbarea vulgaris* R. Br.; *Campe barbarea* (L.) W. Wight ex Piper; *Campe stricta* auct. non (Andrz.) W. Wight ex Piper; *Erysimum arcuatum* Opiz ex C. Presl & J. Presl; *Erysimum barbarea* L.; a few others

GENERAL INFORMATION: Erect *biennial,* sometimes annual or short-lived perennial, to nearly 1 m tall, with *bright yellow flowers* and *linear fruits.* Plants exist as rosettes until flower stems develop at maturity, typically in the second season. Yellow rocket is more common in the eastern half of the United States. Like many species in the mustard family, yellow rocket foliage and especially

Yellow rocket (*Barbarea vulgaris*) plant   J. M. DiTomaso

Yellow rocket

Yellow rocket (*Barbarea vulgaris*) flowering stem. J. M. DiTomaso

Yellow rocket (*Barbarea vulgaris*) fruiting stem. J. M. DiTomaso

the seeds contain glucosinolates and can cause digestive tract irritation when ingested in a large quantity. Thyroid dysfunction can develop when quantities of plant material are ingested over time. Livestock confined to pastures that consist primarily of yellow rocket and other members of the family are most likely to exhibit toxicity problems. Symptoms can include colic, diarrhea, excessive salivation, weight loss, neurological disturbances, and thyroid enlargement. Occasionally humans consume the young leaves as salad greens or as a cooked vegetable. Yellow rocket is a host to several pests and diseases that affect vegetable crops, including certain nematodes, cucumber mosaic virus, curly top virus, and potato yellow dwarf virus. Native to Eurasia.

SEEDLINGS: Cotyledons oval, about 3–6 mm long, on stalks 4–7 mm long, glabrous. Leaves alternate. First and subsequent few leaves ovate, about 2–6 mm long, tip rounded, margins ± slightly wavy, glabrous, on stalks equal to or longer than the blade.

MATURE PLANT: Foliage glabrous. Rosette and lower stem leaves deeply pinnate-lobed, mostly 5–20 cm long, glossy, with 1–4 pairs of mostly opposite small lateral lobes. *Terminal lobe nearly round to ovate, base slightly heart-shaped, much larger than lateral lobes.* Stems coarse, angled, mostly branched in the upper portion. Upper stem leaves alternate, reduced, sessile, *base weakly clasping the stem, margins smooth to coarsely toothed or shallow-lobed.*

ROOTS AND UNDERGROUND STRUCTURES: Taprooted, usually with numerous, highly branched fibrous lateral roots. The taproot can grow to 50 cm deep or more.

Yellow rocket

Yellow rocket (*Barbarea vulgaris*) leaf with fewer lobes than *B. verna*. J. M. DiTomaso

Yellow rocket (*Barbarea vulgaris*) seedling. J. M. DiTomaso

Yellow rocket (*Barbarea vulgaris*) seeds. J. O'Brien

FLOWERS: April–May. Petals 4, *bright yellow*, usually 6–8 mm long. Sepals 4, often yellow. Stamens 6, 2 short, 4 longer. Flower stalks less than 1 mm thick.

FRUITS AND SEEDS: Pods ascending to erect, *linear, 1–3 cm long,* 2–3 mm wide, nearly straight, *cylindrical to slightly 4-angled in cross-section*, with a *slender, beaklike style about 2–3 mm long* at the tip. Stalks 3–5 mm long. Pods open at the bottom by 2 valves, each valve conspicuously 1-veined. Seeds in 1 row per pod chamber, ovoid to oblong-rectangular, 1–1.5 mm long, notched at one end, dull, yellowish to reddish brown or gray, surface finely textured.

POSTSENESCENCE CHARACTERISTICS: Old stems with the central membranes of the pods still attached can persist into winter.

HABITAT: Agronomic and vegetable crop fields, orchards, landscaped areas, nurseries, ditches, roadsides, pastures, gardens, and other disturbed moist places. Often inhabits sites that receive some summer moisture. Grows best on disturbed, moist, fertile soils.

DISTRIBUTION: Great Basin, northern Sierra Nevada, and southern North

Early wintercress (*Barbarea verna*) plant.
J. M. DiTomaso

Early wintercress (*Barbarea verna*) fruit base nearly equal to width of pedicel. J. M. DiTomaso

Early wintercress (*Barbarea verna*) leaf. Note the number of lobes. J. M. DiTomaso

Early wintercress (*Barbarea verna*) seeds.
J. O'Brien

Coast Ranges, to 1000 m. Expected to expand its range in California. Most contiguous states, except a few southern and western states.

PROPAGATION/PHENOLOGY: *Reproduces by seed.* Seed production is often high. Most seeds are ejected a short distance from the parent plant when pods snap open, but some can disperse to greater distances with water, soil movement, agricultural equipment, as a seed and feed contaminant, and by clinging to animals, the shoes and clothing of humans, and vehicle tires. Seeds become sticky with mucilage when moistened. Some seeds survive ingestion by animals. Germination requires light, fluctuating temperature, and a cool, moist period. In California, most germination occurs in fall after the rainy season begins, but some seeds can germinate in spring. Seeds have been reported to survive for up to 10–20 years under field conditions.

MANAGEMENT FAVORING/DISCOURAGING SURVIVAL: Mowing flower stems before fruits develop, repeated cultivation or cultivation at the rosette stage, and smother crops can help control yellow rocket.

SIMILAR SPECIES: Early wintercress [*Barbarea verna* (Miller) Asch., synonyms: early yellowrocket; scurvy grass; *Barbarea praecox* (Smith) R. Br.; *Campe*

Yellow rocket

Early wintercress (*Barbarea verna*) flowering stem. J. M. DiTomaso

American yellow rocket (*Barbarea orthoceras*) plant. J. M. DiTomaso

verna (Miller) Heller; *Erysimum verna* Mill.][BARVE] is a less-common *biennial* or short-lived perennial to nearly 1 m tall, with bright yellow flowers. It is distinguished by having basal leaves with *4–10 pairs of lateral lobes*, flower stalks *more than 1 mm thick*, and pods usually *4.5–8 cm long*. Early wintercress inhabits roadsides, fields, damp places, and other disturbed sites in the North Coast Ranges, southeastern Klamath Ranges (Shasta Co.), Cascade Range foothills, Sierra Nevada foothills, Central Coast, and northeastern South Coast Ranges, to at least 500 m, and it is reported to probably be present in the San Joaquin Valley (Fresno Co.). Early wintercress is expanding range in California. It also occurs in Montana, Oregon, Washington, and most eastern states. Native to Eurasia.

American yellow rocket [*Barbarea orthoceras* Ledeb.] is a native *biennial* or short-lived perennial to about 0.6 m tall that more closely resembles yellow rocket by having *lower leaves usually with 2–3 pairs of lateral lobes* and pods on stalks *less than 1 mm thick*. American yellow rocket is usually considered a desirable component of the vegetation in natural areas and not a weed. Unlike yellow rocket, American yellow rocket has *pale yellow flowers*, pods with a *stout style less than 1 mm long*, and upper stem leaves that are *deeply pinnate-lobed at the base*. Pods are straight, erect to widely ascending, and (1.5)2.5–5 cm long. American yellow rocket inhabits damp places in meadows, woodlands, rocky areas, and riparian sites in the Cascade Range, Klamath Ranges, North and South Coast Ranges, Transverse Ranges, and Peninsular Ranges and coastal regions, mostly from 700–3350 m. It also occurs in all other western states and a few north-central and northeastern states.

Rapeseed mustard [*Brassica napus* L.]
Black mustard [*Brassica nigra* (L.) Koch][BRSNI] [Cal-IPC: Moderate]
Birdsrape mustard [*Brassica rapa* L.][BRSRA] [Cal-IPC: Limited]
Saharan mustard [*Brassica tournefortii* Gouan.] [BRSTO][Cal-IPC: High]

SYNONYMS: **Rapeseed mustard:** canola; rape; rapeseed; rutabaga; Swede rape; Swedish turnip; wild mustard; *Brassica napobrassica* (L.) P. Mill.; *Brassica napus* var. *napobrassica* (L.) Reichb.

Black mustard: *Sinapis nigra* L.

Birdsrape mustard: common mustard; field mustard; wild rutabaga; wild turnip; *Brassica campestris* L.; *Brassica campestris* L. var. *rapa* (L.) Hartm.; *Brassica campestris* L. ssp. *rapifera* (Metzger) Sinsk.; *Brassica rapa* L. ssp. *campestris* (L.) Clapham; ssp. *olifera* DC.; and ssp. *sylvestris* Janchen; *Brassica rapa* L. var. *rapa*

Saharan mustard: African mustard; Asian mustard; long-fruited wild turnip; Mediterranean mustard; Mediterranean turnip; Moroccan mustard

Rapeseed mustard (*Brassica napus*) inflorescence. J. M. DiTomaso

Rapeseed mustard (*Brassica napus*) leaf with clasping base. J. M. DiTomaso

Rapeseed mustard • Black mustard • Birdsrape mustard • Saharan mustard

Rapeseed mustard (*Brassica napus*) fruit.
J. M. DiTomaso

Rapeseed mustard (*Brassica napus*) seeds.
J. O'Brien

GENERAL INFORMATION: Erect winter *annuals*, with *yellow, 4-petaled flowers* and *linear fruits*. Birdsrape and rapeseed mustard are sometimes biennial. All exist as basal rosettes until flowering stems develop at maturity. Mustard species are highly variable, and most consist of regional biotypes. Saharan and black mustard are adapted to periodic fire. Newly burned sites are subject to invasion, and along with annual grasses, Saharan and black mustard stands contribute to increased fuel load and fire frequency. Increasing the fire frequency can lead to the conversion of desert scrub to grassland (Saharan mustard) and of chaparral and coastal sage scrub to annual grassland (black mustard). Desert shrubs are killed by fire, and some sage scrub and chaparral shrubs that normally resprout from the crown after fire do not survive frequent burning. Many cultivars of birdsrape, black, and rapeseed mustard have been developed for oil, condiment, fodder, and vegetable crops. Cultivars that escape cultivation soon revert to the wild type. Wild mustards can harbor diseases and pests that attack closely related crops in the mustard family. The foliage, roots, and especially seeds of *Brassica* and many related species contain glucosinolates, which are sulfur-containing compounds that can irritate the digestive tract and cause thyroid dysfunction when consumed in large quantities over time. Toxicity problems in livestock arise when large quantities of seeds are ingested or when animals are confined to pastures that consist primarily of mustard family species. Symptoms can include colic, diarrhea, excessive salivation, and thyroid enlargement. Refer to table 26 (p. 468) for a quick comparison of important distinguishing characteristics of yellow-flowered mustards.

Rapeseed mustard: To 1.5 m tall. Seeds of certain cultivars are the source of canola oil, and the roots of others are consumed as a vegetable. Rapeseed mus-

Rapeseed mustard • Black mustard • Birdsrape mustard • Saharan mustard

Black mustard (*Brassica nigra*) plants.
J. M. DiTomaso

Black mustard (*Brassica nigra*) leaves.
J. M. DiTomaso

Black mustard (*Brassica nigra*) inflorescence.
J. M. DiTomaso

Black mustard (*Brassica nigra*) fruit.
J. M. DiTomaso

Black mustard (*Brassica nigra*) seedling.
J. O'Brien

Black mustard (*Brassica nigra*) seeds.
J. O'Brien

Rapeseed mustard • Black mustard • Birdsrape mustard • Saharan mustard

Birdsrape mustard *(Brassica rapa)* plants.
J. M. DiTomaso

Birdsrape mustard *(Brassica rapa)* fruiting stem.
J. M. DiTomaso

Birdsrape mustard *(Brassica rapa)* inflorescence.
J. M. DiTomaso

Birdsrape mustard *(Brassica rapa)* leaves with clasping bases.
J. M. DiTomaso

Birdsrape mustard *(Brassica rapa)* seedling.
J. M. DiTomaso

Birdsrape mustard *(Brassica rapa)* seeds.
J. O'Brien

tard is an amphidiploid (doubled chromosome set) species that was originally derived from the hybridization of birdsrape mustard and cabbage [*Brassica oleracea* L.] in the area where the native ranges overlap. Native to Europe.

Black mustard: To 2 m tall. In coastal grasslands, dense stands of black mustard outcompete native vegetation. Black mustard appears to have allelopathic properties. Seeds of cultivars are used to produce mustard oil. Native to Europe.

Birdsrape mustard: To 1 m tall. The turnip is the root of birdsrape mustard cultivars. Native to Europe.

Saharan mustard: To about 1 m tall. Saharan mustard is especially problematic in the Sonoran Desert, including the Imperial Valley. It readily spreads from roadsides and other disturbed places into washes, drainages, desert shrubland, and sensitive dune areas. In the Algodonas Dunes, Saharan mustard threatens to outcompete rare plant species. Saharan mustard is a noxious weed in Nevada. Native to the Mediterranean region.

SEEDLINGS: Cotyledons glabrous, ± kidney-shaped, tip indented, broader than long, on stalks longer than the cotyledons.

Rapeseed mustard: Cotyledons mostly 5–10 mm long, 12–14 mm wide. First and subsequent leaves ± elliptic to oblanceolate with 0–3 lobes at the base, margins irregularly wavy-toothed, sparsely covered with papillae-based hairs.

Saharan mustard: Cotyledons mostly 3–5 mm long, 4–10 mm wide. First leaf obovate, about 5–12 mm long, margin bluntly toothed and ciliate, on a short stalk. Subsequent leaves increasingly larger, margins ± evenly wavy-toothed, moderately covered with short papillae-based hairs.

Saharan mustard (*Brassica tournefortii*) plant.
J. M. DiTomaso

Saharan mustard (*Brassica tournefortii*) inflorescence.
J. M. DiTomaso

Rapeseed mustard • Black mustard • Birdsrape mustard • Saharan mustard

MATURE PLANT: Hairs simple. Stems branched, upper portions usually glabrous. Basal leaves ± pinnate-lobed, with a large, ± rounded terminal lobe (lyrate). Basal and lower stem leaves stalked. Stem leaves reduced, sessile.

Rapeseed mustard: *Foliage green* and *lacking a whitish bloom*, otherwise resembles the foliage of birdsrape mustard.

Black mustard: Basal leaves mostly have 1–2 pairs of distinct lateral lobes at the base, terminal lobe much larger than the lateral lobes. Upper stem leaves oblong to linear, *base tapered*, margins entire to toothed or weakly lobed.

Birdsrape mustard: Foliage covered with a *whitish bloom* (glaucous). Basal leaves usually have 1–4 pairs of distinct lateral lobes at the base, terminal lobe much larger than the lateral lobes. Upper stem leaves ± lanceolate, *base lobed* and *weakly clasping the stem*, margins toothed.

Saharan mustard: Basal leaves deeply pinnate-lobed, with 6–14 pairs of deep, toothed lateral lobes, terminal lobe slightly larger than lateral lobes. Upper stem leaves oblong to linear, base tapered, margins finely toothed to lobed.

ROOTS AND UNDERGROUND STRUCTURES: Rapeseed mustard, black mustard, Saharan mustard: Taproots slender, often long in Saharan mustard.

Saharan mustard (*Brassica tournefortii*) fruit.
J. M. DiTomaso

Saharan mustard (*Brassica tournefortii*) leaves.
J. M. DiTomaso

Saharan mustard (*Brassica tournefortii*) downward-turning hairs at base of stem.
J. M. DiTomaso

Saharan mustard (*Brassica tournefortii*) seedling.
J. M. DiTomaso

Rapeseed mustard • Black mustard • Birdsrape mustard • Saharan mustard

Saharan mustard (*Brassica tournefortii*) seeds.
J. O'BRIEN

Elongated mustard (*Brassica elongata*) infestation.
J. YOUNG

Birdsrape mustard: Taproot usually thick, ± fleshy.

FLOWERS: Petals 4. Sepals 4, erect. Stamens 6, 2 short, 4 long. Insect-pollinated.

Rapeseed mustard: Mostly April–June. Petals ± golden to dull yellow, paler than those of birdsrape mustard, mostly 9–14 mm long, more than 3 mm wide. Sepals about 6–8 mm long. Self-compatible.

Black mustard: April–July. Petals bright yellow, 6–11 mm long. Sepals 3–5 mm long. Self-incompatible.

Birdsrape mustard: January–June, nearly year-round in some areas. Petals yellow, 6–11 mm long, more than 3 mm wide. Sepals mostly 4–5 mm long. Self-incompatible.

Saharan mustard: December–June, depending on yearly weather patterns. Petals pale yellow, mostly 4–8 mm long, *1–2 mm wide*. Sepals ± 3 mm long. Probably self-compatible.

FRUITS AND SEEDS: Fruits (siliques) linear, 2-chambered, with *1 vein per valve* and a ± conical beak at the apex that sometimes contains 1 seed. Each chamber contains 1 row of seeds. Fruits open from the base to release seeds. Fruit length includes the beak.

Rapeseed mustard: Mature fruits *ascending,* mostly 5–10 cm long. Beak 7–20(30) mm long, seedless. Seeds nearly spherical, 1.5–3 mm in diameter, ± coarsely reticulate-veined, often purplish brown.

Black mustard: Mature fruits *erect, usually lying close to stem, 1–2 cm long.* Beak slender, 1–4 mm long, seedless. Seeds *ovoid to ellipsoid,* mostly 1.5–2 mm long, coarsely reticulate-veined, dark reddish brown.

Birdsrape mustard: Mature fruits *spreading to ascending,* 2–7 cm long. Beak 10–15 mm long, often seedless. Seeds nearly spherical, 1.5–3 mm in diameter, weakly fine reticulate-veined, dark.

Saharan mustard: Mature fruits strongly constricted between the seeds and appearing beaded, *spreading to ascending, mostly 3–7 cm long.* Beak thick, 10–15 mm long, usually contains 1 seed. Seeds spherical, mostly *1–1.5 mm in*

diameter, weakly reticulate-veined, purplish brown.

POSTSENESCENCE CHARACTERISTICS: Some fruits may persist on senesced stems through summer.

HABITAT: Roadsides, fields, disturbed waste places.

Rapeseed mustard: Roadsides, fields, and disturbed waste places. Inhabits coastal cliffs and cooler inland areas in its native range.

Black mustard: Roadsides, fields, disturbed waste places, and grasslands, especially in coastal areas. Mostly inhabits areas with a mild winter climate in its native range.

Birdsrape mustard: Roadsides, fields, disturbed waste places, orchards, grain fields, and other agronomic crops. Often inhabits cold-winter inland areas in its native range.

Indian mustard (*Brassica juncea*) inflorescence.
J. M. DiTomaso

Indian mustard (*Brassica juncea*) leaves and stem.
J. M. DiTomaso

Indian mustard (*Brassica juncea*) seedling.
J. M. DiTomaso

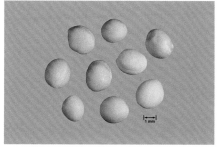

Indian mustard (*Brassica juncea*) seeds.
J. O'Brien

Saharan mustard: Roadsides, washes, annual grassland, coastal sage scrub, dunes, and desert shrubland. Typically grows in arid climate areas on sandy soil and where competing vegetation is sparse. Inhabits coastal and inland dunes in its native range.

DISTRIBUTION: Rapeseed mustard: San Joaquin Valley and South Coast, to

Table 26. Yellow-flowered mustards (*Brassica, Hirschfeldia,* and *Sinapis* spp.)

Species	Mature fruits relative to stem	Upper stem leaf base	Fruit length including beak (cm)	Fruit beak length (cm)	Conspicuous veins per valve (young fruits)	Basal leaf lateral lobe pairs	Other
Brassica fruticulosa Mediterranean cabbage	spreading to ascending	tapered	1.5–4	0.2–0.6	1	5–8	sepals most 3–4 mm lor
Brassica juncea Indian mustard	spreading to ascending	tapered	2–4	0.5–1	1	0–3	sepals most 4–6 mm lor
Brassica napus rapeseed mustard	ascending	weakly lobed, clasping stem	mostly 5–10	0.7–2(–3)	1	1–3	petals most 10–14 mm long; fruits mostly 4–5 wide; seeds 1–1.5 mm diameter
Brassica nigra black mustard	appressed to stem	tapered	1–2	0.1–0.4	1	1–3	beak slende cylindrical
Brassica rapa birdsrape mustard	spreading to ascending	weakly lobed, clasping stem	2–7	1–1.5	1	1–3	petals most 6–11 mm lo fruits mostl 3 mm wide less; seeds] to 2 mm diameter
Brassica tournefortii Saharan mustard	spreading to ascending	tapered	3–7	1–1.5	1	5–8	lower stems densely stiff white-hairy
Hirschfeldia incana shortpod mustard	appressed to stem	tapered	0.8–1.5	0.3–0.6	3–7 ± obscure	(3)4–6	biennial or short-lived perennial; b ± flattened, base abrupt swollen
Sinapis alba white mustard	ascending	tapered	2–3	1.5–3; equal to or longer than fruit body	3, especially on beak of mature fruits	mostly 3–5	fruit body u ally densely long-hairy; beak flatten
Sinapis arvensis wild mustard	ascending	tapered	2–4	0.6–1.2; less than fruit body	3–7, obscure or conspicuous	mostly 0–3	fruit body glabrous or sparsely hai beak angled

500 m. Idaho, Montana, Nevada, Oregon, some central states, and most eastern states.

Black mustard: Throughout California to 1500 m, except Great Basin region, east side of Sierra Nevada, and deserts. All contiguous states, with the possibly exceptions of Arkansas, Georgia, New Mexico, South Carolina, and Wyoming.

Birdsrape mustard: Throughout California to 1500 m, except Great Basin region and deserts. All contiguous states.

Saharan mustard: San Joaquin Valley, Southwestern region, deserts, to about 1000 m. Nevada, Arizona, New Mexico, and Texas.

PROPAGATION/PHENOLOGY: *Reproduce by seed.* Most seeds fall near parent plants when fruits open at maturity. Some seeds disperse to greater distances with water, soil movement, human activities, and possibly by clinging to vehicle tires, shoes, clothing, and to animals. In California, seeds typically germinate in fall or winter after the rainy season begins. Many mustard species develop a large, persistent seedbank. Deeply buried seeds of black and birdsrape mustard can survive for 50 years or more. Seeds nearer to the soil surface are not as long-lived under field conditions.

Rapeseed mustard: Grows slower than birdsrape mustard.

Black mustard: Seeds do not require a cold, moist period to germinate. Germination is typically even and occurs over a period of about 4–6 weeks. Black mustard usually develops a large, persistent seedbank.

Birdsrape mustard: Seeds of some biotypes require a cold, moist period to stimulate germination.

Saharan mustard: Can produce up to 16,000 seeds per plant. Sometimes seeds disperse when dried plant stems break at ground level and tumble under windy conditions. Seeds become slightly sticky with mucilage when moistened with water.

MANAGEMENT FAVORING/DISCOURAGING SURVIVAL: Burning and other kinds of disturbance usually favor the increase of mustard species. Yearly manual removal of plants before seeds mature can eventually deplete the seedbank.

SIMILAR SPECIES: Refer to table 26 (p. 468) for a comparison of important distinguishing characteristics among yellow-flowered mustards.

Elongated mustard [*Brassica elongata* Erhr. ssp. *integrifolia* (Boiss.) Breistr., synonyms: Wilson weed; *Brassica elongata* Erhr. ssp. *persica* (Boiss. & Hohen.) Thell.; *Brassica persica* Boiss. & Hohen.] is a *biennial, short-lived perennial*, or sometimes winter annual that is spreading rapidly in Nevada. It has not yet been established in California or other southwestern states, but may be a potentially invasive weed in these areas. Elongated mustard is distinguished by having *elliptic-oblong basal leaves with margins that are barely round-toothed or slightly wavy*, upper leaves that are *sessile or short-stalked* (not clasping the stem), and *ascending-erect fruits 1.5–2.2 cm long* including a *stalklike base 1.5–4.5 mm long* and a *tapered, seedless beak about 0.5–3 mm long*. In addition to a *stalklike* *base*, fruits are on *conspicuous stalks* ± 8 mm long and the leaves are *glabrous or nearly glabrous*. Flowers have yellow petals mostly 7–10 mm long. Elongated mustard currently inhabits roadsides, juniper/black sagebrush woodlands, pinyon/juniper woodlands, and other desert communities on the margins of salt desert areas in Nevada, especially around the highway near the city of Eureka and about 200 km further west. Native to southeastern Europe and Asia. It was first collected in Portland, Oregon, in 1911, where it did not persist. It was next collected near Eureka, Nevada, in 1968.

Indian mustard [*Brassica juncea* (L.) Czernov] [BRSJU], synonyms: brown mustard; India mustard; oriental mustard; *Brassica integrifolia* Rupr.; *Brassica japonica* Thunb.; *Brassica willdenowii* Boiss.] is a *glabrous or nearly glabrous winter annual* to 1 m tall. It is distinguished by having upper leaves that are linear to lanceolate with a *tapered base* and smooth to toothed or sometimes lobed margins, and by fruits that are *spreading to ascending, 2–4 cm long* including a slender *beak 5–10 mm long*. In addition, the basal leaves *have 0–3 pairs of distinct lateral lobes* and typically *wither early* as flowering stems develop. Sepals are *4–6 mm long*. Seeds are ovoid to nearly spherical, 1.5–2 mm long, finely reticulate-veined, and yellowish to brown. In California, Indian mustard is less common than most other mustards. Indian mustard inhabits fields, roadsides, and disturbed areas in the Central Valley, San Francisco Bay region, and South Coast, to 300 m. It also occurs in all other contiguous states. Indian mustard is an amphidiploid derived from the hybridization of black mustard and birdsrape mustard. Native to Eurasia.

Mediterranean cabbage [*Brassica fruticulosa* Cyr.] is a *biennial or perennial* or sometimes annual to about 0.5 m tall. Recently introduced to California, it is not described in current floras to date. It most closely resembles Indian mustard. Unlike Indian mustard, Mediterranean cabbage has basal leaves with *5–8 pairs of lateral lobes*, fruits that are *strongly constricted between the seeds* and 2–4 cm long with a *beak 2–6 mm long*, and sepals that are *3–4 mm long*. In addition, Mediterranean cabbage has upper stem leaves that are usually *more consistently pinnate-lobed* than those of Indian mustard. Mediterranean cabbage is scattered in disturbed places, washes, and canyons in the San Francisco Bay region (San Mateo Co.) and South Coast (Los Angeles, Riverside, and San Bernardino Cos.), particularly along the southern base of the San Gabriel and San Bernardino Mountains and adjacent valleys to the south, to 700 m. Mediterranean cabbage appears to have spread rapidly in the 1980s and may be overlooked elsewhere in California. It is expected to expand its range in California. Native to Europe.

Shepherd's-purse [*Capsella bursa-pastoris* (L.) Medik.] [CAPBP]

SYNONYMS: case plant or weed; lady's purse; pepperplant; pick-purse; shepherd's bag; shepherd's pouch; St. James' weed; *Bursa bursa-pastoris* (L.) Britt.; *Bursa gracilis* Gren.; *Capsella rubella* Reut.; *Thlaspi bursa-pastoris* L.

GENERAL INFORMATION: Winter *annual* to 0.5 m tall, with small whitish flowers and *flat, heart-shaped fruits*. Plants exist as flat basal rosettes until flower stems develop at maturity. Fruits taste peppery and are sometimes added to salad greens as a spice. However, like many other members of the mustard family, foliage and seeds contain glucosinolates, which can irritate the digestive tract when ingested in quantity. Native to Europe.

SEEDLINGS: Cotyledons ± oblong-elliptic, about 2–5 mm long, tip rounded, base tapered to a stalk ± 1 mm long, glabrous. Leaves alternate, sparsely covered with simple and forked hairs. First and subsequent few leaves oblong-elliptic to obovate, 3–6 mm long, tips rounded, margins smooth to toothed, ± on stalks to 1 mm long. Later leaves often toothed to deeply pinnate-lobed.

MATURE PLANT: Foliage highly variable, sparsely covered with *simple and forked hairs*. Rosette leaves nearly smooth-margined to deeply pinnate-lobed, 3–10 cm long. Terminal lobes elliptic, ovate, or nearly triangular. Flower stems simple or branched. Stem leaves alternate, sparse, reduced in size, sessile, base

Shepherd's-purse (*Capsella bursa-pastoris*) plant. C. Elmore

Shepherd's-purse

Shepherd's-purse (*Capsella bursa-pastoris*) inflorescence.   J. M. DiTomaso

lobed and clasping the stem, margins usually toothed or pinnate-lobed.

ROOTS AND UNDERGROUND STRUCTURES: Taproot slender, often branched, with fibrous lateral roots.

FLOWERS: Mostly late winter or spring, but sometimes year-round under favorable conditions. Racemes often terminal. Petals 4, *white*, sometimes tinged pale pink, ± 2 mm long, narrowed at the base (clawed). Flower stalks spreading to ascending, mostly 10–15 mm long.

FRUITS AND SEEDS: *Pods triangular-heart-shaped, 2-chambered, flattened perpendicular to the narrow septum*, mostly 4–8 mm long, ± 6 mm wide, taste peppery, open to release *numerous seeds per chamber*. Style short. Seeds oblong, ± 1 mm long, 0.5 mm wide, slightly compressed, grooved on each side, otherwise smooth, finely textured at 10× magnification, dull reddish to yellowish brown.

POSTSENESCENCE CHARACTERISTICS: Dead stems with remnants of the pods can persist for a short period.

HABITAT: Agronomic and vegetable crops, nurseries, gardens, turf, landscaped areas, orchards, vineyards, roadsides, waste places, urban sites, and other disturbed places.

DISTRIBUTION: Common. Throughout California, to 2300 m. All contiguous states. Nearly worldwide.

PROPAGATION/PHENOLOGY: *Reproduces by seed.* Seeds fall near parent

Shepherd's-purse (*Capsella bursa-pastoris*) seedling.   J. M. DiTomaso

plants and disperse to greater distances with water, soil movement, and human activities, including landscape maintenance and agricultural work. Seeds mostly germinate fall through early spring, but in coastal areas with moderate temperatures, seeds can germinate year-round. Buried seeds can survive up to 15 years or more.

MANAGEMENT FAVORING/DISCOURAGING SURVIVAL: Repeated cultivation to prevent seeds from maturing can help control shepherd's-purse in cultivated fields. Manually removing plants before seeds develop can control populations in gardens and small areas.

SIMILAR SPECIES: The *flat, triangular-heart-shaped pods* readily distinguish shepherd's-purse from other weedy members of the mustard family.

Shepherd's-purse (*Capsella bursa-pastoris*) seeds.   J. O'Brien

Hairy bittercress [*Cardamine hirsuta* L.][CARHI]
Little bittercress [*Cardamine oligosperma* Nutt. ex Torrey & A. Gray]

SYNONYMS: Hairy bittercress: hoary bittercress

Little bittercress: few-seeded bittercress; lesser-seeded bittercress; little western bittercress; *Cardamine acuminata* Rydb.; *Cardamine bracteata* Suksd.; *Cardamine hirsuta* L. ssp. *oligosperma* (Nutt. ex Torrey & A. Gray) O. E. Schulz; *Cardamine hirsuta* L. var. *acuminata* Nutt. ex Torrey & A. Gray and var. *parviflora* Nutt. ex Torrey & A. Gray; *Cardamine unijuga* Rydb.; others

GENERAL INFORMATION: Winter or summer *annuals*, sometimes biennials, to 0.4 m tall, with *pinnate-compound leaves*, tiny white flowers, and *very slender, linear pods*. Plants exist as rosettes of *pinnate-compound leaves* until flowering stems develop at maturity. These species are often difficult to distinguish from one another.

Hairy bittercress (*Cardamine hirsuta*) plant.
J. M. DiTomaso

Hairy bittercress (*Cardamine hirsuta*) seedling.
J. M. DiTomaso

Hairy bittercress (*Cardamine hirsuta*) flowering stem.
J. M. DiTomaso

Hairy bittercress (*Cardamine hirsuta*) seeds.
J. O'Brien

Hairy bittercress: Sometimes a nuisance weed of container crops in nurseries. Native to Europe.

Little bittercress: A ruderal native of most western states. In California, little bittercress is considered a desirable component of the vegetation in natural areas. However, it is sometimes weedy in landscaped areas, orchards, nurseries, turf, vegetable crops, and other favorable disturbed places. It has recently been introduced to New York and possibly elsewhere in the Northeast, most likely through the transportation of nursery crops. Some botanists currently recognize 2 varieties, of which only variety *oligosperma* occurs throughout the range of the species, including California. Variety *kamtschatica* (Regel) Detling occurs only in a few northwestern states.

SEEDLINGS: Cotyledons ovate to nearly round, tip slightly indented (emarginate), 3–5 mm long, glabrous, on stalks mostly 3–5 mm long. Leaves alternate. First 1–3 leaves semicircular to kidney-shaped, broader than long, 3–8 mm long, 4–9 mm wide, sparsely coarse-hairy, on stalks as long or longer than blades. Subsequent few leaves pinnate-lobed, usually with 1–2 pairs of lateral lobes, coarse-hairy.

MATURE PLANT: Stems erect to ascending, angled, simple or branched. Foliage glabrous to sparsely coarse-hairy. Hairs *simple*, straight. Leaves *pinnate-compound*, mostly 2–10 cm long. Stem leaves alternate, sparse. Rosette leaves usually present at flowering. Leaflets of lower leaves nearly round, kidney-shaped, to obovate. Terminal leaflets larger than lateral lobes. Leaflets of upper leaves usually elongate. Leaflet margins smooth to slightly lobed.

Hairy bittercress: Lateral leaflet pairs often 2–3. Leaflets of rosette leaves *often obovate*, margins usually wavy to slightly lobed.

Little bittercress: Lateral leaflet pairs 2–5. Leaflets of rosette leaves *often round to kidney-shaped*.

ROOTS AND UNDERGROUND STRUCTURES: *Taproot weak, slender, often inconspicuous*, yellowish, usually highly branched, with *numerous fibrous lateral roots*.

FLOWERS: Petals 4, white. Sepals 4, often tinged purplish. Stamens 4 or 6.

Hairy bittercress: January–June. Flowering stems often several to numerous. Outer stems usually spreading and curved upward at the base (decumbent). Petals mostly 1.5–2 mm long.

Little bittercress: March–July. Flowering stems usually single or few, erect from the base. Petals mostly 2–4 mm long.

FRUITS AND SEEDS: Pods (siliques) *linear, flattened*, 12–25 mm long, ascending to erect. *Style (beak) at fruit tip inconspicuous*, thick, to 0.5 mm long. *Fruit valves spring open elastically from the bottom. Seeds in 1 row per chamber*, orange-brown, smooth except for 1 groove on each face.

Little bittercress (*Cardamine oligosperma*) plant.
J. M. DiTomaso

Little bittercress (*Cardamine oligosperma*) seedling.
J. M. DiTomaso

Little bittercress (*Cardamine oligosperma*) seeds.
J. O'Brien

Hairy bittercress: Pods *1 mm wide or less*. Seeds oval to oblong, ± 1 mm long, 0.5 mm wide, mostly *22–36 per fruit*. Upon opening, *fruit valves roll into a tight coil*.

Little bittercress: Pods *1–2 mm wide*. Seeds nearly round to oblong, 0.75–1 mm long, 0.5–0.75 mm wide, *margin often very narrowly winged* (requires magnification), usually *15–24 per fruit*. Upon opening, fruit valves usually curve or coil loosely.

POSTSENESCENCE CHARACTERISTICS: Dead plants with opened fruits can persist for a short period.

HABITAT: Landscaped areas, yards, gardens, nurseries, container plants, greenhouses, occasionally turf, roadsides, ditches, orchards, vineyards, vegetable crops, and other disturbed places. Little bittercress also inhabits moist meadows, woodlands, creek bottoms, and other moist shady places in natural areas.

DISTRIBUTION: Hairy bittercress: Klamath Ranges, North Coast Ranges, San Francisco Bay region, Sacramento Valley, and probably elsewhere, to 800 m. Expanding range in California. New Mexico, Washington, southern states, and most eastern states.

Little bittercress: Throughout California, except Great Basin and deserts, to 1100 m. All western states, except Arizona and New Mexico. New York and probably other eastern states.

PROPAGATION/PHENOLOGY: *Reproduce by seed.* Most seeds fall near the parent plant or are flung to within a few feet when pods snap open. Some seeds disperse to greater distances with water, mud, soil movement, and human activities, especially through the transportation of nursery container crops.

MANAGEMENT FAVORING/DISCOURAGING SURVIVAL: Continually removing plants manually before seeds are produced can help to control both species. In nurseries, scrubbing containers that are to be reused removes seeds that cling to the sides. Disposing of used potting media that may be contaminated with bittercress seeds eliminates a seed source. Composting or fumigating used potting soil helps eliminate the seed source.

SIMILAR SPECIES: Sibara [*Sibara virginica* (L.) Rollins][SIBVI] is a native *annual or biennial* with flower stems to 0.3 m tall that is occasionally weedy in orchards and vegetable crop fields. Unlike the bittercresses, sibara has *deeply pinnate-lobed leaves* (not pinnate-compound) and *linear pods that open slowly and rigidly from the bottom with the valves remaining straight.* Refer to the entry for **Sibara** (p. 527) for more information.

Lens-podded whitetop [*Cardaria chalepensis* (L.) Hand.-Mazz.][CADDC][Cal-IPC: Moderate Alert] [CDFA list: B]

Hoary cress [*Cardaria draba* (L.) Desv.][CADDR] [Cal-IPC: Moderate][CDFA list: B]

Hairy whitetop [*Cardaria pubescens* (C. Meyer) Jarmol.][CADPU][Cal-IPC: Limited][CDFA list: B]

SYNONYMS: Lens-podded whitetop: hoary cress; lens-podded hoary cress; peppergrass; whitetop; whiteweed; *Cardaria draba* (L.) Desv. ssp. *chalepensis* (L.) O.E. Schulz; *Cardaria draba* (L.) Desv. ssp. *repens* (Schrenk) O.E. Schulz;

Lens-podded whitetop (*Cardaria chalepensis*) plant. J. M. DiTomaso

Cardaria draba (L.) Desv. var. *repens* (Schrenk) Rollins; *Lepidium draba* L. var. *repens* Thell; *Lepidium repens* (Schrenk) Boiss.

Hoary cress: globe-podded hoarycress; heart-podded hoary cress; peppergrass; pepperweed; perennial peppergrass; short whitetop; whitetop; whiteweed; *Lepidium draba* L.

Hairy whitetop: ball cress; globe-podded hoarycress; hoary cress; whitetop; *Cardaria pubescens* (C. Meyer) Jarmol. var. *elongata* Rollins; *Hymenophysa pubescens* C.A. Meyer

GENERAL INFORMATION: Erect *perennials* to 0.4(0.5) m tall, with nearly flat-topped to rounded inflorescences of small white flowers, small disk- to heart-shaped pods, and *vigorously creeping horizontal roots*. Clonal colonies usually develop from the creeping roots. *Cardaria* species are problematic in natural areas and crop fields, especially irrigated crops such as alfalfa and sugar beets. In 1933 three distinct *Cardaria* species were recognized in North America. Consequently, taxonomic references to *Cardaria* species in articles published prior to 1933 are unclear as to which species was the subject. Although similar, each species differs in chromosome number and herbicide resistance. All are self-incompatible. Lens-podded whitetop and hoary cress can hybridize, but it appears that only first-generation hybrids survive under natural conditions.

Lens-podded whitetop: Currently, some botanists treat lens-podded whitetop as a subspecies of hoary cress. Besides California, lens-podded whitetop is a state-listed noxious weed in Arizona (prohibited), Oregon (class B), and Utah. Native to Central Asia.

Lens-podded whitetop (*Cardaria chalepensis*) fruiting stem. J. M. DiTomaso

Lens-podded whitetop (*Cardaria chalepensis*) flowers and fruit. J. M. DiTomaso

Hoary cress: Besides California, hoary cress is a state-listed noxious weed in Arizona (restricted), Colorado, Idaho, Iowa (primary), Kansas, Montana (category 1), Nevada, New Mexico (class A), North Dakota, Oregon (class B), South Dakota, Utah, Washington (class C), and Wyoming. Native to Eurasia.

Hairy whitetop: Besides California, hairy whitetop is a state-listed noxious weed in Arizona (prohibited), Oregon (class B), Utah, Washington (class C), and Wyoming. Native to Central Asia.

SEEDLINGS: Within a month, the taproot can grow to a depth of 25 cm and lateral roots can develop shoot buds.

Hoary cress: Cotyledons oval to elliptic, 7–9 mm long, ± 2.5 mm wide, unequal, pale gray-green, with peppery taste. First leaves ovate to oblong, dull, scaly, somewhat larger than the cotyledons, often with slightly wavy margins. Subsequent leaves resemble first leaves, but sometimes have short fine hairs along the margins. First and subsequent *leaf bases taper to a stalk equal to or longer than the length of the blade.*

Lens-podded whitetop, hairy whitetop: Seedlings probably resemble those of hoary cress.

MATURE PLANT: Stems generally erect, sparse to densely covered with short, simple hairs. Leaves alternate, gray-green, highly variable, obovate, (ob)lanceolate, or oblong to elliptic, margin irregularly toothed to entire. Upper and especially the lower blade surface sparse to densely covered with *short, simple white hairs.* Basal leaves short-stalked. Upper leaves sessile, *base lobed and clasping the stem, lobes acute to weakly rounded.*

Hoary cress (*Cardaria draba*) plants.　　　　　　　　　　　　J. M. DiTomaso

Hoary cress (*Cardaria draba*) infestation in a wet area.   J. M. DiTomaso

Hoary cress (*Cardaria draba*) flowering stem.
J. M. DiTomaso

Hoary cress (*Cardaria draba*) fruit.
J. M. DiTomaso

Lens-podded whitetop: Leaves to 8 cm long, to 3 cm wide, often smaller, sparsely hairy. Leaf base lobes often rounded-acute. Very difficult to distinguish from hoary cress in the vegetative state.

Hoary cress: Leaves to 9 cm long, to 4 cm wide, sometimes smaller, sparse to densely hairy. Leaf base lobes often rounded-acute.

Hairy whitetop: Leaves to 8 cm long, to 2 cm wide, usually much smaller, densely hairy. Leaf base lobes often acute.

ROOTS AND UNDERGROUND STRUCTURES: Plants develop *extensive systems of persistent, deep, vertical and horizontal roots that vigorously produce new shoots at irregular intervals. Root fragments can generate new plants.* Vertical roots can penetrate the soil to depths of 2 m or more. Roots can account for 75% of the total plant biomass and, as a result, store considerable amounts of carbohydrates. Carbohydrate reserves typically accumulate to maximum levels by mid-summer and are minimal in early to mid-spring. Roots survive cold winter climates and periods of drought. Roots do not associate with mycorrhizae.

FLOWERS: Inflorescences (compound corymbs) nearly flat-topped to rounded. Flowers fragrant, numerous, *white,* with 4 petals 2–4 mm long. Sepals 4, separate. Insect-pollinated. Self-incompatible.

Hoary cress (*Cardaria draba*) seedling.
J. M. DiTomaso

Hoary cress (*Cardaria draba*) seeds and fruit.
J. O'Brien

Clasping leaf base of hoary cress (*Cardaria draba*). J. M. DiTomaso

Lens-podded whitetop: April–August. Sepals *glabrous*. Petals mostly 3–4 mm long.

Hoary cress: March–July. Sepals *glabrous*. Petals mostly 3–4 mm long.

Hairy whitetop: April–October. Sepals *covered with short, simple hairs*. Petals mostly 2–3.5 mm long.

FRUITS AND SEEDS: Pods (silicles) 2-chambered, variable, *inflated*, with a *persistent style 1–2 mm long at the apex, do not open to release the seeds* or open slowly after a long period. Seeds usually 1–2 per chamber, ovoid, slightly flattened, reddish brown, 1.5–2 mm long, 1–1.5 mm wide, surface minutely granular.

Lens-podded whitetop: Pods ± *disk-shaped, round to broadly (ob)ovate* or barely kidney-shaped in outline with the indentation at the apex, *not constricted at septum or 2-lobed*, 2.5–6(8) mm long, 4–6(7) mm wide, glabrous.

Hoary cress: Pods *upside-down heart-shaped* to broadly ovate in outline, *constricted at septum and often 2-lobed*, 2.5–3.5 mm long, 3–5 mm wide, glabrous.

Hairy whitetop: Pods *strongly inflated, spherical to ovoid*, sometimes (ob)ovate in outline, *not constricted at the septum or 2-lobed*, 3–4.5 mm long, 2.5–4.5 mm wide, *covered with short hairs*.

POSTSENESCENCE CHARACTERISTICS: Foliage dies back during extended periods of freezing temperatures or drought.

HABITAT: Disturbed open sites, fields, pastures, grain and vegetable crops

Hairy whitetop (*Cardaria pubescens*) flowering stem. J. M. DiTomaso

Hairy whitetop (*Cardaria pubescens*) seedling. J. M. DiTomaso

Hairy whitetop (*Cardaria pubescens*) plants late in season with fruits. J. M. DiTomaso

Hairy whitetop (*Cardaria pubescens*) fruit.
J. M. DiTomaso

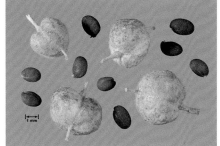

Hairy whitetop (*Cardaria pubescens*) stem and leaf bases.   J. M. DiTomaso

Hairy whitetop (*Cardaria pubescens*) seeds and fruit.   J. O'Brien

(especially irrigated crops such as alfalfa and sugar beets), orchards, vineyards, roadsides, and ditches. Often grows on moderately moist alkaline to saline soils, but tolerates a wide range of soil types and moisture conditions.

DISTRIBUTION: All species are scattered throughout California and uncommon in the desert regions.

Lens-podded whitetop: More frequent in the Sacramento Valley, southern San Joaquin Valley, and northern Siskiyou Co., to 1500 m. Western states except possibly Arizona, many central states.

Hoary cress: More frequent in the Klamath Range, Great Basin, southern Sacramento Valley, San Francisco Bay region, and Southwestern region, to 1200 m. All western states, most central and eastern states.

Hairy whitetop: More frequent in the Sacramento Valley, Southwestern region, and Great Basin, to 2000 m. Western states except New Mexico, scattered in a few central and eastern states

PROPAGATION/PHENOLOGY: *Reproduce vegetatively from creeping roots* and *by seed*. Root fragments generate new plants, but regeneration is generally poor in dry soils. Under favorable conditions, plants often increase vegetatively by more than a 60-cm radius per year. Light stimulates seed germination but is not required. Seed germinates in the fall after the first rains. Plants typically do not flower the first year. One flowering stem of lens-podded whitetop or hoary cress can produce up to 850 mature pods. Lens-podded whitetop and hairy whitetop (and probably hoary cress) compete poorly with shrubs in natural communities.

Lens-podded whitetop: Seedlings recover from injury more readily than those of hoary cress.

Hoary cress: In 1 year, a single plant on open ground without competition can spread vegetatively to cover an area to 3.7 m in diameter and can produce up to 455 shoots. One plant can produce up to 4800 seeds, with a viability of about 85%. Seeds germinate at temperatures ranging from 0.5°–40°C (optimum 20°–35°C). Dry-stored seed can remain viable for up to 5 years, but under field conditions, seeds are probably viable for a much shorter period.

Hairy whitetop: Plants are reported to produce about 30–560 (average 300) pods per plant.

MANAGEMENT FAVORING/DISCOURAGING SURVIVAL: Colonies are difficult to eliminate because of deep, persistent roots. Cultivation can facilitate spread of plants by dispersing root fragments. However, repeated cultivation of at least 1–2 times per month can eliminate colonies in about 2–4 years. Where feasible, flooding an area with 15–25 cm of water for 2 months can eliminate colonies.

SIMILAR SPECIES: Unlike *Cardaria* species, **perennial pepperweed** [*Lepidium latifolium* L.] has *glabrous foliage*, sessile stem leaves that are *tapered at the base*, and *smaller pods* (± 2 mm long) that are *flattened* and *not inflated* and that have a *sessile stigma less than 1 mm long at the apex*. In addition, perennial pepperweed is typically more than 0.5 m tall, and its *pods open at maturity to release the seeds*. Refer to the entry for **Perennial pepperweed** (p. 511) for more information.

Blue mustard [*Chorispora tenella* (Pallas) DC.] [COBTE][CDFA list: B]

SYNONYMS: bead-podded mustard; crossflower; musk mustard; purple mustard; tenella mustard

GENERAL INFORMATION: Winter *annual* to 0.5 m tall, with *showy magenta to bluish violet or pale violet flowers* and *linear pods that are slightly constricted between the seeds*. Foliage is usually *glandular-pubescent* and has an *unpleasant scent*. Plants exist as basal rosettes until flower stems develop in early spring. Blue mustard is a highly competitive species and is especially widespread in the central states and southern Canada. Large populations of blue mustard in grain fields significantly reduce grain yields, and milk from dairy cattle grazing on infested pastures can have an unpleasant flavor. Besides California, blue mustard is a state-listed noxious weed in Colorado. Native to Russia and adjacent regions of Asia.

SEEDLINGS: Cotyledons lanceolate. Rosette leaves oblong to oblanceolate, typically sparsely covered with minute glandular hairs. Subsequent rosette leaf margins entire, wavy, or pinnate-lobed.

MATURE PLANT: Foliage *sparse to moderately covered with simple, minute glandular hairs* that are sticky to touch. Stems leafy, branched mostly from the base. Leaves alternate, elliptic-oblong to (ob)lanceolate. Lower stem leaves stalked, 3–8 cm long, margins wavy-toothed to pinnate-lobed, usually do not form a rosette. Upper stem leaves sessile, margins entire to wavy-toothed.

Blue mustard (*Chorispora tenella*) plant. J. M. DiTomaso

ROOTS AND UNDERGROUND STRUCTURES: Taproot shallow.

FLOWERS: Early spring. Petals 4, *pale violet to bluish violet or magenta*, narrowly clawed, 10–13 mm long. Sepals 4, *erect, separate to the base, but together appear cylindrical*, 6–8 mm long, usually violet with narrow membranous margins. Inner sepals sac-like at the base. Stamens 6 (4 long, 2 short), arrowhead-shaped. Style lacking, stigma minute and entire.

FRUITS AND SEEDS: Pods *linear-lanceolate*, erect to spreading, often *curved upward, nearly round in cross-section, slightly constricted between seeds at maturity*, 2–4 mm wide, mostly 3–4.5 cm long including a *slender tapered beak* 0.7–2 cm long at the apex. Pod stalks thick, ascending, 2–4 mm long. Pods eventually *break apart transversely between the seeds into sections*. Seeds reddish to brown, nearly spherical, ± 1.5 mm in diameter, usually remain within the pod sections.

HABITAT: Dry disturbed sites, winter annual crops (especially winter wheat), roadsides, and waste places. Tolerates a broad range of moisture, temperature, and soil conditions.

DISTRIBUTION: Cascade Range, Central Valley, South Coast Ranges, South Coast, and Great Basin, to 1300 m. All western states, most central states, some northeastern states, and a few southern states.

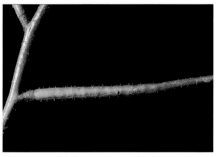

Blue mustard (*Chorispora tenella*) fruit.
J. M. DiTomaso

Blue mustard (*Chorispora tenella*) flowering stem.
J. M. DiTomaso

Blue mustard (*Chorispora tenella*) seedling.
J. M. DiTomaso

PROPAGATION/PHENOLOGY: *Reproduces by seed.* Seeds germinate during the cool season, mostly after the first fall or early winter rain in California. Seeds can mature in as few as 10 days after flowers open.

MANAGEMENT FAVORING/DISCOURAGING SURVIVAL: Large populations can be difficult to control in grain fields because plants are somewhat tolerant to typical 2,4-D use rates. Rotation to spring-planted crops allows additional control options, including cultivation or other registered herbicides.

SIMILAR SPECIES: Malcolm stock [*Malcolmia africana* R.Br.; synonym: African mustard][MAMAF] is an uncommon prostrate to erect *annual* to 0.5 m tall, with simple toothed leaves, pink to rose-violet flowers, and linear pods that are slightly constricted between the seeds. It is distinguished by having foliage and pods that are densely covered with small, *branched, nonglandular hairs.* In addition, pods are ascending, 4–6 cm long, 1–15 mm wide, *lack a beak,* round to weakly 4-angled in cross-section, and *open longitudinally to release the seeds.* Flowers have conspicuous styles with pointed stigma lobes. Malcolm stock mostly inhabits dry disturbed areas and desert shrubland on the east side of the Sierra Nevada, to 2000 m. It occurs in all western states, Illinois, Kansas, and Texas. Native to the Mediterranean region.

Radish [*Raphanus sativus* L.][RAPSN] and **wild radish** [*Raphanus raphanistrum* L.][RAPRA] have linear to linear-lanceolate pods that are ± slightly constricted between the seeds and that break apart between the seeds. Unlike blue mustard, the radishes have *foliage that is sparsely covered with stiff, nonglandular hairs.* Also, radish pods are *4–6 mm wide* with a *lengthwise groove on the lower surface,* and flowers are pale pink or lavender, yellow, or white. Refer to the entry for **Wild radish and Radish** (p. 516) for more information.

Lesser swinecress [*Coronopus didymus* (L.) Smith] [COPDI]

Greater swinecress [*Coronopus squamatus* (Forsskal) Asch.][CDFA list: B]

SYNONYMS: Lesser swinecress: carpet cress; swinecress; swine watercress; wartcress; *Carara didyma* (L.) Britt.; *Lepidium didymum* L.; *Senebiera didyma* (L.) Pers.; *Senebiera incisa* Willd.; *Senebiera pectinata* DC.; *Senebiera pinnatifida* DC.

Greater swinecress: creeping wartcress; Eurasian swinecress; *Carara coronopus* (L.) Medik.; *Cochlearia coronopus* L.; *Coronopus procumbens* Gilib.; *Coronopus verrucarius* (Garsault) Muschler & Thell.; *Lepidium squamatum* Forsskal; *Senebiera coronopus* (L.) Poiret

GENERAL INFORMATION: Prostrate or low-growing winter or summer *annuals*, sometimes biennials, with *deeply pinnate-lobed to pinnate-dissected leaves* and *small 2-lobed fruits*. Both species exist as rosettes until flower stems develop at maturity.

Lesser swinecress: Stems to 0.5 m long. Foliage has an *unpleasant skunklike scent*. Milk from dairy cattle grazing in pastures infested with lesser swinecress is reported to have an unpleasant flavor. The plant is also a problem in newly established alfalfa plantings in Imperial, San Joaquin, and Sacramento Valleys. High concentrations in alfalfa hay reduce feed palatability. Native to Eurasia.

Greater swinecress: Stems to 0.3 m long. Greater swinecress is particularly problematic in vegetable crops in the Imperial Valley. It is an asymptomatic host of tomato spotted wilt virus in the Mediterranean region. Native to Europe.

Lesser swinecress (*Coronopus didymus*) flowering stem. J. M. DiTomaso

SEEDLINGS: Lesser swinecress: Cotyledons narrowly oblanceolate, about 5–12 mm long, tip rounded, base long-tapered, glabrous. First and subsequent few leaves alternate, resemble cotyledons, except margins often have 1 or more rounded coarse teeth and tips often have a few short hairs. Later leaves have margins toothed to pinnate-lobed.

Greater swinecress: Cotyledons ± linear, about 5–15 mm long, tip slightly rounded, glabrous. Early leaves often appear opposite. First leaf pair resembles cotyledons, mostly 8–20 mm long. Later leaves alternate, margin ± toothed.

MATURE PLANT: Foliage glabrous to pubescent with simple hairs. Stems prostrate to decumbent, usually highly branched. Lower leaves stalked. Upper leaves alternate, nearly sessile.

Lesser swinecress: Leaves deeply pinnate-lobed 1–2 times or dissected, 1.5–7 cm long.

Greater swinecress: Leaves deeply pinnate-lobed, lobes coarsely toothed to lobed, to 30 cm long.

ROOTS AND UNDERGROUND STRUCTURES: Taproots slender, simple or branched, with few fibrous roots.

FLOWERS: Racemes dense, mostly *axillary*, about 1–4 cm long. Petals 4, white. Sepals 4, spreading. Stamens 2 or 4.

Lesser swinecress (*Coronopus didymus*) flowering and fruiting stem.   C. ELMORE

Lesser swinecress (*Coronopus didymus*) rosette.   J. M. DiTomaso

Lesser swinecress (*Coronopus didymus*) seedlings.   J. M. DiTomaso

Lesser swinecress (*Coronopus didymus*) seeds.   J. O'Brien

Greater swinecress (*Coronopus squamatus*) plant. J. M. DiTomaso

Greater swinecress (*Coronopus squamatus*) flowers at plant base. J. M. DiTomaso

Greater swinecress (*Coronopus squamatus*) inflorescence. J. M. DiTomaso

Greater swinecress (*Coronopus squamatus*) young rosette. J. M. DiTomaso

Lesser swinecress: February–October. Petals linear, ± *0.5 mm long*. Sepals deciduous.

Greater swinecress: May–October. Petals obovate, *1–1.5 mm long*.

FRUITS AND SEEDS: Pods (siliques) 2-chambered, *broader than long, slightly compressed perpendicular to septum*, on short stalks. Seeds 1 per chamber, oblong to ± kidney-shaped, remain enclosed within the hardened fruit chambers at maturity.

Lesser swinecress: Pods *conspicuously 2-lobed*, ± *1.5 mm long, 2–2.75 mm wide*, surface finely reticulate-wrinkled. Stigma sessile. Pod chambers eventually separate at maturity.

Greater swinecress: Pods *broadly ovoid, lobes inconspicuous, 2.5–3 mm long, 3–4 mm wide, body conspicuously wrinkled* and with *elongated tubercles on the margins*. Style longer than pod lobes. Pod chambers generally do not separate.

POSTSENESCENCE CHARACTERISTICS: Dead plants with fruits can persist for a short period.

HABITAT: Fields, roadsides, gardens, vegetable crops, turf, alfalfa pastures, orchards, ditch banks, and other disturbed places.

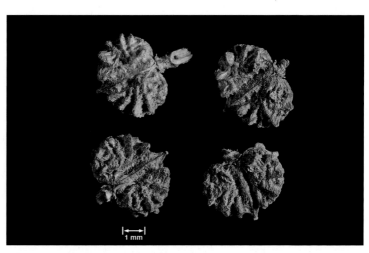

Greater swinecress (*Coronopus squamatus*) fruit.   J. O'Brien

Lesser swinecress: Also nurseries.

Greater swinecress: Grows best in warm regions and on fertile, heavy, compacted soils that periodically receive moisture.

DISTRIBUTION: Lesser swinecress: Throughout California, to 2000 m. Oregon, Washington, Arizona, New Mexico, and southern and eastern states.

Greater swinecress: San Francisco Bay region, Sacramento Valley, and Sonoran Desert (Imperial Valley), to 300 m. Oregon and some southern and northeastern states.

PROPAGATION/PHENOLOGY: *Reproduce by seed.* Fruits fall near the parent plant and disperse to greater distances with water, mud, soil movement, and human activities such as landscape maintenance and agricultural operations.

MANAGEMENT FAVORING/DISCOURAGING SURVIVAL: Manual removal or cutting plants below the crown before fruits mature can control small populations of both species. However, because of its low growth form, successful manual removal by mowing is difficult. Tillage is effective in most crop fields, but herbicides are often necessary where tillage is not practical. Some large plants of greater swinecress can survive uprooting by vertical cultivating implements under moist conditions or if reburied.

SIMILAR SPECIES: Tongue pepperweed [*Lepidium nitidum* Nutt.], **clasping pepperweed** [*Lepidium perfoliatum* L.], and **veiny pepperweed** [*Lepidium oblongum* Small] have prostrate rosettes with leaves that ± resemble those of the swinecresses. Unlike the swinecresses, these pepperweeds have *smooth or obscurely veined fruits that are slightly longer than wide* and *not conspicuously 2-lobed or tubercled*. In addition, inflorescences are primarily terminal. Refer to the entry for **Field pepperweed and Clasping pepperweed** (p. 500) for more information.

Flixweed [*Descurainia sophia* (L.) Webb ex Engler & Prantl][DESSO][Cal-IPC: Limited]

SYNONYMS: herb sophia; pinnate tansymustard; tansymustard; *Sisymbrium sophia* L.; *Sisymbrium tripinnatum* DC.; *Sophia sophia* (L.) Britt.

GENERAL INFORMATION: Erect winter or summer *annual or biennial* to 0.8 m tall, with tiny yellow flowers and leaves that are *pinnate-dissected 2–3 times*. Plants exist as basal rosettes until flower stems develop at maturity. Flixweed can be fatally *toxic* to cattle when flowering plants are ingested in quantity over a period time. Animals develop a condition known as paralyzed tongue. Symptoms include nervousness, blindness, muscular twitching, and the inability to feed or drink. If treated in time, animals generally recover within a short duration. Flixweed appears to be increasing in the Mojave Desert. It is a state-listed noxious weed in Colorado and Minnesota (secondary). Native to Eurasia.

SEEDLINGS: Cotyledons oblong-lanceolate to elliptic, mostly 3–7 mm long, ± 1 mm wide, glabrous. Stalk below cotyledons (hypocotyl) often inconspicuous, covered with minute branched hairs. First leaf pair usually appears opposite. First leaves 3-lobed, 3–7 mm long, covered with minute branched hairs, stalked, lobes narrow. Subsequent leaves alternate, 1–2 pinnate-lobed.

MATURE PLANT: Stems simple or branched near the middle or above. Foliage sparse to densely covered with minute branched hairs, sometimes interspersed with simple nonglandular hairs. *Leaves pinnate-dissected 2–3 times or compound*, 1–10 cm long, *ultimate lobes linear to narrowly obovate*.

Flixweed (*Descurainia sophia*) infestation.　　　　　　　J. M. DiTomaso

ROOTS AND UNDERGROUND STRUCTURES: Taproot simple or branched.

FLOWERS: March–August, depending on the region. Petals *yellow*, erect, 2–3 mm long. Sepals yellow, 2–3 mm long. Basal leaves usually withered at flowering. Racemes elongate in fruit.

FRUITS AND SEEDS: Pods *linear, nearly cylindrical, 10–35 mm long, ± 1 mm wide, straight or slightly curved upward, with a beaklike style less than 1 mm long at the tip* and *seeds in 1 row per chamber*. Pod septum *2–3-veined*. Stalks *ascending to erect*. Seeds mostly 10–20 per pod, oblong, 0.7–1.5 mm long, orange-brown, coated with sticky mucilage when moistened. Usually self-pollinated.

HABITAT: Roadsides, waste places, fields, vineyards, orchards, agronomic crops, gardens, disturbed desert areas, and canyon bottoms. Often grows in sandy or stony soils. Overwintering rosettes tolerate freezing and snow cover.

DISTRIBUTION: Throughout California, to 2600 m. All contiguous states, except possibly Alabama and Florida.

PROPAGATION/PHENOLOGY: *Reproduces by seed.* Seeds fall near the parent plant or disperse to greater distances with water, human activities, as a seed contaminant, and by clinging to animals, the shoes and clothing of humans, and vehicle tires, especially when the seeds are moistened. Seed production is

Flixweed *(Descurainia sophia)* plant.
J. M. DiTomaso

Flixweed *(Descurainia sophia)* flowering stem.
J. M. DiTomaso

Flixweed *(Descurainia sophia)* seedling.
J. M. DiTomaso

typically abundant. Most seeds germinate in fall after the first rain, but a smaller flush of seeds often germinate in spring. Seedlings can flower in about 1 month. Seeds can survive for at least 3 years in the soil seedbank.

MANAGEMENT FAVORING/DISCOURAGING SURVIVAL: Fall and early spring cultivation can usually control seedlings, unless conditions are wet, in which case many seedlings may survive.

SIMILAR SPECIES: *Sisymbrium* mustards have *simple hairs*, and most species have more coarse leaves that are only deeply 1-pinnate-lobed. Refer to the entry for **Tumble mustard, London rocket, and Hedge mustard** (p. 537) for more information.

In most cases native *Descurainia* species are not considered weeds, but are distinguishable by having one or more of the following characteristics: pods that are *oblong, elliptic,* or *club-shaped* and often *less than 12 mm long* or pods that are linear with a *septa that has 0 or 1 indistinct vein*; pod stalks that are *erect to ascending*; and/or upper leaves that are *simple with smooth margins to 1-pinnate-*

Flixweed (*Descurainia sophia*) stem, leaves, and young inflorescences.
J. M. DiTomaso

lobed. **Pinnate tansymustard** [*Descurainia pinnata* (Walter) Britton][DESPI] is a widespread native with several subspecies. It is often considered a problem in Nevada and northeastern California rangelands and can be distinguished from flixweed by its shorter (4–20 mm) oblong to club-shaped fruit, shorter style (< 0.5 mm), and two rows of seeds per chamber.

Pinnate tansymustard (*Descurainia pinnata*) stem with leaves and young inflorescences.
J. M. DiTomaso

Pinnate tansymustard (*Descurainia pinnata*) inflorescence.
J. M. DiTomaso

Flixweed (*Descurainia sophia*) seeds.
J. O'Brien

Dyer's woad [*Isatis tinctoria* L.][ISATI]
[Cal-IPC: Moderate][CDFA list: B]

SYNONYMS: Marlahan mustard

GENERAL INFORMATION: Erect *biennial*, sometimes winter annual or short-lived perennial, to 1.2 m tall, with *bright yellow flowers* and distinctive *dark, pendant fruits*. Plants exist as basal rosettes until flower stems develop at maturity. Dyer's woad is highly competitive and can grow in large, dense colonies that displace desirable rangeland species, crop plants, and the native vegetation of an area. It is most problematic on rangeland, in agronomic crops, and in undisturbed natural areas in the intermountain west region of the northwestern United States. Besides California, dyer's woad is a state-listed noxious weed in Arizona (restricted), Colorado, Idaho, Montana (category 2), Nevada, New Mexico (class A), Oregon (class B), Utah, Washington (class A, plant quarantine), and Wyoming. The foliage contains compounds that appear to have insecticidal and fungicidal properties. However, the native rust [*Puccinia thlaspeos*] can significantly reduce seed production and may have the potential to be used as a biocontrol agent. Dyer's woad has been cultivated for several centuries in Europe as a medicinal herb and source of blue dye, and it was cultivated by the early settlers of the eastern states. Native to Europe.

SEEDLINGS: Cotyledons oblong, glabrous, about 15–30 mm long, tip often truncate and slightly indented, base wedge-shaped and tapered into a stalk about 5–12 mm long. First and subsequent few leaves alternate, elliptic to oblong or obovate, about 15–30 mm long, sparsely covered with long hairs, tip often rounded, base tapered to a hairy stalk about 4–15 mm long.

Dyer's woad (*Isatis tinctoria*) infestation.
J. M. DiTomaso

Dyer's woad (*Isatis tinctoria*) plant in fruit.
J. M. DiTomaso

Dyer's wood

MATURE PLANT: Leaves usually bluish green, with a *conspicuously pale midvein*, covered with a powdery white bloom (glaucous). *Rosette leaves oblanceolate to elliptic*, mostly 3–18 cm long, 1–4 cm wide, tip weakly rounded, base gradually tapered to stalk about one-half to three-quarters of the length of the blade, sparsely covered with simple long hairs, especially on the veins, *margin weakly toothed to slightly wavy*. Stem leaves alternate, sessile, broad to narrowly arrowhead-shaped, sometimes broadest near the tip, base lobed and clasping the stem, margin smooth, glabrous or nearly glabrous.

ROOTS AND UNDERGROUND STRUCTURES: Taproots of rosette and mature plants penetrate the soil to an average depth of about 1 m. Most lateral root growth occurs in the top 30 cm of soil during the second year.

FLOWERS: April–June. Panicles of racemes nearly flat-topped or umbrella-shaped. Petals 4, *bright yellow*, spoon-shaped, mostly 3–4 mm long. Sepals 4, separate to the base, shorter than petals. Stamens 6, 4 long, 2 short. Insect- or self-pollinated.

FRUITS AND SEEDS: *Fruits pendant, black to blue- or purplish black, flattened, oblong to oblanceolate*, 8–18 mm long, 5–7 mm wide, longitudinally ridged at the center of each side, gradually tapered to a slender stalk, *do not open*. Stigma sessile. Seeds 1 per fruit, oblong, nearly round in cross-section, grooved into 2 unequal halves, dull yellowish to orangish brown, about 3–4 mm long.

POSTSENESCENCE CHARACTERISTICS: Dried plants with a few fruits may persist well into winter.

HABITAT: Disturbed and undisturbed sites, roadsides, railroad rights-of-ways, fields, pastures, grain and alfalfa fields, forest, and rangeland. Often grows on dry, rocky or sandy soils.

Dyer's woad (*Isatis tinctoria*) plants in flower. J. M. DiTomaso

Dyer's woad

Dyer's woad *(Isatis tinctoria)* rosette.
J. M. DiTomaso

Dyer's woad *(Isatis tinctoria)* fruits.
J. M. DiTomaso

Dyer's woad *(Isatis tinctoria)* inflorescence.
J. M. DiTomaso

Dyer's woad *(Isatis tinctoria)* seedling.
J. M. DiTomaso

DISTRIBUTION: Klamath Ranges, Cascade Range, North Coast Ranges, northern & central Sierra Nevada, Modoc Plateau, and northern San Francisco Bay region, to 1000 m. Colorado, Idaho, Montana, Nevada, New Mexico, Oregon, Utah, Wyoming, Illinois, New York, New Jersey, Virginia, and West Virginia. Expanding range in the intermountain west region.

PROPAGATION/PHENOLOGY: Reproduces by seed. Most fruits fall near the parent plants, but some fruits disperse short distances with wind and to greater distances with water, soil movement, human activities, as a seed and hay contaminant, and possibly by clinging to animals. Seeds mature about 8 weeks after flower stem initiation. In a Utah study, plants produced an average of 383 fruits per plant. Seeds removed from fruits lack a dormancy period. Fruit coats contain water-soluble inhibitors that prevent many seeds from germinating until leaching occurs and reduce seedling growth of dyer's woad and other species. Some seeds germinate in the presence of the inhibitor. Rupture of fruit coats increases germination. Seeds germinate in fall and early spring. At maturity, fall-germinating plants typically produce more seeds than spring-germinating plants. Seed longevity under field conditions has not been studied.

MANAGEMENT FAVORING/DISCOURAGING SURVIVAL: Plants cut above the crown can grow new shoots and may persist as short-lived perennials. Spring cultivation can control infestations in crop fields.

SIMILAR SPECIES: Dyer's woad is distinguishable from other members of the mustard family by its unique fruits.

Field pepperweed [*Lepidium campestre* (L.) R. Br. ex W.T. Ait.][LEPCA]

Clasping pepperweed [*Lepidium perfoliatum* L.] [LEPPE]

SYNONYMS: Field pepperweed: cow-cress; cream-anther field pepperwort; downy peppergrass; English peppergrass; field cress; field peppercress; field peppergrass; field pepperwort; poor-man's pepper; *Neolepia campestre* (L.) W.A. Weber; *Thlaspi campestre* L.

Clasping pepperweed: clasping cress; clasping-leaved peppercress; perfoliate pepperwort; shield-cress; yellowflower pepperweed

Field pepperweed (*Lepidium campestre*) plant.   J. M. DiTomaso

Field pepperweed • Clasping pepperweed

Field pepperweed *(Lepidium campestre)* fruiting stem. J. M. DiTomaso

Field pepperweed *(Lepidium campestre)* stem and leaf bases. J. M. DiTomaso

Field pepperweed *(Lepidium campestre)* seedling. J. M. DiTomaso

Field pepperweed *(Lepidium campestre)* rosette. J. M. DiTomaso

Field pepperweed *(Lepidium campestre)* seeds, and fruit. J. O'Brien

Clasping pepperweed (*Lepidium perfoliatum*) plant. J. M. DiTomaso

Clasping pepperweed (*Lepidium perfoliatum*) flowering stems. J. M. DiTomaso

Clasping pepperweed (*Lepidium perfoliatum*) stem with clasping leaf bases. J. M. DiTomaso

GENERAL INFORMATION: Erect winter *annuals or biennials*, with flower stems to 0.6 m tall and *lens-shaped fruits*. Plants exist as rosettes until the inflorescence develops at maturity. Both species are widespread in the United States. In California, they primarily inhabit noncrop areas of the northern region and generally do not form dense colonies. Field pepperweed is native to Europe. Clasping pepperweed is native to Eurasia.

SEEDLINGS: First 2 leaves often appear opposite, later leaves alternate.

Field pepperweed: Cotyledons oval to elliptic, tip rounded to slightly truncate, 6–15 mm long, glabrous, base tapered to a stalk. First leaf pair oblong to ovate, margins barely wavy, usually glabrous, blade about 6–17 mm long, on a stalk longer than the blade. Subsequent leaves oval to oblong, margins slightly wavy to toothed and sometimes lobed once or twice at the base, base tapered to a long stalk, often with some spreading hairs, especially on the stalk. Cotyledons and leaves taste peppery or mustardlike.

Clasping pepperweed: Cotyledons narrowly oblanceolate, tip slightly rounded, base long-tapered. First and subsequent leaves deeply pinnate-dissected once or twice into narrow segments.

MATURE PLANT: Rosette and lower leaves stalked. Upper stem leaves sessile, smaller than lower leaves, *base strongly lobed and clasping the stem*, margins weakly toothed.

Field pepperweed: Stems usually 1 per plant, branched in the upper portion, often densely pubescent, especially near the tips. Leaves pubescent to sparsely spreading-hairy. Rosette and lower leaves *oblong to oblanceolate*, sometimes with 1–few lobes near the base, 4–12 cm long, ± 1 cm wide, margins irregularly toothed, on stalks. Upper stem leaves usually narrow, *oblong to oblanceolate, basal lobes usually acute* (sagittate) and not completely surrounding the stem, margin often wavy-toothed.

Clasping pepperweed: Stems simple or branched, minutely hairy in the lower portion, glabrous in the upper portion. Lower leaves *pinnate-dissected 2–3 times into narrow segments*, glabrous to sparsely hairy. Upper stem leaves *broadly ovate to nearly round, basal lobes rounded and completely surrounding the stem*, glabrous, margins smooth or minutely toothed.

ROOTS AND UNDERGROUND STRUCTURES: Taproots simple or branched with fibrous lateral roots.

FLOWERS: Racemes elongate as fruits mature. Petals 4, narrow spoon-shaped. Sepals 4, separate, hairy. Stamens 6.

Field pepperweed: April–August. Petals *white*, ± 2 mm long. Sepals ± 1.5 mm long. Anthers yellow.

Clasping pepperweed: March–July. Petals usually *pale yellow*, ± 1.5 mm long. Sepals ± 1 mm long.

FRUITS AND SEEDS: Fruiting stalks 4–8 m long, ± spreading. Seeds obovoid, ±

Clasping pepperweed (*Lepidium perfoliatum*) seedling. J. M. DiTomaso

Clasping pepperweed (*Lepidium perfoliatum*) seeds and fruit. J. O'Brien

Tongue pepperweed (*Lepidium nitidum* var. *nitidum*) flowering stem. J. M. DiTomaso

Tongue pepperweed (*Lepidium nitidum* var. *nitidum*) fruiting stem. J. M. DiTomaso

2 mm long, slightly flattened, with 1 or 2 grooves on each face, surface dull and ± finely roughened.

Field pepperweed: Pods (silicles) ovate to oblong, body slightly flattened, margins winged, mostly 5–6 mm long, 3–4 mm wide, tip indented (emarginate), surface appears glabrous, but with magnification is *minutely papillate with scale-like hairs.* Style 0.2–1 mm long, *shorter than to slightly longer than fruit wing tips.* Stalks slightly flattened. Seeds dark orange-brown or reddish brown.

Clasping pepperweed: Pods oval to round or slightly diamond-shaped, flattened, ± 4 mm long, 3–4 mm wide, not or slightly winged at apex, tip slightly indented, surface often glabrous or sometimes minutely pubescent. Style about 0.2 mm long, equal to or barely longer than fruit tip. Stalks round in cross-section. Seeds orange-brown, often with a translucent marginal wing about 0.1 mm wide.

POSTSENESCENCE CHARACTERISTICS: Dead stems with some fruits can persist into fall.

HABITAT: Roadsides, pastures, fields, waste places, orchards, vineyards, agronomic crops, other disturbed sites, and sagebrush flats.

DISTRIBUTION: Field pepperweed: Klamath Ranges, Cascade Range, and northern Sierra Nevada foothills, mostly 900–1900 m. Expected to expand its range in California. Nearly all contiguous states, except possibly Alabama, North Dakota, and Texas.

Clasping pepperweed: Great Basin, Cascade Range, Central Valley, eastern South Coast Ranges, and deserts, mostly 600–1500 m. Most contiguous states, except Indiana and some southern and northeastern states.

PROPAGATION/PHENOLOGY: *Reproduce by seed.* Seeds generally fall near the parent plant, but some seeds disperse to greater distances with water, mud, soil movement, animals, vehicle tires, agricultural equipment, and other human activities.

Field pepperweed: Most seeds are ballistically expelled to within 2 m of the parent plant.

MANAGEMENT FAVORING/DISCOURAGING SURVIVAL: Manual removal of scattered plants before seeds develop can help to control the annual and biennial pepperweeds. Cultivation may be necessary to prevent seed production and to kill seedlings of both species in agricultural fields.

SIMILAR SPECIES: Refer to table 27 (p. 509) for a comparison of distinguishing characteristics of pepperweeds. Some of the following species and varieties are ruderal natives of California and other states. These species and varieties are usually considered a desirable component of the vegetation in natural areas, but are occasionally weedy on agricultural lands and elsewhere.

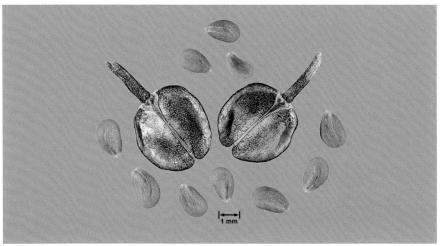

Tongue pepperweed (*Lepidium nitidum* var. *nitidum*) seeds and fruit.   J. O'BRIEN

Field pepperweed • Clasping pepperweed

Typical veiny pepperweed (*Lepidium oblongum* var. *oblongum*) plant.   J. M. DiTomaso

Veiny pepperweed (*Lepidium oblongum* var. *oblongum*) flowering stem.   J. M. DiTomaso

Robust veiny pepperweed (*Lepidium oblongum* var. *oblongum*) plant.   J. M. DiTomaso

Veiny pepperweed (*Lepidium oblongum* var. *oblongum*) seeds and fruit.   J. O'Brien

Tongue pepperweed [*Lepidium nitidum* Nutt. ex Torrey & A. Gray var. *nitidum*, var. *howellii* C. L. Hitchc., and var. *oreganum* (T. J. Howell ex E. L. Greene) C. L. Hitchc., synonyms: shining pepperweed; California tonguegrass] [LEPNI] is a ruderal native *annual* with erect to spreading flower stems to 0.4 m tall and white flowers. It is distinguished by having *lower and upper leaves pinnatedissected into linear segments* and *shiny, glabrous, oval fruits* that have a *shallow, narrow notch at the apex* and a *sessile stigma*. Upper leaves are *not* lobed at the base. Fruit stalks are *strongly flattened*. In addition, fruits are mostly 3–6 mm long and 3–4 mm wide. Tongue pepperweed occurs throughout California, except eastern deserts, to 1500 m. It also occurs in Nevada, Oregon, Washington, and New York.

Featherleaf pepperweed (*Lepidium pinnatifidum*) plant. J. M. DiTomaso

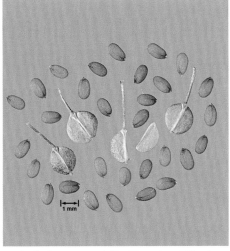

Featherleaf pepperweed (*Lepidium pinnatifidum*) seeds and fruit. J. O'Brien

Veiny pepperweed [*Lepidium oblongum* Small var. *oblongum* and var. *insulare* C. L. Hitchc., synonyms: *Lepidium greenei* Thell.; *Lepidium bipinnatifidum* auth., not Desv.] is a ruderal native *annual* with prostrate to ascending flower stems to 0.3 m tall and *flowers that lack petals or have inconspicuous linear petals that are much shorter than the sepals*. Stems typically branch near the base. Veiny pepperweed is further distinguished by having *upper and lower leaves once or twice pinnate-lobed into narrow segments mostly 1–3 mm long*. Upper leaves are *not* lobed at the base. Fruits are entirely glabrous or hairy on the margins, 2–3 mm long, *weakly veined, elliptic to round or obovate,* with an unwinged margin. The fruit tip has a *broad notch, erect wings,* and a *sessile stigma*. Veiny pepperweed occurs in the Central Valley and Southwestern region, to 500 m. It also occurs in Arizona, New Mexico, Oregon, Texas, and a few other southern and central states.

Featherleaf pepperweed [*Lepidium pinnatifidum* Ledeb.] is a taprooted *annual, biennial,* or occasionally short-lived perennial, with flower stems to about 0.5 m tall and *flowers without petals or with inconspicuous white petals*. It is primarily distinguished by having *narrow-oblong upper leaves* that are *not* lobed at the base and *nearly round fruits about 2 mm long and wide* that are usually *sparsely covered with minute hairs*. Fruits are *barely indented at the tip* and have a *stigma about equal to the tip*. In addition, lower leaves are pinnate-lobed 1–2 times. Featherleaf pepperweed inhabits roadsides, fields, and other disturbed places in the Sacramento Valley, San Francisco Bay region, Central Coast, South Coast, and western South Coast Ranges, to 200 m. It is expanding range and may occur elsewhere. Native to Eurasia.

Virginia pepperweed [*Lepidium virginicum* L. var. *medium* (E. Greene) C. Hitchc., var. *pubescens* (E. Greene) Thell., and var. *virginicum*] is a ruderal *annual* with erect stems to 0.7 m tall that are branched in the upper portion and white flowers with *petals that are equal to or longer than the sepals*. It is also distinguished by having *oblong lower leaves* that are *pinnate-lobed at the base* and ± *elliptic upper leaves with toothed margins*. Upper leaves are *not* lobed at the base. *Fruits are glabrous, round*, 2.5–4 mm long and about as wide, and have a *sessile stigma*. Most Virginia pepperweed varieties occurring in California are native, except for variety *virginicum*, which is thought to be introduced into California from the central and eastern states. Variety *virginicum* is distinguished by having fruit stalks that are sparse to densely hairy and round in cross-section. Varieties of Virginia pepperweed occur throughout California, except in Klamath Ranges and higher elevations of the Sierra Nevada, to 2400 m. Virginia pepperweed also occurs in all other contiguous states. Buried Virginia pepperweed seeds have been reported to survive up to about 40 years.

Virginia pepperweed *(Lepidium virginicum)* plant.   J. M. DiTomaso

Virginia pepperweed *(Lepidium virginicum)* flowering stems.   J. M. DiTomaso

Virginia pepperweed *(Lepidium virginicum)* seedling.   J. M. DiTomaso

Virginia pepperweed *(Lepidium virginicum)* seeds and fruit.   J. O'Brien

Purple-anther field pepperweed [*Lepidium heterophyllum* Bent., synonym: *Lepidium smithii* Hook.] is a less common *perennial*, sometimes biennial or annual, with foliage and fruits that closely resemble that of field pepperweed. Unlike field pepperweed, purple-anther field pepperweed typically has *more than one flowering stem from the crown* and fruits are *smooth or have just a few minute scale- or papilla-like hairs* when viewed under magnification. Purple-anther field pepperweed *often has purple anthers* and a style that is as long as or longer than the fruit tip. Purple-anther field pepperweed inhabits fields, roadsides, and open hillsides in the Cascade Range, Klamath Ranges, northwestern North Coast Ranges, and southeastern San Francisco Bay region, to 1500 m. It is expanding range in California and is reported to be spreading into some undisturbed natural areas (Plumas National Forest). It also occurs in Oregon and a few northeastern states. Native to Europe.

Table 27. Pepperweeds (*Lepidium* spp.) and hoary cress *(Cardaria draba)*

Species	Life cycle	Flowering stems	Upper leaf base (all are sessile)	Upper leaf shape	Lower leaf shape	Fruits	Other
L. campestre field pepperweed	annual; biennial	usually single, erect, to 0.5 m tall	narrow-lobed and clasping stem	± narrow, oblong to oblanceolate	oblong to oblanceolate, base ± with 1–2 lobes	5–6 mm long, ovate-oblong, tip indented, minutely papillate, style shorter to longer than tip	anthers yellow; style shorter to longer than fruit tip
L. heterophyllum purple-anther field pepperweed	perennial; (biennial, annual)	often more than 1 from crown, ± decumbent, to 0.5 m tall	narrow-lobed and clasping stem	± narrow, oblong to oblanceolate	oblanceolate to elliptic	4–7 mm long, ovate-oblong, tip indented, smooth or with a few minute papillae, style ± longer than tip	anthers often purple; style equal to or longer than fruit tip
L. latifolium perennial pepperweed	perennial	erect, usually in colonies, to 2 m tall	tapered	lanceolate to oblong or elliptc	oblong	± 2 mm long, round-ovate, tip rounded, smooth or hairy, stigma sessile	extensive creeping roots; foliage glabrous; upper leaves 1–4 cm wide
L. nitidum tongue pepperweed	annual	erect to spreading, to 0.4 m tall	not lobed	linear or pinnate-dissected into linear segments	pinnate-dissected into linear segments	2.5–6 mm long, round-ovate, very smooth, shiny, tip narrow-notched, stigma sessile	petals ± equal to or longer than sepals; fruit stalks very flat

Species	Life cycle	Flowering stems	Upper leaf base (all are sessile)	Upper leaf shape	Lower leaf shape	Fruits	Other
L. oblongum veiny pepperweed	annual	prostrate to ascending, to 0.3 m long	not lobed	pinnate-lobed 1–2 times into short narrow segments	pinnate lobed 1–2 times into short narrow segments	2–3.5 mm long, ovate, smooth, weakly-veined, tip broad-notched, stigma sessile	petals lacking or much shorter than sepals; fruit stalks slightly flat
L. perfoliatum clasping pepperweed	annual; biennial	erect, to 0.6 m tall	broad-lobed, lobes clasping and ± surrounding stem	broadly ovate to nearly round	pinnate-dissected 2–3 times	± 4 mm long, round-elliptic, tip barely indented, ± smooth, style ± equal to tip	flowers pale yellow; style equal to or barely longer than fruit tip
L. pinnatifidum featherleaf pepperweed	annual; biennial; (perennial)	± erect, to 0.5 m tall	not lobed	uppermost narrow oblong, lower pinnate-toothed or -lobed	pinnate-lobed 1–2 times into acute ± broad lobes or teeth	± 2 mm long, round-elliptic, ± minutely hairy, tip barely indented, stigma ± sessile & equal to tip	petals lacking; foliage unpleasant-scented
L. virginicum Virginia pepperweed	annual	erect, to 0.7 m tall	tapered	elliptic	obovate, base pinnate-lobed	2.5–4 mm long, round-ovate, tip narrow-notched, smooth, stigma ± sessile	petals equal to or longer than sepals
Cardaria draba hoary cress	perennial	erect, to 0.4 m tall	acute to weakly rounded lobes and clasping stem	highly variable, obovate, lanceolate, or oblong to elliptic	similar to upper leaves	2.5–3.5 mm long, 3–5 mm wide, upside-down heart-shaped to broadly ovate, tip lacking a notch, smooth	petals present, large persistent perennial roots

Perennial pepperweed [*Lepidium latifolium* L.] [LEPLA][Cal-IPC: High][CDFA list: B]

SYNONYMS: broadleaf pepperweed; broadleaved pepperweed; giant whiteweed; iron weed; perennial peppercress; perennial peppergrass; slender perennial peppercress; tall whitetop; *Cardaria latifolia* (L.) Spach

GENERAL INFORMATION: Erect *perennial* to 2 m tall, with *glabrous foliage*, rounded to pyramidal inflorescences of small white flowers and an *extensive creeping root system*. Perennial pepperweed is highly competitive and often forms *dense colonies* that displace native vegetation and wildlife. Toxicity problems have not been documented for livestock grazing in areas infested with perennial pepperweed. Goats appear to tolerate heavy consumption of fresh plants. However, there have been reports of horses becoming ill after being fed contaminated hay. Since its introduction in the mid-1930s, perennial pepperweed has spread rapidly throughout the western states, particularly in California and Nevada, where it is a state-listed noxious weed. Perennial pepperweed is also a state-listed noxious weed in Colorado, Idaho, New Mexico (class A), Oregon (class B), South Dakota, Utah, Washington (class B, plant quarantine), and Wyoming. Native to Eurasia.

SEEDLINGS: Cotyledons obovate to oblong, about 3–8 mm long, glabrous, tip rounded, base tapered into a short stalk about 2–3 mm long. First leaves developmentally alternate, but appear opposite, ovate to oblong, about 4–12 mm

Perennial pepperweed (*Lepidium latifolium*) infesting a wetland site. J. M. DiTomaso

long, glabrous, tip ± rounded, base ± wedge-shaped, margin entire to slightly wavy, on a stalk ± 5 mm long. Subsequent leaves resemble first leaves, but increasingly larger.

MATURE PLANT: Crown and lower stems ± weakly woody. Foliage *glabrous*, green to gray-green. Leaves alternate, lanceolate to elliptic or oblong. Basal leaves larger and wider that stem leaves, to 30 cm long and 8 cm wide, margin serrate, on a stalk about 3–15 cm long. Stem leaves reduced, *sessile or nearly sessile, base tapered*, margins entire to weakly serrate.

ROOTS AND UNDERGROUND STRUCTURES: Roots long, thick, minimally branched, vigorously creeping. Most roots occur in the top 60 cm of soil, but some can penetrate to a depth of 3 m or more. Experimental evidence indicates that plants extract salts (e.g., calcium, magnesium, and sodium) from deep soil

Perennial pepperweed (*Lepidium latifolium*) inflorescence.   J. M. DiTomaso

Perennial pepperweed (*Lepidium latifolium*) tapered leaf base.   J. M. DiTomaso

Perennial pepperweed (*Lepidium latifolium*) plant.   J. M. DiTomaso

Perennial pepperweed (*Lepidium latifolium*) basal rosette.   J. M. DiTomaso

and deposit them on the soil surface. This increased soil salt concentration is speculated to inhibit the germination and growth of other species. Carbohydrate reserves are lowest when flowering stems are elongating (bolting stage).

FLOWERS: May–September. Inflorescences ± pyramidal to rounded on top. Petals 4, white, spoon-shaped, ± 1.5 mm long. Sepals oval, less than 1 mm long, sometimes covered with long simple hairs. Stamens 6, 4 long, 2 short. Insect-pollinated. Self-compatible.

FRUITS AND SEEDS: Pods (silicles) 2-chambered, *round to slightly ovate, slightly flattened, lack a notch at the apex*, ± 2 mm long, usually sparse to moderately covered with *long, simple hairs. Stigma sessile, persistent.* Pod stalks much longer than pods, glabrous or sparsely pubescent. Seeds 1 per chamber, ellipsoid, slightly flattened, with a shallow groove on each side, ± 1 mm long, 0.5 mm

Perennial pepperweed (*Lepidium latifolium*) seedling.   J. M. DiTomaso

Perennial pepperweed (*Lepidium latifolium*) seeds and fruit.   J. O'Brien

Perennial pepperweed (*Lepidium latifolium*) young leaves from root.   J. M. DiTomaso

wide, reddish brown, surface minutely granular. Seeds fall from pods irregularly through winter and some may remain in pods until the following season.

POSTSENESCENCE CHARACTERISTICS: Aboveground parts typically die in late fall and winter. The pale, tan dead stems can persist for several years.

HABITAT: Noncrop areas including wetlands, riparian areas, meadows, vernal pools, salt marshes, flood plains, sand dunes, roadsides, and irrigation ditches, as well as ornamental plantings and agronomic crops, including alfalfa, orchards, vineyards, and irrigated pastures. Typically grows on moist or seasonally wet sites. Tolerates saline and alkaline conditions.

DISTRIBUTION: Throughout California, except deserts and northern North Coast and adjacent mountains (Del Norte, Humboldt, northern Mendocino Cos.), to 2200 m. All western states, Texas and a few other central states, a few eastern and northeastern states.

PROPAGATION/PHENOLOGY: *Reproduces vegetatively from creeping roots* and *root fragments* and *by seed*. Roots do not hold soil together very well, allowing erosion of river, stream, or ditch banks.

Root fragments and seeds float and disperse with flooding, soil movement, and agricultural and other human activities. Seeds can also cling to tires, shoes, and to animals and can contaminate hay or crop and pasture seed. Large fragments can survive extreme desiccation on the soil surface for extended periods. Fragments as small as 1–2 cm long and 2–8 mm in diameter can develop into new plants. New shoots begin to grow from roots in late winter. Fluctuating temperatures appear to stimulate seed germination. Plants usually produce abundant, often highly viable seed, but seedlings are seldom detected in the field. In wet years, seed production is sometimes limited by white rust (*Albugo* sp.) infection. Seedlings are not often encountered but appear to emerge midwinter through mid-spring.

MANAGEMENT FAVORING/DISCOURAGING SURVIVAL: Dense infestations are difficult to control. Cleaning agricultural or earth-moving machinery after use in infested areas and curtailing movement or use of soil, hay, and crop or pasture seed contaminated with perennial pepperweed root fragments and/or seed and can help prevent new infestations. Techniques such as repeated mowing, hand-digging, cultivation, grazing, and burning do not adequately control established perennial pepperweed infestations, except in the early stages of establishment. Cultivation typically increases infestations by dispersing root fragments. Seasonal flooding for an extended period during the growing season or mowing at the flower bud stage and treating the regrowth at the bolting to flowering stage with systemic herbicide can significantly reduce populations.

SIMILAR SPECIES: Hoary cress [*Cardaria draba* (L.) Desv.], **lens-podded whitetop** [*Cardaria chalepensis* (L.) Hand.-Mazz.], and **hairy whitetop** [*Cardaria pubescens* (C. Meyer) Jarmol.] are noxious *perennials with creeping roots* that resemble perennial pepperweed. **Field pepperweed** [*Lepidium campestre* (L.) R.Br.] and **clasping pepperweed** [*Lepidium perfoliatum* L.] are related *annual/biennials*. Unlike perennial pepperweed, all of these species *grow only to about 0.5 m tall*, have stem leaves with *bases that are lobed and clasping the stem*, and foliage that is at least sparsely *covered with short, simple hairs*. In addition, *Cardaria* species typically have *inflated pods greater than 2 mm long* with persistent a *style 1–2 mm long*. Field pepperweed and clasping pepperweed *do not grow in dense colonies*. Refer to the entries for **Lens-podded whitetop, Hoary cress, and Hairy whitetop** (p. 478) and **Field pepperweed and Clasping pepperweed** (p. 500), and also table 27 (p. 509), for more information about these species.

Wild radish [*Raphanus raphanistrum* L.][RAPRA]
Radish [*Raphanus sativus* L.][RAPSN][Cal-IPC: Limited]

SYNONYMS: Wild radish: cadlock; jointed charlock; jointed wild radish; white charlock; wild kale; wild turnip

Radish: cultivated radish; wild radish; *Raphanus raphanistrum* L. var. *sativus* (L.) G. Beck

GENERAL INFORMATION: Erect winter or summer *annuals*, sometimes biennials, to 1.2 m tall, with white, yellow, or pale purplish pink flowers and *elongate pods that do not open to release seeds*. Plants exist as rosettes until flower stems develop at maturity. Wild radish and radish readily hybridize, and introgression is common in populations where both species occur. In California, both species are widespread and many populations exhibit a range of species characteristics, making identification to species problematic. Wild radish and radish are susceptible to several pests and diseases that affect a variety of crops, including flea beetles, thrips, blackleg of brassicas (*Leptosphaeria* sp.), and turnip yellow mosaic tymovirus (TYMV). Conversely, plants also attract beneficial predators of crop pests, and the flowers attract bees. Livestock that consume large quantities of wild radish or radish seeds can develop irritation of the digestive tract. Wild radish is a state-listed noxious weed in Minnesota (secondary) and south-

Wild radish (*Raphanus raphanistrum*) plant.　　　　　　　　　　J. M. DiTomaso

ern Australia, where it is a highly competitive crop and pasture weed. Native to the Mediterranean region.

SEEDLINGS: Cotyledons kidney- to broadly heart-shaped, 1–2 cm long, length about equal to the width, tip broadly notched, base abruptly tapered to a stalk 1–2.5 cm long, glabrous. Leaves alternate, sparsely covered with stiff flattened (appressed) simple hairs. First and subsequent few leaves elliptic-oblong, about 1–2.5 cm long, margins irregularly round-toothed, tip ± rounded, base tapered or deeply lobed once or twice, on long stalks.

MATURE PLANT: Foliage typically sparsely covered with stiff flattened (appressed) simple hairs, especially near the base. Stems usually branched in the upper portion. Lower leaves elliptic to oblanceolate, deeply pinnate-lobed to

Wild radish (*Raphanus raphanistrum*) seedling.
J. M. DiTomaso

Wild radish (*Raphanus raphanistrum*) seeds.
J. O'Brien

Wild radish (*Raphanus raphanistrum*) inflorescence.
J. M. DiTomaso

Wild radish (*Raphanus raphanistrum*) fruit.
J. M. DiTomaso

compound, mostly 6–20 cm long. Lobe margins irregularly round- or sharp-toothed, sometimes weakly lobed. Terminal lobes ± oval to round, larger than lateral lobes. Upper stem leaves reduced in size, ± elliptic, toothed, sometimes lobed at the base, sessile or short-stalked.

ROOTS AND UNDERGROUND STRUCTURES: Taproot *radish scented and flavored.*

Wild radish: Taproot usually slender, ± branched, can grow to 1.6 m deep, usually with numerous spreading fibrous lateral roots in the top 20 cm of soil.

Radish: Taproot generally thick, elongate, with fibrous lateral roots.

FLOWERS: Introgressive populations typically exhibit all flower colors. Petals 4, long-clawed, mostly 15–25 mm long, ± with dark violet veins. Sepals 4, erect, separate. Bracts below flowers lacking.

Wild radish: April–July. Petals white, yellow fading to white, or pale purplish pink.

Radish: February–July. Petals usually pale purplish pink, occasionally white.

FRUITS AND SEEDS: Pods *elongate, nearly straight, ± round in cross-section, initially ± fleshy,* corky when dry. *Beak about 1–2 or 3 cm long, taper to a narrow apex,* seedless. *Pods do not open to release seeds.* Seeds nearly round to ovoid, 2–4 mm long, ± 2 mm wide, surface minutely pitted in a reticulate pattern, brown to reddish or yellowish brown.

Wild radish: Dried pods *strongly constricted between seeds* (often not apparent in fresh pods), *longitudinally grooved below, body nearly uniform in diameter for most of its length,* mostly 4–6 mm wide, mostly 4–8 cm long including a beak

Radish (*Raphanus sativus*) infestation.   J. M. DiTomaso

about 1–2 cm long, *often break apart horizontally* into 1-seeded segments. Seeds mostly 4–12 per pod.

Radish: Dried pods *straight-sided or slightly constricted between seeds*, weakly longitudinally grooved or smooth below, *body diameter largest below the middle and tapering toward the apex*, mostly 6–8 mm wide, 3–6 cm long including a beak about 1–3 cm long, generally do not break apart horizontally. Seeds mostly 1–3(–5) per pod.

POSTSENESCENCE CHARACTERISTICS: Dead stems with pods can persist into winter.

HABITAT: Roadsides, pastures, fields, crop fields, orchards, vineyards, old gardens, playgrounds, parks, and other disturbed places.

DISTRIBUTION: Wild radish: Throughout California, except deserts, to 800 m. Arizona, Colorado, Idaho, Montana, Nevada, Oregon, Washington, some central states, and most southern and eastern states.

Radish: Throughout California, except deserts, Great Basin, and some mountain areas, to 1000 m. Most contiguous states.

PROPAGATION/PHENOLOGY: *Reproduce by seed.* Fruits and seeds disperse with water, soil movement, animals, human activities and agricultural operations, and as a contaminant of crop seed and hay.

Wild radish: Plants can produce large quantities of seeds under favorable conditions. A proportion of seeds are dormant at maturity. A wide fluctuation of daily temperature stimulates germination, and light is not required. Most germination

Radish (*Raphanus sativus*) plant.
J. M. DiTomaso

Radish (*Raphanus sativus*) inflorescence.
J. M. DiTomaso

Radish *(Raphanus sativus)* seedling. J. M. DiTomaso

Radish *(Raphanus sativus)* fruit. J. M. DiTomaso

occurs in fall after the first significant rain, but some seeds continue to germinate throughout spring or at other times when conditions are favorable. Most seedlings emerge from a shallow soil depth. Buried seeds can survive up to 20–30 years or more.

MANAGEMENT FAVORING/DISCOURAGING SURVIVAL: Manual removal of plants before seed production can help to control the radishes in small areas or in larger areas where there are small numbers of plants. Cultivation as needed to prevent seed production can help to control these species in agricultural fields.

SIMILAR SPECIES: Blue mustard [*Chorispora tenella* (Pallas) DC.][COBTE] [CDFA list: B] is a winter *annual* to 0.5 m tall, with pale purple to bluish purple flowers and fruits that resemble those of wild radish. Unlike wild radish, blue mustard typically has *glandular-pubescent foliage that is sticky to touch* and has an unpleasant scent. In addition, the *pods curve upward* and are *smooth on the lower surface*. Refer to the entry for **Blue mustard** (p. 486) for more information about this species. The yellow-flowered radish forms can be confused with a variety of *Brassica* (mustard) species. However, all radish species have conspicuous dark veins in the petals and indehiscent fruit with constrictions between the seeds. Unlike mustards, the roots of radish have a hot, peppery flavor.

Radish *(Raphanus sativus)* seeds. J. O'Brien

Austrian fieldcress [*Rorippa austriaca* (Crantz) Spach] [RORAU][CDFA list: B]

Yellow fieldcress [*Rorippa sylvestris* (L.) Bess.] [RORSY][CDFA list: Q]

SYNONYMS: Austrian fieldcress: Austrian yellowcress; *Nasturtium austriacum* Crantz; *Radicula austriaca* (Crantz) Small

Yellow fieldcress: creeping fieldcress; creeping yellowcress; kiek; *Nasturium sylvestre* (L.) R.Br.; *Radicula sylvestris* (L.) Druce; *Sisymbrium sylvestre* L.

GENERAL INFORMATION: Mustardlike *perennials* with *yellow flowers* and *creeping rootstocks*. Both species can form dense clonal patches. Plants typically

Austrian fieldcress (*Rorippa austriaca*) plant. No viable seed produced. J. M. DiTomaso

grow on wet or frequently irrigated soils. Both species are native to Europe.

Austrian fieldcress: To 1 m tall. Seeds sometimes enter California with bulbs from the Netherlands. Besides California, Austrian fieldcress is a state-listed noxious weed in Arizona (prohibited), Nevada, and Washington (class B).

Yellow fieldcress: Seeds and rhizomes are sometimes transported with interstate nursery stock. Yellow fieldcress is common in the northeastern states. It is state-listed noxious weed in Oregon (class B) and North Carolina (class B).

SEEDLINGS: Austrian fieldcress: Cotyledons oval, 2–6 mm long, 1–2 mm wide, on stalks about equal in length, often withered by the 4-leaf stage. Early rosette leaves alternate, ovate, usually at least 4 times larger than the cotyledons, on stalks equal to or longer than the blades, surfaces often sparse to mod-

Austrian fieldcress (*Rorippa austriaca*) flowering and fruiting stem. J. K. CLARK

Austrian fieldcress (*Rorippa austriaca*) leaf and stem. J. K. CLARK

Austrian fieldcress (*Rorippa austriaca*) rosette. J. K. CLARK

erately covered with *short, unicellular hairs.* Margins of early leaves smooth to *irregular-toothed.* Later rosette leaves oblong to oblanceolate, *margin unequally toothed* (serrate to dentate) but not lobed.

Yellow fieldcress: To 0.5 m tall. Young seedlings resemble those of *Austrian fieldcress*, except first and subsequent leaves are *glabrous.* Later rosette leaves deeply pinnate-lobed nearly to midrib.

MATURE PLANT: Plants exist as basal rosettes until flowering stems develop in spring. Stems ascending to nearly erect, simple or branched near the top. Stem leaves alternate, progressively reduced.

Austrian fieldcress: Rosette and stem leaves dull bluish green, glabrous, *narrow oblong to oblanceolate,* 3–10 cm long, *entire to unequally sharp-toothed* (dentate to serrate). Upper stem leaves taper to a narrow lobed base that clasps the stem.

Yellow fieldcress: Rosette and stem leaves glabrous, *deeply pinnate-lobed nearly to the midrib,* mostly 5–20 cm long. *Lobes narrow, often sharp-toothed,* sometimes smooth or nearly smooth.

ROOTS AND UNDERGROUND STRUCTURES: Both species initially have a taproot but also develop an *extensive system of shallow and deep, laterally creeping rhizomes* that produce new shoots. Rhizome fragments can develop into new plants.

Austrian fieldcress: Mature taproots are often deep, thick, and fleshy, and resemble those of horseradish [*Armoracia rusticana* P. Gaertner, Meyer & Scherb.].

Yellow fieldcress: Rhizomes slender, reported to grow to about 1 m deep.

FLOWERS: Racemes short, terminal and axillary. Petals 4, *yellow, longer than sepals.* Stamens 6.

Austrian fieldcress: June–August. Flowers mostly 2–3 mm wide.

Yellow fieldcress: May–August. Flowers 4–5 mm wide.

FRUITS AND SEEDS: Both species appear to rarely develop viable seeds in California and some other places.

Austrian fieldcress: Pods (silicles) generally *ovoid, ± 3 mm long, 2.4–3 mm wide*, with a *persistent style 2–3 mm long.* Pod stalk 7–15 mm long, spreading to ascending. Seeds 0–several per chamber, flattened, nearly oval to heart-shaped, 0.5–1 mm long, reddish to orange-brown, covered with minute bumps under high magnification.

Yellow fieldcress: Pods (siliques) *linear, cylindrical, straight, 10–15 mm long, ± 1.5 mm wide*, with a *persistent style 0.5–1 mm long.* Pod stalk mostly 5–10 mm long, spreading to slightly arched downward. Seeds 0.6–0.8 mm long.

HABITAT: Austrian fieldcress: Moist disturbed and cultivated sites, roadsides, fields (especially hay fields), mud flats, and nurseries. Typically inhabits areas where the soil is wet from 6–8 months during the year.

Yellow fieldcress: Nursery container stock, nursery grounds, moist landscaped and waste areas, stream margins.

DISTRIBUTION: Austrian fieldcress: Modoc Plateau, generally 1200–2000 m. Idaho, Montana, Utah, Washington, some north-central states, and a few northeastern states.

Yellow fieldcress: Central Coast and western South Coast Ranges (Monterey Co.). Discovered in the container stock of various California nurseries in 1997–98. Expected to expand its range.

PROPAGATION/PHENOLOGY: *Reproduce vegetatively from rhizomes and rhizome fragments.* Stems and sometimes leaves can develop adventitious roots in water or moist substrates. Rhizome fragments disperse with soil movement and agricultural and landscaping activities, such as cultivation and transportation of ornamental plants in containers. Both species appear to rarely develop viable seed in California.

MANAGEMENT FAVORING/DISCOURAGING SURVIVAL: As soon as detected, manually removing individual plants and rhizomes can help prevent the spread of Austrian and yellow fieldcress. Improving drainage of wet soils, repeated cultivation, and crop rotation can help control Austrian fieldcress in agricultural fields.

Yellow fieldcress (*Rorippa sylvestris*) foliage.
C. ELMORE

Yellow fieldcress (*Rorippa sylvestris*) flowering stem.
J. M. DITOMASO

Yellow fieldcress (*Rorippa sylvestris*) rhizome.
J. M. DITOMASO

Yellow fieldcress *(Rorippa sylvestris)* seedling. J. M. DiTomaso

SIMILAR SPECIES: Marsh yellowcress [*Rorippa palustris* (L.) Besser vars. *occidentalis* (S. Watson) Rollins and *hispida* (Desv.) Rydb., synonyms: bog marshcress; bog yellowcress; common yellowcress; marshcress; hispid yellow cress (var. *hispida*); western bog yellowcress (var. *occidentalis*); yellow watercress; *Rorippa islandica* (Oeder) Borbás][RORIS] is a glabrous to hairy native *annual, biennial*, or short-lived perennial with a *slender taproot* and variable

Marsh yellowcress *(Rorippa palustris)* flowering stem. J. M. DiTomaso

Marsh yellowcress *(Rorippa palustris)* fruiting stem. J. M. DiTomaso

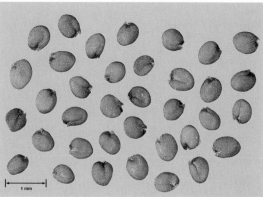

Marsh yellowcress (*Rorippa palustris*) seedling. J. M. DiTomaso

Marsh yellowcress (*Rorippa palustris*) seeds. J. O'Brien

leaves that are irregularly toothed to deeply pinnate-lobed. Unlike Austrian and yellow fieldcress, marsh fieldcress *lacks laterally creeping rhizomes* and *reproduces primarily by seed*. In addition, it typically has one dominant stem from the base to 1.4 m tall, *petals equal to or shorter than the sepals* (1–3.5 mm long), and *pods more than 2 mm wide* that are *round to ovate and 2–6 mm long* (var. *hispida*) or *oblong and 7–15 mm long* (var. *occidentalis*). Both varieties have a persistent style 0.2–1 mm long. Seeds resemble those described for Austrian fieldcress. Seeds typically germinate summer through winter. Marsh yellowcress usually inhabits moist or wet places in natural communities, where it is a desirable component of the native vegetation. It is designated as an obligate wetland indicator species in most regions of the United States. Marsh yellowcress is primarily a weed of nursery operations, but it is sometimes also weedy in ditches, irrigation canals, and irrigated crops. It occurs throughout California (var. *occidentalis*) and the Modoc Plateau (var. *hispida*), to 2000 m. Varieties of marsh yellowcress occur in all contiguous states. Marsh yellowcress was mistakenly called *Rorippa islandica*. However, *Rorippa islandica* (Oeder) Barbas is a separate species and is not known to occur in North America.

Sibara [*Sibara virginica* (L.) Rollins][SIBVI]

SYNONYMS: common rock-cress; Virginia rock cress; Virginia winged rock cress; *Arabis virginica* (L.) Poir.; *Arabis ludoviciana* (Hook.) C. A. Meyer; *Cardamine ludoviciana* Hook.; *Cardamine parviflora* L. ssp. *virginica* (L.) O.E. Schulz; *Cardamine virginica* L.; *Planodes virginicum* (L.) E. L. Greene

GENERAL INFORMATION: Native winter *annual*, sometimes biennial, with a basal rosette of pinnate-lobed leaves, flowering stems to 0.3 m tall, and *flat, linear pods*. Sibara is generally considered a desirable component of the vegetation in natural areas. However, it is a ruderal species and is sometimes a minor weed in agronomic fields, vegetable crops, orchards, and vineyards.

SEEDLINGS: Cotyledons oval, ~ 4–7 mm long, 0.5–1 mm wide, glabrous, on stalks that elongate to a length ~ equal to or slightly longer than the cotyledons. Leaves alternate. First and subsequent few leaves oblanceolate, pinnate-dissected, ~ 10–20 mm long, ± 4 mm wide, glabrous. Terminal lobe ovate, much larger than lateral lobes. Lateral lobes narrow, ± oblong, well-spaced, often abruptly tapered to a minute bristlelike point at the apex (mucro).

MATURE PLANT: Stems 1–several, ascending to decumbent, branched, sparsely stiff-hairy in the lower portion, glabrous to sparsely pubescent in the upper portion. *Hairs mostly simple, a few sometimes forked.* Lower leaves oblong to oblanceolate, *deeply pinnate-lobed to the midrib*, mostly 2–10 cm long, stalked, glabrous to sparsely stiff-hairy, especially on the stalk. Terminal lobe larger than

Sibara (*Sibara virginica*) plant. J. M. DiTomaso

the lateral lobes, oblong to ovate, sometimes shallowly 3-lobed. Lateral lobes narrow, ± oblong, widely spaced. Upper leaves resemble lower leaves, but with shorter stalks and more often glabrous.

ROOTS AND UNDERGROUND STRUCTURES: Taproots slender.

FLOWERS: February–May. Petals 4, white, spoon-shaped, about 2–3 mm long, longer than sepals. Sepals 4, erect, ± 1.5 mm long, often purple-tinged.

FRUITS AND SEEDS: Pods ascending to nearly erect, glabrous, *linear, flattened parallel to septum, straight,* 1.5–2.5 cm long, mostly 1.5–2 mm wide, with an *inconspicuous beak* (style) 0.5 mm long or less. Pods *open by 2 valves to release seeds. Valves only have a visible midvein at the base, remain straight upon opening.* Pod stalk ascending, 2–4 mm long, width nearly equal to pod width. Seeds in 1 row per chamber, nearly round with a shallow notch at one end, slightly flat-

Sibara *(Sibara virginica)* stems   J. M. DiTomaso

Sibara *(Sibara virginica)* inflorescence.
J. M. DiTomaso

Sibara *(Sibara virginica)* seedling.
J. M. DiTomaso

Sibara *(Sibara virginica)* seeds.   J. O'Brien

Mouse-ear cress (*Arabidopsis thaliana*) plant.
J. M. DiTomaso

Mouse-ear cress (*Arabidopsis thaliana*) inflorescence.
J. M. DiTomaso

Mouse-ear cress (*Arabidopsis thaliana*) seedling.
J. M. DiTomaso

Mouse-ear cress (*Arabidopsis thaliana*) seeds.
J. O'Brien

tened, ± 1 mm in diameter, orange-brown, margin narrow-winged all around, surface minutely granular with magnification.

POSTSENESCENCE CHARACTERISTICS: Dead plants with opened fruits can persist for a short period.

HABITAT: Agronomic fields, vegetable crops, orchards, vineyards, open disturbed places, stream banks, ditch banks, and vernal pool margins. Usually inhabits moist places.

DISTRIBUTION: Central Valley, southern Southwestern region (Riverside, Orange, and San Diego Cos.), and western North Coast Ranges, to 300 m. Southern states and adjacent eastern and central states.

PROPAGATION/PHENOLOGY: *Reproduces by seed.* Most seeds fall near the parent plant. Some seeds disperse to greater distances with water, mud, by clinging to animals, and through human activities.

MANAGEMENT FAVORING/DISCOURAGING SURVIVAL: Manual removal or cultivation before seeds mature can help to control sibara.

SIMILAR SPECIES: Mouse-ear cress [*Arabidopsis thaliana* (L.) Heynh., synonyms: arabidopsis; thale cress; *Arabis thaliana* L.; *Sisymbrium thalianum* (L.) J. Gay & Monn.][ARBTH] is an *annual* rosette with flower stems to 0.4 m tall, small white flowers, and straight, linear pods 1–1.5 cm long. Unlike sibara, mouse-ear cress has leaf margins with a *few shallow teeth*, leaf stalks with *simple hairs and 2–4-branched hairs*, cylindrical to slightly flattened pods *0.5–1 mm wide*, and oblong seeds that lack a narrow-winged margin. Mouse-ear cress inhabits disturbed open areas in the North Coast Ranges, North Coast, northern Sierra Nevada foothills, northern Sacramento Valley, San Francisco Bay region, and possibly elsewhere, to 1000 m. It also occurs in most northwestern states and is common in the eastern half of the United States. Mouse-ear cress is commonly used for experimental studies in genetics, biochemistry, and physiology. Native to Europe.

Wild mustard [*Sinapis arvensis* L.][SINAR] [Cal-IPC: Limited]

Shortpod mustard [*Hirschfeldia incana* (L.) Lagr.-Fossat][HISIN][Cal-IPC: Moderate]

SYNONYMS: Wild mustard: canola; charlock mustard; common mustard; crunch-weed; field kale; field mustard; kaber mustard; kedlock; rapeseed; *Brassica arvensis* (L.) Rabenh.; *Brassica kaber* (DC.) Wheeler; *Brassica kaber* (DC.) Wheeler var. *pinnatifida* (Stokes) Wheeler; *Brassica kaber* (DC.) Wheeler var. *schkuhriana* (Reichenb.) Wheeler

Shortpod mustard: hairy brassica; hairy mustard; hoary mustard; Mediterranean mustard; *Brassica geniculata* (Desf.) J. Ball; *Sinapis incana* L.

GENERAL INFORMATION: Erect, yellow-flowered mustards to 1 m tall. Refer to table 26 (p. 468) for a comparison of distinguishing features.

Wild mustard: Winter or summer *annual*. Seeds contain allylisothiocyanate, the alkaloid sinapine, and the alkaloidal glucoside sinalbin. Ingestion of large quantities of seed can be fatally *toxic* to livestock. Symptoms of poisoning resemble severe gastroenteritis. Wild mustard is a state-listed noxious weed in Colorado, Iowa (secondary), Ohio (prohibited), and Minnesota (secondary). Native to Europe.

Wild mustard (*Sinapis arvensis*) plant.
J. M. DiTomaso

Wild mustard (*Sinapis arvensis*) inflorescence.
J. M. DiTomaso

Shortpod mustard: *Biennial or short-lived perennial,* sometimes winter annual. Shortpod mustard is becoming more problematic in wildland areas of southern California. Native to the Mediterranean region.

SEEDLINGS: Leaves alternate.

Wild mustard: Cotyledons kidney-shaped, 5–10 mm long, slightly broader than long, glabrous, stalked. First and subsequent few leaves hairy, obovate, about 10–20 mm long, tapered to a hairy stalk ± 10 mm long, margins wavy or rounded-toothed.

Shortpod mustard: Cotyledons ± heart-shaped, tip indented, 4–6 mm long, longer than broad, glabrous, stalked. First and subsequent few leaves hairy, elliptic-oblong, about 5–10 mm long, tapered to a hairy stalk 2–5 mm long, margins slightly wavy to rounded-toothed.

Wild mustard *(Sinapis arvensis)* fruit.
J. M. DiTomaso

Wild mustard *(Sinapis arvensis)* seedling.
J. K. Clark

Wild mustard *(Sinapis arvensis)* young rosette.
J. M. DiTomaso

Wild mustard *(Sinapis arvensis)* seeds. J. O'Brien

MATURE PLANT: Lower leaves obovate, irregularly pinnate-lobed and toothed, about 5–20 cm long, terminal lobe larger than lateral lobes, on a long stalk. Leaf lobes rounded to ± acute. Upper leaves ± sessile, reduced, ± lanceolate, not clasping the stem, margin irregularly toothed.

Wild mustard: Stem bases sparsely covered with short, stiff, downward directed hairs. *Basal leaves do not form a rosette.* Leaves sparsely hairy.

Shortpod mustard: Stem bases moderate to densely covered with stiff, downward directed hairs. *Basal leaves usually form a flat rosette.* Leaves moderate to densely covered with stiff, grayish hairs.

ROOTS AND UNDERGROUND STRUCTURES: Both species have a ± slender taproot with fibrous lateral roots.

Shortpod mustard: Taproot often deep, with lateral roots in the upper 30 cm of soil.

FLOWERS: Racemes elongate in fruit. Sepals yellowish, spreading. Insect-pollinated. Mostly self-incompatible.

Wild mustard: February–May, occasionally to October. Petals pale to bright yellow, 8–12 mm long. Sepals 4–5 mm long.

Shortpod mustard: May–October, sometimes nearly year-round. Petals pale yellow to white, mostly 5–6 mm long. Sepals ± 3 mm long.

FRUITS AND SEEDS: Pods (siliques) linear, 2-valved, with a persistent beak at the tip, *on a stalk that is shorter than the pod. Immature pods have 3–7 faint veins on each valve.* Seed surface minutely reticulate at 10× magnification.

Wild mustard: Pods *ascending, not closely appressed to stem,* glabrous, cylindri-

Shortpod mustard (*Hirschfeldia incana*) plant. J. M. DiTomaso

Shortpod mustard (*Hirschfeldia incana*) inflorescence. J. M. DiTomaso

Shortpod mustard (*Hirschfeldia incana*) rosette. J. M. DiTomaso

Shortpod mustard (*Hirschfeldia incana*) seedling. J. M. DiTomaso

Shortpod mustard (*Hirschfeldia incana*) seeds. J. O'Brien

cal, angled or ± flat, *20–35 mm long including beak*, often slightly constricted between seeds. Beak ± *conical or angled, 0.6–1.2 cm long*, usually contains 1–2 seeds. Seeds nearly spherical, 1.5–2 mm in diameter, dull reddish brown to nearly black.

Shortpod mustard: Most pods *erect, appressed to stem*, glabrous, ± cylindrical, *8–15 mm long including beak*. Beak ± *flattened, 3–6 mm long, slender, abruptly swollen at the base*, often contains 1–2 seeds. Seeds ovoid, ± 1 mm long, dull reddish brown.

POSTSENESCENCE CHARACTERISTICS: Brown stems with pod remnants can persist for a few months.

HABITAT: Disturbed places, roadsides, fields, pastures, agronomic crops, orchards, vineyards, ditch banks, and dry washes.

White mustard *(Sinapis alba)* plants.
J. M. DiTomaso

White mustard *(Sinapis alba)* stem and leaves.
J. M. DiTomaso

White mustard *(Sinapis alba)* seedling.
J. M. DiTomaso

White mustard *(Sinapis alba)* inflorescence.
J. M. DiTomaso

White mustard *(Sinapis alba)* fruit.
J. M. DiTomaso

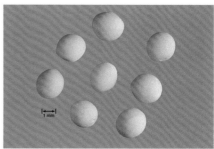
White mustard *(Sinapis alba)* seeds.   J. O'Brien

DISTRIBUTION: Wild mustard: Throughout California, except deserts and Great Basin, to 500 m. All contiguous states, except possibly Alabama.

Shortpod mustard: North Coast, Sierra Nevada foothills, Central Valley, Central-western region, and South Coast, to 1600 m. Expanding range in southern California. Nevada and Oregon.

PROPAGATION/PHENOLOGY: *Reproduce by seed.* Seeds fall near parent plants or disperse to greater distances with water, mud, agricultural activities, as seed and feed contaminants, and animals. Seed production can be high.

Wild mustard: Seeds can survive at least 11 years under field conditions. Seed dormancy is variable.

Shortpod mustard: Seeds appear to require an after-ripening period of about 1 month. Fruiting stems die when plants run short of soil moisture at the end of summer or early fall. New foliage grows from the rootstock after the first fall rain.

MANAGEMENT FAVORING/DISCOURAGING SURVIVAL: Where possible, manual removal or cultivation before seeds develop, particularly during the seedling stage, can control populations. Control methods implemented over a period of years will eventually exhaust the seedbank.

SIMILAR SPECIES: White mustard [*Sinapis alba* L., synonyms: *Brassica alba* Rabench.; *Brassica hirta* Moench.][SINAL] is a winter or summer *annual* to about 0.75 m tall. White mustard is distinguished by having *upper leaves on a short stalk* and *densely hairy pods* 2–3 cm long including the ± glabrous, *flattened beak that is equal to or longer than the body.* Pod stalks are ascending to spreading and often *equal to or longer than the pods.* In addition, white mustard seeds are pale yellow, nearly spherical, and 1.5–2 mm in diameter. White mustard is commercially cultivated for the seeds and greens, but, similar to wild mustard, white mustard seeds contain isothiocyanates and ingestion of large quantities can be *toxic* to livestock. White mustard has escaped cultivation and inhabits disturbed places along the Central and South Coast to 1000 m. It is expected to expand its range in California. White mustard is naturalized in all other western states, except possibly Nevada, most central and eastern states, and some southern states. Native to Eurasia. Shortpod mustard is commonly found along with black mustard (*Brassica nigra* (L.) Koch). See table 26 (p. 468) for specific differences between these two and other similar species.

Tumble mustard [*Sisymbrium altissimum* L.][SSYAL]
London rocket [*Sisymbrium irio* L.][SSYIR]
[Cal-IPC: Moderate]
Hedge mustard [*Sisymbrium officinale* (L.) Scop.] [SSYOF]

SYNONYMS: Tumble mustard: Jim Hill mustard; tall mustard; tall tumble-mustard; tumble mustard; tumbleweed mustard; *Norta altissima* (L.) Britt.

London rocket: desert mustard; *Norta irio* (L.) Britt.

Hedge mustard: hedge weed; *Erysimum officinale* L.; *Sisymbrium officinale* (L.) Scop. var. *leiocarpum* DC.

GENERAL INFORMATION: Erect mustards, usually with *pale yellow flowers* and *nearly cylindrical or ± 4-angled pods*. All three species exist as rosettes with *pinnate-dissected or pinnate-lobed to pinnate-compound leaves* until flower stems develop at maturity. Native to Europe.

Tumble mustard: Winter or summer *annual*, sometimes biennial, to 1.5 m tall, with *ascending linear pods*.

London rocket: Winter *annual* to 0.5 m tall, with *ascending linear pods*.

Hedge mustard: Winter or summer *annual* to 1 m tall, with *erect, awl-shaped pods that typically lay close to the stem* (appressed).

Tumble mustard (*Sisymbrium altissimum*) flowering stem. J. M. DiTomaso

Tumble mustard • London rocket • Hedge mustard

Tumble mustard (*Sisymbrium altissimum*) plant.
J. M. DiTomaso

Tumble mustard (*Sisymbrium altissimum*) stem and leaves.
J. M. DiTomaso

Tumble mustard (*Sisymbrium altissimum*) rosette.
J. M. DiTomaso

Tumble mustard (*Sisymbrium altissimum*) seedling.
J. M. DiTomaso

Tumble mustard (*Sisymbrium altissimum*) seeds.
J. O'Brien

SEEDLINGS: Leaves alternate.

Tumble mustard: Cotyledons ovate to oblong, about 2–6 mm long, 1–2 mm wide, glabrous, on short stalks. First leaf oblong-elliptic, slightly larger than cotyledons, bristly-hairy, margin smooth or with a few weak teeth, base tapered to a broad stalk ± equal to the blade length. Subsequent leaves similar to first leaf, except increasingly larger and with shallow-toothed margins.

London rocket: Cotyledons oval, about 2–6 mm long, 1–3 mm wide, glabrous, on stalks ± equal to the blade length. First leaf oval, slightly larger than cotyledons, margins smooth or with a few weak teeth, initially sparsely short-hairy, on a long stalk that is initially short-hairy and later glabrous. Subsequent few leaves resemble first leaf, but with shallow-toothed margins.

Hedge mustard: Cotyledons ovate to nearly round, about 3–7 mm long, 3–6 mm wide, glabrous, tip ± slightly indented, on stalks that elongate with age. First leaf ovate to nearly round, about 5–25 mm long, 3–20 mm wide, initially spreading-hairy, margins weakly toothed, on a hairy stalk 1–3 cm long. Subsequent 2 leaves resemble first leaf, but usually have 1–2 small lobes at the base and more pronounced teeth on the margins.

MATURE PLANT: Hairs *simple* (unbranched). Stems erect, branched. Rosette leaves *deeply pinnate-dissected or -lobed to pinnate-compound in the lower part*. Upper leaves *sessile, not clasping the stem*, smaller than lower leaves.

Tumble mustard: Foliage sparsely long-hairy. Lower leaves to about 15 cm long, terminal lobe ± deltoid-lanceolate, lateral lobes lanceolate. *Upper leaves dissected into long linear lobes.*

London rocket (*Sisymbrium irio*) plant.　　　　　　　　　　J. M. DiTomaso

Tumble mustard • London rocket • Hedge mustard

London rocket (*Sisymbrium irio*) flowering stem.
J. M. DiTomaso

London rocket (*Sisymbrium irio*) inflorescence.
J. M. DiTomaso

London rocket (*Sisymbrium irio*) rosette.
J. M. DiTomaso

London rocket (*Sisymbrium irio*) seedling.
J. M. DiTomaso

London rocket (*Sisymbrium irio*) seeds.
J. O'Brien

London rocket: Foliage nearly glabrous. Stems mostly branched near the base. Lower leaves to 15 cm long, terminal and lateral lobes ovate to lanceolate. Upper leaves ± oblong, mostly with 2 spreading lobes at the base (± hastate).

Hedge mustard: Foliage sparsely stiff-hairy. Lower leaves to 20 cm long, ter-

Hedge mustard (*Sisymbrium officinale*) plant. J. M. DiTomaso

Hedge mustard (*Sisymbrium officinale*) flowering stem. J. M. DiTomaso

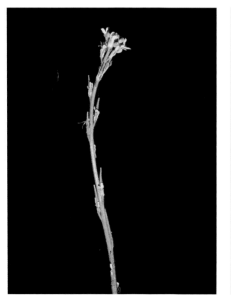
Hedge mustard (*Sisymbrium officinale*) inflorescence. J. M. DiTomaso

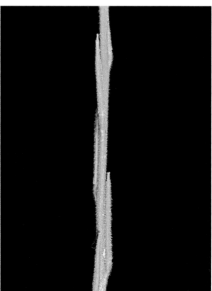
Hedge mustard (*Sisymbrium officinale*) fruit. J. M. DiTomaso

minal and lateral lobes ovate to lanceolate. *Upper leaves ± oblong, mostly with 2 spreading lobes at the base* (± hastate).

ROOTS AND UNDERGROUND STRUCTURES: All have a ± slender, often branched taproot, with fibrous lateral roots.

FLOWERS: Racemes lack bracts. Petals 4, *pale yellow,* rarely white, narrowed at the base (clawed). Sepals 4, separate, erect to slightly spreading.

Tumble mustard: April–September. Petals mostly 6–8 mm long. Sepals ± 4 mm long, *outer 2 with small, erect, whitish horns at the tip.*

London rocket: January–May. Petals mostly 2.5–4 mm long, barely longer than sepals.

Hedge mustard: April–September. Petals mostly 3–4 mm long. Sepals ± 2 mm long.

FRUITS AND SEEDS: Pods *open by 2 valves* to release numerous seeds, *have 3 veins on each face* that are easiest to see with magnification. Pods of most *Sisymbrium* species lack a conspicuous beak, but are tipped with a small persistent style that is about as wide as the pod. Seeds in 1 row per chamber.

Tumble mustard: Pods ± ascending, *linear, cylindrical,* 5–10 cm long, ± 1 mm

Hedge mustard (*Sisymbrium officinale*) basal leaves. J. M. DiTomaso

Hedge mustard (*Sisymbrium officinale*) seedling. J. M. DiTomaso

Hedge mustard (*Sisymbrium officinale*) seeds. J. O'Brien

wide, *lack a conspicuous beak* (style ± 1 mm long), ± rigid, straight, ± glabrous. Midvein on valves more prominent than lateral veins. Pod stalks ± ascending, mostly 4–10 mm long, *equal to or greater than pod width*. Seeds oblong, ± 1 mm long, 0.5 mm wide, brownish yellow to dull pale orange.

London rocket: Pods ± ascending, *linear, cylindrical*, 3–4 cm long, ± 1 mm wide, *lack a conspicuous beak* (style ± 0.5 mm long), *flexible*, straight or slightly curved, glabrous, *usually extend above flowers*. Pod stalks ascending, mostly

Mediterranean rocket (*Sisymbrium erysimoides*) inflorescence. J. M. DiTomaso

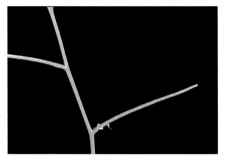

Mediterranean rocket (*Sisymbrium erysimoides*) fruit. J. M. DiTomaso

Mediterranean rocket (*Sisymbrium erysimoides*) basal leaves. J. M. DiTomaso

Mediterranean rocket (*Sisymbrium erysimoides*) seeds. J. O'Brien

Tall hedge mustard (*Sisymbrium loeselii*) plant.
J. M. DiTomaso

Tall hedge mustard (*Sisymbrium loeselii*) foliage.
J. M. DiTomaso

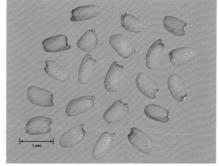

Tall hedge mustard (*Sisymbrium loeselii*) seeds.
J. O'Brien

Tall hedge mustard (*Sisymbrium loeselii*) inflorescence.
J. M. DiTomaso

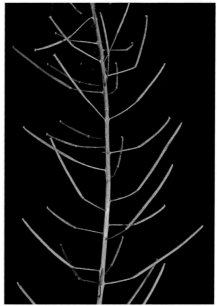

Tall hedge mustard (*Sisymbrium loeselii*) fruiting stem.
J. M. DiTomaso

5–10 mm long, *less than pod width*. Seeds oblong, less than 1 mm long, ± minutely papillate.

Hedge mustard: Pods ± *erect, most lay close to stems* (appressed), *narrowly awl-shaped, tip tapered into an inconspicuous beak* (style 1–2 mm long), ± *4-angled*, 0.8–1.5 cm long, ± 2 mm wide at the base, rigid, ± minutely pubescent. Style 1–2 mm long. Pod stalks erect (± appressed), mostly 2–4 mm long, thick, club-shaped. Septum of open pods distinctly corrugated. Seeds ± ovoid, variable, ± 1 mm long, dark brown.

POSTSENESCENCE CHARACTERISTICS: Dead stems with the remnants of pods can persist through fall.

Oriental mustard (*Sisymbrium orientale*) plant. J. M. DiTomaso

Oriental mustard (*Sisymbrium orientale*) flowers.
J. M. DiTomaso

Oriental mustard (*Sisymbrium orientale*) cauline leaf.
J. M. DiTomaso

Oriental mustard (*Sisymbrium orientale*) fruiting stem.
J. M. DiTomaso

HABITAT: Roadsides, fields, orchards, vineyards, gardens, agronomic and vegetable crops, landscaped areas, waste places, and other disturbed sites. Hedge mustard is often associated with shady places, particularly under oak trees.

DISTRIBUTION: Tumble mustard: Throughout California, to 2500 m. All contiguous states, except possibly Alabama.

London rocket: Central Valley, eastern San Francisco Bay region, Southwestern

Oriental mustard (*Sisymbrium orientale*) seeds. J. O'BRIEN

region, southwestern Great Basin, and White, Inyo, and desert mountains, to 800 m. Arizona, Nevada, New Mexico, Texas, Utah, and scattered in a few northeastern and southern states.

Hedge mustard: Throughout California, except deserts and Great Basin, to 2200 m. All contiguous states, except possibly Arizona.

PROPAGATION/PHENOLOGY: *Reproduce by seed.* Seeds fall near the parent plant or disperse to greater distances with water, mud, soil movement, animals, human activities, and as crop seed contaminants. Under favorable conditions, seed production can be high.

Tumble mustard: Seeds also disperse when senesced stems break off at ground level and tumble with the wind. Buried seeds survive up to about 10 years.

Hedge mustard: Germination requires light and the presence of nitrate. Field studies of buried seeds suggest that cool temperatures stimulate germination in seeds that have been dormant for an extended period, while newly matured seeds and seeds that have been dormant for a few months germinate best under warm conditions.

MANAGEMENT FAVORING/DISCOURAGING SURVIVAL: Manual removal or cultivation as needed to prevent seed production can control these species.

SIMILAR SPECIES: Refer to table 28 (p. 548) for a comparison of distinguishing features of *Sisymbrium* species. The following species inhabit roadsides, fields, and other disturbed places and are most similar to London rocket. All species have pods that lack a conspicuous beak, but are tipped with a small, persistent style about as wide as the pod.

Mediterranean rocket [*Sisymbrium erysimoides* Desf.] is an uncommon, *nearly glabrous annual* to 0.8 m tall, with *rigidly straight pods 2–4(–5) cm long*. Unlike London rocket, Mediterranean rocket has *pod stalks with a width that is about* equal to the pod width. In addition, *petals are 1.5–3 mm long*, and the *upper leaf lobe margins are often dentate*. Pods have a style ± 1 mm long. Mediterranean rocket inhabits fields and disturbed places in the South Coast, to 300 m. Native to the western Mediterranean region.

Tall hedge mustard [*Sisymbrium loeselii* L., synonym: small tumbleweed mustard] is a *sparsely long-hairy annual*, sometimes biennial, to 1.2 m tall. Unlike London rocket, tall hedge mustard has *stems that branch primarily in the upper portion, pods 2–3.5 cm long* that usually do *not* extend above the flowers, and *long hairs on the lower stems*. Pods have a style ± 0.5 mm long. Tall hedge mustard occurs in the Great Basin and Intermountain states, primarily from 1000–2000 m. It also occurs in Colorado, Idaho, Montana, Nevada, New Mexico, Oregon, Washington, Wyoming, South Carolina, and most central and northeastern states. Native to Europe.

Oriental mustard [*Sisymbrium orientale* L., synonyms: Indian hedge mustard; *Brassica kaber* (DC.) L. C. Wheeler var. *orientalis* (L.) Scoggan] is a *sparsely hairy annual* to about 0.3 m tall. It is distinguished by having *straight pods 3–10 cm long on stalks with a width that is about equal to the pod width* and *petals mostly 8–10 mm long*. Pods have a club-shaped style 1–3 mm long. Oriental mustard occurs throughout much of California to 1000 m, except the Northwestern region, Cascade Range, and Sierra Nevada. It also occurs in Arizona, Nevada, New Mexico, Oregon, South Carolina, Texas, and a few northeastern states. In Australia, a few Oriental mustard populations have developed tolerance to acetolactate synthase (ALS)-inhibiting herbicides. Seeds appear to survive for only a few years under field conditions. Native to Europe.

Table 28. *Sisymbrium* species

Sisymbrium spp.	Pod shape; length (cm)	Pod stalk length and width	Petal length (mm)	Other
S. altissimum tumble mustard	linear; 5–10	4–10 mm, ≥ pod width	6–8	upper leaves dissected into long, linear segments; outer 2 sepals with small erect whitish horns at tip
S. erysimoides Mediterranean rocket	linear; 2–4(–5)	2–4 mm, ≈ pod width	1.5–3	upper leaf lobes often dentate toothed
S. irio London rocket	linear; 3–4	5–10 mm, < pod width	2.5–4	foliage nearly glabrous; stems branched near base; pods usually extend above flowers
S. loeselii tall hedge mustard	linear; 2–3.5	10–20 mm, < pod width	6–8	stem bases sparsely long-hairy; stems branched mostly in upper portion; pods usually not extending above flowers
S. officinale hedge mustard	narrow awl-shaped; 0.8–1.5	2–4 mm, ≈ pod width	3–4	pods ± erect, appressed to stem
S. orientale Oriental mustard	linear; 3–10	2–6 mm, = pod width	8–10	style in fruit club-shaped, 1–3 mm long

Field pennycress [*Thlaspi arvense* L.][THLAR]

SYNONYMS: bastard cress; fanweed; Frenchweed; mithridate mustard; pennycress; stinkweed; *Teruncius arvensis* (L.) Lunnell

GENERAL INFORMATION: Winter or summer *annual* with flower stems to 0.5 m tall, white flowers, and nearly circular, flat pods. Plants exist as rosettes until flowering stems develop at maturity. Foliage has an unpleasant odor, especially when crushed. Field pennycress is most problematic in the agricultural regions of Canada and the northern United States. Foliage, flowers, and especially seeds contain sinigrin, a compound that releases the digestive tract irritant allylthiocyanate when ingested by cattle. Allythiocyanate is mostly released

Field pennycress (*Thlaspi arvense*) plant.   J. M. DiTomaso

Field pennycress

Field pennycress (*Thlaspi arvense*) flowering stem. J. M. DiTomaso

Field pennycress (*Thlaspi arvense*) fruit. J. M. DiTomaso

Field pennycress (*Thlaspi arvense*) seedlings. J. M. DiTomaso

Field pennycress (*Thlaspi arvense*) seeds. J. O'Brien

under neutral to alkaline pH conditions, such as that found in the rumen of cattle. Animals with acidic stomachs are less affected. Ingestion of field pennycress can also taint the milk of dairy cattle and cause photosensitization in cattle and pigs. Native to Europe.

SEEDLINGS: Foliage light green, with an unpleasant odor when crushed and a pungent flavor resembling mustard and garlic. Cotyledons oval, 4–10 mm long, 2–4 mm wide, glabrous, stalk below (hypocotyl) usually long. Leaves alternate, but first 2 may appear opposite. First leaf ± oval, slightly larger than cotyledons, usually glabrous, margin weakly wavy-toothed, teeth ± thickened at the tip, lower surface paler than upper surface, stalked. Subsequent few leaves resemble first leaf, except increasingly larger.

MATURE PLANT: Foliage *glabrous* or nearly glabrous. Hairs simple when present. Stems simple or branched. Leaves alternate. Lower leaves early-deciduous, absent or few at maturity, oblanceolate, mostly 2–6 cm long, *margins smooth to shallow-toothed*, on a short stalk. Upper leaves sessile, oblong-lanceolate, 2–5 cm long, *base lobed and clasping the stem*, margins smooth to shallow-toothed.

ROOTS AND UNDERGROUND STRUCTURES: Taproot slender, with fibrous lateral roots.

FLOWERS: April–August. Racemes initially dense, elongate with maturity. Petals 4, white, 3–4 mm long. Sepal 4, ± oval, white-margined, 1.5–2 mm long.

FRUITS AND SEEDS: Pods (silicles) *nearly circular to round-oblong, flattened perpendicular to the septum, 1–1.8 cm long, 1–1.5 cm wide*, 2-chambered, *margin broadly winged*, tip notch 1.5–2.5 mm deep. Stigma nearly sessile. Pod stalks slender, 7–15 mm long, spreading to curved upward. Seeds *2–8 per chamber*, ± oval, flattened, 1.5–2.3 mm long, concentrically striate, brown to dark brown.

POSTSENESCENCE CHARACTERISTICS: Stems with membranous old pods can persist through fall.

HABITAT: Roadsides, fields, agronomic crops, irrigation ditches, orchards, pastures, open places in cultivated forests, waste places, and other disturbed sites. Often associated with fertile agricultural soil.

DISTRIBUTION: Probably throughout California, but primarily found in the Northwestern region, Cascade Range, Modoc Plateau, northern Sacramento Valley, northern Sierra Nevada foothills, and South Coast, to 500 m. All contiguous states, except possibly Alabama.

PROPAGATION/PHENOLOGY: *Reproduces by seed*. Seeds fall near the parent plant, disperse short distances with wind, and disperse to greater distances with water, mud, soil movement, animals, agricultural operations, and as seed and feed contaminants. Fluctuating temperature and exposure to light stimulates germination in a large proportion of seeds. Scarification also promotes germination. Some buried seeds can survive up to about 20 years. Seed production is often prolific, and a large persistent seedbank can develop.

MANAGEMENT FAVORING/DISCOURAGING SURVIVAL: Manual removal or cultivation as needed to prevent seed production can control field pennycress.

SIMILAR SPECIES: The *distinctive pods, clasping upper leaves*, and *smooth to shallow-toothed margins of the lower leaves* make field pennycress unlikely to be confused with other weedy members of the mustard family.

Marijuana [*Cannabis sativa* L.][CNISA]

SYNONYMS: bhang; cannabis; dagga; ganja; grass; hashish; hemp; kif; marihuana; Mary Jane; pot; spleef; weed; *Cannabis indica* Lam.; *Cannabis sativa* L. ssp. *indica* (Lam.) E. Small & Cronq.; *Cannabis sativa* L. ssp. *sativa*

GENERAL INFORMATION: Highly variable, stout summer *annual* to 4 m tall, with *distinctively aromatic foliage* and *palmate-compound leaves*. Marijuana has been cultivated since prehistoric time. It is used for fiber, oil, birdseed, medicine, and as a psychoactive recreational drug. Plants contain numerous chemical constituents, including the cannabinoids, a large group of compounds of which 3 are primarily responsible for the neurological effects produced in humans and animals that ingest dried or fresh leaves and flower buds. Ingestion of large quantities of plant material can cause extreme intoxication, but is rarely fatal to animals or humans. Symptoms of marijuana intoxication can include drowsiness, depression, impaired coordination and reaction time, visual disturbance, increased appetite or vomiting, hypersensitivity to touch and sound, periods of arousal, hyperactivity, unusual behavior, tremors, disorientation, and dizziness. As a medicine, marijuana is used to treat glaucoma and to alleviate some of the side effects of cancer chemotherapy. The numerous cultivated varieties of marijuana vary widely in the proportion of cannabinoids they contain. Plants grown for fiber generally contain very small quantities of the neurotoxic principles. Handling plants for prolonged periods (as for fiber production) can cause dermatitis in sensitive individuals. Marijuana is a state-listed noxious weed in Illinois, Minnesota (primary), Missouri, North Dakota, Pennsylvania, and West Virginia. Its cultivation is illegal in most states, including California. Some tax-

Marijuana (*Cannibis sativa*) infestation.   J. M. DiTomaso

Marijuana (*Cannibis sativa*) plant.
J. M. DiTomaso

Marijuana (*Cannibis sativa*) foliage.
J. M. DiTomaso

Marijuana (*Cannibis sativa*) seedling.
J. M. DiTomaso

onomists currently recognize 2 subspecies, of which only *Cannabis sativa* L. ssp. *sativa* is reported to be naturalized in the United States. Probably native to Central Asia.

SEEDLINGS: Cotyledons spoon-shaped to oblanceolate, mostly 9–12 mm long, ± 3 mm wide, upper surface and margins covered with short stiff hairs. Leaves opposite. First leaf pair simple, lanceolate to elliptic, much larger than cotyledons, margin coarsely toothed, surfaces short stiff-hairy. Second leaf pair consists of palmate-compound leaves with 3 leaflets. Leaflets resemble the first leaf pair. Leaves of the third leaf pair usually consist of 5 leaflets.

MATURE PLANT: Foliage typically covered with *short, ± resinous hairs*. Stems erect or twining, opposite and/or alternately branched, often hollow, angled, ± ridged, inner bark layer fibrous. *Lower leaves opposite, upper leaves often alternate.* Leaves *palmate-compound*, on long stalks. Leaflets (3–)5–11(–16), *narrowly lanceolate to elliptic*, mostly 5–15 cm long, *margin coarsely toothed, tip long-tapered and slightly pinched-in* (acuminate), base tapered, lower surface paler than upper surface.

ROOTS AND UNDERGROUND STRUCTURES: Taproot sturdy, branched, can grow to 0.6 m deep.

FLOWERS: Flowers *inconspicuous*, greenish. *Male and female flowers develop*

Marijuana

Marijuana (*Cannibis sativa*) seeds.   J. O'Brien

separately, usually on different plants (dioecious). Flower clusters axillary. Male clusters open, ± panicle-like, usually more than 15 cm long. Female clusters spikelike, erect to spreading, 2 cm long or more. Male flowers consist of 5 drooping stamens and 5 separate perianth parts (sepals or petals). Female flowers consist of a superior ovary tipped with 2 slender feathery stigmas and surrounded by a tubular calyx. Each female flower is enclosed by a leafy bract. Wind-pollinated.

FRUITS AND SEEDS: Achenes ovoid, about 2–4 mm long, smooth, fine-veined with magnification, yellowish tan, gray, brown, or greenish gray or greenish brown, sometimes mottled with darker brown or black, typically remain enclosed by the bract.

HABITAT: Roadsides, waste places, and disturbed sites; sometimes illegally cultivated in rural and wildland areas. Tolerates a wide range of soil and climatic conditions.

DISTRIBUTION: Sporadic throughout much of California, except deserts and Great Basin, to about 600 m. Naturalized in most contiguous states, except possibly Arizona, Nevada, and Mississippi.

PROPAGATION/PHENOLOGY: *Reproduces by seed.* Seeds disperse primarily with human activities, but possibly also with water, mud, and animals.

MANAGEMENT FAVORING/DISCOURAGING SURVIVAL: Manual removal can control marijuana.

SIMILAR SPECIES: Marijuana is unlikely to be confused with other weedy species.

Mouseear chickweed [*Cerastium fontanum* Baumg. ssp. *vulgare* (Hartman) Greuter & Burdet][CERVU]
Sticky chickweed [*Cerastium glomeratum* Thuill.] [CERGL]
Common chickweed [*Stellaria media* (L.) Vill.][STEME]

SYNONYMS: Mouseear chickweed: big chickweed; big mouseear chickweed; common mouseear chickweed; large mouseear chickweed; mouse-ear; perennial mouse-ear chickweed; *Cerastium adsurgens* Greene; *Cerastium caespitosum* Gilib.; *Cerastium holosteoides* Fries var. *vulgare* (Hartm.) Hyl.; *Cerastium triviale* Link; *Cerastium vulgare* Hartm.; *Cerastium vulgatum* L. var. *hirsutum* Fries; others

Sticky chickweed: annual mouse-ear chickweed; *Cerastium acutatum* Suksd.; *Cerastium consanguineum* Wedd.; *Cerastium vulgatum* L.; a few others. *Cerastium viscosum* L. has been incorrectly applied in some references.

Common chickweed: starwort; starweed; satin flower; tongue-grass; winterweed; *Alsine media* L.; *Stellaria media* (L.) Vill. ssp. *media*; ssp. *pallida* (Dumort.) Aschers. & Graebn., and ssp. *neglecta* (Weihe) Murb.; others

GENERAL INFORMATION: The chickweeds have *opposite leaves without stipules at the base*, and small white flowers with *deeply 2-lobed petals*. The following species are generally minor weeds of lawns, yards, gardens, crop fields,

Mouseear chickweed (*Cerastium fontanum* ssp. *vulgare*) plant. J. M. DiTomaso

Mouseear chickweed • Sticky chickweed • Common chickweed

Mouseear chickweed (*Cerastium fontanum* ssp. *vulgare*) flower.　　J. M. DiTomaso

Mouseear chickweed (*Cerastium fontanum* ssp. *vulgare*) fruit capsule.　　J. M. DiTomaso

Mouseear chickweed (*Cerastium fontanum* ssp. *vulgare*) seedling.　　J. K. Clark

Mouseear chickweed (*Cerastium fontanum* ssp. *vulgare*) seeds.　　J. O'Brien

orchards, and elsewhere, but they can be more problematic in coastal regions.

Mouseear chickweed: Hairy *perennial* to 0.5 m tall, with *creeping matlike vegetative stems that root at the nodes,* erect flower stems, and *cylindrical capsules that have 10 teeth at the tip.* Plants often flower in the first season and are sometimes biennial or annual in disturbed places. Native to Europe.

Sticky chickweed: Ascending to erect winter or summer *annual* to 0.5 m tall, with *cylindrical capsules that have 10 teeth at the tip.* Foliage is *often glandular-hairy* and slightly sticky to touch. In most regions of California, sticky chickweed is primarily a winter annual. Native to Europe.

Common chickweed: Prostrate to erect winter or summer *annual,* rarely biennial or short-lived perennial, to 0.5 m tall, with sparsely hairy foliage and *ovoid capsules that open by 6 valves.* The *hairs on the stems are aligned in 2 longitudinal rows.* Plants are sometimes matlike with prostrate stems that root at the nodes. In most areas of California, common chickweed is primarily a winter annual. However, in foggy coastal areas it occurs year-round. Common chickweed can harbor several viruses and other pests that affect a variety of vegetable crops. The foliage is sometimes used medicinally and is consumed as a salad green. A variety of wildlife feed on the seeds and foliage. Common chickweed occurs nearly worldwide. Some botanists recognize 3 subspecies of

Mouseear chickweed • Sticky chickweed • Common chickweed

Sticky chickweed (*Cerastium glomeratum*) plant. J. M. DiTomaso

Sticky chickweed (*Cerastium glomeratum*) flowering stem. J. M. DiTomaso

Sticky chickweed (*Cerastium glomeratum*) flower. J. K. Clark

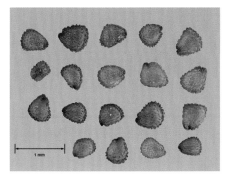

Sticky chickweed (*Cerastium glomeratum*) seeds. J. O'Brien

common chickweed in the United States and include **pale starwort** [*Stellaria pallida* (Dumort.) Crepin] (refer to pale starwort in the "Similar Species" section below). Of the 3 subspecies, only ssp. *media* and ssp. *pallida* (Dumort.)

Mouseear chickweed • Sticky chickweed • Common chickweed

Common chickweed (*Stellaria media*) plants.
J. M. DiTomaso

Common chickweed (*Stellaria media*) hairs along one side of stem.   J. M. DiTomaso

Common chickweed (*Stellaria media*) flowering stem.   J. M. DiTomaso

Common chickweed (*Stellaria media*) seedling.
J. K. Clark

Common chickweed (*Stellaria media*) seeds.
J. O'Brien

Aschers. & Graebn. occur in California, and only subspecies *media* is reported to occur in other western states. Native to southwestern Europe.

SEEDLINGS: Leaves opposite.

Mouseear chickweed: Cotyledons oval-elliptic to lanceolate, tip rounded to ± acute, 2–7 mm long, 0.5–2 mm wide, glabrous or with a few hairs on the upper surface, on a stalk 2–4 mm long. First and subsequent few leaf pairs oval- to oblong-elliptic, 2–10 mm long, long-hairy, tip abruptly pinched in to a small point (acuminate), on a stalk about equal in length to the blade.

Sticky chickweed: Cotyledons oval-elliptic, 1–3 mm long, tip ± acute, glabrous or with a few hairs, on a stalk ± 1 mm long. First and subsequent few leaf pairs oval-elliptic, mostly 2–4 mm long, long-hairy, tip ± abruptly pinched in to a small point, on a stalk ± 1 mm long.

Common chickweed: Cotyledons lanceolate-elliptic, 2–12 mm long, 0.25–1.5 mm wide, tip acute, glabrous, on a short, ± sparsely hairy stalk. First and subsequent few leaf pairs oval-elliptic, 2–12 mm long, tip ± abruptly pinched-in to a small point, glabrous, on a stalk 2–10 mm long with long hairs aligned in a longitudinal row on each side.

MATURE PLANT: Stems mostly forked. Leaves opposite.

Mouseear chickweed: Foliage covered with *nonglandular hairs*. Leaves sessile or on a short stalk. Lower leaves oblong-elliptic to oblanceolate or obovate, 2–25 mm long. Flower stem leaves ± oblong-elliptic, mostly 8–25 mm long, tip rounded to obtusely acute.

Sticky chickweed: Foliage moderate to densely covered with nonglandular and *glandular hairs, slightly sticky to touch*. Leaves sessile or on a short stalk, (ob)lanceolate to (ob)ovate, 5–35 mm long, tip rounded to obtusely acute.

Common chickweed: Stem internodes and leaf stalks have *hairs aligned in a longitudinal row on each side*. Leaves evenly spaced, broadly ovate, tip acute, *mostly glabrous* or with hairy margins at the base. Lower leaves mostly 3–25 mm long, sessile. Upper leaves to 45 mm long, on a stalk.

ROOTS AND UNDERGROUND STRUCTURES: Mouseear chickweed: Initial taproot slender. Roots from nodes fibrous, shallow.

Sticky chickweed, common chickweed: Taproot slender.

Little starwort (*Stellaria graminea*) flowering stem. R. Uva

Mouseear chickweed • Sticky chickweed • Common chickweed

Little starwort (*Stellaria graminea*) seeds.
J. O'Brien

Little starwort (*Stellaria graminea*) flower.   R. Uva   Pale starwort (*Stellaria pallida*) seeds.   J. O'Brien

FLOWERS: Cyme branches forked (dichotomous) and opposite. Petals 5, *white, deeply 2-lobed*. Sepals 5, separate, margins membranous.

Mouseear chickweed: March–August. Sepals 4–7 mm long, 0.2–0.6 mm wide, *nonglandular hairs shorter than the tip*, membranous margin 0.2–0.6 mm wide. Petals about 4–8 mm long, slightly less to slightly longer than sepals. Stamens 10(5).

Sticky chickweed: February–June. Sepals 3.5–5 mm long, *glandular and nonglandular hairs slightly longer than the tip*, membranous margin 0.1 mm wide or less. Petals slightly less to slightly longer than sepals or sometimes lacking. Stamens 10(5). Primarily self-pollinating.

Common chickweed: Mostly February–September, but flowering can occur year-round under favorable conditions. Sepals 3–5 mm long (to 6 mm long in fruit), membranous margin 0.1–0.4 mm wide, glabrous or sparsely covered with short, glandular and nonglandular hairs. Petals slightly shorter than sepals. Stamens 3–10.

FRUITS AND SEEDS: Capsules contain numerous seeds. Seed surface characteristics are visible with magnification.

Mouseear chickweed: Capsules *cylindrical*, 6.5–11 mm long, about 2 times longer than sepals, open at the tip by *10 teeth*. Seeds broadly triangular-ovoid, ± irregular, 0.5–1 mm long or wide, ± notched at one end, reddish brown, covered with minute, elongate papillae.

Sticky chickweed: Capsules *cylindrical*, 3.5–9 mm long, about 2 times longer than sepals, often *slightly curved*, open at the tip by *10 teeth*. Seeds broadly triangular-ovoid, ± irregular, ± 0.5 mm long or wide, ± shallow-notched at one end, reddish brown, covered with minute papillae.

Common chickweed: Capsules *ovoid*, about 4–8 mm long, slightly longer than sepals, open about *halfway or nearly to the base by 6 valves*. Seeds ± circular, wedge-shaped in cross-section, ± 1 mm in diameter, pale tan to pale reddish brown, with a short groove at one end and concentric circles of minute, low tubercles on each face.

POSTSENESCENCE CHARACTERISTICS: Dead plants generally degrade rapidly, especially in moist places.

HABITAT: Yards, turf, gardens, landscaped areas, agronomic and vegetable crops, orchards, vineyards, grasslands, managed forests, nurseries, roadsides, urban sites, and other disturbed places.

Mouseear chickweed: Also inhabits wet soil near marshes, moist woodlands.

Sticky chickweed: Also inhabits chaparral.

Common chickweed: Especially common in urban areas. Grows best on moist fertile soil with some shade.

Thymeleaf sandwort (*Arenaria serpyllifolia*) seeds.   J. O'BRIEN

DISTRIBUTION: Mouseear chickweed: Northwestern region, Cascade Range, and Sierra Nevada, to 2200 m. Probably also occurs in the South Coast Ranges, southern Central Coast, and southern South Coast. All contiguous states.

Sticky chickweed: Throughout most of California, except deserts, Great Basin, and higher regions of the southern Sierra Nevada, to 1600 m. New Mexico, Idaho, Montana, Oregon, Washington, southern states, most northeastern states, and many central states, except a few north-central states.

Common chickweed: Throughout most of California, except Mojave Desert, to 1300 m. All contiguous states.

PROPAGATION/PHENOLOGY: *All reproduce by seed.* Mouseear chickweed and sometimes common chickweed also reproduce vegetatively from creeping prostrate stems that root at the nodes. Most seeds fall within a short distance of the parent plant. Some seeds disperse to greater distances with wind, water, mud, soil movement, animals, human activities, and as commercial seed contaminants. Most seeds germinate in fall after the first rains, but some germination can occur in spring and at other times when conditions are favorable. For mouseear chickweed and common chickweed, newly matured seeds do not require light to germinate, but seeds that have been buried for some time require exposure to light to germinate. Buried seeds are long-lived and a persistent seedbank can develop. Buried mouseear chickweed seeds can survive up to 68 years. Under typical field conditions, seeds of common chickweed can survive at least 10 years and are reported to survive for up to 30 years or more.

MANAGEMENT FAVORING/DISCOURAGING SURVIVAL: Manual removal, cultivation, and close mowing can limit seed production and control the chickweeds.

SIMILAR SPECIES: Little starwort [*Stellaria graminea* L., synonyms: grass-leaved starwort; grasslike starwort; grassy starwort; *Alsine graminea* (L.) Britt.; *Stellaria graminea* L. var. *latifolia* Peterm.][STEGR] is an erect to spreading *perennial* to 0.6 m tall, with mostly *glabrous foliage, slender white rhizomes*, small white flowers with 2-lobed petals, and *ovoid capsules that open by 6 valves*. It is distinguished from other *Stellaria* species by having *sepals mostly 4–5.5 mm long,* some with *densely ciliate margins, petals equal to or slightly longer than the sepals,* and dark brown *seeds ± 1 mm in diameter with elongate tubercles.* Little starwort is an infrequent weed that inhabits turf, gardens, landscaped areas, and other disturbed places in the San Joaquin Valley, San Francisco Bay region, and South Coast, to 50 m. It also occurs in Colorado, Idaho, Montana, Nevada, Oregon, Washington, northeastern states, some central states, and a few southern states. Native to Europe.

Pale starwort [*Stellaria pallida* (Dumort.) Crepin, synonyms: lesser chickweed; *Alsine pallida* Dumort.; *Stellaria media* (L.) Thuill. ssp. *pallida* (Dumort.) Aschers. & Graebn.] is an erect to prostrate *annual* to 0.5 m tall, with *hairs aligned in a longitudinal row on each side of the stems and leaf stalks* and *ovoid capsules that open by 6 valves.* Unlike common chickweed, pale starwort has *sepals 2–3 mm long* (to 4 mm long in fruit), flowers that *lack petals,* and *seeds 0.7–0.8 mm in diameter.* Pale starwort inhabits grassland, oak woodlands, stream banks, and disturbed places in the Sacramento Valley, central Sierra Nevada foothills, Central Coast, southern Channel Islands, and Peninsular Ranges, to 450 m. It also occurs in Texas and some eastern and central states. Pale starwort is encountered less frequently than common chickweed. Native to southwestern Europe.

Thymeleaf sandwort [*Arenaria serpyllifolia* L. ssp. *serpyllifolia*; synonyms: *Arenaria leptoclados* (Reichenb.) Guss.; *Arenaria serpyllifolia* L. ssp. *leptoclados* (Reichenb.) Nyman; *Arenaria serpyllifolia* L. var. *tenuior* Mert. & Koch] [ARISE] is a ± tufted *annual* to 0.25 m tall, with *minutely pubescent foliage* and small white flowers. Unlike the chickweeds and little starwort, thymeleaf sandwort has *ovate leaves that are 3–5-veined from the base* and flowers with *entire petals* (not deeply 2-lobed). Thymeleaf sandwort is a minor weed that inhabits sand bars, dry woodland, and other disturbed places in the Northwestern region, central Sierra Nevada foothills, and South Coast, to 1800 m. It also occurs in all contiguous states, except possibly Arizona, Colorado, and North Dakota. Native to Europe.

Thymeleaf sandwort (*Arenaria serpyllifolia*) plants.   J. M. DiTomaso

Babysbreath [*Gypsophila paniculata* L. var. *paniculata*][GYPPA][CDFA list: B]

SYNONYMS: baby's breath; babysbreath gypsophila; bachelor's button; maiden's breath; tall gypsophyll

GENERAL INFORMATION: Herbaceous *perennial* to 1 m tall, with a thick rhizome, sparse foliage, and small white flowers on slender, openly branched stems. Babysbreath has escaped cultivation as an ornamental and become an invasive weed in some areas of California and elsewhere the Pacific Northwest. Besides California, babysbreath is a state-listed noxious weed in Washington (class C). Native to Eurasia.

SEEDLINGS: Cotyledons narrowly oblong, about 3–15 mm long, 1.5–5 mm wide, glabrous. First and subsequent leaves opposite, sessile, joined at the base, narrowly oblanceolate, mostly 10–25 mm long, 3–6 mm wide, arched downward, tips round, margins and lower midvein minutely ciliate.

MATURE PLANT: Stems slender, erect to spreading, swollen at the nodes (point of leaf attachment). Leaves *opposite*, narrowly lanceolate, *2–9 mm wide, glabrous*.

ROOTS AND UNDERGROUND STRUCTURES: Crown and rhizome *thick*, not creeping. Roots deep, can penetrate soil to a depth of 4 m. *New shoots grow from the crown and rhizome, but not the roots*. Severed crown and rhizome pieces can generate new plants. Root fragments do not develop new shoots.

FLOWERS: July–October. Inflorescences open, panicle-like, *lack bracts below the flowers*, branches slender and mostly forked. Flowers numerous, mostly

Babysbreath (*Gypsophila paniculata*) plant. J. M. DiTomaso

Babysbreath (*Gypsophila paniculata*) flowers.
J. M. DiTomaso

Babysbreath (*Gypsophila paniculata*) stem with leaves.
J. M. DiTomaso

6–8 mm wide. Petals 5, white, usually 2–4 mm long. Sepals 5, *fused and cuplike with 5 teeth, glabrous, 1.5–2.1 mm long, membranous between the teeth*. Styles 2. Stamens 10.

FRUITS AND SEEDS: Capsules spherical to oblong, slightly longer than the fused sepals, *open by 4 valves* to release 2–several seeds. Seeds, black, 1–2 mm long, disk- to bean-shaped, with longitudinal rows of minute tubercles.

POSTSENESCENCE CHARACTERISTICS: Aerial parts die back in cold winter climates, but the crown and roots survive.

HABITAT: Disturbed sites, especially on sandy soils and in open, grassy places. Tolerates considerable variation in temperature and moisture, including Mediterranean, semidesert, cold winter climates, and soils with a permafrost.

DISTRIBUTION: Cascade Range, northern Sierra Nevada, Great Basin, eastern North Coast Ranges, western South Coast Ranges, San Joaquin Valley, South Coast, and Mojave Desert, to 2000 m, but mostly above 1200 m. Colorado, Idaho, Montana, Nevada, Oregon, Washington, Utah, Wyoming, Florida, middle central and north-central states, and most northeastern states.

PROPAGATION/PHENOLOGY: *Reproduces by seed*. Most seeds fall near the parent, but plants can break off at ground level and tumble with the wind, dispersing seeds to greater distances. One plant can produce several thousand seeds. Newly matured seeds typically lack or have a short dormancy period. Most germination occurs from 10°–30°C. Seedlings emerge from soil depth of 0–5 cm, with 0–1 cm optimal.

MANAGEMENT FAVORING/DISCOURAGING SURVIVAL: Severing the crown from the roots by cultivation or hand-cutting to several inches below the soil surface usually kills babysbreath. Mowing or clipping does not appear to reduce plant vigor.

SIMILAR SPECIES: Annual babysbreath [*Gypsophila elegans* Bieb. var. *elegans*, synonyms: showy babysbreath] is an uncommon *annual or biennial* to 0.5 m tall, with a *slender taproot* and fine, openly branched stems with a few small flowers

Babysbreath (*Gypsophila paniculata*) seedling.
J. M. DiTomaso

Babysbreath (*Gypsophila paniculata*) seeds.
J. O'Brien

that have white to pink petals with purple veins. Annual babysbreath is further distinguished by having flowers on a *glabrous stalk* and with a *glabrous calyx 3–5 mm long*. Petals are about 2–5 times as long as the calyx. Leaves are mostly linear-lanceolate and *2–5 mm wide*. Annual babysbreath inhabits pine-fir forests in the northern Sierra Nevada (Placer County), mostly at ± 2000 m. It also occurs in Colorado, Kansas, North Carolina, Utah, many northeastern states, and a few north-central states. Native to southwestern Europe and southeastern Asia.

Glandular babysbreath [*Gypsophila scorzonerifolia* Ser., synonym: garden babysbreath] is an uncommon *perennial* to nearly 1 m tall, with a *thick rhizome* and small white to pale pink flowers. Unlike babysbreath, glandular babysbreath has *glandular-hairy flower stalks* and flowers with a *glandular-hairy calyx*. Petals are about 1.5–2 times the length of the calyx. In addition, glandular babysbreath has leaves that are linear-oblong to lanceolate and mostly *10–35 mm wide*, but sometimes as narrow as 3 mm wide. Glandular babysbreath inhabits disturbed sites on the eastern side of the southern Sierra Nevada (Inyo County), mostly at ± 1200 m. It also occurs in Colorado, Nebraska, Nevada, New York, Utah, Wyoming, and a few other states in the Great Lakes region. Glandular babysbreath is cultivated as a garden ornamental. Native to eastern Europe.

Dwarf pearlwort [*Sagina apetala* Ard.]
Corn spurry [*Spergula arvensis* L.][SPRAR]

SYNONYMS: Dwarf pearlwort: sticky pearlwort; *Alsinella ciliata* Greene; *Sagina apetala* Ard. var. *barbata* Fenzl; *Sagina ciliata* Heller

Corn spurry: corn spurrey; devil's-gut; pickpurse; sandweed; starwort; stickwort; yarr; *Spergula arvensis* L. ssp. *sativa* (Boenn.) Celak.; *Spergula arvensis* L. var. *sativa* (Boenn.) Mert. & Koch; *Spergula sativa* Boenn.; *Spergula ramosissima* Dougl. ex Torrey & A. Gray

GENERAL INFORMATION: Refer to table 29 (p. 575) for a comparison of distinguishing characteristics of pearlworts and associated species.

Dwarf pearlwort: Diminutive, ± matlike native *annual* to 8 cm tall, with *opposite linear leaves* and small, solitary flowers that *lack petals*. Dwarf pearlwort is a desirable component of the vegetation in natural areas, but it is sometimes a weed in gardens, landscaped areas, nursery containers, turf, orchards, and elsewhere.

Corn spurry: Erect to spreading summer or winter *annual* to about 0.5 m tall, with numerous *linear leaves in fascicles at the nodes* that *appear whorled* and panicle-like cymes of small white flowers. Corn spurry is more common in coastal regions. The foliage may contain oxalates. In New Zealand, dairy cattle that consumed a large quantity of corn spurry developed hypocalcemia. Native to Europe.

SEEDLINGS: Foliage slightly fleshy.

Dwarf pearlwort (*Sagina apetala*) plant.   J. M. DiTomaso

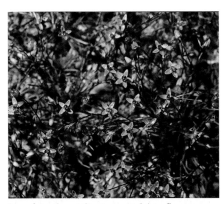
Dwarf pearlwort *(Sagina apetala)* in flower.
J. M. DiTomaso

Dwarf pearlwort *(Sagina apetala)* juvenile plants.
J. M. DiTomaso

Dwarf pearlwort: Cotyledons linear, mostly 3–10 mm long, tip rounded. Leaves opposite, linear, 3–12 mm long, usually have a few short glandular hairs, tip acute, often with a minute bristle.

Corn spurry: Cotyledons linear, ± 1 cm long, fused at the base, tip rounded. First few leaves opposite, linear, mostly 1–2 cm long, tip rounded, glabrous or with short glandular hairs. Later leaves resemble first leaves, but appear whorled and can be longer. Stipules small, membranous.

MATURE PLANT: Foliage slightly fleshy, flexible.

Dwarf pearlwort: Upper foliage ± covered with minute glandular hairs. Stems *threadlike*, erect to prostrate with upturned tips (decumbent). *Stipules lacking.* Leaves *opposite*, sessile, linear, 3–12 mm long, often tipped with a minute bristle. *Upper leaf margins minutely ciliate near the base.* Plants lack sterile (non-flowering) rosettes of leaves at the base.

Dwarf pearlwort *(Sagina apetala)* seeds.
J. O'Brien

Corn spurry (*Spergula arvensis*) plant.
J. M. DiTomaso

Corn spurry (*Spergula arvensis*) flowering stem.
J. M. DiTomaso

Corn spurry (*Spergula arvensis*) seedling.
C. Elmore

Corn spurry (*Spergula arvensis*) seeds. J. O'Brien

Corn spurry: Foliage glabrous to short glandular-hairy, sometimes sticky to touch. Stems erect to spreading. Main stems mostly branched at the base. *Stipules minute, membranous,* fused. Leaves sessile, linear, *appear whorled in fascicles of mostly 10–30 at the nodes,* 1–5 cm long, margin often rolled under.

ROOTS AND UNDERGROUND STRUCTURES: Dwarf pearlwort: Taproot threadlike to slender, shallow, with fine fibrous lateral roots.

Corn spurry: Taproot slender, with fibrous lateral roots.

FLOWERS: Sepals separate, alternate with styles. Numbers of flower parts can vary on the same plant.

Dwarf pearlwort: (March) April–July. Year-round in nurseries. Flowers terminal and axillary, *solitary* on *straight,* threadlike stalks 2–12 mm long. *Sepals 4(5),* ovate, 1.5–2 mm long, tip rounded. *Petals usually lacking,* occasionally 4, min-

Boccone's sandspurry (*Spergularia bocconii*) plant. J. M. DiTomaso

Boccone's sandspurry (*Spergularia bocconii*) flowers. J. M. DiTomaso

Boccone's sandspurry (*Spergularia bocconii*) seedling. J. M. DiTomaso

Boccone's sandspurry (*Spergularia bocconii*) seeds. J. O'Brien

ute. Stamens mostly 4. *Styles 4(5).*

Corn spurry: February–October. Flower clusters (cymes) terminal, *panicle-like.* Flower stalks (pedicels) slender, about 2–4 cm long or more, initially erect to ascending, in fruit spreading to reflexed. Petals 5, white, equal to or slightly longer than sepals. *Sepals 5,* ovate, 3–5 mm long, tip rounded or acute, margin broadly membranous. Stamens 10(5). *Styles 5.* Flowers open only in sunlight.

FRUITS AND SEEDS: Capsules ovoid, smooth, 1-chambered, contain numerous seeds, open by spreading or down-curved valves.

Dwarf pearlwort: Capsules mostly 1.5–3 mm long, open by *4(5) valves,* on a straight stalk, resemble straw-colored flowers when open. Seeds ± kidney-shaped to triangular, slightly flattened, 0.2–0.3 mm long, back grooved, covered with minute papillae, brown.

Corn spurry: Capsules mostly 4–6 mm long, open by *5 valves.* Seeds circular to oval lens-shaped, 1–2 mm in diameter. Margins narrow-winged with a small notch, wing pale. Surfaces dull black to gray, granular, sometimes speckled with pale papillae.

POSTSENESCENCE CHARACTERISTICS: Dead plants do not persist for long.

HABITAT: Dwarf pearlwort: Disturbed sites, roadsides, gardens, paths, walkways, landscaped areas, turf, orchards, vineyards, waste places, river bars, and stream banks. Often grows on sandy soil.

Corn spurry: Disturbed sites, roadsides, waste places, fields, crop fields, orchards, vineyards, gardens, landscaped areas, coniferous woodland, sand dunes, open slopes, coastal scrub, and vernal pool margins. Tolerates serpentine soil.

DISTRIBUTION: Dwarf pearlwort: Northwestern region, Cascade Range, northern and central Sierra Nevada, Central Valley, Central Coast, South Coast, and northern Channel Islands, to 700 m. Oregon, Washington, Colorado, Louisiana, Illinois, Maryland, and New Jersey.

Corn spurry: North Coast, western North Coast Ranges, San Francisco Bay region, Central Coast, Central Valley, western South Coast Ranges, South Coast, and possibly elsewhere, to 730 m. Colorado, Idaho, Montana, Oregon, Washington, Wyoming, and most central, eastern, and southern states.

PROPAGATION/PHENOLOGY: *Reproduce by seed.* Seeds fall near the parent plant and disperse to greater distances with water, mud, soil movement, agricultural operations, and landscape maintenance.

Dwarf pearlwort: Some seeds can germinate immediately following maturation, while the rest remain dormant.

Corn spurry: Germination requirements are complex and highly variable. Like many species in the pink family, buried corn spurry seeds can survive for many years. Some corn spurry seeds that were buried for ± 1700 years in the soil beneath an ancient church still germinated when placed under favorable conditions.

Red sandspurry (*Spergularia rubra*) weedy area. J. M. DiTomaso

Red sandspurry (*Spergularia rubra*) flowers.
J. M. DiTomaso

Red sandspurry (*Spergularia rubra*) plant.
J. M. DiTomaso

Red sandspurry (*Spergularia rubra*) seeds.
J. O'Brien

MANAGEMENT FAVORING/DISCOURAGING SURVIVAL: Manual removal or cultivation before fruits develop can help control both species. Cultivation followed by favorable moisture conditions can stimulate seed germination.

SIMILAR SPECIES: Refer to the table below for a comparison of distinguishing characteristics.

Boccone's sandspurry [*Spergularia bocconii* (Scheele) Foucaud ex Merino, synonyms: purple sandspurry; *Spergularia bocconei* (Scheele) Foucaud; *Alsine bocconei* Scheele; *Tissa luteola* Greene; *Alsine luteola* House] is a low to ± mat-like *annual*, with stems to 0.3 m tall or long and with small, *white to rosy pink flowers*. Foliage is slightly fleshy and glabrous to short glandular-hairy. Unlike dwarf pearlwort and corn spurry, Boccone's sandspurry has *0–2 leaves in the axils, triangular stipules* 1.5–4.5 mm long that are *membranous and dull white, 3 styles*, and capsules 2.5–5.5 mm long that *open by 3 valves*. In addition, the flowers have *8–10 stamens*. Seeds are *light brown, ± 0.5 mm long*, rounded kidney-shaped, barely flattened, and covered with minute papillae. Plants usually grow on sandy soil. Boccone's sandspurry inhabits roadsides, paths, ditch banks, crop fields, sand dunes, salt marsh areas, and other alkaline and saline places in the Central Valley, central Sierra Nevada foothills, Central Coast, western South Coast Ranges, South Coast, southern Channel Islands, and probably elsewhere, to 400 m. It also occurs in Oregon. Native to southwestern Europe and the Mediterranean region.

Red sandspurry [*Spergularia rubra* (L.) J. & C. Presl, synonyms: *Arenaria rubra* L.; *Tissa rubra* Britton][SPBRU] is a matlike *annual to short-lived perennial*, with stems to 0.3 m long, *fine foliage*, and small *pink flowers*. Unlike Boccone's sandspurry, red sandspurry has opposite leaves with *fascicles of 2–4 or more leaves per axil*, *stipules* 3.5–5 mm long that are ± *lanceolate* and *shiny white*, and *dark brown to reddish brown seeds*. Foliage is glabrous to short glandular-hairy and barely fleshy. Red sandspurry inhabits disturbed sites, roadsides, waste places, fields, meadows, open forest, and mud flats in the Northwestern region, Cascade Range, northern and central Sierra Nevada, Sacramento Valley, Central-western region, South Coast, San Gabriel Mountains, Peninsular Ranges, and possibly San Joaquin Valley, to 2400 m. Red sandspurry also occurs in most western states except Arizona, most northeastern states and adjacent central states, and a few southern states. Native to Europe.

Birdseye pearlwort [*Sagina procumbens* L., synonyms: arctic pearlwort; birdeye pearlwort; procumbent pearlwort; *Sagina procumbens* L. var. *compacta* Lange] is a ruderal native *perennial* to about 18 cm tall, with *glabrous foliage*, decumbent to ascending stems that are *not* threadlike, and *opposite linear leaves*. Flowers have *4(5) sepals* and *4(5) petals that are one quarter to one-half as long as the sepals*. Sometimes flowers lack petals. After the capsules open, the sepals are *spreading to ascending*. Unlike dwarf pearlwort, flowering plants of birdseye pearlwort usually have *rosettes of leaves at the base that do not develop flower stems*, and fruit stalks usually *curve downward*. Birdseye pearlwort inhabits roadsides, waste places, nursery containers, sidewalk cracks, and other disturbed places, usually on wet gravelly to sandy soil, in the North Coast and Central Coast, to 15 m. It also occurs in Colorado, Idaho, Montana, Oregon, Utah, Washington, Texas, and most eastern and adjacent central states.

Four-leaved polycarp [*Polycarpum tetraphyllum* (L.) L., synonyms: four-leaved allseed; fourleaf manyseed; *Mollugo tetraphylla* L.; *Polycarpum tetraphyllum*

Birdseye pearlwort (*Sagina procumbens*) plant. J. M. DiTomaso

Dwarf pearlwort • Corn spurry

Four-leaved polycarp (*Polycarpum tetraphyllum*) flowering stem. J. M. DiTomaso

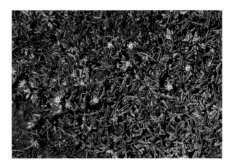

Salt sandspurry (*Spergularia marina*) is a native, nonweedy plant that resembles non-native weedy species. J. M. DiTomaso

Sticky sandspurry (*Spergularia marcrotheca*) is a showy native, nonweedy plant that resembles non-native weedy species. J. M. DiTomaso

Four-leaved polycarp (*Polycarpum tetraphyllum*) plant. J. M. DiTomaso

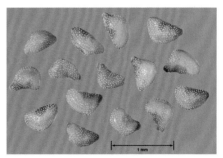

Four-leaved polycarp (*Polycarpum tetraphyllum*) seeds. J. O'Brien

(L.) L. ssp. *tetraphyllum*] is a matlike or tufted *annual, biennial, or perennial* to 18 cm tall, with glabrous foliage and *nearly sessile, obovate leaves* 4–12 mm long that are *opposite*, but often *appear to be in whorls of 4* in conjunction with the stipules. Plants are usually annual in California. Four-leaved polycarp is further distinguished by having *papery stipules that are lanceolate to broadly triangular and 1.8–2.8 mm long*. Flowers have 5 separate sepals ± 2 mm long, with white, papery margins and that are *keeled along the midline* and *tipped with a bristle* ± *0.5 mm long*, 5 linear to elliptic petals 0.5–1 mm long, 3–5 stamens that are fused at the base, and *1 style with 3 branches*. Capsules are ovoid to nearly round and open by 3 valves with margins that roll inward after opening. Seeds are obliquely triangular, ± 0.5 mm long, and minutely granular. Four-leaved polycarp inhabits roadsides, shady waste places, and other disturbed sites in the western North Coast Ranges, northern Sierra Nevada foothills, Sacramento Valley, Central Coast, South Coast, San Gabriel Mountains, and possibly elsewhere, to 450 m. It also occurs in Pennsylvania, Texas, Alabama, Georgia, Florida, and South Carolina. Native to southern Europe.

There are a few other non-native *Spergularia* species that are less commonly encountered than the species previously described. Refer to the appendix, **Nonnative Naturalized Plants** (p. 1601, vol. 2), for more information.

Table 29. Pearlworts, spurry, sandspurrys, and polycarp (*Sagina, Spergula, Spergularia,* and *Polycarpum* spp.)

Species	Leaf arrangement	Leaf shape	Stipules	Style number	Fruit valves	Other
Polycarpum tetraphyllum four-leaved polycarp	opposite or appear in whorls of 4 (with the stipules)	obovate	yes	1, with 3 branches	3	stipules papery, broadly triangular to lanceolate; 1.8–2.8 mm long; sepals 5, keeled, tipped with a bristle ± 0.5 mm long
Sagina apetala dwarf pearlwort	opposite	linear	no	4(5)	4–5	annual; upper leaves minutely ciliate near base; flower stalks and sepals often glandular; petals 0
Sagina procumbens birdseye pearlwort	opposite	linear	no	4(5)	4–5	glabrous perennial with sterile rosettes at base; fruit stalks curve downward
Spergula arvensis corn spurry	appear whorled with 10–30 leaves in fascicles at the nodes	linear	yes	5	5	flowers in terminal clusters (cymes); foliage glabrous or glandular; stamens 10(5)
Spergularia bocconii Boccone's sandspurry	opposite, often with fascicles of smaller leaves in the axils	linear	yes	3	3	leaves mostly 0–2 per fascicle; stipules dull white; seeds light brown
Spergularia rubra red sandspurry	opposite, often with fascicles of smaller leaves in the axils	linear	yes	3	3	leaves mostly 2–4 (sometimes more) per fascicle; stipules shiny white; seeds reddish to dark-brown

Bouncingbet [*Saponaria officinalis* L.][SAWOF] [Cal-IPC: Limited]

SYNONYMS: Fuller's herb; hedge pink; scourwort; soapwort; sweet betty; *Lychnis saponaria* Jessen

GENERAL INFORMATION: Erect *perennial* to 1 m tall, with *vigorous rhizomes* and showy pink to white flowers. Colonies often develop from the rhizomes. Bouncingbet is cultivated as an ornamental, but it has escaped cultivation and become weedy in various regions throughout the United States. In California, bouncingbet is especially problematic in the meadows and riparian areas of Mono Lake basin. It is a state-listed noxious weed in Colorado. Bouncingbet is potentially *toxic* to animals and humans if ingested in quantity. The seeds and, to a lesser degree, the foliage, contain large amounts of a saponin, a soaplike compound that foams in water. However, animals typically avoid eating the plant because of its unpleasant flavor. In Europe, bouncingbet has been cultivated as a garden and medicinal plant for hundreds of years. Various cultivars are available in the nursery trade. Native to southern Europe.

SEEDLINGS: Cotyledons narrowly elliptic-lanceolate, about 9–12 mm long, 1–3 mm wide, glabrous, bases fused, weakly sheathing. First and subsequent

Bouncingbet (*Saponaria officinalis*) plant. J. M. DiTomaso

Bouncingbet

Bouncingbet (*Saponaria officinalis*) seedling.
J. M. DiTomaso

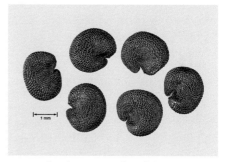

Bouncingbet (*Saponaria officinalis*) flowering stem. J. M. DiTomaso

Bouncingbet (*Saponaria officinalis*) seeds. J. O'Brien

leaves opposite, glabrous, elliptic to oblong, about 7–12 mm long, 2–5 mm wide, base tapered to a short stalk 2–4 mm long. Stalks fused at base, ± sheathing stem.

MATURE PLANT: Foliage glabrous. Stems leafy, nodes swollen, simple or oppositely branched in the upper portions. Leaves opposite, ovate to elliptic-lanceolate, mostly 3–10 cm long, with 3 main veins from the base.

ROOTS AND UNDERGROUND STRUCTURES: Rhizomes short, thick, white, can spread rapidly.

FLOWERS: June–September. *Flowers densely clustered at stem tips,* nearly sessile, showy, 8–12 mm wide. *Calyx tube cylindrical,* usually 15–22 mm long, 4–8 mm wide, with 5 lanceolate lobes about 2–5 mm long. Petals 5 (sometimes doubled), pink to white, about 3.5–4 cm long including the claw within the calyx tube, spreading to reflexed, tip sometimes indented, with *2 appendages at the junction of the petal blade and claw.* Styles 2. *Stamens 10,* fused with petals to the ovary. Ovary superior.

FRUITS AND SEEDS: Capsules oblong-cylindrical, mostly 15–22 mm long, surrounded by a persistent calyx tube, open at the apex by 4 short valves, contain numerous seeds. Seeds kidney-shaped, flattened, usually 1.5–2 mm long, covered with minute shiny tubercles, dark brown to black.

POSTSENESCENCE CHARACTERISTICS: Stems typically die in winter and regenerate from the rhizomes in spring. Dead stems are oppositely branched, often 4-angled, and have conspicuously swollen nodes. Sometimes a few capsules remain intact. Stems usually do not persist for long periods.

HABITAT: Riparian areas, oak woodlands, grasslands, roadsides, waste areas, and other disturbed places. Tolerates some shade and winter temperatures as low as −40° C or less.

DISTRIBUTION: Northwestern region, Cascade Range, Modoc Plateau, northern Sierra Nevada foothills, Great Basin area east of the central Sierra Nevada, San Francisco Bay region, western South Coast Ranges, South Coast, and Peninsular Ranges, to 2000 m. All contiguous states.

PROPAGATION/PHENOLOGY: *Reproduces by seed* and *vegetatively from rhizomes*. Seeds disperse with water, soil movement, and human activities. The biology of bouncingbet is poorly understood. Under experimental conditions, newly matured seeds and seeds stored for a few years required a cool, moist period of about 1 week before germination could occur. Seeds then germinated readily at 20°C.

SIMILAR SPECIES: Bouncingbet is readily distinguished from weedy catchfly and campion species (*Silene*, *Lychnis*) by having *2 styles* and *4-valved capsules*, and from cow-cockle species (*Vaccaria*) by having a *perennial life cycle, dense flower clusters, cylindrical flower tubes* that are *round in cross-section*, and petals with *2 appendages at the junction of the blade and claw*.

English catchfly [*Silene gallica* L.][SILGA]

SYNONYMS: common catchfly; French catchfly; small-flowered catchfly; windmill pink; *Silene anglica* L.

GENERAL INFORMATION: Erect to decumbent *annual* to 0.5 m tall, with bristly-hairy foliage, *opposite leaves*, and white to pink or lavender flowers that have a *glandular-hairy calyx*. Plants exist as rosettes until flower stems develop at maturity. English catchfly is the most common of the non-native *Silene* species, but is generally not a significant weed problem. Native to Europe.

SEEDLINGS: Cotyledons ovate, 3–8 mm long, 2–5 mm wide, glabrous, light green. Stalk below (hypocotyl) with a few minute transparent down-curved barblike hairs (20× magnification). Leaves opposite. First leaf pair oblanceolate, about 15–30 mm long, 6–12 mm wide, light green. Upper surface and margins have minute barblike hairs.

MATURE PLANT: Foliage covered with *short, bristly hairs*. Stems simple or branched, erect to decumbent, sometimes also glandular. Leaves *opposite*, sparse, sessile, 1–3.5 cm long, to ± 0.5 cm wide. Lower leaves oblanceolate to oblong, tip rounded with a minute nipplelike tip (mucro). *Upper leaves lanceolate to narrowly oblong*.

English catchfly (*Silene gallica*) seedling.
J. M. DiTomaso

English catchfly (*Silene gallica*) flowering stem.
J. M. DiTomaso

English catchfly (*Silene gallica*) seeds.   J. O'Brien

English catchfly

ROOTS AND UNDERGROUND STRUCTURES: Taproot slender, with fibrous lateral roots.

FLOWERS: February–June. *Racemes often one-sided.* Flowers *1 per node, sessile or on stalks to 5 mm long,* ascending to erect. Calyx cylindrical in flower, expanded and ± urn-shaped in fruit, 6–10 mm long, *10-veined,* with 5 lanceolate teeth ± 1 mm long at the top and covered with *bristly hairs and glandular hairs.* Petals 5, white to pale pink or lavender, bottom portion (enclosed within the calyx) narrow (clawed) with *2 linear appendages ± 1 mm long, blade 2–5 mm long, entire or slightly notched,* usually slightly twisted. Styles 3. Stamens 10. *Stamens and styles included within the narrow portion of the flower.*

FRUITS AND SEEDS: *Capsules ovoid,* mostly 6–8 mm long, enclosed by the expanded ± urn-shaped calyx. Seeds numerous per capsule, kidney-shaped, slightly flattened, ± 1 mm long, gray-black, covered with concentric rows of tubercles and ridges. Capsule teeth 6.

POSTSENESCENCE CHARACTERISTICS: Dead stems with stiff hairs and remnants of capsules and sepal tubes still attached can remain intact for a period.

HABITAT: Roadsides, fields, orchards, vineyards, landscaped areas, agronomic crops, gardens, waste places, and other disturbed places. Often grows on dry sandy sites.

White campion *(Silene latifolia* ssp. *alba)* flower.
J. M. DiTomaso

White campion *(Silene latifolia* ssp. *alba)* flowering stems.
J. M. DiTomaso

White campion *(Silene latifolia* ssp. *alba)* seeds.
J. O'Brien

English catchfly

Nightflowering catchfly (*Silene noctiflora*) flower. J. M. DiTomaso

Nightflowering catchfly (*Silene noctiflora*) fruit with distinctive vein pattern. J. M. DiTomaso

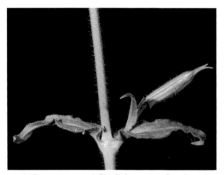

Nightflowering catchfly (*Silene noctiflora*) node and leaves. J. M. DiTomaso

Nightflowering catchfly (*Silene noctiflora*) seedling. J. M. DiTomaso

Nightflowering catchfly (*Silene noctiflora*) seeds. J. O'Brien

English catchfly

DISTRIBUTION: Throughout California, except deserts and Great Basin, to 1000 m. Arizona, Colorado, Oregon, Washington, and Texas. Scattered in the eastern and southern states.

PROPAGATION/PHENOLOGY: *Reproduces by seed.* Most seeds probably fall near the parent plant, but some seeds most likely disperse to greater distances with water, mud, soil movement, animals, and human activities.

MANAGEMENT FAVORING/DISCOURAGING SURVIVAL: Manual removal, cultivation, or mowing before seed is produced can control English catchfly.

SIMILAR SPECIES: The following species are less common than English catchfly. They inhabit fields, roadsides, and other disturbed places. Refer to table 30 (p. 585) for a comparison of important distinguishing characteristics of catchflys and campion. All are native to Europe.

White campion [*Silene latifolia* Poiret ssp. *alba* (Miller) Greuter & Burdet, synonyms: bladder campion; white cockle; *Silene alba* (Mill.) E. H. L. Krause; *Silene pratensis* (Rafn.) Godrn. & Gren.; *Lychnis alba* Mill.; *Melandrium album* (Mill.) Garcke; others][MELAL] is a *biennial or short-lived perennial* to 1 m tall, with coarse foliage that is covered with bristly hairs and some glandular hairs and showy white flowers in clusters of 2–3 on *stalks more than 5 mm long*. Petals are *deeply 2-lobed at the tip*. Flowers appear bisexual, but *male and female flowers develop on separate plants* (dioecious). White campion is further distinguished by having *calyx tubes 12–20 mm long* including lobes 3–6 mm long. Male flowers *lack styles* and have calyx tubes with ± *10 veins*. Female flowers *lack stamens* and have *5 protruding styles* and ± *20-veined calyx tubes that are inflated in fruit*. *Capsules open by 10 valves, with 10 teeth at the top of the capsule.* Biennial plants have a taproot, while perennial plants often develop a ± thick rootstock. White campion inhabits fields and roadsides in the San Francisco Bay region, Central Valley, North Coast, western North Coast Ranges, southern Sierra Nevada foothills, and probably elsewhere, to 1000 m. It occurs in most contiguous states, except Arizona, New Mexico, Texas, and a few other southern states. White campion is a state-listed noxious weed in Minnesota (secondary) and Washington (class C).

Nightflowering catchfly [*Silene noctiflora* L., synonyms: nightflowering silene; sticky cockle; *Melandrium noctiflorum* (L.) Fries][MELNO] is a coarse *annual* to about 0.8 m tall that is often confused with white campion. Like white campion, nightflowering catchfly has foliage that is covered with bristly hairs and glandular hairs, some nodes with 2–3 flowers, some flowers on stalks *more than 5 mm long*, and white to yellowish petals that are *deeply 2-lobed at the tip*. Nightflowering catchfly is distinguished by having *bisexual flowers with 3 styles*, calyx tubes with *netlike venation above the middle* and *long-tapered lobes* that are mostly *6–13 mm long*. Calyx tubes are ± *10-veined* and inflated in fruit, and flowers open at night. Capsules with 6 teeth. Nightflowering catchfly inhabits fields and open, disturbed sites in the Cascade Range, Sierra Nevada foothills, and South Coast, to 1900 m, and is expected to expand its range. It occurs in most contiguous states, except Arizona, Nevada, Texas, and a few other scattered southern states. Nightflowering

English catchfly

Bladder campion (*Silene vulgaris*) flowers, node, and leaves.   J. M. DiTomaso

Bladder campion (*Silene vulgaris*) fruit with vein pattern.   J. M. DiTomaso

Bladder campion (*Silene vulgaris*) stem and leaves.   J. M. DiTomaso

Bladder campion (*Silene vulgaris*) seedling.
  J. M. DiTomaso

Bladder campion (*Silene vulgaris*) seeds.
  J. O'Brien

catchfly is a common agronomic weed in the northern United States and Canada. It is a state-listed noxious weed in Minnesota (secondary).

Bladder campion [*Silene vulgaris* (Moench) Garcke, synonyms: cowbell; maidenstears; maiden's tears; rattleweed; *Oberna commutata* (Guss.) Ikonn.; *Silene cucubalus* Wibel; *Silene inflata* (Salisb.) Sm.; *Silene latifolia* (Mill.) Britt. & Rend.][SILVU] is a *glabrous perennial* to 0.8 m tall, with *short rhizomes*, white flowers that have *deeply 2-lobed petals*, and inflated calyx tubes that are usually *glabrous* and become *papery in fruit*. Unlike white campion, bladder campion has *bisexual flowers* and *sepal tubes 7–10 mm long*, including *lobes 1–3 mm long*. In addition, calyx tubes have 10–15 faint veins, and the stamens and 3 styles protrude from the flower. Petals lack appendages or have 2 minute appendages at the base. Bladder campion inhabits fields and open places in the Cascade Range, Central Valley, San Francisco Bay region, and South Coast, to 1200 m. It probably also occurs in the North Coast Ranges and is expected to expand its range in California. It also occurs in Colorado, Idaho, Montana, New Mexico, Nevada, Oregon, Washington, Wyoming, central states, northeastern states, and a few southern states.

Cone catchfly [*Silene conoidea* L., synonyms: weed silene; *Conosilene conica* (L.) Fourr. ssp. *conoidea* (L.) A. Löve & Kjellq.; *Pleconax conoidea* (L.) Sourkova][SILCD] is an uncommon *annual* to about 0.8 m tall, with the lower foliage minutely hairy, the *upper foliage minutely glandular-hairy*, and white, pink, reddish, or purplish flowers. It is distinguished by having *calyx tubes mostly 18–26 mm long with ± 30 veins*. In addition, stamens and *3 styles* protrude beyond the flower tube, and petals are entire, toothed, or notched at the tip and have *2 bilobed appendages 2–5 mm long at the base*. Capsules are ovoid to cone-shaped and open by *3 valves*. Cone catchfly inhabits disturbed, open sites in the northern Sierra Nevada foothills, South Coast, and Sonoran Desert, to 500 m. It also occurs in Colorado, Delaware, Idaho, Maryland, Missouri, Montana, Nevada, Oregon, and Washington.

Hairy catchfly [*Silene dichotoma* Ehrh., synonym: dichotoma silene][SILDI] is an *annual* to 0.8 m tall, with *densely short-hairy foliage* and *slightly inflated sepal*

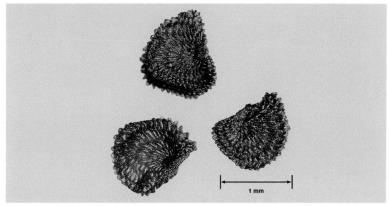

Cone catchfly (*Silene conoidea*) seeds. J. O'BRIEN

tubes that have ± 10 veins. Hairy catchfly resembles English catchfly, but unlike English catchfly, hairy catchfly has calyx tubes with *stiff hairs on the veins* and that *lack glandular hairs*, flowers with white to reddish petals that are *deeply 2-lobed*, petal blades mostly *5–12 mm long*, and *stamens and 3 styles that protrude from the flower*. In addition, the *appendages at the base of the petals are about as wide as they are long*. Hairy catchfly inhabits fields, roadsides, and other disturbed places in the San Francisco Bay region, Cascade Range, and South Coast, to 1000 m. It also occurs in Colorado, Idaho, Montana, Oregon, Washington, Wyoming, northeastern states, many north-central states, and a few southern states.

Table 30. Catchflys and campions (*Silene* spp.)

Species	Life cycle	Calyx tube veins	Calyx tube length (mm)	Calyx tube surface	Petal tip; color	Flower stalk length (mm)	Other
Silene conoidea cone catchfly	annual	~ 30	18–26	minutely hairy, ± glandular	smooth to finely toothed or notched; white, pink, reddish or purplish	10–30	petal appendages at base 2–5 mm long
Silene dichotoma hairy catchfly	annual	10	9–14	short, stiff-hairy	deeply 2-lobed; white to pink or red	1–3	stamens and styles extend beyond flower; styles 3
Silene gallica English catchfly	annual	10	6–10	glandular-hairy	smooth to notched; white, pale pink, or lavender	2–5	stamens and styles included within the flower; styles 3
Silene latifolia ssp. *alba* white campion	biennial, perennial	male flowers 10; female flowers ± 20	12–20	short-hairy	deeply 2-lobed; white	some > 5	dioecious; styles 5 (female flowers); capsules 10-toothed; calyx lobes mostly 3–6 mm long
Silene noctiflora nightflowering catchfly	annual	10	14–22	glandular-hairy	deeply 2-lobed; white or pale pink	3–12, some > 5	styles 3 (all flowers); capsules 6-toothed; calyx lobes mostly 6–13 mm; calyx tube with veins netlike in upper half
Silene vulgaris bladder campion	perennial with short rhizomes	10–15, faint	7–10	glabrous or nearly so	deeply 2-lobed; white	10–30	foliage glabrous; calyx tubes inflated and papery in fruit

Red orach [*Atriplex rosea* L.][ATXRO]
Australian saltbush [*Atriplex semibaccata* R. Br.] [ATXSE][Cal-IPC: Moderate]

SYNONYMS: Red orach: red orache; redscale; tumbling orach; tumbling oracle; tumbling saltweed; *Atriplex spatiosa* A. Nels.

Australian saltbush: berry saltbush; creeping saltbush; scrambling berry saltbush; *Atriplex denticulata* Moq. in DC.; *Atriplex flagellaris* Wooton & Standl.

GENERAL INFORMATION: Red orach: Erect summer *annual* to 1.5 m tall, with scurfy, gray-green to reddish, *triangular-ovate leaves* and small dense clusters of *firm, ± triangular fruits with toothed margins*. Plants shed large quantities of pollen in late summer and can cause allergy symptoms in sensitive individuals. Red orach can harbor the beet leafhopper, which is a vector of beet curly top virus in western states. Besides sugarbeets, the virus affects a variety of crops, including tomatoes, peppers, spinach, cucurbits, beans, and alfalfa. Native to Eurasia.

Australian saltbush: Spreading, *semi-shrubby perennial* to 0.4 m tall and about 1.25 m in diameter, with scurfy, gray-green, *oblong leaves* and small, *fleshy, ± diamond-shaped fruits that turn reddish at maturity*. This species has escaped cultivation and become regionally invasive in coastal grassland, scrub, and on the higher ground of salt marshes, especially in southern California. Australian saltbush was first cultivated in the early 1920s in Tulare County as a livestock forage plant for alkaline and saline areas. More recently it has been promoted as a fire-resistant

Red orach (*Atriplex rosea*) plant. J. M. DiTomaso

plant tolerant of drought, salt and alkali conditions, to be used for groundcover or erosion control and as a component of reclamation vegetation for the restoration of mined sites in the southwestern states. Australian saltbush is susceptible to the

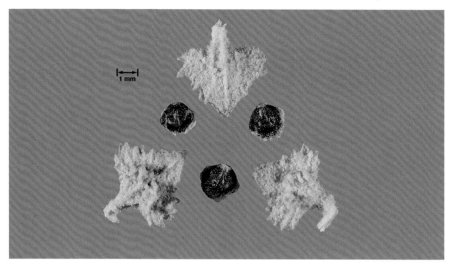

Red orach (*Atriplex rosea*) seeds, with and without surrounding bracts. J. O'Brien

Red orach (*Atriplex rosea*) fruit. J. M. DiTomaso

Red orach (*Atriplex rosea*) flowering stem. J. M. DiTomaso

Red orach (*Atriplex rosea*) seedling. J. M. DiTomaso

Australian saltbush (*Atriplex semibaccata*) plant. J. M. DiTomaso

beet western yellows virus, which can affect a variety of crop plants and is transmitted by aphids. Native to Australia.

SEEDLINGS: First few leaves opposite.

Red orach: Cotyledons linear-lanceolate, about 7–11 mm long, ± 2 mm wide, glabrous, upper surface dull green, lower surface and stalk below (hypocotyl) tinged magenta. First few leaves ovate, mostly 5–12 mm long, 2–5 mm wide, margins weakly wavy-toothed, surfaces evenly covered with minute salt-excreting glands that resemble glistening balls.

Australian saltbush: Cotyledons resemble those of red orach. First few leaves narrow-oblong, ± finely scurfy, nearly the same size as the cotyledons.

MATURE PLANT: Red orach: Branches ascending. Stems pale, scurfy to glabrous. Leaves alternate, triangular-ovate, mostly 1–6 cm long, 0.5–3 cm wide, firm, base ± abruptly tapered to a stalk, margin coarsely wavy-toothed, *surfaces densely white-scurfy, especially the lower surface,* upper surface pale greenish gray to reddish, persistent.

Australian saltbush: Stems usually *weakly woody at the base, spreading or decumbent to ascending,* branched, initially white-scurfy, later ± glabrous. Leaves alternate, mostly short-stalked, oblong to narrow-elliptic, 0.8–3(–5) cm long, 0.3–2 cm wide, tip acute to rounded, margin smooth or wavy-toothed, gray-green, surfaces ± scurfy, especially the lower surface.

ROOTS AND UNDERGROUND STRUCTURES: Both species have a taproot with fibrous lateral roots.

Australian saltbush: Crown and upper root ± woody. Root system typically deep.

Australian saltbush (*Atriplex semibaccata*) flowering stems. J. M. DiTomaso

Australian saltbush (*Atriplex semibaccata*) fruiting stem. J. M. DiTomaso

Australian saltbush (*Atriplex semibaccata*) seedling. J. K. Clark

Australian saltbush (*Atriplex semibaccata*) seeds with and without surrounding bracts.
J. O'Brien

FLOWERS: Male and female flowers develop on the same plant (monoecious). Male flowers consist of a 3–5-lobed calyx and 3–5 stamens. *Female flowers consist of 2 bracts* that enclose the ovary and 2 stigmas. The fruiting bract characteristics described in the fruits and seeds section are important for species identification.

Red orach: July–October. Male and female flowers develop in mixed clusters in the leaf axils and on terminal spikes. Primarily wind-pollinated.

Australian saltbush: April–December. Male and female flowers develop in separate small clusters. Male flower clusters develop at stem tips. One or a few female flowers develop in leaf axils below the male flower clusters.

FRUITS AND SEEDS: Seeds 1 per fruit, vertical in most *Atriplex* species, except where specified.

Red orach: Bracts usually *triangular*, sometimes ± diamond-shaped, *firm*, 4–8 mm long, *fused ± to the middle, margins acute-toothed* (dentate), *surface with a few toothlike projections and densely white-scurfy*. Seeds circular, flattened, 2–2.5 mm in diameter, mostly dull brown.

Garden orach (*Atriplex hortensis*) flowering stems with leaves. J. M. DiTomaso

Australian saltbush: Bracts *ovate to diamond-shaped, slightly fleshy, 3–6 mm long, reddish at maturity, fused from the base to middle or slightly above, margins smooth or with 1–2 small teeth on each side at or slightly above the middle*, venation netlike. Seeds circular, flattened, 1.5–2 mm in diameter.

POSTSENESCENCE CHARACTERISTICS: Red orach: Senesced plants with persistent fruits can persist into winter.

Australian saltbush: Plants are evergreen, unless the tops are killed by severe frost.

HABITAT: Both species inhabit disturbed places, roadsides, waste places, and fields on many types of soil, but often grow under alkaline or saline conditions.

Red orach: Also inhabits orchards, vineyards, and crop fields, including the margins of rice fields.

Australian saltbush: Also inhabits the margins of cultivated fields, grassland, scrub, shrubland, woodland, and salt marsh areas. Can survive below freezing temperatures in winter.

DISTRIBUTION: Red orach: Throughout California, to 2400 m. Western states, most central states, many northeastern states, and South Carolina.

Australian saltbush: Throughout California, except Modoc Plateau, Cascade Range, and northern and central Sierra Nevada, to 1000 m. Nevada, Arizona, New Mexico, Utah, and Texas.

PROPAGATION/PHENOLOGY: *Reproduce by seed.* Fruits persist on the parent plant, fall near it, and disperse to greater distances with water, mud, soil movement, human activities, and animals, including birds. Seeds germinate under saline and alkaline conditions. Most germination occurs in spring.

Red orach: Fruits also disperse when senesced stems detach near the ground and plants tumble with the wind. Under certain photoperiod conditions, plants produce 2 types of seeds: larger, dull brown seeds and smaller, glossy black seeds. Brown seeds have a softer seed coat and germinate more readily than black seeds. In northern areas such as Canada, only brown seeds develop. Light enhances germination, especially of black seeds, and the bracts appear to contain a germination inhibitor. Germination of both seed types generally occurs around 20°C. Red orach uses the C4 photosynthetic pathway.

Australian saltbush: Most seeds germinate when temperatures fluctuate between 5–25°C. Plants are often short-lived (about 2–5 years), but under favorable conditions can survive up to 10 years. The type of photosynthetic pathway used by Australian saltbush is uncertain.

MANAGEMENT FAVORING/DISCOURAGING SURVIVAL: Manual removal or cultivation before seeds develop can help control both species. Close mowing before seeds develop can also help to control red orach.

SIMILAR SPECIES: Garden orach [*Atriplex hortensis* L., synonyms: garden oracle; garden orach saltweed; sea purslane; mountain spinach][ATXHO] is a variable, erect summer *annual* to 1.8 m tall that has escaped cultivation as a garden

Garden orach (*Atriplex hortensis*) flowering stem. J. M. DiTomaso

Garden orach (*Atriplex hortensis*) seedling. J. M. DiTomaso

Garden orach (*Atriplex hortensis*) seeds with and without surrounding bracts. Note two seed types. J. O'Brien

Triangle orach (*Atriplex prostrata*) plant.
J. M. DiTomaso

Triangle orach (*Atriplex prostrata*) flowering stem.
J. M. DiTomaso

Triangle orach (*Atriplex prostrata*) leaves.
J. M. DiTomaso

ornamental and leafy vegetable. Unlike triangle orach, garden orach has *ovate to nearly round fruit bracts 8–18 mm long* that have *smooth margins* and *netlike veins*. Leaves are *lanceolate to ovate or ± triangular, green,* and *glabrous or nearly glabrous on both surfaces*. In addition, plants produce 2 types of seeds. Dull brown seeds about 2–4 mm in diameter develop in flowers that lack bracts and have a 3–5-lobed calyx. Glossy black seeds about 1–2 mm in diameter develop *horizontally* within the bracts in flowers that lack a calyx. Garden orach inhabits open disturbed sites in San Francisco Bay region, North Coast, South Coast, and Modoc Plateau, to 1300 m, and is expected to expand its range in California. It also occurs in all western states except possibly Arizona, most north-central states, and some northeastern states. Native to Eurasia.

Triangle orach [*Atriplex prostrata* Boucher ex DC., synonyms: fat-hen; halberdleaf orach; spearscale; *Atriplex latifolia* Wahlenb.; *Atriplex patula* L. var. *triangularis* (Willd.) Thorne & Welsh and var. *salina* Wallr.; *Atriplex triangularis* Willd; the following names have been misapplied to triangle orach: *Atriplex hastata* L.; *Atriplex patula* L. ssp. *hastata* (L.) Hall & Clements and ssp. *patula*; *Atriplex patula* L. var. *hastata* (L.) A. Gray] [ATXHA] is an ascending to erect native summer *annual* to about 1 m tall, with lanceolate to *triangular-hastate leaves* that are *green* and *glabrous to sparsely scurfy on both sides*. Triangle orach is a widespread variable ruderal native that often inhabits disturbed, moist saline and alkaline places. In natural areas triangle orach is considered a desir-

Red orach • Australian saltbush

Triangle orach (*Atriplex prostrata*) seedling. J. K. Clark

Triangle orach (*Atriplex prostrata*) seeds with and without surrounding bracts. J. O'Brien

able component of the native vegetation. The foliage and fruits are consumed by livestock and a variety of wildlife. However, it is occasionally weedy in and around crop fields and orchards. Besides its *annual life cycle* and ± *triangular-hastate leaves*, triangle orach is distinguished by having *green fruit bracts* that are usually *triangular to deltate* (sometimes ovate), *fused and truncate at the base*, and that have *weakly wavy-toothed margins*. Bracts are 3–7 mm long, and the faces often have a few small toothlike projections. Only black seeds are produced. Triangle orach typically inhabits moist or seasonally moist places and marshes in the North Coast, southwestern North Coast Ranges, Central Valley, Central Coast, and South Coast, to 300 m. It occurs in most contiguous states, except some southern states and possibly Nevada. Triangle orach uses the C3 photosynthetic pathway.

Twoscale saltbush [*Atriplex heterosperma* Bunge, synonym: *Atriplex micrantha* C.A. Mey. in Ledeb. may be the more correct name] is an erect *annual* to 1.5 m tall, with *green triangular-hastate leaves* and fruits and seeds that resemble those of garden orach. Unlike garden orach, twoscale saltbush has *fruit bracts 2–7 mm long that lack netlike veins* and the glossy black seeds develop *vertically* within the bracts. Twoscale saltbush occasionally inhabits open disturbed places in the Central Valley, North Coast, western North Coast Ranges, Cascade Range foothills, and possibly elsewhere, to 1200 m. It is expected to expand its range in California. Twoscale saltbush also occurs in Colorado, Minnesota, Missouri, Montana, Nebraska, Nevada, North Dakota, South Dakota, Utah, Washington, and Wyoming. Native to Eurasia.

A few other introduced *Atriplex* species occur in California, but they are uncommon compared to the species previously described. Refer to the appendix, **Nonnative Naturalized Plants** (p. 1601, vol. 2), for more information. There are more than 30 native *Atriplex* species and varieties in California, a few of which are rare or endangered. For this reason, it is important to correctly identify species before any control measures are initiated.

Fivehook bassia [*Bassia hyssopifolia* (Pall.) Kuntz] [BASHY][Cal-IPC: Limited]

Kochia [*Kochia scoparia* (L.) Schrad.][KCHSC] [Cal-IPC: Limited]

SYNONYMS: Fivehook bassia: bassia; fivehorn smotherweed; hyssop-leaved echinopsilon; smotherweed; thorn orache; *Echinopsilon hyssopifolius* (Pall.) Moq.; *Kochia hyssopifolia* (Pall.) Schrad.; *Salsola hyssopifolia* Pall.

Kochia: belvedere; belvedere-cypress; fireball; fireweed; Mexican burningbush; Mexican-fireweed; mock cypress; morenita; poor man's alfalfa; red belvedere; summer cypress; *Bassia scoparia* (L.) A.J. Scott; *Bassia sieversiana* (Pall.) W.A. Weber; *Chenopodium scoparia* L.; *Kochia alata* Bates; *Kochia parodii* Aellen; *Kochia sieversiana* (Pall.) C.A. Mey.; *Kochia trichophila* Stapf.; *Kochia virgata* Kostel.

GENERAL INFORMATION: Erect summer *annuals* to about 1.2 m tall, with slender leaves and inconspicuous flowers and fruits. Both are similar in form to **common lambsquarters** [*Chenopodium album* L.] (refer to the entry for **Common lambsquarters and Nettleleaf goosefoot** (p. 602) for more information). In their vegetative state, these species are very difficult to distinguish from one another.

Fivehook bassia: Fruits have 5 *hooked spines*. Fivehook bassia is considered to be fair forage for sheep in small amounts. However, the foliage contains variable amounts of potassium oxalate and can be fatally *toxic* to livestock when plant

Fivehook bassia *(Bassia hyssopifolia)* foliage. J. M. DiTomaso

Fivehook bassia • Kochia

Fivehook bassia (*Bassia hyssopifolia*) plant.
J. M. DiTomaso

Fivehook bassia (*Bassia hyssopifolia*) flowers.
J. M. DiTomaso

Fivehook bassia (*Bassia hyssopifolia*) flowering stem.
J. M. DiTomaso

Fivehook bassia (*Bassia hyssopifolia*) seedling.
J. M. DiTomaso

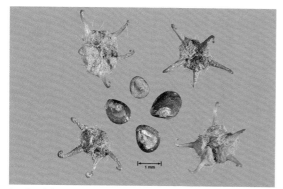

Fivehook bassia (*Bassia hyssopifolia*) seeds with and without surrounding hooked calyx.
J. O'Brien

Kochia (*Kochia scoparia*) plant. J. M. DiTomaso

material is ingested in large amounts within a short period. Fivehook bassia also contains an unidentified substance that can cause digestive tract disturbances. Toxicity symptoms can include diarrhea, uncoordinated gait, tetany, decreased blood calcium, neurological effects, coma, and photosensitization. Animals generally avoid consuming excessive amounts of this species if more palatable forage is available. Native to Eurasia and possibly introduced as a contaminant of alfalfa seed.

Kochia: Fruits typically have 5 *thickened or knoblike lobes* (tubercles) *or short, horizontal wings*. Kochia can provide good livestock forage in small amounts. However, the foliage contains saponins, small amounts of oxalates and nitrates, and possibly a thiamine-destroying substance and can be *toxic* to livestock when more than 50% of the diet consists of kochia. Toxicity symptoms include poor weight gain, mild liver dysfunction, photosensitivity, and neurological symptoms. Kochia is common in the northern plains states and is a state-listed noxious weed in Colorado, Minnesota (secondary), Oregon (class B), and Washington (class B, noxious plant quarantine). Populations of kochia have developed resistance to certain classes of herbicides in several states, Canada, and the Czech Republic. Kochia is sometimes cultivated for livestock forage and as an ornamental. In Asia, it is used as a potherb and seed crop. Kochia plant litter appears to have allelopathic properties that affect certain crop plants and kochia seedlings. Native to Asia.

SEEDLINGS: Fivehook bassia: Cotyledons linear to narrowly lanceolate, about 3–6 mm long, 1–2 mm wide, glabrous. Subsequent leaves narrowly lanceolate, gray-green, similar in size to cotyledons, covered with long soft white hairs, form a small rosette.

Kochia (*Kochia scoparia*) leaves and flowers.
J. K. Clark

Kochia (*Kochia scoparia*) seedlings.
J. M. DiTomaso

Kochia (*Kochia scoparia*) flowering stem.
J. M. DiTomaso

Kochia (*Kochia scoparia*) foliage.
J. M. DiTomaso

Kochia (*Kochia scoparia*) seeds.
J. O'Brien

Kochia: Seedlings are indistinguishable from those of fivehook bassia. Tolerates frost.

MATURE PLANT: Foliage generally gray-green, usually *covered with soft hairs* (fivehook bassia is hairier than kochia). Leaves mostly *alternate, flat, linear-lanceolate to lanceolate, 5–60 mm long, margins smooth.*

Fivehook bassia: Leaves mostly 1–5 mm wide, often sparse when plants are in fruit.

Kochia: Stems sometimes reddish late in the season. Leaves sometimes fascicled, mostly 1–10 mm wide, usually *3–5-veined from the base in the lower portion of the plant.* Upper leaf surfaces are sometimes nearly glabrous. Kochia often appears leafier than fivehook bassia.

ROOTS AND UNDERGROUND STRUCTURES: Both species have a taproot, usually with few to several branched, fibrous lateral roots.

Kochia: Roots can grow to nearly 2 m deep and horizontally to about 3 m.

FLOWERS: July–October. *Spikes consist of sessile clusters of inconspicuous, mostly bisexual flowers* in the axils of reduced leaves or bracts. Flowers lack petals. Calyx (sepals as a unit) *radial, 5-lobed,* mostly 1–2 mm wide (excluding spines on fivehook bassia).

Fivehook bassia: *Calyx lobes densely covered with tan to golden, soft, furry hairs. Each lobe has a yellowish incurved or hooked spine ± 1 mm long on the back.*

Kochia: *Calyx lobes glabrous to ciliate or sparsely hairy, usually with knoblike tubercles or sometimes with short horizontal wings in fruit.* Tubercles or wings less than 1 mm long or wide.

FRUITS AND SEEDS: Fruits (utricles) contain 1 horizontal seed and remain enclosed within the calyx. Seeds flattened, ovate, mostly 1–2 mm long, grooved on each face, contain an *annular embryo.*

HABITAT: Roadsides, fields, disturbed places, crop fields (especially newly planted cropland), ditch margins, and seasonal wetlands. Both species tolerate alkaline or saline soil and drought.

Fivehook bassia: A facultative wetland indicator species in California and many other states.

Kochia: Predominately inhabits upland sites.

DISTRIBUTION: Fivehook bassia: Throughout California, except the Northwestern region and high-elevation areas of the Sierra Nevada, to 1200 m. All western states, Kentucky, South Dakota, Texas, and a few northeastern states.

Kochia: Central Valley, San Francisco Bay region, Central Coast, South Coast, Mojave and Sonoran deserts, and Great Basin, to 1500 m. Expanding range in California. Most contiguous states, except possibly Maryland, and a few southern states.

PROPAGATION/PHENOLOGY: *Reproduce by seed.* Both species produce abundant seed. Seeds germinate in spring. Kochia and probably fivehook bassia use the C4 photosynthetic pathway.

Kochia: Senesced plants break off at the base and scatter fruits as they tumble with the wind. Shallow burial of seed typically reduces seedling emergence. Seeds near the soil surface generally survive about 1–2 years, but buried seed may survive 3 or more years.

MANAGEMENT FAVORING/DISCOURAGING SURVIVAL: In natural areas, manually removing plants before seed is produced can help control small populations and is less likely to disturb nearby desirable vegetation. Mowed or grazed plants typically resprout from the base. Shallow tillage can control populations on cropland.

SIMILAR SPECIES: Fivehook bassia and kochia are distinguishable from other weedy members of the Chenopodiaceae by the combination of having *flat, generally lanceolate leaves, hairy foliage,* and *radial fruits with hooked spines or tubercles.*

Wild beet [*Beta vulgaris* L.]

SYNONYMS: common beet; beetroot; wild beet; *Beta maritima* L.; *Beta vulgaris* L. var. *maritima* (L.) Moq.; many varieties have been named

GENERAL INFORMATION: *Biennial, sometimes annual or perennial*, with a thick, fleshy taproot and flower stems to 1.2 m tall. Plants exist as a loose rosette until flower stems develop at maturity, usually in the second year. Varieties of this species are commonly cultivated for their edible roots and greens. Plants occasionally escape cultivation, but seldom persist as a naturalized population. Native to southern Europe.

MATURE PLANT: Foliage glabrous. Stems decumbent to erect, simple or branched. Leaves ovate or diamond-shaped to oblong, base tapered or truncate, margins weakly wavy, often slightly succulent. Rosette leaves about 10–20 cm long, on stalks about as long as the blades. Stems leaves alternate, smaller than rosette leaves, on short stalks.

ROOTS AND UNDERGROUND STRUCTURES: Taproot thick, fleshy.

FLOWERS: July–October. Inflorescences 10–50 cm long, spikelike or in panicles with spikelike branches. Flowers bisexual, in sessile clusters of 5–11. Petals lacking. Sepals 5, curved inward, 2–2.5 mm long, margins membranous, *persist*

Wild beet (*Beta vulgaris*) plant. J. M. DiTomaso

in fruit. Stamens 5. Ovary half inferior, sunken into receptacle. Stigmas 2(3).

FRUITS AND SEEDS: Fruits hard, 3–5 mm in diameter, about 2.5 mm long, *fall in clusters* of 5–11, eventually open irregularly like a lid at the top to release the seed. Sepals thick and hardened at the base, curved, back keeled. Seeds 1 per fruit, horizontally oriented, nearly round, smooth, dark brown.

HABITAT: Disturbed places, often near crop fields and gardens.

DISTRIBUTION: San Francisco Bay region, Central Coast, South Coast, and Channel Islands, to 200 m. Oregon, Montana, Utah, a few central and southern states, and several eastern states.

PROPAGATION/PHENOLOGY: *Reproduces by seed.* Buried seeds can survive for up to about 20 years.

MANAGEMENT FAVORING/DISCOURAGING SURVIVAL: Manual removal or cultivation can readily control wild beet.

SIMILAR SPECIES: Wild beet is unlikely to be confused with other weedy members of the goosefoot family. However, wild beet is distinguished by having flowers with an *ovary that is half inferior and sunken into the receptacle, sepals that are thick and hardened at the base in fruit,* and *fruits that fall in clusters*.

Wild beet (*Beta vulgaris*) rosette.

J. M. DiTomaso

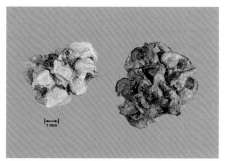

Wild beet (*Beta vulgaris*) flowering stem.

J. M. DiTomaso

Wild beet (*Beta vulgaris*) fruit clusters.

J. O'Brien

Common lambsquarters [*Chenopodium album* L.] [CHEAL]

Nettleleaf goosefoot [*Chenopodium murale* L.] [CHEMU]

SYNONYMS: Common lambsquarters: bacon-weed; fat-hen; frost-blite; meal-weed; pigweed; poor man's spinach; white goosefoot; white pigweed; wild spinach; *Chenopodium album* L. var. *album*; *Chenopodium album* L. var. *lanceolatum* (Muhl. ex Willd.) Coss. & Germ.; *Chenopodium amaranticolor* Coste & Reyn.; *Chenopodium browneanum* Roem. & Schult.; *Chenopodium centrorubrum*

Common lambsquarters *(Chenopodium album)* plant.   J. K. CLARK

(Makino) Nakai; *Chenopodium giganteum* D. Don; *Chenopodium lanceolatum* Muhl. ex Willd.; *Chenopodium suecicum* J. Murr.; others. This species is highly variable, with many named varieties. Currently, some botanists treat a few taxa that have been recognized as distinct species as varieties of *Chenopodium album*.

Nettleleaf goosefoot: Australian spinach; salt-green; sowbane; swine-bane; wall goosefoot; *Chenopodium biforme* Nees; *Chenopodium congestum* Hook f.; *Chenopodium lucidum* Gilib.; *Chenopodium triangulare* Forssk.; a few others

GENERAL INFORMATION: Common lambsquarters: Erect, primarily summer *annual* to 1.5 m tall, with *dull, pale gray green foliage* and at least a few *triangular-ovate leaves*. Foliage, especially new growth, is covered with a fine, *white, powdery scurf*. Common lambsquarters is a nearly ubiquitous weed of cultivated lands and other disturbed places. It is generally considered an edible species, and certain garden cultivars are grown for the tender leaves and shoots, which are consumed as a vegetable. However, under certain conditions, common lambsquarters can contain enough oxalates to be fatally *toxic* to livestock when a large quantity of the foliage is ingested within a short period. Ingestion of large amounts of oxalates causes the amount of calcium in the blood to drop to a low level. Toxicity symptoms include weakness, depression, tremors, labored breathing, seizures, and coma. Sheep are most often affected. Plants can also accumulate *toxic* levels of nitrates, but nitrate poisoning by ingestion of

Common lambsquarters (*Chenopodium album*) flowering stem. J. M. DiTomaso

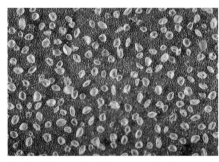

Swollen hairs on leaf of common lambsquarters (*Chenopodium album*). J. K. Clark

Common lambsquarters (*Chenopodium album*) seedling. J. M. DiTomaso

this species is uncommon. Common lambsquarters is susceptible to numerous viruses that affect many crops and ornamentals, including beet curly top; potato viruses X, M, and S; ringspot viruses of passionfruit, tomato, pepper, potato, *Prunus* species, and mulberry; and mosaic viruses of beet, bean, barley, lettuce, alfalfa, cucumber, squash, eggplant, hops, primula, watermelon, and wisteria. A variety of birds and small mammals consume the seeds. Probably native to Europe. Because many other *Chenopodium* species are often misidentified as common lambsquarters, the distribution of this species, particularly in southern California, is probably overestimated. See the "Similar Species" section below for comparisons.

Nettleleaf goosefoot: Erect *annual* to about 1 m tall, with ovate to deltate leaves that have toothed margins and slightly glossy, dark green upper surface. Like common lambsquarters, nettleleaf goosefoot can occasionally contain potentially *toxic* amounts of oxalates or nitrates under certain conditions. It is also susceptible to numerous crop and ornamental viruses, including plum pox; cherry leaf roll; beet curly top; ringspot viruses of hibiscus, pepper, potato, and tomato; and mosaic viruses of apple, beet, cucumber, hop, lettuce, parsnip, poplar, primula, radish, tomato, turnip, and watermelon. Small animals consume the seeds. Native to Europe.

SEEDLINGS: First leaves appear opposite, about equal to or slightly larger than cotyledons. Subsequent leaves alternate, increasingly larger.

Common lambsquarters: Cotyledons narrow oblong-lanceolate, 4–15 mm long, 0.7–2 mm wide, dull green. Stalk below cotyledons (hypocotyl) often tinged purplish red. Leaves oblong-ovate to triangular-ovate, often folded upward

Common lambsquarters (*Chenopodium album*) seeds, some with calyx attached. Note black and brown seed types.   J. O'BRIEN

Common lambsquarters • Nettleleaf goosefoot

Nettleleaf goosefoot (*Chenopodium murale*) flowering stem.   J. M. DiTomaso

Nettleleaf goosefoot (*Chenopodium murale*) habit.   J. K. Clark

Nettleleaf goosefoot (*Chenopodium murale*) seedling.   J. K. Clark

Nettleleaf goosefoot (*Chenopodium murale*) seeds, some with calyx attached.   J. O'Brien

along the midvein, margins smooth to weakly wavy-toothed, initially covered with glistening translucent granules that later become a white-powdery scurf, especially on the lower surfaces; lower surface often tinged purplish red.

Nettleleaf goosefoot: Cotyledons narrow oblong-lanceolate, 6–15 mm long, ± 1–2 mm wide, dark green. Leaves ± oblong-ovate, margins weakly toothed, ± glossy dark green, upper surface and especially the lower surface sparsely covered with minute white dotlike scales, lower surface and stalks sometimes purplish red.

MATURE PLANT: Leaves generally triangular-ovate to lanceolate, on slender stalks about half as long at the blade. Lower leaves mostly 3-veined from the base.

Common lambsquarters: Stems simple or sparsely branched, angled, ridged, sometimes striated purplish red. Leaves 1–5 cm long, *dull, pale gray-green*, ± covered with a *fine, white, powdery scurf* especially on the lower surface and new growth, margins smooth to coarsely toothed or sometimes weakly 3-lobed. Length of lower leaves usually less than 1.5 times the width.

Nettleleaf goosefoot: Leaves 1.5–6 cm long, margins coarsely toothed, *upper surface glossy dark green*, lower surface sparsely covered with a fine, white, powdery scurf.

ROOTS AND UNDERGROUND STRUCTURES: Both species have a simple or branched taproot with fibrous lateral roots.

FLOWERS: Inflorescences ± *panicle-like*, consist of *dense*, ± *spikelike clusters* of sessile green flowers. Flowers ± 1.5 mm in diameter, lack petals, covered with a fine, white, powdery scurf. *Sepals 5*, ovate, separate nearly to base, *narrowly keeled down the center*, usually enclose the fruit. Wind-pollinated.

Common lambsquarters: May–November. Self-compatible.

Nettleleaf goosefoot: Mostly spring, sometimes nearly year-round.

FRUITS AND SEEDS: Fruits (utricles) 1–1.5 mm in diameter, enclosed by sepals, contain 1 seed. Outer fruit wall (pericarp) thin, *adherent to and difficult to separate from the seed*, smooth to irregularly textured at 20× magnification. Seeds circular disk-shaped, 1–1.5 mm in diameter, glossy, black to dark brown, mostly oriented *horizontally* in the fruit.

Common lambsquarters: Seed equator *rounded to broadly angled*. Surface nearly smooth.

Nettleleaf goosefoot: Seed equator *acutely angled*. Surface minutely pitted.

POSTSENESCENCE CHARACTERISTICS: Fruit clusters persist on dried stems into winter.

HABITAT: Roadsides, fields, pastures, agronomic and vegetable crops, gardens, orchards, vineyards, landscaped areas, waste places, and other disturbed sites. Common lambsquarters inhabits many soil types, but grows best in cultivated calcium-rich soil.

Common lambsquarters • Nettleleaf goosefoot

Missouri goosefoot *(Chenopodium album* var. *missouriense)* stem and leaves. J. M. DiTomaso

Missouri goosefoot *(Chenopodium album* var. *missouriense)* flowering stem. J. M. DiTomaso

Missouri goosefoot *(Chenopodium album* var. *missouriense)* seedling. J. M. DiTomaso

Missouri goosefoot *(Chenopodium album* var. *missouriense)* seeds. J. O'Brien

DISTRIBUTION: Common lambsquarters: Common throughout California, to 1800 m. All contiguous states. Nearly worldwide in temperate regions.

Nettleleaf goosefoot: Throughout California, except Great Basin, to 2900 m. Common, except in desert regions. Most contiguous states, except some central states.

PROPAGATION/PHENOLOGY: *Reproduce by seed.* Fruits fall near the parent plant and disperse to greater distances with water, soil movement, mud, animals, agricultural operations, other human activities, and as a crop seed contaminant. Seed production can be high.

Common lambsquarters: Plants produce 2 types of seeds, depending on photoperiod. Small black seeds typically are dormant at maturity and develop when the daily photoperiod is long. Large nondormant brown seeds develop when the

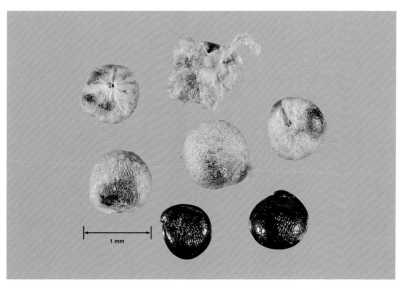

Oakleaf goosefoot (*Chenopodium glaucum*) seeds. J. O'Brien

photoperiod is short. Most seeds germinate in spring. Seeds survive ingestion by animals. Buried seeds can survive up to about 40 years or more.

Nettleleaf goosefoot: Seeds often germinate during late winter or early spring.

Nuttall povertyweed (*Monolepis nuttalliana*) plant. J. M. DiTomaso

Common lambsquarters • Nettleleaf goosefoot

Nuttall povertyweed (*Monolepis nuttalliana*) flowering stem.   J. M. DiTomaso

Nuttall povertyweed (*Monolepis nuttalliana*) fruiting stem.   J. M. DiTomaso

Nuttall povertyweed (*Monolepis nuttalliana*) seedling.   J. M. DiTomaso

Nuttall povertyweed (*Monolepis nuttalliana*) seeds.   J. O'Brien

MANAGEMENT FAVORING/DISCOURAGING SURVIVAL: Manual removal or cultivation before plants produce seed can help to control both species. Common lambsquarters seedlings usually do not survive clipping or trampling.

SIMILAR SPECIES: The following *Chenopodium* species and varieties are generally encountered less frequently than lambsquarters and nettleleaf goosefoot.

Missouri goosefoot [*Chenopodium album* L. var. *missouriense* (Aellen) I. J. Bassett & C. W. Crompton, synonyms: Missouri lambsquarters; *Chenopodium missouriense* Aellen; *Chenopodium paganum* auct. non Reichb.][CHEMI] is a sprawling *annual* to 2 m tall, with ovate to deltate leaves that have irregularly wavy-toothed to deeply toothed margins. It was previously recognized as a distinct species, but many botanists now consider it to be a variety of common lambsquarters. Unlike common lambsquarters, Missouri goosefoot is usually *densely branched* and has *arching to pendant inflorescence branches*. In addition, fruits are *1–1.3 mm in diameter*, and the *length of the lower leaves is often 1.5–2 times the width*. Most seeds are horizontal in the fruit. Missouri goosefoot inhabits disturbed sites in the northern Sierra Nevada foothills, southern Sierra Nevada, San Joaquin Valley, South Coast, and Peninsular Ranges, to 1000 m. It also occurs in New Mexico, Nevada, southern states, and most central and eastern states. Native to the central and eastern states.

Oakleaf goosefoot [*Chenopodium glaucum* L., synonyms: Rocky Mountain goosefoot; *Blitum glaucum* Koch; *Chenopodium glaucum* L. var. *pulchrum* Aellen; *Chenopodium glaucum* L. ssp. *salinum* (Standl.) Aellen; *Chenopodium glaucum* L. var. *salinum* (Standl.) Boivin; *Chenopodium salinum* Sandl.][CHEGL] is a *prostrate to ascending annual* to about 0.3 m tall, with *oblong to ± ovate leaves* that have entire to toothed margins and are *densely covered with white, powdery scurf below*. Currently, some botanists considered the taxon in California and other western states to be a distinct species known as **Rocky Mountain goosefoot** [*Chenopodium salinum* Sandl.]. Oakleaf goosefoot is distinguished by having *3 sepals* that are *membranous in fruit*, and *the number of fruits with the seed oriented vertically is roughly equal to the number of fruits with the seed oriented horizontally*. Vertical seeds are not enclosed by a sepal tube. Oakleaf goosefoot often grows in moist sandy soil. It inhabits disturbed open sites, pond margins, stream banks, and saline places in the Cascade Range, Great Basin, eastern slope of the Sierra Nevada, and the San Bernardino Mountains, to 2200 m, and is expected to expand its range in California. It also occurs in all western states, most north-central and central states, and a few northeastern states. Like common lambsquarters, oakleaf goosefoot can occasionally contain potentially *toxic* levels of oxalates or nitrates. Native to Eurasia.

Stinking goosefoot [*Chenopodium vulvaria* L.] is a *prostrate to decumbent annual* to about 0.3 m tall, with foliage that has an *unpleasant fishy odor*. Leaves are ± white powdery, especially below. Stinking goosefoot is further distinguished by having *± broadly elliptic to ovate leaves 0.5–3 cm long and nearly as wide*, with a *rounded base*, smooth margins, and on a *stalk ± equal to the blade*. In addition, the sepal tube is much longer than the lobes and encloses the fruit. The fruit wall adheres to the seed, and seeds are usually horizontal in the fruit. Stinking goosefoot sporadically inhabits roadsides, fields, grassland, cropland, and other disturbed places in the Central-western region, Central Valley, southern Sierra Nevada, Cascade Range foothills, and western Transverse Ranges, to 1400 m. It also occurs in Oregon, Texas, Florida, Mississippi, and several northeastern states. Native to the Mediterranean region.

There are a few other similar, non-native *Chenopodium* species in California that are uncommon or occur in localized areas. Refer to the appendix, **Non-native Naturalized Plants** (p. 1601, vol. 2), for more information.

Nuttall povertyweed [*Monolepis nuttalliana* (Schultes) E. Greene, synonyms: Nuttall's monolepis; *Blitum chenopodioides* Nutt.; *Blitum nuttallianum* Schult. ex

Rowm. & Sch.; *Chenopodium trifidum* Trev.] [MOPNU] is a winter or summer *annual* to 0.4 m tall, with *fleshy foliage*. Stems have *simple ascending branches from near the base*. Leaves are lanceolate, 1–4.5 cm long, and have a *tooth or lobe near the base on each side* (± hastate). Flower clusters are nearly sessile in the leaf axils and consist of 5–15 or more flowers. Mature foliage is nearly glabrous, and new foliage is often sparsely covered with a white powder. Unlike *Chenopodium* species, Nutall povertyweed flowers mostly have *1 ± membranous sepal* (some flowers can have 2–3) and *0–1 stamen*. In addition, fruits are ovoid, 1.5–2 mm long, with a thin minutely pitted wall that adheres to the seed. Nuttall povertyweed is a ruderal native and is generally not considered a weed in natural areas. However, it can be weedy in crop fields and other dry to wet, open, disturbed sites throughout much of California, except the Northwestern region, to 3500 m. It also occurs in all western states, most central states, and a few northeastern and southern states.

Mexicantea [*Chenopodium ambrosioides* L.][CHEAM]

SYNONYMS: American wormseed; apazote; bitter weed; Chilean tea; epazote; epazote; erva de Santa Maria; Jerusalem tea; paico; pasote; Spanish tea; strong-scented pigweed; wormseed; wormseed goosefoot; yerba de Santa Maria; *Ambrina ambrosioides* (L.) Spach; *Atriplex ambrosioides* (L.) Crantz; *Blitum ambrosioides* (L.) Beck; *Botrys ambrosioides* (L.) Nieuwl.; *Chenopodium ambrosioides* L. var. *ambrosioides*; *Chenopodium ambrosioides* L. var. *anthelminticum* A. Gray; *Chenopodium ambrosioides* L. var. *chilensis* Speg.; *Chenopodium ambrosioides* L. var. *vagans* (Standl.) J. T. Howell; *Chenopodium anthelminticum* L.; *Chenopodium chilense* Schrad.; *Chenopodium vagans* Standl.; many others

GENERAL INFORMATION: *Erect annual to perennial* to 1.3 m tall, with coarse, *distinctively strong-scented, glandular foliage* and terminal clusters of inconspicuous greenish flowers. Mexicantea occurs nearly worldwide as a weed of agronomic fields, vegetable crops, orchards, and pastures. The leaves are used as a folk medicine in many countries and as a spice in Latin America. Seeds, and to a much smaller extent, leaves, contain a highly *neurotoxic* terpene peroxide, ascaridole, in the essential oil. Ascaridole has antihelmintic and antifungal properties. In the United States, mexicantea was cultivated for the seed oil, known as oil of Chenopodium, which was widely used to treat intestinal parasites in livestock and humans. However, the margin of safety between ingestion of an effective dose and toxicity symptoms is narrow and variable. Neurotoxic symptoms resulting from ingestion of the oil include vomiting, headache, weakness, sleepiness, hallucinations, convulsions, heart irregularities, and coma. Under certain conditions, mexicantea foliage can contain levels of oxalates that are

Mexicantea (*Chenopodium ambrosioides*) plant. J. M. DiTomaso

Mexicantea (*Chenopodium ambrosioides*) seedling. J. M. DiTomaso

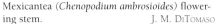

Mexicantea (*Chenopodium ambrosioides*) flowering stem. J. M. DiTomaso

Mexicantea (*Chenopodium ambrosioides*) seeds, two with calyx attached. J. O'Brien

toxic to livestock when a large quantity of plant material is ingested in a short period. Acute toxicity symptoms include depression, weakness, tremors, labored breathing, and low levels of blood calcium. However, animals generally avoid consuming the strong-scented foliage unless more palatable forage is unavailable. Studies suggest that plant residues incorporated into the soil can reduce local populations of plant-parasitic nematodes. and in tropical climates, they can limit the amount of seedling damping-off due to certain fungi. Mexicantea also appears to have allelopathic properties. In some tropical regions, farmers allow mexicantea to grow in established crop fields to suppress the growth of other weeds. However, it is a weed in alfalfa because of the undesirable odor it gives to the hay. Native to tropical America.

SEEDLINGS: Cotyledons lanceolate, about 2–8 mm long, 1.5–2 mm wide, glabrous. First 2 leaves often appear opposite, subsequent leaves alternate. First few leaves ± oblong, short-stalked, margins barely wavy to shallow-lobed, pale green, ± pubescent, lower surfaces dotted with sessile golden-brown resin glands, upper surface veins depressed.

MATURE PLANT: Stems typically *erect*, glabrous to short stiff-hairy, ± weakly woody at the base, ± angular-ridged. Upper branches mostly erect to ascending. Lower branches sometimes spreading with upturned tips (decumbent). Leaves alternate, mostly short-stalked, ± elliptic-lanceolate, 1.5–10 cm long, nearly

Jerusalem-oak goosefoot (*Chenopodium botrys*) seedling. J. M. DiTomaso

Jerusalem-oak goosefoot (*Chenopodium botrys*) plant. J. M. DiTomaso

Jerusalem-oak goosefoot (*Chenopodium botrys*) flowering stem. J. M. DiTomaso

Jerusalem-oak goosefoot (*Chenopodium botrys*) seeds, some with calyx attached. J. O'Brien

glabrous to pubescent, margins barely toothed to shallow pinnate-lobed, lobes acute. *Lower surface dotted with minute, sessile, ± golden-brown resin glands (visible with magnification).*

ROOTS AND UNDERGROUND STRUCTURES: Taproot fleshy to weakly woody, often branched, with numerous lateral and fibrous roots.

FLOWERS: June–December. Inflorescences terminal, *panicle-like*, consist of *dense spikelike clusters of green flowers on ± straight branches, ± pyramidal*. Bractlike leaves narrow-oblong, about 1–2 cm long. Flowers *sessile*, lack petals. Sepals 5, separate to base, ovate, ± 1 mm long, *glabrous or sparsely dotted with sessile glands*, persistent in fruit. Stamens 5. Styles usually 2.

FRUITS AND SEEDS: Fruits (utricle) ± spherical, ± 0.7 mm in diameter, gland-dotted, enclosed by the calyx, contain one seed. *Most seeds oriented horizontally* *in the fruit.* Outer fruit wall (pericarp) very thin, *separates easily from the seed*, deciduous. Seeds circular-disk-shaped, glossy, dark brown to nearly black, ± 0.7 mm in diameter.

POSTSENESCENCE CHARACTERISTICS: Dead plants with a few persistent fruits can persist into spring.

HABITAT: Agronomic and vegetable crops, orchards, vineyards, gardens, roadsides, fields, pastures, waste places, and other disturbed noncrop sites. Usually grows in moist places and peat soils, often found on floodplains or sandbanks along lowland streams. Tolerates saline soil.

DISTRIBUTION: Throughout California, except deserts and Great Basin, to 1400 m. All contiguous states, except Idaho, Minnesota, Montana, North Dakota, and Wyoming. Nearly worldwide.

PROPAGATION/PHENOLOGY: *Reproduces by seed.* One plant can produce several thousand to more than 200,000 seeds. Seeds fall near the parent plant and disperse to greater distances with water, soil movement, mud, agricultural operations and other human activities, and possibly as a crop seed contaminant. Newly matured seeds are typically dormant. A cool period followed by fluctuating temperature with light exposure stimulates germination. In California, seeds typically germinate in mid-spring through early summer.

Cutleaf goosefoot *(Chenopodium multifidum)* flowering stem.   J. M. DiTomaso

Cutleaf goosefoot *(Chenopodium multifidum)* plant.   J. M. DiTomaso

Cutleaf goosefoot *(Chenopodium multifidum)* seeds, some with calyx attached.   J. O'Brien

Cutleaf goosefoot *(Chenopodium multifidum)* seedling.   J. M. DiTomaso

Clammy goosefoot (*Chenopodium pumilo*) flowering stem. J. M. DiTomaso

Clammy goosefoot (*Chenopodium pumilo*) seedling. J. M. DiTomaso

Clammy goosefoot (*Chenopodium pumilo*) plant. J. M. DiTomaso

MANAGEMENT FAVORING/DISCOURAGING SURVIVAL: Manual removal or cultivation before plants produce seed can control mexicantea.

SIMILAR SPECIES: The following *Chenopodium* species have *glandular-pubescent foliage* and lower leaves with *wavy-toothed to pinnate-lobed margins*.

Jerusalem-oak goosefoot [*Chenopodium botrys* L., synonyms: feather geranium; *Botrydium botrys* (L.) Sm.; *Teloxys botrys* (L.) W. A. Weber] [CHEBO] is an *erect, strong-scented annual* to 0.7 m tall. Unlike mexicantea, Jerusalem-oak goosefoot has foliage and sepals with *gland-tipped hairs*, inflorescence branches that are spreading and *down-curved to ± coiled* with nearly flat-topped cymelike clusters of *flowers on short stalks*, and fruits with an *outer wall that adheres to the seed*. Jerusalem-oak goosefoot inhabits agricultural land and disturbed noncrop areas throughout California, to 2100 m. It occurs in most contiguous states, except a few southern states. Native to Europe.

Clammy goosefoot (*Chenopodium pumilo*) seeds, some with attached calyx. J. O'Brien

Cutleaf goosefoot [*Chenopodium multifidum* L., synonyms: *Roubevia multifida* (L.) Moq.; *Teloxys multifida* (L.) W.A. Weber] is a *spreading annual* to 0.3 m tall, with *gland-tipped hairs* on the foliage and flowers. Cutleaf goosefoot is distinguished by *having sepals fused into an obovoid tube with 3–5 minute teeth at the apex*. The sepal tube has a *netlike surface*, is covered with *gland-tipped hairs*, and *encloses the fruit*. The seed is *oriented vertically* in the fruit. Cutleaf goosefoot inhabits disturbed places in the Northwestern region, Sierra Nevada foothills, Central Valley, Central-western region, southwestern Transverse Ranges, South Coast, Santa Rosa Island, San Nicolas Island, and probably elsewhere, to 1400 m. It also occurs in Oregon and many eastern states. Native to South America.

Clammy goosefoot [*Chenopodium pumilio* R. Br., synonyms: Australian goosefoot; Tasmanian goosefoot; *Chenopodium carinatum* auct. non R. Br.; *Teloxys pumilio* (R. Br.) W. A. Weber] is an erect to prostrate *annual* to 0.3 m tall, with *strong-scented foliage*. Unlike mexicantea, clammy goosefoot has *spherical flower clusters 3–6 mm in diameter that are mostly in the leaf axils*, and the *sepals and lower surfaces of the leaves are dotted with nearly sessile, opaque, pale yellow glands* (easiest to see with magnification). In addition, *all leaves are less than 3 cm long*, and *most seeds are oriented vertically in the fruit*. Clammy goosefoot inhabits agricultural land and other disturbed places in the Northwestern region, Modoc Plateau, Sierra Nevada, Central Valley, San Francisco Bay region, Transverse Ranges, South Coast, and western Peninsular Ranges, to 3000 m. It also occurs in Arizona, Nevada, Oregon, Washington, southern states, many eastern states, and a few central states. Native to Australia.

Halogeton [*Halogeton glomeratus* (M. Bieb.) C. Meyer] [HALGL][Cal-IPC: Moderate][CDFA list: A]

SYNONYMS: barilla; saltlover; *Anabasis glomeratus* Bieb.

GENERAL INFORMATION: Erect winter or summer *annual* to 0.5 m tall, with small fleshy leaves. Halogeton typically invades disturbed arid and semiarid sites with alkaline to saline soils. Plant tissues accumulate salts from lower soil horizons. The salts leach from dead plant material, increasing topsoil salinity and favoring halogeton seed germination and seedling establishment. The foliage contains variable amounts of soluble sodium oxalates and can be fatally *toxic* to livestock, especially sheep, when ingested in quantity. Soluble oxalates can cause an acute reduction in bloodstream calcium (hypocalcemia). Symptoms of poisoning can include depression, frothy salivation, weakness, staggering, muscular spasms, and coma. Toxicity of plant material depends on environmental conditions, plant maturity, and the condition of livestock. As little as 340 g of foliage can be fatal to poorly nourished animals, and although younger plants are most toxic, mature plants are probably also poisonous. Livestock supplemented with calcium-fortified feeds are less susceptible to the toxic effects. Animals generally avoid consuming the bitter-tasting foliage if more palatable forage is available. Besides California, halogeton is a state-listed noxious weed in Arizona (restricted), Colorado, Hawaii, New Mexico (class B), and Oregon (class B). Native to the cold, desert regions of Eurasia.

SEEDLINGS: Cotyledons cylindrical, gradually narrowed to the slightly blunt apex, mostly 3–6 mm long, 1–2 mm wide, glabrous. First leaves appear oppo-

Halogeton (*Halogeton glomeratus*) plant. J. M. DiTomaso

Halogeton

Halogeton (*Halogeton glomeratus*) senesced plants. J. M. DiTomaso

Halogeton (*Halogeton glomeratus*) flowering stem. J. M. DiTomaso

Halogeton (*Halogeton glomeratus*) stems with fruit. J. M. DiTomaso

Halogeton (*Halogeton glomeratus*) seeds and bractlike sepals. J. O'Brien

site, cylindrical, usually broadest near the tip, with tufts of *long white interwoven hairs in the axils, tip rounded with a short bristle at the apex.*

MATURE PLANT: Stems branched, often curved at the base, ascending to erect, often somewhat fleshy, usually tinged reddish or purple. Leaves alternate, sessile, dull green to bluish green, *fleshy, cylindrical, 4–22 mm long, 1–2 mm wide, broadest at the tip, tip bluntly rounded with a stiff bristle 1–2 mm long. Foliage glabrous, except for tufts of long white interwoven hairs in the leaf axils.* Leaves deciduous or shriveled in fruit.

ROOTS AND UNDERGROUND STRUCTURES: The taproot grows slowly and can penetrate the soil to a depth of up to 0.5 m. Lateral roots may spread up to about 0.5 m in all directions.

Halogeton

Halogeton (*Halogeton glomeratus*) fruiting stem.   COURTESY OF ROSS O'CONNELL

FLOWERS: June–September. Flower clusters numerous and dense in most leaf axils, small, headlike, with 0–3 bractlets 1.5–2 mm long below each cluster. Flowers bisexual and female (pistillate). Both flower types lack petals and have 5 sepals. Most sepals petal-like, with a *narrow base 1–2 mm long* and a *membranous, fan-shaped tip 2–3.5 mm long* that is greenish yellow to red-tinged and conspicuously veined. Some flowers have bractlike sepals 2–3 mm long. Female flowers lack stamens. Bisexual flowers have 2–5 stamens.

FRUITS AND SEEDS: August–October. Utricles (thin-walled one-seeded fruits) densely numerous, 1–2 mm long, *enclosed by the sepals. Fruits with sepals typically conceal the stems.* Utricles loosely enclosed by fan-shaped sepals contain a blackish brown seed that are collectively referred to as black seeds in the literature. Utricles tightly enclosed by adherent, brown, bractlike sepals contain a brown seed, and the entire structure is referred to as a brown seed. Seeds ± teardrop-shaped, often with 2 points, *flattened,* 1–2 mm long, have a coiled embryo.

POSTSENESCENCE CHARACTERISTICS: Plants turn straw-colored when cool-season frosts begin. Plants with some fruits, particularly those enclosed by bractlike sepals, may remain intact through winter.

HABITAT: Disturbed open sites, dry lakebeds, shrublands, and roadsides, typically where native vegetation is sparse. Inhabits arid and semiarid regions, especially where winters are cold. Grows on many soil types, but is adapted to alkaline and saline soils with at least 5800 ppm of sodium chloride.

DISTRIBUTION: Abundance depends on yearly rainfall amounts. Great Basin (e Modoc, e Lassen, Mono, and n Inyo Cos.), Mojave Desert (s Inyo, e Kern, ne Los Angeles, and w and ne San Bernardino Cos.), and northern Sierra Nevada (c Placer, e Nevada Cos.), to 1800 m. Previous infestations now eradicated occurred in the Cascade range (c Siskiyou Co.). Occurs in all western states, Nebraska, and South Dakota.

PROPAGATION/PHENOLOGY: *Reproduces by seed.* Plants typically produce enormous quantities of seed (average is ± 75 seeds per inch of stem). Seeds disperse with wind, water, human activities, seed-gathering ants, animals, and when dry plants break off at ground level and tumble with the wind. Many seeds survive ingestion by animals, including sheep and rabbits. Plants produce 2 types of seeds depending on photoperiod. Black seeds typically develop after mid-August, lack or have a short after-ripening period, and remain viable for about 1 year. Brown seeds usually develop before mid-August, are dormant at maturity, and can survive for about 10 years or more under field conditions. Experimental evidence suggests that the bractlike sepals enforce dormancy of brown seeds. Cool, moist vernalization appears to enhance germination of brown seeds by decomposing the adherent sepals. Plants typically produce more black seeds than brown, but the ratio varies according to environmental conditions. Most black seeds are shed by early November. Brown seeds may remain on plants until February. Most seeds germinate in late fall to early spring in cold-winter areas, but some germination can occur year-round when conditions become favorable. Black seeds can imbibe water and germinate in less than 1 hour.

MANAGEMENT FAVORING/DISCOURAGING SURVIVAL: Halogeton competes poorly with established perennial vegetation. Overgrazing, human disturbance, and fire typically reduce desirable vegetation and increase open sites with bare soil, encouraging halogeton invasion and establishment. Fire disturbance often enhances seed germination and favors the growth of dense stands.

SIMILAR SPECIES: Before flowering, halogeton resembles immature **barbwire Russian thistle** [*Salsola paulsenii* Litv.], **Russian thistle** [*Salsola tragus* L.] or **kochia** [*Kochia scoparia* (L.) Schrader]. Unlike halogeton, immature Russian thistles have *linear leaves 0.5–1 mm wide* and *lack hairs in the axils.* In addition, Russian thistle seeds are *cone-shaped.* Kochia has *pubescent leaves that do not have a stiff bristle at the tip.* Refer to the entries for **Barbwire thistle, Russian thistle,** and **Spineless Thistle** (p. 622) and **Five-hook bassia and Kochia** (p. 594) for more information.

Barbwire Russian thistle [*Salsola paulsenii* Litv.] [SASPA][Cal-IPC: Limited][CDFA list: C]

Russian thistle [*Salsola tragus* L.][SASKR] [Cal-IPC: Limited] [CDFA list: C]

Spineless Russian thistle [*Salsola collina* Pallas] [CDFA list: Q]

SYNONYMS: Russian thistle: common saltwort; prickly Russian thistle; Russian tumbleweed; tumbleweed; tumbling weed; windwitch; witchweed; *Salsola australis* R.Br.; *Salsola iberica* (Sennen & Pau) Botsch; *Salsola kali* ssp. *austroafricana* Aellen (proposed change to *S. austroafricana* (Aellen) Hrusa); *Salsola kali* L. var. *tenuifolia* Tausch.; *Salsola kali* L. ssp. *ruthenica* (Iljin) Soó; ssp. *tenuifolia* Moq.; and ssp. *tragus* (L.) Celakovsky; *Salsola pestifer* Nelson; *Salsola ruthenica* Iljin. The synonymy of this taxon is complex, and correct use of some scientific names is currently uncertain. Only frequently encountered synonyms are listed.

Spineless Russian thistle: slender Russian thistle; tumble thistle

GENERAL INFORMATION: Bushy, large noxious summer *annuals* with rigid branches and reduced, stiff, prickly upper stem leaves (bracts) at maturity. Under certain conditions, these species can accumulate oxalates to levels that are *toxic* to livestock, especially sheep. Diarrhea and toxicity problems associated with high levels of oxalate ingestion most often occur when sheep have

Barbwire Russian thistle (*Salsola paulsenii*) plant. J. M. DiTomaso

Barbwire Russian thistle *(Salsola paulsenii)* flowering stem. J. M. DiTomaso

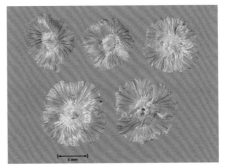

Barbwire Russian thistle *(Salsola paulsenii)* seeds with bractlike sepals. J. O'Brien

Barbwire Russian thistle *(Salsola paulsenii)* fruiting stem. J. M. DiTomaso

been foraging almost exclusively on Russian thistle or other *Salsola* species for several weeks. Symptoms of oxalate poisoning include abrupt onset of depression, weakness, labored breathing, prostration, seizures, and sometimes coma and death. All *Salsola* species are native to Eurasia. Russian thistles can also impede traffic, create fire hazards, and are the alternate host for the beet leafhopper *(Circulifer tenellus)* that can carry the virus causing curly top of sugarbeets, tomatoes, melons, and other crops. The taxonomy of *Salsola* in the western United States is complex and not fully understood.

Barbwire Russian thistle: This species can resemble Russian thistle and is often confused with it. To make matters more complex, the two species can hybridize. The intermediate has been named *Salsola* × *gobicola* Iljin (in Asia). In California, the intermediate acts like a separate species, easily recognized (unlike the parents, most plants have a yellow patch at the perianth base that fades when dried). Although this hybrid has some variability, it is not nearly as diverse as *S. tragus*. It occurs with both, either, or neither parent, and looks more like *S. paulsenii* than *S. tragus*. Aside from the uncommon *S.* × *gobicola*, there appear to be two more common forms of barbwire Russian thistle. One form has a more lax perianth and a chromosome count of $2n = 54$. This form, called *Type* "*lax*," is a complex hybrid involving *S. tragus*, *S. paulsenii*, and *S. kali* ssp. *austroafricana* Aellen (previously referred to as *Type B*, but proposed as a change to *S. austroafricana* (Aellen) Hrusa). *Salsola paulsenii* has a more

Barbwire Russian thistle • Russian thistle • Spineless Russian thistle

Russian thistle *(Salsola tragus)* plant.
J. M. DiTomaso

Russian thistle *(Salsola tragus)* flowering stem.
J. M. DiTomaso

Russian thistle *(Salsola tragus)* flower in leaf axil.
J. M. DiTomaso

Russian thistle *(Salsola tragus)* juvenile plant.
J. M. DiTomaso

Russian thistle *(Salsola tragus)* seedling.
J. M. DiTomaso

Russian thistle *(Salsola kali* ssp. *austroafricana)* fruit in leaf axil.
J. M. DiTomaso

Russian thistle (*Salsola tragus*) seeds.   J. O'Brien

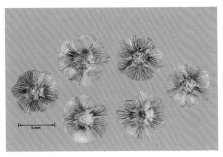

Russian thistle (*Salsola kali* ssp. *austroafricana*) seeds.   J. O'Brien

spinose perianth form than *Type "lax"* and has a chromosome count of 2n = 36. Barbwire Russian thistle is rarely a host of the beet leafhopper. It was probably introduced to the southwestern United States at the turn of the century, but was unrecognized until 1967.

Russian thistle: Although it is often a noxious weed, immature plants in moderation can provide an extra source of nutritious forage for livestock on arid rangeland. Russian thistle consists of 3 variants or types in California. Currently designated *Salsola tragus*, *Salsola kali* ssp. *austroafricana*, and *Type C*, the three types differ in chromosome number (*S. tragus*: tetraploid 2n = 36; *Salsola kali* ssp. *austroafricana*: diploid 2n = 18; *Type C*: hexaploid 2n = 54), calyx wing development, and their shapes and sizes, pubescence, and anther length. *Type C* is an interspecific hybrid between the tetraploid *S. tragus* and the diploid *S. kali* ssp. *austroafricana*. Russian thistle was introduced into South Dakota around 1874 with flax seed from Russia and quickly spread throughout the United States in the following years. The leaf mining moth *Coleophora klimeschiella* and the stem boring moth *Coleophora parthenica* have been released as biocontrol agents in California in the 1970s, but control of Russian thistle with these insects has been poor. Additional biocontrol efforts are underway, but may depend on a better understanding of the species complexes that exist in the western United States. Besides California, Russian thistle is a state-listed noxious weed in Colorado and Ohio (prohibited).

Spineless Russian thistle: No infestations are currently known in California, but spineless Russian thistle seed has been found as a contaminant of commercial birdseed. Spineless Russian thistle is a widespread weed in the Midwest and appears to be expanding its range. It is also a state-listed noxious weed in Colorado.

Key to *Salsola paulsenii, S. tragus,* and their various types

(Key developed by Dr. Fred Hrusa, CDFA).

1. Mature perianth (not the wings) forming a spinose tip. All (±) fruits forming wings. Desert regions
   Salsola paulsenii

1. Mature perianth (not the wings) soft, generally crumpled, not forming a spinose tip. All fruits winged or some fruits not winged; deserts and cismontane regions.
    2. Many mature perianths not forming distinct wings. Winged fruit generally found only on the upper one-half of stems, or in northern California some plants flowering but not forming winged fruits
        Salsola tragus
    2. Generally all mature perianths with wings, rarely a few flowers on lowermost stems with reduced wings
        3. All fruits winged; individual minor perianth fruit wings greater than 1.8 mm in width, often more than 2.5 mm wide; gen. greater than 1.5 mm in length, (the ratio of length to width about 1); stems usually glabrous, rarely minutely papillose, or thinly hairy
            Salsola kali ssp. *austroafricana*
        3. A few lowermost fruits not or only slightly winged. Minor wing length to width ratio generally > 2; stem vestiture variable, ± glabrous to hairy, papillose or papillose-hairy
            4. Stems generally papillose hairy, occasionally sparsely hairy, minor perianth wings generally highly reduced or obsolete, sometimes one (rarely both) long and linear with a slightly broadened or sharply acuminate tip; venation of a minor wing generally not visible, except sometimes a single central vein may be seen; plants stiff and spiny at maturity, similar to *Salsola paulsenii*
                Type "lax"
            4. Stems sparsely hairy or glabrous; minor perianth wings with broadened distal end, broadly oblanceolate or obovate; venation of the minor wings visible, with several veins in the distal portion
                Type C

Glasswort (*Salsola soda*) plant. J. M. DiTomaso

Glasswort (*Salsola soda*) seedling.
J. M. DiTomaso

Glasswort (*Salsola soda*) seeds, some with calyx attached.
Jim O'Brien

Glasswort (*Salsola soda*) fruiting stem.
J. M. DiTomaso

SEEDLINGS: Cotyledons and subsequent leaves needlelike. Leaves alternate, but often appear opposite because of short internodes.

Barbwire Russian thistle: Cotyledons similar to those of Russian thistle. Subsequent leaves resemble those of mature plants. Stems thick, ± rigid, without purple striations. *Main stem shorter than lateral branches.* First 4 lateral branches often form a cross-shaped pattern and are nearly prostrate, with the tips curved upward. *Immature plants prickly, often wider than tall and appear mounded at an early stage.*

Russian thistle: Cotyledons 10–35 mm long. Subsequent leaves fleshy, soft, weakly spine-tipped. Stems slender, flexible, often with reddish purple longitudinal striations. *Immature plants usually taller than wide, with lateral branches ascending and shorter than the main stem.*

Spineless Russian thistle: Subsequent leaves pliable, ± fleshy, with a soft bristle at the apex.

MATURE PLANT: Leaves alternate, sessile, linear to needlelike, gradate into rigid, spine-tipped bracts in the inflorescences.

Barbwire Russian thistle: Typically to 0.5 m tall, wider than tall. Foliage ± yellowish green, covered with minute papillae or rarely glabrous. Leaves fleshy, thick, stiff, spine-tipped, 5–32 mm long, 1–1.5 mm wide, often curved away from the stem (recurved) near the tip, covered with short hairs or minute papillae. Bracts resemble those of Russian thistle.

Mediterranean saltwort (*Salsola vermiculata*) infestation. D. GRIFFIN

Russian thistle: To 1 m tall, usually height ± equal to width or taller than wide. Stems rigid, typically curved upward, purple-striated or green. Foliage ± bluish green, glabrous (mostly *S. kali* ssp. *austroafricana*) or covered with short, stiff hairs. Leaves fleshy to leathery, 8–52 mm long, mostly 0.5–1 mm wide, tip sharp-pointed to spine-tipped. Bracts ± *awl-shaped, reflexed, not overlapping at maturity*, margins membranous and minutely barbed.

Spineless Russian thistle: To 1 m tall. Stems green and white striated, straight, erect to ascending. Foliage ± glabrous, covered with minute papillae. Leaves to 5 cm long, pliable, ± succulent to leathery, tipped with a soft bristle, base expanded and extending a short distance down the stems (slightly decurrent). *Bracts lanceolate, appressed to stems, strongly overlap one another.*

ROOTS AND UNDERGROUND STRUCTURES: All have a taproot. Do not form mycorrhizal associations.

Russian thistle: Taproot to 1.5 m deep, with lateral roots spreading to 1.8 m. Plants can extract deep soil moisture that is not available to winter wheat.

FLOWERS: Bisexual, axillary, mostly solitary. Petals lacking. *Calyx usually 2.5–3.5 mm long, 4–5-parted, persistent in fruit.* Russian thistle and barbwire Russian thistle have calyxes with *winglike appendages that appear petal-like.* Stamens 5, exserted beyond sepals. Style branches 2, exserted. Wind-pollinated. Out-crossing and self-fertile.

Barbwire Russian thistle: June–September. Often flowers 2–3 weeks earlier than Russian thistle. *Sepal tips often stiff, erect, usually spinelike in the spiny form, or less stiff and not spiny in the lax form.* The lax form is often pinkish. Sepal wings similar to those of Russian thistle except larger, *2.5–4.5 mm long*, typically flat, margin smooth to minutely irregular-toothed.

Barbwire Russian thistle • Russian thistle • Spineless Russian thistle

Mediterranean saltwort *(Salsola vermiculata)* plant. J. M. DiTomaso

Mediterranean saltwort *(Salsola vermiculata)* flowering stem. J. M. DiTomaso

Mediterranean saltwort *(Salsola vermiculata)* leaves and stem. J. M. DiTomaso

Mediterranean saltwort *(Salsola vermiculata)* fruiting stem. R. Breckenridge

Russian thistle: July–October. Male flowers often develop in early July. Bisexual flowers develop from about mid-July *(Salsola kali* ssp. *austroafricana)* to early October. Sepal tips acute, ± *lax, not spinelike*. Sepal wings 0.5–2.5 mm long, fan-shaped, margin usually minutely toothed to scalloped, translucent, often pinkish to deep red with conspicuous veins, and dark brownish in *Salsola kali* ssp. *austroafricana*.

Spineless Russian thistle: Summer/fall. Stems sometimes have gall-like flowers in leaf axils in the lower portion of the plant. *Flowers ± hidden by 2 large, fleshy, spine-tipped bracts.* Sepal wings lacking or narrow, mostly less than 1 mm long, margin minutely irregular-toothed.

FRUITS AND SEEDS: Utricles (fruiting structures) ± spherical, 1-seeded, enclosed by the persistent calyx. Seeds compressed-round to slightly conical, about 1.5–2 mm in diameter. Seed coat thin, translucent, gray to brown, with a dark greenish brown, coiled embryo visible beneath.

Barbwire Russian thistle: *Fruits 7–9 mm in diameter, including sepal wings.*

Russian thistle: *Fruits to 8 mm in diameter, including sepal wing.* Fruits are generally smaller (3–5 mm) than *S. kali* ssp. *austroafricana* fruits (8 mm). Sepal wings flat and open (*Salsola kali* ssp. *austroafricana*) or folded over in the true *S. tragus* form. Seeds typically 2–2.5 mg/seed, but in *Salsola kali* ssp. *austroafricana* seeds are usually 1.25–1.75 mg/seed.

Spineless Russian thistle: Fruits 3–5(7) mm in diameter including the narrow sepal wings when present, *hidden by 2 large, fleshy, spine-tipped bracts.*

POSTSENESCENCE CHARACTERISTICS: Plants become gray to brown. Main stems of Russian thistle break off at ground level under windy conditions, allowing plants to disperse numerous seeds as they tumble with the wind. Skeletons usually persist for at least 1 year and are typically found along fences and other structures. *Salsola kali* ssp. *austroafricana* can occasionally act as a perennial.

HABITAT: Disturbed sites, waste places, roadsides, fields, cultivated fields, and disturbed natural and seminatural plant communities. Grows best on loose, sandy soils.

DISTRIBUTION: Barbwire Russian thistle: True barbwire Russian thistle is found throughout the Mojave Desert, mostly below 500 m, Arizona, Colorado, Nevada, New Mexico, Oregon, and Utah. *Salsola* × *gobicola* distribution is somewhat intermediate in scope and is the only "*paulsenii*" type in the San Joaquin Valley, where it is restricted to the area from Bakersfield to Taft. The *Type "lax"* form is found in limited regions of western Utah and extensively in Nevada and in the southern San Joaquin Valley (Bakersfield area) and western Mojave Desert of California (northern Western Transverse Ranges), below 1300 m. Also occurs in western Nevada where it is common and mixed with *S. tragus.*

Russian thistle: The species is common throughout California, especially in the southern region of the state, to 2700 m. *Salsola tragus* is widespread in the central Valley and Coast Ranges. *Salsola kali* ssp. *austroafricana* is found in the San Joaquin Valley and in the south Coast Range, as well as in Arizona. **Type C** is found in the southern San Joaquin Valley (Kern County), around Coalinga in Kings County, and around Yuba City in Sutter County. The species as a group is found in all contiguous states except Florida.

Spineless Russian thistle: Currently not known to occur in California. Arizona, Colorado, Montana, New Mexico, Utah, Wyoming, most central states, and some northeastern states.

PROPAGATION/PHENOLOGY: *Reproduce by seed.* Seed appears to require an after-ripening period. Cotyledons are photosynthetic upon emergence.

Barbwire Russian thistle: Most seed falls near the parent plant, and plants typically do not become "tumbleweeds." Seed appears to require less of an after-ripening period before germination can occur, and germination occurs over a broader range of temperature. Otherwise, the reproductive biology of barbwire Russian thistle is similar to that of Russian thistle.

Russian thistle: Seeds disperse when plants break off at ground level and tumble with the wind. Seeds require an after-ripening period before germination can occur. Most seeds germinate the spring following maturation. Germination can occur when night temperatures are below freezing and daytime temperatures reach 2°C. Optimal temperature for germination is generally 7°–35°C. Seeds require very little moisture (about 7.5 mm of rainfall) to germinate, and germination occurs rapidly within in a few hours. Seedlings require loose soil for successful establishment. Seedlings that germinate on firm soil seldom survive because the young root or radicle is unable to penetrate the soil. Under field conditions, most seeds survive about 1 year, and a few seeds may survive up to 3 years. Plants about 0.5 m tall can produce about 1500–2000 seeds, and large plants can produce up to 100,000. Optimal emergence occurs from litter or soil depths to 1 cm, but some seedlings may emerge from soil depths to about 6 cm.

MANAGEMENT FAVORING/DISCOURAGING SURVIVAL: Seedlings cut just above the cotyledons seldom survive. Properly timed cultivation of seedlings prevents seed production and can control infestations, but cultivation must be repeated until the short-lived soil seedbank ($\leq$ 2 years) becomes depleted.

SIMILAR SPECIES: Glasswort [*Salsola soda* L., synonym: oppositeleaf Russian thistle] is a slender, erect to rounded summer *annual* to 0.5 m tall, with glabrous, fleshy, narrowly oblong leaves 6–55 mm long that have a rigid, ± acute tip, a narrow-winged base, and a ± whitish translucent margin. Unlike the other Russian thistles, glasswort *foliage remains fleshy in fruit,* and the fruits have a *calyx 3.5–5 mm long* with *tubercled inner sepals* (facing the stem) and *outer sepals that have wings less than 1.5 mm long*. Glasswort inhabits mudflats and saltmarshes in the San Francisco Bay region. Glasswort seedlings have glabrous, succulent, linear cotyledons that are about 10–15 mm long, ± 0.5 mm wide, and ± oval in cross-section. Seedlings are usually initially brownish, but develop green cotyledons with a reddish stalk (hypocotyl) within a short period. The subsequent 2 leaves resemble cotyledons, but may be up to 70 mm long. Native to southern Europe.

Mediterranean saltwort [*Salsola vermiculata* L., synonyms: Damascus saltwort; shrubby Russian thistle; wormleaf salsola][SASVE][Federal Noxious Weed][CDFA list: A] is a *shrubby perennial* to 1 m tall or more, with *minutely hairy, oblong to ovate leaves that have a rounded* tip and with *leaflike inflorescence bracts*. Flowers and fruits resemble those of Russian and barbwire Russian thistle. Mediterranean saltwort is an uncommon weed of disturbed rocky slopes and flats, often on clay soils, in the Temblor Range (se San Luis Obispo and possibly cw Kern Cos.), to 1000 m. Besides California, Mediterranean saltwort is a state-listed noxious weed in Florida and North Carolina. There remains some confusion as to which scientific name is the correct name to apply to the species in California. Currently, most taxonomists are using *Salsola vermiculata* L. However, California material is sometimes called *Salsola damascena* Bochantsev. Native to Syria and introduced in 1969 as an experimental range plant.

Immature **halogeton** [*Halogeton glomeratus* (M. Bieb.) C. Meyer] is distinguished from immature Russian thistles by having *fleshy, cylindrical leaves that are broadest near the tip* and *tufts of long white hairs in the leaf axils*. Refer to the entry for **Halogeton** (p. 618) for more information.

Common St. Johnswort [*Hypericum perforatum* L.] [HYPPE][Cal-IPC: Moderate][CDFA list: C]

SYNONYMS: goatweed; klamathweed; St. John's-wort; tipton weed. The genus *Hypericum* is sometimes segregated into the Hypericaceae, also known as the St. Johnswort family.

GENERAL INFORMATION: Erect *perennial* to 1.2 m tall, with *rhizomes* and *showy, bright yellow flowers* that have *numerous stamens*. Foliage is *dotted with tiny translucent and black oil glands* that contain hypericin, a fluorescent red pigment that is *toxic* to livestock when consumed in quantity, especially to animals with light-colored skin. Toxicity symptoms include skin photosensitivity of light-colored areas and loss of condition. Most animals graze plants only when more desirable forage is unavailable. In herbal medicine, hypericin is the antidepressant ingredient in St. Johnswort remedies, and is sometimes cultivated as a crop. There are several regional varieties of common St. Johnswort. The variety in the Pacific Northwest is aggressively competitive and can spread rapidly by seed and rhizomes. By 1940, more than 2 million hectares (about 1 million ha in California) of rangeland were infested. Several years later, the leaf-feeding flea beetles *Chrysolina quadrigemina* and *C. hyperici* and the root-boring beetle *Agrilus hyperici* were successfully introduced as biocontrol agents. Since beetles survive better at elevations below 1500 m, infestations are more common at higher elevations. Today infestations of the weed have been reduced by 97 to 99%. When control of common St. Johnswort is high, the beetles

Common St. Johnswort (*Hypericum perforatum*) infestation.   J. M. DiTomaso

Common St. Johnswort

populations may decline so low that they are unable to respond as the weed populations increase. Under these situations, the beetles must be reintroduced. Localized outbreaks of the plant sometimes occur after disturbances such as logging, fire, or during low population cycles of the flea and root-boring beetles. St. Johnswort has been used as a medicinal remedy for centuries in Europe. Hypericin concentrations can vary depending on the biotype of common St. Johnswort. Native to Europe.

SEEDLINGS: Cotyledons lanceolate to ovate, 1.5–3 mm long, 1–2 mm wide. Subsequent leaves opposite, oval to elliptic, increasingly larger. Underside of leaf margins dotted with a few elevated black glands.

MATURE PLANT: Stems highly branched near the top, glabrous, often reddish, with black glands along 2 longitudinal ridges. The main stem or stems usually

Common St. Johnswort *(Hypericum perforatum)* plant.   J. M. DiTomaso

Common St. Johnswort *(Hypericum perforatum)* sprout from rhizome.   J. M. DiTomaso

Common St. Johnswort *(Hypericum perforatum)* seedling.   J. M. DiTomaso

Common St. Johnswort *(Hypericum perforatum)* flower.   J. M. DiTomaso

have *numerous sterile shoots 2–10 cm long in the lower leaf axils*. Leaves opposite, sessile, *elliptic-oblong* to linear, 1–3 cm long, *flat, tip rounded*, glabrous, green, 3–5-veined from the base, *dotted with numerous tiny translucent glands* that are visible when a leaf is held up to the light, margin *rolled under* (revolute). Undersurface of leaf margins lined with *elevated black glands*.

ROOTS AND UNDERGROUND STRUCTURES: Taproot stout, with many branched lateral roots, to about 1.5 m deep. Rhizomes develop just below the soil surface from the crown and can extend outward to about 0.5 m. New shoots grow from the crown and rhizomes in early spring. Fragmented rhizomes can develop into new plants. Under favorable conditions, roots grow deeper and fewer rhizomes develop. Specialized corky tissue (polyderm), found only in a few plant families, protects the roots and crowns.

FLOWERS: June–September. Flowers bright yellow, ± 2 cm in diameter, clustered at the stem tips. Petals 5, separate, *8–12 mm long*, typically *dotted with black glands along the margins*. Sepals 5, *linear-lanceolate, 4–5 mm long, much shorter than petals*. Stamens yellow, *numerous*, in 3 fascicles. Styles 3, *3–10 mm long*. Plants typically do not flower the first year. Insect-pollinated and apomictic (seed develops without pollination).

FRUITS AND SEEDS: Capsules ovoid, *3-chambered* but *not lobed*, sticky-glandular, 5–10 mm long, with 3 persistent styles 3–10 mm long, open longitudinally to release seed. Seed shiny black to brown, nearly cylindrical, ± 1 mm long, densely pitted, often coated with gelatinous material from the capsule that aids with dispersal and that may inhibit germination until it breaks down or leaches out in about 4–6 months.

Common St. Johnswort (*Hypericum perforatum*) seeds. J. O'Brien

Common St. Johnswort

Canary Island hypericum (*Hypericum canariense*) infestation. J. WADE

POSTSENESCENCE CHARACTERISTICS: Aerial growth dies back during the cold season. In forested or wildland areas, dry flower stems can contribute to fire hazard risks.

HABITAT: Rangeland areas and pastures (especially those that are poorly managed), fields, roadsides, and forest clearings or burned areas in temperate regions with cool, moist winters and dry summers. Grows best on open, disturbed sites on slightly acidic to neutral soils. Does not tolerate water-saturated soils.

DISTRIBUTION: Northwestern region, Cascade Range, northern and central Sierra Nevada, Sacramento Valley, San Francisco Bay region, Central Coast, and Peninsular Ranges, to 1500 m. Most contiguous states except Arizona, New Mexico, Utah, Alabama, and Florida.

PROPAGATION/PHENOLOGY: *Reproduces by seed* and *vegetatively from rhizomes*. Plants can develop seed with or without pollination (facultative apomixis). Seed and capsules disperse with water and adhere to machinery, tires, shoes, clothing, and to animals. Seeds are hard-coated, and most of the seeds ingested by animals remain intact and viable. Seeds from plants in the Pacific Northwest usually do not require an after-ripening period. Germination occurs in fall through spring. Brief exposure to fire (100°–140°C) often stimulates germination. Calcium ions in water appear to inhibit germination of some biotypes. Seed can remain viable for about 10 years or more in the soil and for at least 5 years when submerged in fresh water. Plants typically produce an average of 15,000–33,000 seeds per plant. Seedlings survive best on disturbed open sites.

MANAGEMENT FAVORING/DISCOURAGING SURVIVAL: Releasing the leaf-feeding beetle *Chrysolina quadrigemina* on uninfested plant populations can

Common St. Johnswort

Canary Island hypericum (*Hypericum canariense*) plant. J. WADE

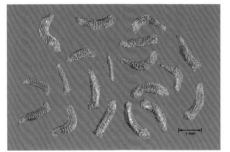

Canary Island hypericum (*Hypericum canariense*) seeds. J. O'BRIEN

Canary Island hypericum (*Hypericum canariense*) flowers. J. WADE

eventually give excellent long-term control. However, insects should be released under adequate moisture condition or reproduction will be greatly reduced. Cultivation readily controls plants on agricultural lands. Improving soil fertility and reseeding with desirable vegetation can help reduce or eliminate infestations in pastures. Mowing and overgrazing reduces seed production but promotes vegetative spread from the rhizomes. Systemic herbicides in spring can be effective. Burning stimulates seed germination and vegetative reproduction.

SIMILAR SPECIES: Canary Island hypericum [*Hypericum canariense* L.] [Cal-IPC: Moderate Alert] is an uncommon ornamental *shrub to 4(5) m tall* with large, showy, yellow flowers that has escaped cultivation in some areas. It is distinguished by having *narrow oblong-lanceolate to elliptic leaves 2–7 cm long* that are *often folded along the midvein* and have an *acute tip* and a *wedge-shaped base*. In addition, *petals are much longer than the sepals*, and the *sepals are lined with hairs on the margin* (ciliate). The 3 styles are 3–6 mm long. *Petals and stamens persist after flowering*. Capsules are *leathery, 3-chambered, and open at maturity*. Canary Island hypericum has become invasive in disturbed places, coastal sage scrub, and grassland in the South Coast (especially San Diego area) and southern San Francisco Bay region, to 100 m. Native to the Canary Islands.

Small-flowered St. Johnswort [*Hypericum mutilum* L., synonym: dwarf St. Johnswort] is an uncommon *erect annual to perennial to 0.6 m tall* that is distinguished by having *petals ± 2 mm long, sepals equal to or slightly longer than the petals, styles ± 1 mm long*, and *1-chambered capsules*. Leaves are elliptic to ovate,

Common St. Johnswort

Small-flowered St. Johnswort (*Hypericum mutilum*) plant.   J. M. DiTomaso

Small-flowered St. Johnswort (*Hypericum mutilum*) flowers.   J. M. DiTomaso

Small-flowered St. Johnswort (*Hypericum mutilum*) flowering stem.   J. M. DiTomaso

mostly 10–25 mm long, and dotted with flat, translucent glands. Small-flowered St. Johnswort inhabits moist places or riparian woodlands in the central-eastern Sacramento Valley and adjacent Sierra Nevada foothills, to 300 m. It is native to eastern North America and expected to expand its range in California.

In addition, three native *Hypericum* species occur in roughly the same regions as common St. Johnswort. However, the native species have at least one of the following characteristics: *± prostrate habit with matted stolons, linear to lanceolate leaves with acute tips and usually folded along the midrib, sterile shoots to 2 cm long at the base of the main stem(s), or 3-lobed capsules.*

Small-flowered St. Johnswort (*Hypericum mutilum*) seedlings.   J. M. DiTomaso

Small-flowered St. Johnswort (*Hypericum mutilum*) seeds.   J. O'Brien

Field bindweed [*Convolvulus arvensis* L.]
[CONAR][CDFA list: C]

SYNONYMS: cornbind; creeping charlie; creeping jenny; European morningglory; field morningglory; greenvine; lovevine; orchard morningglory; small bindweed; small-flowered morningglory; wild morningglory; *Convolvulus ambigens* House; *Convolvulus incanus* auct. non Vahl.; *Strophocaulos arvensis* (L.) Small

GENERAL INFORMATION: Herbaceous *perennial*, with vinelike stems and an *extensive system of deep, creeping roots*. Field bindweed is considered one of the most noxious weeds of agricultural fields throughout temperate regions of the world. Plants typically develop large patches and are difficult to control. Field bindweed is troublesome in numerous crops, but is especially difficult in cereals, beans, and potatoes. Heavy infestations in cereal crops can reduce harvest yields 30–40% or more. In California, it is estimated that 770,000 hectares of agricultural land were infested in 1981. Field bindweed can harbor the viruses that cause potato X disease, tomato spotted wilt, and vaccinium false bottom. Foliage contains tropane alkaloids and can cause intestinal problems in horses grazing on heavily infested pastures. Two biocontrol agents, the bindweed gall mite (*Aceria malherbae*) and bindweed moth (*Tyta luctuosa*), are cleared for release in the United States. However, these biocontrol agents are not registered for use in California because uncommon native morningglory (*Calystegia* spp.) may also be susceptible to attack. Native to Europe.

SEEDLINGS: Cotyledons unequal, *generally rectangular, square to kidney-shaped, indented at the tip*, about 8–20 mm long, 3–10 mm wide, glabrous, dull green

Field bindweed (*Convolvulus arvensis*) in field. J. K. Clark

Field bindweed

Field bindweed (*Convolvulus arvensis*) white flower form.   J. M. DiTomaso

Field bindweed (*Convolvulus arvensis*) pink flower form.   J. M. DiTomaso

Field bindweed (*Convolvulus arvensis*) seedling.   J. M. DiTomaso

with whitish veins, on stalks about 10–20 mm long. First leaves ± arrowhead-shaped, blunt at the tip, similar in size to the cotyledons. Subsequent leaves increasingly larger, generally resemble mature leaves. New leaves loosely creased along the main vein in bud. The taproot grows deep rapidly. By 6 weeks, creeping lateral roots have developed, mostly in the top 30 cm of soil.

MATURE PLANT: *Stems twine around and over other plants or trail along the ground.* Leaves alternate, short-stalked, dull green, *arrowhead-shaped* to ± oblong or nearly round, typically 2–4 cm long, tip often rounded, basal lobes ± pointed and often *flared outward* (hastate), surface glabrous or sparsely covered with short hairs, sometimes covered with a whitish powdery bloom. Leaf size and

Field bindweed (*Convolvulus arvensis*) foliage.   J. M. DiTomaso

Field bindweed (*Convolvulus arvensis*) seeds.   J. O'Brien

Field bindweed

Hollyhock bindweed (*Convolvulus althaeoides*) flower and leaf. J. M. DiTomaso

Hollyhock bindweed (*Convolvulus althaeoides*) seeds. J. O'Brien

Hollyhock bindweed (*Convolvulus althaeoides*) seedling. J. M. DiTomaso

Western morningglory (*Calystegia occidentalis*) (right) and field bindweed (*Convolvulus arvensis*) (left). J. M. DiTomaso

Western morningglory (*Calystegia occidentalis*) seeds. J. O'Brien

shape vary greatly depending on environmental conditions such as light intensity, soil moisture, and frequency of cultivation or defoliation.

ROOTS AND UNDERGROUND STRUCTURES: Roots cordlike, white, fleshy, brittle. Root systems consist of a vigorous, extensive network of primary and secondary taproots, numerous short-lived lateral feeder roots, and long-lived horizontal creeping roots that develop vertical rhizomes from endogenous buds. Rhizomes grow to the soil surface to produce new shoots. Roots can penetrate soil to a depth of 3 m or more depending on the availability of soil moisture. Most horizontal creeping roots develop in the top 60 cm of soil. Root systems

Field bindweed

Kidney-weed dichondra (*Dichondra micrantha*) plant.   J. M. DiTomaso

competitively extract soil moisture and can survive extended periods of drought and repeated cultivation.

FLOWERS: April–October or until the first frost. Flowers axillary, solitary, or in cymes of 2–4, on stalks (peduncles) about 2–6 cm long. *Corolla white or pinkish, funnel-shaped, 1.5–3 cm long*, pleated and spiraled in bud. Calyx ± bell-shaped, usually *less than 5 mm long. Stigmas 2, linear, cylindrical*, not flattened. Bracts 2, linear to narrowly lanceolate, 1–10 mm long, *attached mostly 1–2.5 cm below the flower.* Flowers open for 1 day. Insect-pollinated. Self-incompatible.

FRUITS AND SEEDS: Capsules spherical, ± inflated, ± 8 mm in diameter. Seeds few per capsule, variable in shape, but typically obovate, slightly compressed, ± 3-sided in cross-section, 3–4 mm long, dull dark gray-brown, covered with small, rough, irregular tubercles.

POSTSENESCENCE CHARACTERISTICS: Foliage typically dies back during the cool season.

HABITAT: Cultivated fields, orchards, vineyards, gardens, pastures, abandoned fields, roadsides, and waste places. Grows best on moist, fertile soils. Tolerates poor, dry, gravelly soils, but seldom grows in wet soils. Inhabits regions with temperate, Mediterranean, and tropical climates.

DISTRIBUTION: Abundant throughout California, to 1500 m. All contiguous states.

PROPAGATION/PHENOLOGY: *Reproduces by seed* and *vegetatively from deep, creeping roots and rhizomes*. Most seeds fall near the parent plant, but some seeds may disperse to greater distances with water, agricultural activities, and animals. Seeds are hard-coated and can survive ingestion by birds and other animals. Most seeds can imbibe water and germinate 10–15 days after pollination. However, seed coats mature 15–30 days after pollination, and about 80% of seeds become impermeable to water. Impermeable seeds require scarification or degradation of the seed coat by microbial action to imbibe water and germinate. Seeds germinate throughout the growing season, but peak germination usually occurs in mid-spring through early summer. Germination can occur under various temperature regimes from 5°–40°C, but it is highest and most rapid when temperatures fluctuate between 20°–35°C. A 3–6 week period of chilling to ± 5°C appears to increase germination. Light is not required. A large portion of the seedbank remains dormant from year to year. Buried seeds can survive for 15–20 years or more. A high percentage of seed under dry storage can survive at least 50 years. Seed production is highly variable. Dry, sunny conditions and calcareous soils favor seed production. Frequent cultivation, rain, or heavy wet soils can inhibit seed set. One plant can produce up to 500 seeds. In the field, young plants seldom produce seed the first season. Root starch reserves are highest from mid-summer through early fall, but then decline rapidly with conversion to sugars. Root carbohydrates are lowest in mid-spring before flowers develop. Maximum translocation of carbohydrates from shoots to roots occurs from the bud to full flower stages. Acclimated roots can survive temperatures as low as –6°C. Most new shoots appear in early spring. Undisturbed patches can expand their radius up to 10 m per year. Root fragments as small as 5 cm can generate new shoots.

Kidney-weed dichondra *(Dichondra micrantha)* seedling. J. K. CLARK

MANAGEMENT FAVORING/DISCOURAGING SURVIVAL: Deep cultivation before flowering and repeated cultivation when new shoots appear for one to several years, followed by rotation to competitive crops such as winter wheat or alfalfa, can control troublesome infestations in agricultural fields. Where practical, flooding fields with water to a depth of 15–25 cm for 60 to 90 days can effectively eliminate most field bindweed plants. Cultivation to a depth of at least 10 cm within 3 weeks of emergence can control seedlings. Stems and leaf surfaces have a texture that is difficult to wet. The addition of a wetting agent to certain herbicides may make them more effective. Some field bindweed biotypes are tolerant to certain herbicides.

SIMILAR SPECIES: Hollyhock bindweed [*Convolvulus althaeoides* L., synonym: mallow bindweed] is a *showy perennial* with *purple to deep pink, funnel-shaped flowers*. Hollyhock bindweed is further distinguished by having *some upper leaves deeply 3–many-lobed*. It grows in localized populations on disturbed sites in the northern Sierra Nevada foothills (Nevada Co.), Transverse Ranges, Peninsular Ranges, and Southwest region, to 1000 m. Hollyhock bindweed is a state-listed noxious weed in Utah. Native to the Mediterranean region.

Western morningglory [*Calystegia occidentalis* (A. Gray) Brummitt ssp. *fulcrata* (A. Gray) Brummitt and ssp. *occidentalis*, synonyms: chaparral false bindweed; *Convolvulus occidentalis* A. Gray; many others specific to each subspecies] is a variable native *perennial* that closely resembles *field bindweed*. It is a desirable component of the vegetation in natural communities, but is sometimes weedy in agricultural or managed forest systems. Unlike field bindweed, western morningglory typically has *flowers 2.5–4 cm long* with a *calyx more than 7 mm long, flattened stigma lobes,* and *bracts usually attached less than 1 cm below the flowers*. In addition, western morningglory foliage is *hairy throughout*, and the *basal lobes of leaves are often squared to slightly indented or 2-lobed*. Western morningglory inhabits dry slopes in chaparral and pine forests throughout California, except the Mojave and Sonoran deserts, to 2700 m. Subspecies *occidentalis* also occurs in Oregon.

Kidney-weed dichondra [*Dichondra micrantha* Urb., *Dichondra repens* J.R. & G. Forst.][DIORE] is *creeping, matlike perennial* that is cultivated as a lawn grass substitute. It sometimes escapes cultivation and becomes weedy in nearby landscaped areas and turf. Kidney-weed dichondra is easily recognized by its *creeping, matlike habit* and *small, kidney-shaped to nearly circular leaves with rounded lobes at base*. Leaves often form a shallow cup shape and are on stalks. Kidney-weed dichondra occurs as a localized weed throughout California. It is a widespread native of tropical and subtropical regions.

Alkaliweed [*Cressa truxillensis* Kunth][CSVTR]

SYNONYMS: alkali clover; cressa; spreading alkaliweed; *Cressa australis* R. Br.; *Cressa cretica* var. *truxillensis* (Kunth) Choisy (American authors, non L.); *Cressa depressa* Goodd.; *Cressa erecta* Rydb.; *Cressa insularis* House; *Cressa minima* Heller; *Cressa pumila* Heller; *Cressa vallicola* Heller; a few others

GENERAL INFORMATION: Native *perennial*, with prostrate to ascending stems to 0.25 m long or tall, *silvery gray-green foliage* covered with silky hairs, and small white flowers. Alkaliweed is a desirable component of the native vegetation in natural areas. However, it is occasionally weedy in agricultural fields, orchards, and associated margins and ditch banks, especially where the soil is saline or alkaline.

SEEDLINGS: Cotyledons nearly round, covered with short soft gray-white hairs. Leaves alternate, ± ovate.

MATURE PLANT: Stems ± tufted, prostrate or spreading to ascending, mostly branched, weakly woody at the base, densely covered with silky gray-white hairs. Leaves alternate, *nearly sessile*, ± *elliptic, 5–10 mm long, silvery gray-green*, covered with *silky gray-white hairs*.

ROOTS AND UNDERGROUND STRUCTURES: Taproot slender. Crown and upper taproot ± woody.

FLOWERS: May–October. Flowers single in upper leaf axils, 5–8 mm long, short-stalked. Corolla white, 5-lobed, sometimes appearing slightly bilateral. Lobes ± 2 mm long, spreading or curved backward, tip acute, silky-hairy on the outside. Sepals 5, separate, erect, 4–5 mm long, equal. Corolla and sepals persist

Alkaliweed (*Cressa truxillensis*) plants on railroad track slope. J. M. DiTomaso

Alkaliweed

Alkaliweed (*Cressa truxillensis*) plant.
J. M. DiTomaso

Alkaliweed (*Cressa truxillensis*) seeds.
J. O'Brien

Alkaliweed (*Cressa truxillensis*) flowering stem.
J. M. DiTomaso

in fruit. Stamens 5, anthers reddish. *Styles* 2, each with a globular white stigma. Stamens and styles extend beyond the corolla. *Ovary 2-chambered.*

FRUITS AND SEEDS: Capsules ± spherical, about 2–3 mm long, with long wavy hairs at the tip, remain enclosed by the persistent sepals and corolla, usually contain only 1 developed seed. Sepals and corolla orange-brown when dry. Seeds irregularly shaped and wrinkled, slightly flattened, 1.5–2 mm long.

POSTSENESCENCE CHARACTERISTICS: Foliage turns brown and does not persist.

HABITAT: Flood plains, salt marshes, roadsides, margins of agricultural fields, and ditch banks, especially on land recently converted to agriculture. Typically inhabits low-lying saline or alkaline sites in many plant communities, including seasonal wetlands. Designated a facultative wetland indicator species.

DISTRIBUTION: Throughout California, except Great Basin, to 1200 m. Arizona, New Mexico, Nevada, Oklahoma, Oregon, Utah, and Texas.

PROPAGATION/PHENOLOGY: *Reproduces by seed.* Capsules probably fall near the parent plant and disperse to greater distances with water, soil movement, and mud.

MANAGEMENT FAVORING/DISCOURAGING SURVIVAL: Manual removal or cultivation can control alkaliweed.

SIMILAR SPECIES: Alkaliweed is unlikely to be confused with other weedy species.

Japanese morningglory [*Ipomoea nil* (L.) Roth][IPOEH]
Tall morningglory [*Ipomoea purpurea* (L.) Roth] [PHBPU]

SYNONYMS: Japanese morningglory: ivyleaf morningglory; whiteedge morningglory; *Convolvulus hederaceus* L.; *Convolvulus hederaceus* var. *zeta* L.; *Convolvulus nil* L.; *Convolvulus tomentosus* Vell.; *Ipomoea cuspidata* Ruiz & Pav.; *Ipomoea githaginea* A. Rich.; *Ipomoea hederacea* auct. non Jacq.; *Ipomoea scabra* Forssk.; *Pharbitis nil* (L.) Choisy. This species has been incorrectly referred to as *Ipomoea hederacea* Jacq., which is a different species, and plants with heart-shaped leaves have been called *Ipomoea hederacea* Jacq. var. *integriuscula* Gray.

Tall morningglory: common morningglory; *Convolvulus hederaceus* var. *epsilon* L.; *Convolvulus hederaceus* var. *gamma* L.; *Convolvulus purpureus* L.; *Ipomoea diversifolia* Lindl.; *Ipomoea hirsutula* Jacq. f.; *Ipomoea purpurea* (L.) Roth. var. *diversifolia* (Lindl.) O'Donell; *Pharbitis nil* (L.) Choisy var. *diversifolia* (Lindl.) Choisy ex DC.

GENERAL INFORMATION: Variable, summer *annual vines*, with large, *heart-shaped to 3-lobed leaves* and *showy, funnel-shaped flowers*. Both species can have either 3-lobed leaves or heart-shaped leaves, and in tall morningglory, both leaf types can occur on the same plant. Morningglories are often cultivated as garden ornamentals. However, they can be troublesome weeds under favorable conditions. Japanese morningglory and other morningglories can be problematic in agronomic and vegetable crops, especially cotton fields. Established plants are difficult to control when the twining stems become entangled with

Japanese morningglory (*Ipomoea nil*) with lobed leaves.   J. M. DiTomaso

Japanese morningglory • Tall morningglory

Japanese morningglory (*Ipomoea nil* var. *integriuscula*) with unlobed leaves. Also called *Ipomoea hederacea* var. *integriuscula*. J. M. DiTomaso

Japanese morningglory (*Ipomoea nil*) flower. J. K. Clark

Japanese morningglory (*Ipomoea nil* var. *integriuscula*) seedling. J. K. Clark

Japanese morningglory (*Ipomoea nil*) seedling. J. M. DiTomaso

crop plants. Seeds of *Ipomoea* species contain an array of alkaloids, including ergot-type indole alkaloids that are *neurotoxic* to humans and animals when ingested. Japanese and tall morningglory contain low quantities of indole alkaloids, and usually many seeds must be ingested to produce neurological symptoms in adult humans. Symptoms can include nausea, purgative effects, sluggishness, excitement, and hallucinations. Digestive tract disturbances, weight loss, and death can occur in animals that consume small quantities (about

Japanese morningglory (*Ipomoea nil*) seeds. J. O'Brien

Tall morningglory (*Ipomoea purpurea*) seeds. J. O'Brien

Japanese morningglory • Tall morningglory

Tall morningglory (*Ipomoea purpurea*) flower.
J. M. DiTomaso

Tall morningglory (*Ipomoea purpurea*) seedling.
J. M. DiTomaso

Tall morningglory (*Ipomoea purpurea*) vine.
J. M. DiTomaso

3–8% of the diet) of seeds over a long period of time. However, contaminated grain typically does not contain enough morningglory seed to cause problems in livestock. The alkaloids are unstable with heat. Baking contaminated wheat flour at 121°C for an hour can destroy up to 40% of the alkaloid content. Both species are susceptible to the sweet potato feathery mottle virus. Japanese morninglory is also susceptible to apple, cucumber, and watermelon mosaic viruses. Japanese morningglory is a state-listed noxious weed in Arizona (prohibited). Tall morningglory is a state-listed noxious weed in Arizona (prohibited) and Utah. Japanese morningglory is native to the southeastern states and was introduced into Asia in the early 1500s. It has a long history of cultivation in Japan. Tall morningglory is native to tropical America.

SEEDLINGS: Cotyledons butterfly-shaped, length and width nearly equal, tip deeply 2-lobed, base ± truncate to slightly round-lobed, glabrous, ± glossy, upper surface dotted with imbedded translucent glands (punctate), on long stalks. Stalk below cotyledons (hypocotyl) usually purplish red at the base. Leaves alternate, hairy. First leaf usually heart-shaped, sometimes barely 3-lobed. Subsequent leaves heart-shaped to 3-lobed.

Japanese morningglory: Hairs ascending to erect, mostly on the stems, leaf stalks, and veins. Most hairs have a minutely papillate base.

Tall morningglory: Hairs minute, usually lay flat (appressed), ± evenly distributed on leaf surfaces.

Japanese morningglory • Tall morningglory

Blue morningglory (*Ipomoea indica*) vine. J. M. DiTomaso

MATURE PLANT: Stems branched, *twining*. Foliage hairy, visible with magnification. Leaves alternate, heart-shaped to deeply 3-lobed, lobes ± ovate to lanceolate, sinuses rounded.

Japanese morningglory: Leaves 3–8 cm long, 3-lobed or heart-shaped. Hairs short, stiff, *erect to ascending*, most with a minutely papillate base, especially on the stems and stalks. Stems and stalks ± densely hairy. Blades ± sparsely hairy, with most hairs on the veins.

Tall morningglory: Leaves mostly 7–12 cm long, often heart-shaped, sometimes 3-lobed. Both leaf types can occur on the same plant. Blades usually evenly covered with *minute, appressed hairs*, sometimes nearly glabrous.

ROOTS AND UNDERGROUND STRUCTURES: Taproots slender, with shallow spreading lateral roots.

FLOWERS: June–November. Flowers showy, funnel-shaped, axillary, stalked. Sepals 5. Sepal characteristics are important for species identification. Self-compatible.

Japanese morningglory: *Sepals mostly 10–20 mm long, ovate at the base, hairy, with a long, slender, straplike tip.* Flowers mostly 2–4 cm long, usually pale blue to purple, single or in clusters of 2–3.

Tall morningglory: *Sepals 10–16 mm long, lanceolate to oblong, hairy, tip acute.* Flowers mostly *4–6 cm long,* single or in clusters of 2–5, purple, blue, pink, white, or bi-colored.

FRUITS AND SEEDS: Capsules spherical, ± 1 cm in diameter, open by 2–4 valves, contain 4–6 seeds. *Sepals persist in fruit.* Seeds shaped like a wedge of a sphere, dark brown to black, 4–6 mm long, dull, surface granular and ± minutely hairy at low magnification.

POSTSENESCENCE CHARACTERISTICS: Dead vines with capsules can persist through winter.

HABITAT: Agronomic and vegetable crops, gardens, fields, orchards, vineyards, landscaped areas, waste places, and other disturbed sites. Grow best in moist fertile soil. Do not tolerate freezing conditions.

DISTRIBUTION: Japanese morningglory: San Joaquin Valley and Southwestern region, to 250 m. Texas, Louisiana, Oklahoma, Georgia, Florida, and Massachusetts.

Tall morningglory: Central Valley, Sierra Nevada foothills, Central-western region, and South-western region, to 100 m. Arizona, Nevada, New Mexico, Utah, central states, eastern states, and southern states.

PROPAGATION/PHENOLOGY: *Reproduce by seed.* Seed production can be high. Seeds fall near the parent plant or disperse to greater distances with water, mud, soil movement, agricultural operations, other human activities, and as a seed contaminant. Seeds typically germinate throughout spring and summer when moisture conditions are favorable. Some germination can occur in fall, but fall-germinating plants often do not survive through winter. Seedlings emerge from soil depths to 10 cm (4 in). Buried seeds can survive for long periods.

Blue morningglory (*Ipomoea indica*) flowers.   J. M. DiTomaso

MANAGEMENT FAVORING/DISCOURAGING SURVIVAL: Manual removal or cultivation of seedlings or mature plants before plants begin seed production can control the morningglories.

SIMILAR SPECIES: Refer to table 31 (p. 653) for a quick comparison of morningglorys.

Cairo morningglory [*Ipomoea cairica* (L.) Sweet, synonyms: five-leaved morningglory; mile-a-minute vine; messina creeper][IPOCA] is a *perennial vine* with *glabrous foliage, deeply palmate-lobed or compound leaves,* and white to purple flowers 5–9 cm long. Leaves are *compound or deeply lobed to the base* and typically have *5(–7) elliptic leaflets*, with the central leaflet usually less than 5 cm long. Sepals are ± ovate with *rounded-acute tips* and mostly 6–10 mm long. Cairo morningglory is uncommon and inhabits disturbed sites in the western South Coast Ranges, especially Santa Margarita Lake in San Luis Obispo County, to ± 300 m. It also occurs in some southern states. Initially Cairo morningglory was believed to be native to Africa. Native to South America.

Blue morningglory [*Ipomoea indica* (Burm. f.) Merr., synonyms: oceanblue morningglory; *Ipomoea mutabilis* Lindl.; *Ipomoea acuminata* (Vahl) Roemer & Schultes; *Ipomoea cathartica* Poir.; *Ipomoea congesta* R. Br.; others][IPOAC] is a *perennial vine*, with rose-lilac flowers 4–7 cm long that age to blue and 3-lobed to heart-shaped leaves. *Foliage is evenly covered with short, stiff, flattened, ± gray hairs,* and the *lower surfaces of leaves are densely hairy.* Unlike Japanese morningglory, blue morningglory has *sepals 20–30 mm long* that *gradually taper into a long, slender tip* and have *inconspicuous hairs ± 1 mm long*. Blue morningglory inhabits riparian sites and disturbed places in the Central-western and Southwestern regions, to about 50 m. Native to Mexico.

Threelobe morningglory [*Ipomoea triloba* L., synonym: littlebell][IPOTR] is a *glabrous* summer *annual to perennial vine*, with *narrowly ovate sepals 6–10 mm long* that have a *long pinched-in tip* (acuminate). Leaves are heart-shaped or 3-lobed. Unlike tall morningglory, threelobe morningglory has *narrowly bell-shaped flowers 1–2 cm long*. Flower color varies from shades of pink or violet to white. Threelobe morningglory inhabits fields, crop fields (especially cotton, asparagus, and alfalfa fields, orchards), and other disturbed places in the Sonoran Desert, primarily in the Imperial Valley, to 50 m. Like Japanese and tall morningglory, seeds of three-lobe morningglory contain alkaloids that are *toxic* to humans and animals when ingested. Threelobe morningglory is a noxious weed of crop fields in many tropical and subtropical regions of the world. It is state-listed as a noxious weed in Arizona (restricted), Florida, and North Carolina (class A). Native to tropical America.

Table 31. Morningglorys (*Ipomoea* spp.)

Ipomoea spp.	Life cycle	Leaf shape	Sepal length (mm); tips	Flower length (cm)	Foliage hairiness
I. cairica Cairo morningglory	perennial	lobed to base or palmate compound, leaflets 5–7	6–10; rounded-acute	5–9	glabrous
I. indica blue morningglory	perennial	3-lobed or heart-shaped	20–30; slender, long-tapered	4–7	appressed minute-hairy; lower blade surface densely hairy
I. nil Japanese morninglory	annual	3-lobed or heart-shaped	10–20; slender, straplike	2–4	erect-ascending short-hairy, mainly stems, stalks, and veins
I. purpurea tall morningglory	annual	3-lobed or heart-shaped, sometimes both on same plant	10–16; acute	4–6	appressed minute-hairy; blades ± evenly hairy
I. triloba threelobe morningglory	annual, perennial	3-lobed or heart-shaped	mostly 6–10; tapered, pinched-in (acuminate)	1–2	glabrous

Citronmelon [*Citrullus lanatus* (Thunb.) Mansf. var. *citroides* (Bailey) Mansf.][CITLC]

SYNONYMS: citron; preserving melon; wild melon; *Citrullus colocynthis* (L.) Schrader var. *citroides* (Bailey) Mansf.; *Citrullus vulgaris* Schrad. var. *citroides* Bailey

GENERAL INFORMATION: Prostrate summer *annual vine*, with stems to a few meters long, yellow bowl-shaped flowers, and nearly spherical fruits that resemble small watermelons. Citronmelon is a variety of cultivated watermelon [*Citrullus lanatus* (Thunb.) Matsum & Nakai var. *lanatus*]. The fruit rind is used to make preserves, but the flesh is solid, white, and inedible. Citronmelon is especially undesirable in commercial watermelon fields because it readily crosses with commercial watermelon cultivars, resulting in the production of inferior fruit. A few botanists believe the correct name for this species is *Citrullus colocynthis* (L.) Schrader var. *citroides* (Bailey) Mansf. Native to southern Africa.

SEEDLINGS: Resemble watermelon seedlings. Cotyledons ± oval, light green. First leaf ± round, margins wavy, stalk long and hairy. Second and third leaves often resemble the first leaf. Subsequent leaves resemble mature plant leaves.

MATURE PLANT: Foliage hairy. Stems prostrate, branched, with *branched tendrils*. Leaves alternate, 3–8 cm long, *deeply pinnate-lobed, often appearing ± palmate-lobed*. Lobes pinnate-lobed again, tip ± rounded and broad.

ROOTS AND UNDERGROUND STRUCTURES: Taproot stout, with numerous shallow lateral roots.

Citronmelon (*Citrullus lanatus* var. *citroides*) plant in flower.   J. M. DiTomaso

Citronmelon

Citronmelon (*Citrullus lanatus* var. *citroides*) fruit. J. M. DiTomaso

Citronmelon (*Citrullus lanatus* var. *citroides*) seedling. J. M. DiTomaso

Paddymelon (*Cucumis myriocarpus*) fruit and decayed stems. T. Fuller

FLOWERS: Summer. Male and female flowers develop singly at different nodes on the same plant (monoecious). Flowers yellow, shallow bowl-shaped, ± 3 cm wide. Petals 5, fused near the base. Anthers 3. Stigmas 3, kidney-shaped. Ovary inferior. Insect-pollinated, especially by bees.

FRUITS AND SEEDS: Mature fruits resemble a *small, nearly spherical watermelon. Rind hard, green with pale green or whitish blotchy stripes, hairy. Flesh solid, white, inedible.* Seeds oval, flattened, ± 1 cm long, greenish, tan or tan with dark stripes, to black.

POSTSENESCENCE CHARACTERISTICS: Dead stems with fruits can persist into winter.

HABITAT: Agronomic and vegetable crops, orchards, vineyards, sites near gardens, roadsides, dry river and creek beds, and other disturbed places. Grows best on sandy fertile soil. Killed by frost.

DISTRIBUTION: Sporadic in the San Joaquin Valley, and probably elsewhere, to

Citronmelon

Smellmelon (*Cucumis melo* var. *dudaim*) infestation in asparagus. L. Kreps

Smellmelon (*Cucumis melo* var. *dudaim*) fruit. Courtesy of R. O'Connell

300 m. Nevada, New Mexico, Texas, Florida, Illinois, possibly some other states.

PROPAGATION/PHENOLOGY: *Reproduces by seed.* Seeds usually remain within the dried fruits. Dried fruits decompose in place or disperse with water, human activities such as agricultural operations and road grading, and possibly animals. Seeds germinate in mid- to late spring. Fruits mature in mid-summer through early fall.

MANAGEMENT FAVORING/DISCOURAGING SURVIVAL: Citronmelon is generally undistinguishable from cultivated watermelon until fruits develop. Removing plants before fruits mature can limit the quantity of seed available to germinate in subsequent years.

SIMILAR SPECIES: Paddymelon [*Cucumis myriocarpus* E. Meyer ex Naud.; synonyms: bitter apple; gooseberry gourd; prickly paddy melon][CDFA list: A] is a summer *annual vine* with foliage that resembles that of watermelon and citronmelon. Paddymelon is distinguished by having *unbranched tendrils* and *round fruits 2–3 cm wide that are covered with weak prickles*. Fruits are initially pale green striped with dark green and turn yellow at maturity. Fruits, especially seeds, sometimes contain cucurbitacins, and can be *toxic* to livestock or humans when ingested. Toxicity problems from the ingestion of paddymelon occur in Australia and South Africa, where livestock sometimes develop digestive tract irritation. However, livestock and deer appear to consume the fruits without ill effect in California. Paddymelon does not hybridize with commercial melons. It is susceptible to several viruses that affect tomatoes and potatoes. Paddymelon inhabits fields and disturbed sites in the central and southern San Joaquin Valley and southwestern South Coast Ranges (Santa Barbara Co.), to 300 m. Native to South Africa.

Smellmelon [*Cucumis melo* L. var. *dudaim* (L.) Naud., synonyms: dudaim melon; pomegranate melon; Queen Anne's pocket melon; *Cucumis melo* L. var. *chito* (C. Morren) Naud.; *Cucumis odoratissimus* Moench.][CUMMD][CDFA list: A] is a weedy summer *annual variety* of muskmelon [*Cucumis melo* L.], with prostrate stems to 10 m long or more and *unbranched tendrils*. Fruits are more or less edible, but plants are most often grown as ornamentals or for the fragrance of the fruits. Smellmelon readily hybridizes with commercial cantaloupe cultivars and is a stated-listed noxious weed for this reason. Unlike paddy melon, smellmelon has leaves that are *palmately shallow-lobed to angled* and *cylindrical to round fruits mostly 2–3 cm wide that lack prickles*. Fruits are orange at maturity. Currently, smellmelon is considered eradicated in California. It previously occurred in the southeastern Sonoran Desert (c and e Imperial Co.), to 200 m. Weedy varieties of *Cucumis melo* L. are reported to occur in Nevada, New Mexico, Utah, most southern and eastern states, and adjacent central states. Besides California, smellmelon is a state-listed noxious weed in Arizona (prohibited). Native to tropical Africa.

Buffalo gourd [*Cucurbita foetidissima* Kunth, synonyms: calabazilla; stinking gourd; chili coyote; Missouri gourd; wild gourd; wild pumpkin; *Pepo foetidissima* (Kunth) Britt.][CUUFO] is a widespread native *perennial vine*, with *course foliage* and a *large, tuberlike root*. It is distinguished by having *triangular-ovate leaves mostly 15–30 cm long* and *deeply cup-shape, yellow flowers mostly 9–12 cm long*

Buffalo gourd (*Cucurbita foetidissima*) plant.
J. M. DiTomaso

Buffalo gourd (*Cucurbita foetidissima*) flower.
J. M. DiTomaso

Buffalo gourd (*Cucurbita foetidissima*) fruit.
J. M. DiTomaso

and as wide. In addition, the stiff-hairy foliage has an *unpleasant scent* and *tendrils are branched* 1 cm or more above the base. Fruits are round, 7–9 cm in diameter, and green with blotchy white stripes. Buffalo gourd is generally considered a desirable component of the vegetation in natural areas, but it is occasionally weedy in agronomic fields. The fruits contain cucurbitacins and can be *toxic* to livestock and humans when ingested. Symptoms include severe digestive tract irritation. However, animals rarely consume the bitter-tasting fruits. Buffalo gourd inhabits roadsides, fields, including agronomic crop fields, waste places, and other

Wild cucumber *(Marah macrocarpus)* vine in flower. A native nonweedy species.   J. M. DiTomaso

Wild cucumber *(Marah macrocarpus)* fruit. A native nonweedy species.   J. M. DiTomaso

Wild cucumber *(Marah oreganum)* seed. A native nonweedy species. J. M. DiTomaso

sandy sites in the Central Valley, Central-western and Southwestern regions, and deserts, to 1300 m. It also occurs in Arizona, Colorado, New Mexico, Nevada, Utah, Wyoming, Texas, Florida, and most central states, except a few of the northernmost.

Wild cucumbers [*Marah* spp., synonym: manroot] are glabrous to sparsely minute-hairy native *perennial vines* with large tuberlike roots that may be confused with the preceding species. Wild cucumbers are a desirable component of the vegetation in natural areas and are usually *not* considered weeds. On rare occasions, wild cucumber may persist for a period on land that has been recently developed or converted to agriculture. Wild cucumber is distinguished by having *white flowers to 2 cm in diameter, male flowers in clusters,* and round to oblong fruits 2–20 cm long that are *sparse to densely covered with straight or hooked prickles.* Leaves are palmate-lobed, usually with 5–7 ± acute lobes that are sometimes toothed or shallow-lobed again. Wild cucumbers inhabit open sites in shrubland, woodland, forests, and riparian areas throughout most of California, except the Great Basin and Mojave and Sonoran deserts, to 1800 m. *Marah* species occur in all contiguous western states except Colorado and Utah.

Western juniper [*Juniperus occidentalis* Hook. var. *occidentalis*]

SYNONYMS: western cedar; *Juniperus californica* Carr. var. *siskyouensis* Henderson; *Sabina occidentalis* (Hook.) Heller; *Sabina occidentalis* (Hooker) Antoine

GENERAL INFORMATION: Strongly aromatic native *tree* to 20 m tall, with *small, scalelike leaves*. Western juniper is usually considered a desirable component of the natural vegetation in most areas. It is an important source of shelter and food for a variety of animal species. However, western juniper has greatly expanded range on rangeland in the Great Basin region, including the Modoc Plateau. In some shrub-grassland communities, tree density has increased to the point where desirable forage and browse species have severely declined. In addition, soil erosion appears to have accelerated in such areas. Ecologists believe that the introduction of large numbers of livestock in the past century and the reduction of fire frequency have contributed much to the expansion of western juniper. Because the trees are extremely longed-lived, simply reducing grazing pressure cannot reduce tree density where they are established. There are two varieties of western juniper in California. Variety *occidentalis* is most often associated with problematic density increase and range expansion in the Great Basin. Variety *australis* (Vasek) A. & N. Holmgren, also known as Sierra

Western juniper (*Juniperus occidentalis* var. *occidentalis*) tree.   J. M. DiTomaso

Western juniper

Western juniper (*Juniperus occidentalis* var. *occidentalis*) branch with male flower clusters and fruit.
J. M. DiTomaso

juniper, occurs in the Sierra Nevada and adjacent region of Nevada, Yolla Bolly Mountains, San Gabriel Mountains, San Bernardino Mountains, and desert mountains. It does not appear to be significantly expanding its range. Pregnant cows or ewes that ingest large quantities of juniper foliage may abort or give premature birth to weak or stillborn young. In northeastern California, western juniper is sometimes harvested and chipped as a biomass energy source.

SEEDLINGS: Cotyledons 2–4, linear.

MATURE PLANT: Evergreen rounded to pyramidal tree, usually with a single trunk to ± 2 m in in diameter. Foliage and wood strongly aromatic. Branches often large, spreading. Bark shredding, gray-brown to brown, deeply furrowed into broad flat scaly ridges. *Leaves scalelike*, closely appressed, *whorled in 3's or opposite*, ± 3 mm long.

ROOTS AND UNDERGROUND STRUCTURES: Woody roots near the soil surface can extend 2 to 3 times beyond the canopy.

REPRODUCTIVE STRUCTURES: *Male and female cones develop at the stem tips and mostly on the same tree* (± monoecious). Pollen cones oblong, 2–4 mm long, yellowish brown. *Seed cones fleshy, berrylike*, nearly round to ellipsoid, 5–12 mm long, *blue-black at maturity*, covered with a whitish film (glaucous), pulp resinous, mature in the fall of the second season. Seeds 2–3 per cone, ovoid, ± 6 mm long, tip acute.

POSTSENESCENCE CHARACTERISTICS: Dead trees can persist for many years.

HABITAT: Dry mountain slopes, flats, sagebrush communities.

DISTRIBUTION: Modoc Plateau, higher elevations of the Cascade Range, mostly 700–2300 m. Idaho, Nevada, Oregon, and Washington.

PROPAGATION/PHENOLOGY: *Reproduces by seed.* Fruits disperse primarily by animals, especially birds. Seeds are hard-coated and survive ingestion by animals. Most germination occurs in early spring.

MANAGEMENT FAVORING/DISCOURAGING SURVIVAL: The lower branching pattern of mature trees makes mechanical control difficult. Trees killed by fire do not resprout from the roots. Historically, western juniper has been controlled with prescribed fire. When there is enough fuel present to carry fire, small controlled burns can kill young trees. Often the seedbank of desirable species is depleted, especially in areas with many large trees. Immediate reseeding of burned areas with desirable species can help to prevent invasion by unwanted weed species and to reduce soil erosion. Selective herbicides have also been effective for control of these species.

SIMILAR SPECIES: Western juniper is distinguished from other juniper species with appressed scalelike leaves by having *blue-black seed cones.* Singleleaf pinyon (*Pinus monophylla* Torr. and Frem.) has caused simple problems as western juniper in eastern Nevada and some areas of California in the Great Basin. It is easiy distinguished from juniper and most other pines by its single needle, 2–3 cm long.

Dodder [*Cuscuta* spp.][CDFA list: C]

Giant dodder [*Cuscuta reflexa* Roxb.] [CDFA list: A][Federal Noxious Weed]

SYNONYMS: angel's hair; devil's gut; love vine; strangle gut; tangle gut; witch's hair; witch's shoelaces

GENERAL INFORMATION: Annual plants that parasitize the stems or leaves of other plants. The dodders have leafless, threadlike, *orange, red, or yellow stems* that twine over other plants. Some dodder species are problematic in agricultural crops, especially alfalfa and tomatoes. Dodder seed is difficult to exclude from commercial alfalfa, clover, or flax seed.

Dodder: Of the 9 dodder species in California, 8 are native, and several are uncommon. Species identification is difficult and beyond the scope of this publication. Native species usually grow on various herbs and shrubs in most natural communities and are *not* considered weeds under these conditions. In agricultural fields, field dodder [*Cuscuta pentagona* Engelm.], large-seeded dodder [*C. indecora* Choisy], and California dodder [*C. californica* Hook. & Arn.] are the most commonly encountered native species, in the order listed. Alfalfa dodder [*Cuscuta approximata* Bab.] also parasitizes alfalfa and other crops, but is less common than the previous species. Alfalfa dodder is native to the Old World. *Fusarium tricinctum* and *Alternaria conjuncta* cause fungal diseases that are specific to dodder. Both fungi are used as biocontrol agents in some states to help control problematic dodder populations in crop fields. However, neither

Dodder (*Cuscuta* spp.) infestation in a tomato field. J. K. CLARK

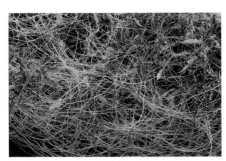

Dodder (*Cuscuta pentagona*) on a tomato plant.
J. M. DiTomaso

Dodder (*Cuscuta pentagona*) flowers.
J. M. DiTomaso

Young dodder (*Cuscuta* spp.) plant parasitizing a tomato plant.
J. K. Clark

Dodder (*Cuscuta pentagona*) seeds.
J. O'Brien

agent is currently registered for use in California. In addition, the effectiveness of these organisms is somewhat limited by the dry climate in California.

Giant dodder: The only known population of giant dodder in California has been eradicated. Giant dodder is a larger, more-vigorous species than the other dodders. It typically attacks woody plants and is a serious pest of citrus, coffee, peach, forest trees, and many other tree species in Asia. Native to Asia.

SEEDLINGS: Lack cotyledons. Develop a small temporary root to support a threadlike shoot, 4–10 cm long. The stem detaches from the root upon penetration into the host tissues. Shoot moves slowly in a circular pattern as it grows until it touches a support. Upon contact with a suitable host, knob-shaped organs (haustoria) develop to penetrate the host stem. Seedlings that do not contact a host plant die within 10 days to several weeks, depending on the species.

MATURE PLANT: Stems 1–2 mm thick, glabrous, lack leaves or have appressed, scalelike leaves about 2 mm long; red, yellow, or orange. Growing stems branch and attach to new host stems with haustoria. Each dodder branch obtains nutrients from the host independent of older branches. One plant can cover 0.9–1.4 m^2.

ROOTS AND UNDERGROUND STRUCTURES: Dodders lack typical underground roots. Stems are *modified into specialized knob-shaped organs (haustoria)*

Dodder (*Cuscuta subinclusa*) flowers. J. M. DiTomaso

that penetrate the stems or leaves of a host plant. Once a host attachment occurs the soil connections dry up and die.

FLOWERS: Generally May–October, but varies according to species and location. Flowers bell-shaped, in axillary clusters. Calyx 4–5 lobed, usually persistent. Corolla 4–5-lobed, usually with an equivalent number of appendages on the inside.

Dodder: Corolla *less than 6 mm long*, white to pinkish. *Styles 2, separate or fused at the base, with 1 stigma per style.* Stigmas linear (alfalfa dodder) or headlike (native dodders).

Giant dodder: Corolla *6–10 mm long*, usually white with a purplish rim. Calyx ± ¼ corolla length. *Style 1, very short, with 2 linear stigmas that are longer than the style.*

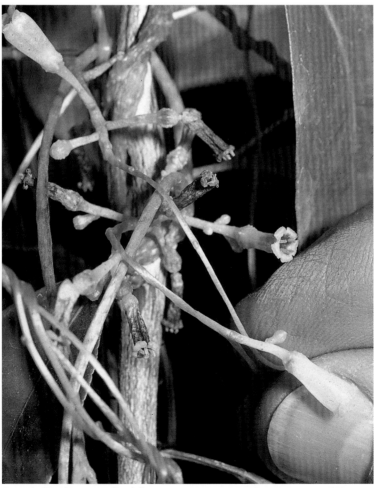

Giant dodder (*Cuscuta reflexa*) flower.   COURTESY OF ROSS O'CONNELL

FRUITS AND SEEDS: Capsules spherical, ovoid, or conical; open irregularly or like a lid at the top (circumscissile). Seeds 1–4, spherical, oblong, ovoid, or angular.

Dodder: Capsules mostly *less than 5 mm long*. Seeds 1–2 mm long.

Giant dodder: Capsules conical, 5–8 mm long. Seed 3–3.5 mm long.

POSTSENESCENCE CHARACTERISTICS: Stems generally do not persist through winter, but haustorial tissue of some species can overwinter in host stems and develop new stems the following spring. Frost kills plants.

HABITAT: Agricultural crops (especially alfalfa, clovers, and tomatoes) and most natural plant communities. Occasionally nursery crops and landscaped sites. Host plant range is broad, but monocots, excluding asparagus and onions, are seldom affected.

DISTRIBUTION: Dodder: Throughout California, to 2500 m. All contiguous states.

Giant dodder: Currently not known to occur in California or any other contiguous state. The one population that has been eradicated occurred in the South Coast (se Los Angeles Co.).

PROPAGATION/PHENOLOGY: *Reproduces by seed and vegetatively from the stems.* Broken stems can develop new haustoria. Seeds disperse with water, animals, and human activities, including agricultural equipment. Seeds germinate in spring. A proportion of the seed has a hard coat that must be weakened by scarification, microbial decomposition, or winter chilling before germination can occur. Germination does not require the presence of a host plant. Under favorable conditions, seeds can germinate in the fruits. Seeds generally remain viable for at least 10 years in the soil seedbank. Most seedlings emerge from the top 5 cm of soil. In most years, the period of emergence ends by mid-May in the Central Valley.

MANAGEMENT FAVORING/DISCOURAGING SURVIVAL: Many options are available to help decrease crop infestations. Useful methods include hand-cultivation, spot or field burning, close mowing, later planting time (mid-May or later), and crop rotation to cereals or corn. Prevent the spread of dodder seed by thoroughly cleaning agricultural machinery immediately after it has been used in an infested field. Hand-cultivation is most efficient at 30 days after planting a crop and repeated at 50 days.

SIMILAR SPECIES: The distinctive orange, red, or yellow twining stems of dodder species are unlikely to be confused with any other plant.

Japanese dodder [*Cuscuta japonica* Choisy [CDFA list: A]; Federal Noxious Weed] is a recent introduction to California (2004). It differs from most other dodder species in that it parasitizes primarily woody plants, including economically important fruit trees. Its stems are about twice as thick as native *Cuscuta* species. Native to Asia.

Smallflower umbrella sedge [*Cyperus difformis* L.] [CYPDI]

Tall flatsedge [*Cyperus eragrostis* Lam.]

SYNONYMS: Smallflower umbrella sedge: variable flatsedge; *Cyperus lateriflorus* Torr.

Tall flatsedge: tall cyperus; tall umbrellaplant; tall umbrellasedge; *Cyperus monandrus* Roth; *Cyperus serrulatus* Vahl; *Cyperus vegetus* Willd.

GENERAL INFORMATION: Erect, grasslike plants, with *closed sheaths, 3-sided stems,* and *clusters of spikelets in compound umbels.*

Smallflower umbrella sedge: Tufted summer *annual* to about 0.4 m tall, with inflorescences consisting of small, spherical clusters of greenish brown to purplish spikelets at maturity. Smallflower umbrella sedge is typically a weed of rice fields in tropical to warm temperate regions nearly worldwide. Some biotypes have developed a tolerance to certain herbicides. Plants use the C3 photosynthetic pathway. Native to the subtropical regions of Asia and Africa.

Tall flatsedge: Erect *perennial* to about 1 m tall, with *rhizomes* and inflorescences consisting of compact clusters of flattened green spikelets that turn tannish with age. Tall flatsedge is a common, widespread native and a desirable component of the vegetation in natural areas. However, plants can be weedy in moist agricultural fields, including rice fields, orchards, landscaped areas, and irrigation ditches.

Smallflower umbrella sedge (*Cyperus difformis*) plant. J. K. Clark

Smallflower umbrella sedge (*Cyperus difformis*) inflorescence. J. K. Clark

SEEDLINGS: Smallflower umbrella sedge: Cotyledon ± elliptic, ± 1 mm long, glabrous, translucent. First leaf blade usually 4–8 mm long, 0.25–0.5 mm wide, 3-veined, bright green. Sheath about 1–2 mm long. Roots sometimes pinkish.

Tall flatsedge: Similar to yellow nutsedge seedlings.

MATURE PLANT: Flower stems (culms) smooth. Basal leaves about equal in length to the culms or less.

Smallflower umbrella sedge: Culms sharply triangular in cross-section, sides often concave. Basal leaves typically 2–4 per culm, mostly 1–4 mm wide, slightly rough to touch (scaberulous) on the margins near the tips.

Tall flatsedge: Flowering stems slightly swollen at the base, *triangular in cross-section, angles smooth and often ± rounded or blunt.* Basal leaves usually 6–10 per culm, mostly (5–)10 mm wide, slightly rough to touch on the margins and lower midvein near the tips.

ROOTS AND UNDERGROUND STRUCTURES: Smallflower umbrella sedge: All roots fibrous, sometimes reddish.

Tall flatsedge: Rhizomes short, thick. Fibrous roots coarse.

SPIKELETS/FRUITS: May–November. *Spikelets persistent,* composed of 2 rows of overlapping *deciduous scales* (tiny bracts) that conceal the achenes. Rachis (stalk to which scales are attached) *straight, lacks narrow wings.* Style 3-branched. *Achenes 3-sided,* obovoid.

Smallflower umbrella sedge (*Cyperus difformis*) on edge of rice field. J. M. DiTomaso

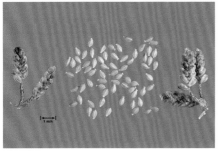

Smallflower umbrella sedge (*Cyperus difformis*) achenes and spikelets. J. K. Clark

Smallflower umbrella sedge (*Cyperus difformis*) seedling. J. K. Clark

Tall flatsedge (*Cyperus eragrostis*) plant.
J. M. DiTomaso

Tall flatsedge (*Cyperus eragrostis*) inflorescence.
J. M. DiTomaso

Tall flatsedge (*Cyperus eragrostis*) spikelet clusters.
J. M. DiTomaso

Tall flatsedge (*Cyperus eragrostis*) achenes and spikelets.
J. K. Clark

Smallflower umbrella sedge: Inflorescence leaves 2–3. Spikelet clusters *dense, spherical*, typically consist of *50–100 spikelets* or more, sessile and/or on unequal stalks (rays) to 7 cm long. Spikelets ± oblong, *slightly flattened*, mostly 4–8 mm long. *Scales rounded at the tips*, membranous, mostly 0.5–1 mm long, green with brown to purplish sides, readily deciduous. Stamens 1–2. Achenes obovoid, 0.5–1 mm long, with a *minute, nipplelike tip* (mucronulate), pale greenish brown, glossy, surface finely cellular under magnification.

Tall flatsedge: Inflorescence leaves 4–8, 3–50 cm long, slightly rough to touch on the margins and midvein. Spikelet clusters dense, ± headlike, consist of about 20–70 spikelets, nearly sessile or on stalks to 10 cm long. Spikelets narrowly ovate, *strongly flattened*, mostly 10–20 mm long, *3–3.5 mm wide*. Scales ovate, 2–2.3 mm long, tip acute, keeled on the backs, 3-veined, midvein minutely roughened at the tip, straw-colored, enfold achene, deciduous with achene. Stamens 1. Achenes obovoid, *1–1.5 mm long and nearly as wide*, with a *short beak at the tip* (mucronate) and *short stalk at the base* (stipitate), black to dark brown, surface covered with minute sunken dots under magnification.

HABITAT: Rice fields, ditches, pond margins.

Smallflower umbrella sedge: Typically inhabits shallow water or wet soils. Grows best on fertile soils. Does not tolerate deep water.

Tall flatsedge: Also stream sides and vernal pools. Inhabits shallow water to moist soil.

DISTRIBUTION: Smallflower umbrella sedge: Central Valley and Southwestern region, to 500 m. Arizona, New Mexico, Nebraska, Pennsylvania, and most southern states.

Tall flatsedge: Throughout California, except Great Basin and desert regions, to 700 m. Oregon, Washington, Texas, Louisiana, Mississippi, New Jersey, Pennsylvania, and South Carolina.

Redroot flatsedge (*Cyperus erythrorhizos*) on bank of irrigation canal.   J. M. DiTomaso

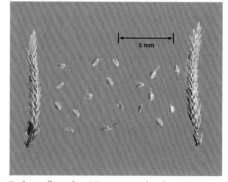

Redroot flatsedge (*Cyperus erythrorhizos*) achenes and spikelets.   J. K. Clark

Redroot flatsedge (*Cyperus erythrorhizos*) inflorescence.   J. M. DiTomaso

PROPAGATION/PHENOLOGY: Seeds disperse with water, soil movement, agricultural activities, and possibly animals.

Smallflower umbrella sedge: *Reproduces by seed.* Plants typically complete their life cycle (from seed to seed production) in about 2 months. Seed production is typically high. Germination requires light. Observations suggest that seed can survive at least 5 years in the field. Deeply buried seeds may survive for many years (about 50 years in one case).

Tall flatsedge: *Reproduces by seed* and *vegetatively from rhizomes.*

SIMILAR SPECIES: Redroot flatsedge [*Cyperus erythrorhizos* Muhl., synonym: *Cyperus halei* Torr. ex Britt][CYPET] is a common, widespread native *annual* to 1 m tall with *reddish roots* that resembles tall flatsedge. Redroot flatsedge is distinguished by having more *open inflorescences* with *barely flattened, linear*

Whitemargined flatsedge (*Cyperus flavicomus*) inflorescence. J. M. DiTomaso

Whitemargined flatsedge (*Cyperus flavicomus*) achenes and spikelets. J. K. Clark

Whitemargined flatsedge (*Cyperus flavicomus*) spikelet clusters. J. M. DiTomaso

spikelets 3–10 mm long and *1–1.5 mm wide*. In addition, the light brown to light reddish brown scales are usually *1–1.5 mm long* and ± *rounded at the tip with a minute, abrupt point* (mucronulate). Achenes are glossy, *light gray to brown*, ± oblong, *unequally 3-sided*, 0.5–1 mm long, nearly as wide, with a blunt nipple-like tip and finely cellular surface under magnification. Like the other sedges described above, redroot flatsedge flowers have a *3-branched style*. Redroot flatsedge is sometimes weedy in wet agricultural fields and ditches and occurs throughout California to 150 m. It occurs in all contiguous states, except possibly Vermont.

Whitemargined flatsedge [*Cyperus flavicomus* Michx.; synonym: *Cyperus albomarginatus* (Mart. & Schrad. ex Nees) Steud.] is an *annual* to 1 m tall that looks somewhat like redroot flatsedge. Whitemargined flatsedge is distinguished by having *flowers with a 2-branched style, (ob)ovate flower scales 1.5–2 mm long with a broad translucent-white margin*, and *glossy black, lens-shaped achenes that are nearly as large as the associated scale*. Whitemargined flatsedge is not included in most current California floras at publication time. It is generally associated with rice fields in the northern Sacramento Valley (Butte Co.), to 50 m. Whitemargined flatsedge has also been collected in the southeastern San Joaquin Valley (Tulare Co.) and may occur elsewhere in California. Native to the southern and eastern states.

Brown flatsedge [*Cyperus fuscus* L.] is an *annual* to 0.3 m tall that resembles smallflower umbrella sedge. Unlike smallflower umbrella sedge, brown flatsedge has *open spikelet clusters that consist of about 3–15 spikelets*. Brown flatsedge inhabits disturbed wet places in the San Joaquin Valley, to 50 m. It also occurs in South Dakota, Nebraska, and a few eastern states. Native to temperate Eurasia.

Yellow nutsedge [*Cyperus esculentus* L.][CYPES] [CDFA list: B]

Purple nutsedge [*Cyperus rotundus* L.][CYPRO] [CDFA list: B]

SYNONYMS: Yellow nutsedge: chufa; chufa flatsedge; earth almond; edible galingale; northern nutgrass; rush nut; watergrass; yellow nutgrass; many varieties scientifically named

Purple nutsedge: chaguan humatag; coco sedge; cocograss; kili'o'opu; pakopako; purple nutgrass

GENERAL INFORMATION: *Perennials* with grasslike leaves, *triangular stems*, and *slender rhizomes with small tubers* attached. The nutsedges are among the most noxious of agricultural weeds in temperate to tropical regions worldwide. They often form dense colonies, can greatly reduce crop yields, and are difficult to control. In California, the nutsedges are especially problematic in annual and perennial crops that receive summer irrigation.

Yellow nutsedge: This species is a widespread, highly variable native of Eurasia and North America, including California and other western states. Some botanists recognize 4 varieties in the contiguous states, of which only *Cyperus esculentus* L. var. *leptostachyus* Boeckl. occurs in California and most other western states. Variety *macrostachyus* Boekl. also occurs in New Mexico and most south-

Yellow nutsedge (*Cyperus esculentus*) plant.
J. M. DiTomaso

Yellow nutsedge (*Cyperus esculentus*) young plant.
J. M. DiTomaso

Yellow nutsedge • Purple nutsedge

Yellow nutsedge *(Cyperus esculentus)* inflorescence. J. M. DiTomaso

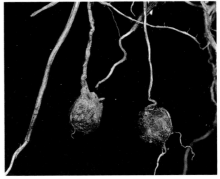

Yellow nutsedge *(Cyperus esculentus)* tubers. J. M. DiTomaso

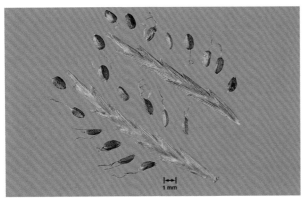

Yellow nutsedge *(Cyperus esculentus)* seeds and spikelets. J. O'Brien

ern states. Both of these varieties are native to North America. Variety *sativus* Boekl., which occurs in Maryland and Virginia, is considered an introduction from Eurasia. Besides California, yellow nutsedge is a state-listed noxious weed in Colorado, Hawaii, Minnesota (secondary), Oregon (class B), and Washington (class B). In some countries, yellow nutsedge is cultivated for its edible, earthy, almond-flavored tubers used to make a milky drink.

Purple nutsedge: This species has been dubbed the "world's worst weed," but it is not as widespread in California as yellow nutsedge. Besides California, it is a state-listed noxious weed in Oregon (class A). The tubers taste bitter and are used medicinally in China and India. Native to Eurasia.

SEEDLINGS: Seedlings not often found. Leaves similar to those of mature plants but smaller. Stem base slightly *triangular*. Midvein region often pale. First 2–3 leaves emerge simultaneously and are folded lengthwise.

MATURE PLANT: Stems erect, simple, glabrous, *triangular in cross-section.* Leaves *3-ranked*, mostly basal, glossy, glabrous, often creased lengthwise, margin finely serrate. Sheath *closed*, membranous, pale green. *Leaves lack ligules, auricles, and collar regions.*

Purple nutsedge (*Cyperus rotundus*) plant.
J. M. DiTomaso

Purple nutsedge (*Cyperus rotundus*) tubers.
J. M. DiTomaso

Purple nutsedge (*Cyperus rotundus*) inflorescence.
J. M. DiTomaso

Purple nutsedge (*Cyperus rotundus*) plant with tubers and rhizomes.
J. M. DiTomaso

Yellow nutsedge: Stems to 0.9 m tall. Leaves with pointed tips and keeled, *about equal to or longer than flower stems,* light green, 4–9 mm wide.

Purple nutsedge: Stems to 0.4 m tall. Leaves rounded at the tip and less keeled, *shorter than flower stems,* dark green, 3–6 mm wide.

ROOTS AND UNDERGROUND STRUCTURES: Plants develop an *extensive system of rhizomes, tubers, and roots.* Rhizomes slender, fleshy when young, covered with scales, develop tubers and basal bulbs that bear aerial shoots. Tubers store starch and have several buds that produce new rhizomes, which develop more basal bulbs and new plants. Roots often grow to a depth greater than that of the tubers and rhizomes.

Yellow nutsedge: Rhizomes *terminate with a tuber*, break easily. Tubers generally *irregularly round*, 0.3–1.5 cm in diameter, brown to black, hard, *smooth* at maturity (with scales when young), with active *buds only at the apex, taste mildly almondlike*.

Purple nutsedge: Rhizomes develop *chains of up to 15 tubers at intervals of 5–25 cm along their length*. Rhizomes can penetrate root vegetables and become wiry and fibrous over time. Tubers *irregular-shaped or oblong to nearly round*, to 2.5 cm long, usually less than 1 cm thick, *covered with red-brown papery scales*, with *buds scattered on the surface, taste bitter*. Roots can grow to 1.3 m deep in heavy clay.

SPIKELETS/FRUITS: Inflorescences loose, *umbel-like*, with leaflike bracts at the base. Spikelets consist of several to ± 40 florets. Achenes ± elliptic, triangular in cross-section, 1–1.5 mm long, seldom mature.

Yellow nutsedge: June–October. *Spikelets straw-colored to gold-brown*, 5–30 mm long, mostly 2–3 mm wide. Longest leaflike bract much longer than the inflorescence. Achenes light brown.

Purple nutsedge: July–November. *Spikelets reddish to purplish brown*, 8–25 mm long, mostly 1–2 mm wide. Longest leaflike bracts often about equal to the inflorescence. Seeds not typically produced in the United States. When present, achenes dark brown or black.

POSTSENESCENCE CHARACTERISTICS: Foliage dies back with cool temperatures in fall, but tubers survive and sprout the following spring.

HABITAT: Disturbed areas, crop fields (especially those irrigated in summer), irrigation ditches, orchards, vineyards, gardens, turf, and landscaped areas. Both species often grow where moisture is plentiful; tolerate many soil conditions, including periods of drought and flooding; and are designated as facultative wetland indicator species in California. Yellow nutsedge often grows on sandy, well-drained soils.

DISTRIBUTION: Nearly worldwide.

Yellow nutsedge: Throughout California, to 1000 m. All contiguous states, except possibly Montana and Wyoming. Occurs in most regions of the world with temperate to tropical climates.

Purple nutsedge: Central Valley, South Coast, Sonoran Desert, to 250 m. Arizona, southern states, a few northeastern and central states. Occurs in most regions of the world with warm temperate to tropical climates, and in California primarily in the low desert areas.

PROPAGATION/PHENOLOGY: *Reproduce vegetatively from tubers* in both species and by seed in yellow nutsedge. Tubers and seeds disperse with agricultural and nursery activities, soil movement, and in water, especially flooding. Seeds also disperse short distances with wind. Seeds and tubers germinate in spring. Tubers develop to soil depths of about 32 cm, but most are in the top 20 cm. Under favorable conditions, one plant can produce hundreds to thousands of tubers in one season. Seed production can be high in yellow nutsedge, but viability is variable. In California, seed viability is typically low. One plant can develop into a dense colony 3 m or more in diameter. Patch boundaries can increase by more than 1 m per year.

Yellow nutsedge: Tubers planted to a soil depth of 80 cm can produce a new plant. Tubers survive soil temperatures as cold as −5°C and sprout more readily following a period of chilling to break dormancy. Tubers germinate when soil temperatures remain above 6°C. Under field conditions, tubers typically survive up to about 3–4 years.

Purple nutsedge: Tubers tolerate high temperatures, but not freezing. Tubers germinate when soil temperatures remain above 15°C. Typically only 1 tuber in a chain germinates, unless the chain is severed. Low oxygen and high carbon dioxide levels appear to promote tuber dormancy. Tuber dormancy is high in undisturbed soils and at deeper soil levels. Tubers can remain dormant for long periods and can become dormant after sprouting. Tuber dormancy often increases with age. New tubers are initiated when flowers develop, often around 4–8 weeks after shoots emerge. Tubers planted at a depth of 90 cm are usually unable to produce aerial shoots. Tubers desiccate quickly when detached from the rhizome-root system under dry conditions and can survive flooded soils for at least 200 days. Tuber longevity is variable and depends on environmental conditions. In most cases, tubers survive about 3–4 years, but under certain conditions they can remain viable for up to 10 years or more. Plants use the C4 photosynthetic pathway and appear to have allelopathic properties.

MANAGEMENT FAVORING/DISCOURAGING SURVIVAL: Limiting tuber production and draining tuber energy reserves by repeatedly removing small plants before the 6-leaf stage (every 2–3 weeks in summer) can eventually control populations. Mature tubers can resprout up to about 12 times. Shading or solarization can help to limit populations by weakening shoots and decreasing new tuber formation, but mature tubers may not be eliminated. Cultivation can worsen an infestation if not repeated often enough to exhaust tuber reserves and prevent new tuber formation.

Yellow nutsedge: Pigs relish tubers and can search out, uncover, and consume most tubers in an area within a period of days.

Purple nutsedge: A thorough, deep cultivation to about 30 cm deep that fragments tuber chains and covers green parts, followed by persistent repeated cultivation to about 15 cm deep to kill new shoots and limit new tuber formation, especially when the soil is dry, can help to control purple nutsedge.

SIMILAR SPECIES: Yellow and purple nutsedge are the only sedges in California that develop tubers. Unlike the nutsedges, grass species have 2-ranked leaves with ligules and round or flattened stems.

Green kyllinga [*Kyllinga brevifolia* Rottb.][KYLBR]

SYNONYMS: shortleaf spikesedge; *Cyperus brevifolius* (Rottb.) Endl. ex Hassk.

GENERAL INFORMATION: Warm-season *perennial* to 0.5 m tall but is generally much shorter, with *glossy, grasslike leaves* and *creeping rhizomes*. Green kyllinga is most problematic in turf and ornamental plantings. Plants typically form dense mats from a prolific network of rhizomes that can crowd out bermudagrass. In the warm season, green kyllinga grows more rapidly than most turfgrasses, creating an uneven surface texture within a short period of mowing. Green kyllinga is used as a folk medicine in Paraguay. Native to tropical America.

MATURE PLANT: Foliage glabrous. Leaves usually 1–3, ± dark green, glossy, flat, (2–)6–15(–21) cm long, 1.5–3.5 mm wide, tip acute, base sheathing, sheath closed. Margins and lower midvein sparsely minute-barbed and ± rough to touch, especially near the tip. Flower stems erect, triangular in cross-section, base slender.

ROOTS AND UNDERGROUND STRUCTURES: Rhizomes (1–)3–12(–30) cm long, 0.5–2 mm in diameter, covered with reddish brown, lanceolate scales 6–13 mm long, with fibrous roots and usually a shoot at each node, internodes (2–)5–12(–30) mm long. During the warm season, rhizomes can grow more than 2.5 cm per day. Plants with short rhizomes and short internodes sometimes appear tufted.

Green kyllinga (*Kyllinga brevifolia*) infestation in turf. J. K. CLARK

SPIKELETS/FRUITS: May–November, sometimes earlier in warm locations. *Spikelet heads dense, ovoid, 4–7 mm long, with 3–4 leaflike bracts (1.5–)4–12 (–18) cm long just below the numerous sessile spikelets. Longest bract usually erect*, resembles an extension of the stem. Shorter bracts spreading to ascending. Spikelets flat, oblong-lanceolate, ± 2.5–3 mm long, consist of *3–4(–5) two-ranked scales and 1 or 2 flowers. Scales membranous, pale brownish to greenish.* Lowest 2 scales sterile, much smaller than others. The third scale subtends *one fertile, bisexual flower.* Upper 1–2 scales sterile or subtend staminate flowers. Achenes flattened, 2-sided, elliptic, ± 1 mm long, body *light to medium brown*, base ± short-stalked (stipitate) and whitish.

POSTSENESCENCE CHARACTERISTICS: Green kyllinga is dormant during the cool season. In mild-winter areas green kyllinga foliage remains green during the cool season, while in colder regions, the foliage turns brown or purplish brown and is darker than senesced bermudagrass.

HABITAT: Turf, ditches, landscaped areas, and ornamental plantings. Inhabits cropland elsewhere but is not usually associated with agricultural land in California. Often grows on over-watered or poorly drained sites. Established plants tolerate some shade and drying.

DISTRIBUTION: Central Valley from Sacramento to southern California, South and Central coast regions from San Diego County probably to southeastern San Francisco Bay, and deserts, to 300 m. Southern and some eastern states.

PROPAGATION/PHENOLOGY: *Reproduces by seed and vegetatively from rhizomes.* Plants often produce large quantities of highly viable seed. Long-distance dispersal occurs primarily by seed. Seeds disperse with water, soil movement, mud, vehicle tires, landscape maintenance such as mowing and leaf blowing, and by clinging to animals and to the shoes and clothing of humans. Most seedlings emerge from the soil surface to a depth of 0.5 cm. Green kyllinga seed longevity is undocumented. However, many members of the sedge family have long-lived seeds. Under favorable conditions, new plants can establish from short rhizome fragments with at least 1 node. Rhizome fragments disperse with

Green kyllinga *(Kyllinga brevifolia)* rhizomes.
J. K. Clark

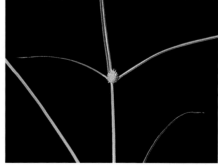

Green kyllinga *(Kyllinga brevifolia)* flowering stem.
J. M. DiTomaso

Green kyllinga (*Kyllinga brevifolia*) flower cluster.
J. M. DiTomaso

mud, soil movement, and landscape maintenance and alteration. Most shoots emerge from rhizome fragments buried to 4 cm deep.

MANAGEMENT FAVORING/DISCOURAGING SURVIVAL: Green kyllinga can survive and flower when mowed to a height as low as 2 cm. In healthy turf, mowing to a height of 2.5 cm or less favors the spread of green kyllinga. Applying nitrogen to turf in poor condition can to help to slow its spread. Manually removing small patches, including all of the rhizomes, can also help to limit the spread of green kyllinga. However, control of large patches often requires treatment with an appropriate herbicide. The spread of green kyllinga is slow when in competition with healthy turf. Large treated patches are susceptible to reinvasion if left vegetation-free.

SIMILAR SPECIES: During the growing season, green kyllinga may resemble ryegrass (*Lolium* spp.) turf. Unlike grasses, green kyllinga leaves *lack a ligule* or a conspicuous junction between the blade base and sheath. **Yellow nutsedge** [*Cyperus esculentus* L.] and **purple nutsedge** [*Cyperus rotundus* L.] sometimes grow in the same habitats as green kyllinga and have similar foliage, but are distinguished by having *small, round or oblong tubers* (~ 0.5–2.5 cm in diameter or long) in rhizomatous chains or at the ends of slender rhizomes. Refer to the entry for **Yellow nutsedge and Purple nutsedge** (p. 674) for more information.

Green kyllinga (*Kyllinga brevifolia*) seeds, with and without surreounding bracts.
J. O'Brien

Western brackenfern [*Pteridium aquilinum* (L.) Kuhn var. *pubescens* Underw.][PTEAP]

SYNONYMS: adderspit; bracken; brake; common brake; eagle fern; female fern; fiddlehead; hairy brackenfern; hog brake; pasture-brake; *Pteridium aquilinum* (L.) Kuhn var. *lanuginosum* (Bong.) Fern; *Pteridium aquilinum* (L.) Kuhn ssp. *lanuginosum* (Bong.) Hultén; *Pteris aquilinum* L. var. *lanuginosum* Bong. *Pteridium aquilinum* (L.) Kuhn is a nearly worldwide native species with many regional, named varieties; *Pteridium* is sometimes placed in the family Pteridaceae.

GENERAL INFORMATION: Native *perennial fern*, with *long, creeping rhizomes* and *triangular fronds to 1.5 m tall*. Western brackenfern is usually considered a desirable component of the vegetation in undisturbed natural areas, but can be problematic in managed forests, pastures, rangeland, orchards, agronomic crops, and other disturbed places. Plants typically form dense colonies and can prevent the reestablishment of desirable trees or grasses on disturbed sites. All plant parts of all brackenfern varieties contain an unstable glycoside and possibly other compounds that are *toxic* and *carcinogenic* to humans and animals when ingested over time. Ingestion of large quantities of fronds over a period of several weeks to months can cause various problems in cattle and sheep, including bone marrow depression, hemorrhagic disease, enzootic hematuria (cyclic urinary tract inflammation in cattle), retinal degeneration, and digestive and urinary tract cancers. Horses can develop potentially lethal neurological problems caused by the destruction of thiamine (vitamin B1) within 2–3 weeks. Humans consume the fresh, cooked, or preserved young coiled fronds (croziers) in some

Western brackenfern (*Pteridium aquilinum* var. *pubescens*) plant. J. M. DiTomaso

Western brackenfern

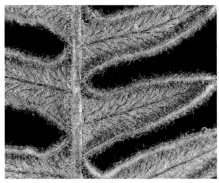

Western brackenfern (*Pteridium aquilinum* var. *pubescens*) developing frond.   J. M. DiTomaso

Western brackenfern (*Pteridium aquilinum* var. *pubescens*) sporangia.   Courtesy of Jim McHenry

regions of the world, particularly Japan and Brazil, where the croziers are considered a delicacy. However, the consumption of croziers or milk from cows that ingest brackenfern has been implicated with an increased incidence of digestive tract cancers. Western brackenfern is a state-listed noxious weed in Minnesota (secondary).

MATURE PLANT: Frond stalks *coarse* (more than 3 mm wide), *strongly grooved on the upper side, u-shaped in cross-section, base nearly black and densely covered with dark feltlike hairs, upper portion straw-colored to greenish* and ± glabrous. Fronds triangular, arched, lower portion usually pinnate-divided 3 times, upper portion mostly pinnate-divided 2 times. Primary divisions (pinnae) ± opposite, sometimes with nectary glands in the axils of the main axis (rachis). Lowest primary divisions usually form a ± 45° angle with the main axis. Secondary divisions (pinnules) alternate. Ultimate divisions oblong, *some toothed or lobed*, tip rounded, mostly 0.5–2 cm long, 0.3–0.6 cm wide, somewhat leathery, margins rolled under, lower surface densely covered with straight or woolly translucent hairs, upper surface glabrous to sparsely hairy. New leaves (croziers) coiled.

ROOTS AND UNDERGROUND STRUCTURES: True roots lacking. Rhizomes *long, creeping*, branched, black to dark brown, lack scales, *covered with dark, feltlike hairs*. Deep rhizomes mostly 1–2.5 cm in diameter, develop few fronds, primarily function in carbohydrate storage. Surface rhizomes slender, especially hairy, develop more fronds. Most rhizome growth occurs in fall.

SPORE PRODUCING STRUCTURES: Vary depending on conditions, but usually appear in summer. Spore-producing structures (sporangia) brown, less than 1 mm wide, slender-stalked, develop in a ± *continuous line along the leaflet margins on the lower surface* and *partly covered by the rolled margin*. Not all fronds develop sporangia. Spores about 25–30 micrometers wide.

POSTSENESCENCE CHARACTERISTICS: Fronds often turn yellow before they are killed by frost. Dead brown fronds can persist into winter.

HABITAT: Forest, meadows, open woodland, rangeland, fields, pastures, agronomic crops, and orchards. Tolerates poor, rocky soil, but not poorly drained sites.

Western brackenfern

DISTRIBUTION: Throughout California, except Central Valley, deserts, and Great Basin, to 3200 m . All western states, South Dakota, and Texas. Different varieties occur in various regions nearly worldwide.

PROPAGATION/PHENOLOGY: *Reproduces by rhizomes and spores.* Rhizomes survive frost or fire, while fronds do not. New fronds grow from the rhizomes in spring. Rhizome fragments can develop into new plants. Removal of overstory can often lead to prolific sprouting from rhizomes. One frond can produce hundreds of thousands of spores, but not all fronds develop sporangia. Spores disperse with wind and water. Spore germination, gametophyte survival, sexual reproduction, and young sporophyte (equivalent to a seedling) survival requires sufficiently moist conditions over a period of several weeks. Spores can survive extended dry periods.

MANAGEMENT FAVORING/DISCOURAGING SURVIVAL: Rhizomes survive fire. In some cases regular cultivation or mowing can eventually exhaust rhizome energy reserves. However, in many situations this has been unsuccessful. Systemic foliar herbicide treatments can control western brackenfern.

SIMILAR SPECIES: Western brackenfern is distinguishable from other native ferns by having all of the following characteristics: rhizomes that are *dark, long, creeping, and hairy*; *triangular fronds to about 1.5 m tall*; sporangia that develop in a *± continuous line along the leaflet margins on the lower surface* and *partly covered by the rolled margin*; fronds with *some ultimate divisions pinnate-toothed or -lobed*; and frond stalks that are *u-shaped in cross-section, much more than 3 mm wide,* densely covered with *dark feltlike hairs,* and that are *black to dark brown at the base* and *straw-colored to greenish above.*

Western brackenfern (*Pteridium aquilinum* var. *pubescens*) gametophytes.

COURTESY OF JIM MCHENRY

Common teasel [*Dipsacus fullonum* L.][DIWSI]
[Cal-IPC: Moderate]

Fullers teasel [*Dipsacus sativus* (L.) Honck.][DIWSA]
[Cal-IPC: Moderate]

SYNONYMS: Common names that have been used for both species include card teasel, card thistle, fullers teasel, gypsy-combs, teasel, and Venus-cup.

Common teasel: wild teasel; *Dipsacus fullonum* L. ssp. *fullonum*; *Dipsacus sylvestris* Huds.

Fullers teasel: cultivated teasel; Indian teasel; *Dipsacus fullonum* L. ssp. *sativus* (L.) Thellung. In the past, this species was not recognized or distinguished from

Common teasel (*Dipsacus fullonum*) plants.   J. M. DiTomaso

Common teasel • Fullers teasel

Common teasel (*Dipsacus fullonum*) leaf base and stem.
J. M. DiTomaso

Common teasel (*Dipsacus fullonum*) seedling.
J. M. DiTomaso

Common teasel (*Dipsacus fullonum*) flower cluster.
J. M. DiTomaso

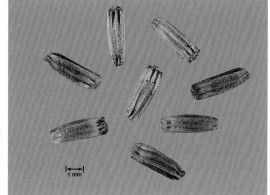

Common teasel (*Dipsacus fullonum*) seeds.
J. O'Brien

common teasel, and many older references mistakenly refer to fullers teasel as *Dipsacus fullonum* L. or *Dipsacus sylvestris* Huds.

GENERAL INFORMATION: Coarse *biennials*, occasionally short-lived perennials, with erect flower stems to 2 m tall and *large, spiny flower heads densely covered with small, lavender to white flowers*. Plants exist as rosettes until flowering stems develop, usually during the second season. Teasel is most problematic when stands become dense and impenetrable to humans or livestock. Both species are state-listed noxious weeds in Iowa (secondary). Common teasel is also a state-listed noxious weed in Colorado and New Mexico (class B). Common teasel is thought to be the wild ancestor of fullers teasel. Before metal carding combs were created, fullers teasel was cultivated for the dried flower heads, which were used to process wool. Dried teasel flower heads are sometimes used in floral arrangements. Native to Europe.

SEEDLINGS: Cotyledons oval to spatula-shaped, about 10–13 mm long, 5–6 mm wide, tip often truncate and slightly indented. Leaves opposite, elliptic-oblong, hairy, margins entire to scalloped, ± long-tapered to a stalk, stalks fused at the base. First leaf pair about 1–2 cm long. Subsequent pairs increasingly larger.

MATURE PLANT: Stems coarse, straight, ridges lined with stiff prickles. Branches opposite, ascending. Rosette and stem leaves to 50 cm long, margins ± entire to scalloped or toothed, often with stiff prickles on the lower midvein and sometimes smaller prickles on the upper surface. Rosette leaves lanceolate to oblanceolate. Stem leaves opposite, sessile, ± fused at the base, lanceolate.

Common teasel: Stem leaves slightly fused at the base.

Fullers teasel: Stem leaves ± strongly fused at the base.

ROOTS AND UNDERGROUND STRUCTURES: Taproot thick, fleshy, yellowish, can grow to a depth of about 0.8 m.

Fullers teasel (*Dipsacus sativus*) flower clusters.
J. M. DiTomaso

Fullers teasel (*Dipsacus sativus*) leaf base and stem.
J. M. DiTomaso

Fullers teasel (*Dipsacus sativus*) seedling.
J. M. DiTomaso

Fullers teasel (*Dipsacus sativus*) seeds. J. O'Brien

FLOWERS: Heads terminal, *ovoid to cylindrical, 3–10 cm long, densely flowered,* on long stalks. *Bracts below flower heads several, linear, unequal, some about as long or longer than the flower head, prickly.* Just below each flower is a *stiff receptacle bract with a spinelike tip that is longer than the flower.* Corolla ± funnel-shaped, unequally 4-lobed, lavender to white. Calyx *cup-shaped, 4-lobed,* green, persistent. *Stamens 4. Ovary inferior.* Flowers around the middle of the head mature first, and two rings of flowers mature sequentially from the middle ring, one ring moving upward and the other ring moving downward. Insect-pollinated. Primarily out-crossing.

Common teasel: April–September. Flower bract tips *straight,* slightly flexible when the whole flower head is gently compressed with the fingers.

Fullers teasel: May–August. Flower bract tips *curve downward slightly or hooked,* rigid when the whole flower head is gently compressed with the fingers.

FRUITS AND SEEDS: Achenes rectangular, square in cross-section, 3–6(8) mm long, ± 1 mm wide, minutely hairy, gray-brown, usually with 8 pale ribs, one at each angle and on each face.

POSTSENESCENCE CHARACTERISTICS: In some areas, rosette leaves die during winter and regrow from the crown in spring. Dead gray-brown stems with old flower heads can persist for a year or more.

HABITAT: Fields, pastures, roadsides, waste places, ditches, riparian sites, and other disturbed sites.

DISTRIBUTION: Common teasel: North Coast, Klamath Ranges, central and southern Sierra Nevada foothills, and San Francisco Bay region, to 1700 m. Most contiguous states, except some southern and north-central states.

Fullers teasel: North Coast, western North Coast Ranges, San Francisco Bay region, South Coast Ranges, and Peninsular Ranges, to 800 m. A few northeastern states.

PROPAGATION/PHENOLOGY: *Reproduce by seed.* Seeds mature September through November. Most seeds eventually fall near the parent plant. Seeds float for 2 weeks or more and can disperse to greater distances with water, mud, soil movement, human activities, and possibly animals. In Canada, most common teasel seeds germinate in spring, while in Australia, most germination occurs in fall. Common teasel seeds survive for at least 6 years under field conditions. Once established, dense populations can persist for 25 years or more.

MANAGEMENT FAVORING/DISCOURAGING SURVIVAL: Manual removal of plants to a few inches below the crown can control small populations. On agricultural land, cultivation of rosettes can control larger populations. Mowing flower stems before flowers develop prevents seed production.

SIMILAR SPECIES: The teasels are unlikely to be confused with other species.

Russian-olive [*Elaeagnus angustifolia* L.] [Cal-IPC: Moderate]

SYNONYMS: narrow-leafed oleaster; oleaster; silverberry; wild olive. The original spelling was changed to *Elaeagnus angustifolius* L. in a few references.

GENERAL INFORMATION: Fast-growing, *deciduous tree* to 7 m tall, with *silvery foliage*. Russian-olive is cultivated as a hardy landscape ornamental and windbreak tree, but it has escaped cultivation in many areas of the United States. In California and other western states, it is especially invasive in seasonally wet riparian areas and may eventually replace stands of native willows (*Salix* spp.) and cottonwoods (*Populus* spp.) at some locations. Although Russian-olive fruits provide food for wildlife, trees are used to a lesser degree than the native vegetation. Russian-olive is most problematic in the Southwest, Intermountain West, and Great Plains regions. Native to the temperate regions of Asia.

MATURE PLANT: Small branches dark reddish brown, ± smooth. Twigs and branches sometimes thorny. Leaves *alternate, simple, narrowly lanceolate or elliptic, mostly 4–8 cm long,* margin smooth, *upper surface gray-green* and moderately covered with *silvery star-shaped hairs and scales. Twigs, leaf stalks, and lower leaf surfaces silvery gray and densely covered with silvery shield-shaped (peltate) scales.*

ROOTS AND UNDERGROUND STRUCTURES: Root system deep, with many well-developed laterals. Cut trees typically resprout from the crown and roots. Depending on the location, roots sometimes associate with nitrogen-fixing bacteria (*Frankia* spp.).

Russian-olive (*Elaeagnus angustifolia*) infestation in a riparian area.   J. M. DiTomaso

Russian-olive

Russian-olive *(Elaeagnus angustifolia)* tree.
J. M. DiTomaso

Russian-olive *(Elaeagnus angustifolia)* in flower.
J. M. DiTomaso

Russian-olive *(Elaeagnus angustifolia)* in fruit.
J. M. DiTomaso

FLOWERS: May–June. Flower clusters umbel-like, axillary on current years growth. Flowers highly *fragrant, bisexual*, mostly 5–10 mm long and wide, consist of a *narrow, bell-shaped calyx* (sepals as a unit) with *4 acute lobes, lack petals*. Calyx *dark yellow inside, silver-scaly outside*. Ovary superior, but appears inferior. Stamens 4, open by slits. Insect-pollinated.

FRUITS AND SEEDS: Fruits *drupelike* (with a fleshy outer layer covering 1 seed), ovoid, about 10–20 mm long, covered with *silvery scales*, internal fleshy layer yellow at maturity. Most fruits mature September–November. Seeds oblong to ± football-shaped with a few longitudinal ridges, slightly smaller than fruits.

POSTSENESCENCE CHARACTERISTICS: Dormant trees without leaves usually retain a few fruits.

HABITAT: Riparian areas, flood plains, grasslands, roadsides, fencerows, seasonally moist pastures, ditches, and other disturbed sites. Often inhabits seasonally moist areas and sites near farmlands. Grows under a wide range of environmental conditions, including clay, sandy, and fairly alkaline or saline soils. Grows best in inland areas with warm summers and cold winters. Tolerates drought, high water tables, and temperatures well below freezing (to –45° C) to as high as 46°C.

DISTRIBUTION: San Joaquin Valley, San Francisco Bay region, eastern Sierra Nevada, and Mojave Desert, mostly to 1500 m. Western states, central states, most northeastern and eastern states, and a few southern states.

PROPAGATION/PHENOLOGY: *Reproduces by seed.* Most fruits remain on trees until distributed by animals, especially birds. Seeds survive ingestion by animals. Seeds are dormant at maturity and require a cool, moist stratification period of about 2–3 months. Some seeds are hard-coated and may require scarification as well as stratification. Seeds germinate in many soil types and over a variable period of time depending on the environmental conditions. Stored seeds survive up to ± 3 years, but longevity in the field is undocumented. Seedlings grow best under moist, slightly alkaline conditions.

ADDITIONAL ECOLOGICAL ASPECTS: Seeds germinate under a broader range of conditions than native willows and cottonwoods. Russian-olive seedlings can survive under a canopy of mature willows and cottonwoods, and then grow quickly when the loss of a tree creates an opening in the canopy. Conversely, willow and cottonwood seedlings seldom survive under a canopy of Russian-olive trees.

MANAGEMENT FAVORING/DISCOURAGING SURVIVAL: Manually removing seedlings and saplings with roots before they mature is a more effective method of control than removing mature trees. Cut trees typically resprout from the roots and crown. Cutting trees before fruits mature and immediately painting

Glossy privet (*Ligustrum lucidum*) leaves and inflorescence. J. M. DiTomaso

Glossy privet (*Ligustrum lucidum*) in fruit. J. M. DiTomaso

Olive (*Olea europaea*) in a riparian site. J. M. DiTomaso

Olive (*Olea europaea*) in flower. J. M. DiTomaso

Olive (*Olea europaea*) with fruit. J. M. D<small>I</small>T<small>OMASO</small>

the stump with systemic herbicide is more effective than cutting alone. Choosing other landscape ornamentals to plant at sites where seedlings are likely to invade nearby natural areas can help prevent the spread of Russian-olive.

SIMILAR SPECIES: Glossy privet [*Ligustrum lucidum* Ait.] and **olive** [*Olea europaea* L.][Cal-IPC: Limited] are other members of the olive family that sometimes escape cultivation. Both species are ornamental *evergreen trees* with *opposite leaves*. Leaves of olive resemble those of Russian-olive, but unlike Russian-olive, olive has *white flowers* with *2 stamens* and *oily fruits that are green to glossy black*. Olive occasionally inhabits disturbed places in the Central Valley, San Francisco Bay region, southern North Coast Ranges, South Coast, and Santa Cruz Island, to 200 m. In southwestern Australia, olive is an invasive weed of parklands where grazing is prohibited. Native to western Asia. Glossy privet has flowers that are similar to those of olive. However, glossy privet is distinguished by having leathery, *glabrous leaves more than 2.5 cm wide* that are *dark glossy green above* and pale green below. Like olive, birds primarily disperse the small purplish to bluish black, berrylike fruits of glossy privet. The naturalized distribution of glossy privet in California is uncertain. At publication time glossy privet is spreading rapidly in natural areas along the Mendocino coast. Glossy privet seedlings often emerge in landscaped areas near mature trees. Glossy privet has naturalized in most southern states from Texas to North Carolina. Native to Asia.

Field horsetail [*Equisetum arvense* L.][EQUAR]

Scouringrush [*Equisetum hyemale* L. ssp. *affine* (Engelm.) Calder & R.H. Taylor][EQUHY]

SYNONYMS: Field horsetail: bottlebrush; common horsetail; foxtail-rush; horsepipes; horsetail fern; meadow-pine; pinegrass; scouringrush; shavegrass; snake-grass; *Equisetum arvense* L. var. *alpestre* Wahenb., var. *boreale* (Bong.) Rupr., var. *campestre* Wahlenb., and var. *riparium* Farw.; *Equisetum boreale* Bong.; *Equisetum calderi* Boivin; *Equisetum saxicola* Suksd.

Scouringrush: common scouringrush; scouringrush horsetail; shavegrass; western scouringrush; *Equisetum affine* Engelm.; *Equisetum hiemale* L. var. *californicum* in some older California references; *Equisetum hyemale* L. var. *affine* (Engelm.) A.A. Eat.; *Equisetum praealtum* Raf.; *Equisetum robustum* A. Braun; *Hippochaete hyemalis* (L.) Bruhin

GENERAL INFORMATION: Primitive native *perennials* with *rhizomes, whorled branches*, and *tubelike, sheathing leaves*. In natural areas, *Equisetum* species are a desirable component of the ecosystem. However, dense colonies can be problematic in agricultural fields, pastures, and controlled aquatic systems. Both species contain alkaloids that destroy thiamine and are *toxic* to livestock, especially horses, when ingested. Stem surfaces accumulate much silica, and for this reason, both species were historically used for scrubbing purposes.

Field horsetail (*Equisetum arvense*) infertile stem and branches. J. NEAL

Field horsetail (*Equisetum arvense*) fertile stem. J. NEAL

Scouringrush (*Equisetum hyemale* ssp. *affine*) stem. J. M. DiTomaso

Field horsetail: Plants produce *2 types of stems: green, annual, vegetative stems* to about 0.6 m tall and *pale tan, short-lived, fertile stems* to 0.4 m tall. Under certain conditions, field horsetail can accumulate heavy metals, such as cadmium, lead, gold, zinc, and copper. It is a state-listed noxious weed in Oregon (class B) and a noxious weed in parts of Australia. Field horsetail is a highly variable, widespread native of North America and Eurasia.

Scouringrush: *All stems green, perennial*, to about 2 m tall. Scouringrush is a widespread native of North America.

MATURE PLANT: Main stems longitudinally ribbed, hollow except at the nodes. Branches whorled. Leaves small, united into a node-sheathing tube that is toothed along the upper margin (referred to as leaf sheath).

Field horsetail: *Vegetative stems green, slender, mostly 2–5 mm wide, 6–14-ribbed*, with *many slender, evenly whorled, 3–4-angled branches* at most nodes. *Lowest branch internodes are longer than the leaf sheath just below.* Leaf sheaths ± 5 mm long, with *6–14 teeth. Fertile stems unbranched, lack chlorophyll, pale tan to pinkish*, ± succulent, slightly thicker than vegetative stems, with dark leaf sheaths mostly 10–20 mm long with *6–10 teeth.*

Scouringrush: *All stems green ± thick, mostly 5–15 mm wide*, mostly 20–40-ribbed, *unbranched or with a few scattered branches at some nodes.* Leaf sheaths mostly *10–15 mm long and wide*, usually with *2 dark bands* (top and bottom) and about *22–50 deciduous teeth.*

ROOTS AND UNDERGROUND STRUCTURES: Rhizomes extensive, creeping, branched, dark, root at the nodes. Rhizomes typically store starch in shortened swollen internodes (tubers). Some tubers readily detach from the parent rhizomes and develop into new plants. The rhizome system can grow to about 1.5 m deep or more.

SPORE-BEARING STRUCTURES: Mostly March–July. Cones (strobilus) terminal, consist of numerous small ± flower-shaped bracts (sporophylls) on short stalks, with spore-bearing structures (sporangia) on lower surfaces.

Field horsetail: Spikes *lanceolate-ovoid*, about 20–30 mm long, 3–10 mm wide, *tip bluntly tapered*.

Scouringrush: Spikes *ovoid*, about 10–30 mm long, mostly 10–15 mm wide, *tip abruptly pointed* (apiculate).

POSTSENESCENCE CHARACTERISTICS: Field horsetail: Vegetative stems typically die at the end of the growing season, turn blackish, and usually do not persist into winter.

HABITAT: Field horsetail: Moist, disturbed places, pastures, orchards, nursery crops, agricultural fields, and irrigation ditches; also, moist meadows and riparian sites in natural areas. Grows best in sandy neutral to slightly alkaline soils on sites with a high water table and poor drainage.

Scouringrush: Moist, sandy sites, riparian areas, marshy places, and ditches.

DISTRIBUTION: Field horsetail: Throughout California, except deserts and

Smooth scouringrush (*Equisetum laevigatum*) fertile stems. J. M. DiTomaso

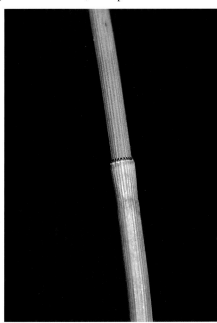

Smooth scouringrush (*Equisetum laevigatum*) node with one dark band at tip. J. M. DiTomaso

Giant horsetail (*Equisetum telmateia* ssp. *braunii*) along creek in native habitat.   J. M. DiTomaso

Giant horsetail (*Equisetum telmateia* ssp. *braunii*) stems.   J. M. DiTomaso

region east of the Sierra Nevada, to 3000 m. All contiguous states, except possibly Louisiana and Florida.

Scouringrush: Throughout California to 3000 m. All contiguous states.

PROPAGATION/PHENOLOGY: *Reproduce vegetatively from rhizomes and tubers,* and less importantly by spores. Rhizome fragments and tubers can develop into new plants. Fragments and tubers disperse with water, soil movement, and human activities. Plants produce large numbers of spores, but few spores survive because they are short-lived and grow best on mud that has recently been flooded and is rich in nutrients. Haploid gametophytes develop from the spores. These produce gametes that need water for fertilization. The transition through these vulnerable stages limits the survival of young plants.

Field horsetail: Fertile stems typically emerge in late winter/early spring and usually wither by the time vegetative stems emerge.

MANAGEMENT FAVORING/DISCOURAGING SURVIVAL: Colonies are difficult to control because of the extensive rhizome systems. Mowing and burning do not kill the rhizomes or tubers. Cultivation fragments, rhizomes, and tubers and can spread horsetail infestations. Mulching field horsetail with leaf compost can reduce the horizontal growth of rhizomes. Mulching with black plastic sheeting for 3–4 years can kill rhizomes in the upper 60 cm of soil. Field horsetail tolerates most herbicides used in agriculture.

SIMILAR SPECIES: Smooth scouringrush [*Equisetum laevigatum* A. Braun, synonyms: *Equisetum funstoni* A.A. Eaton; *Equisetum kansanum* J. Schaffner] is a widespread native *perennial* that is seldom weedy but is sometimes confused with scouringrush. Unlike scouringrush, smooth scouringrush has *annual stems*, spikes with a *rounded tip*, and leaf sheaths that are *longer than wide* and have *1 dark band at the tip*. Smooth scouringrush occurs primarily in natural communities throughout California to 3000 m, including desert mountains but excluding areas east of the Sierra Nevada and low elevation areas of the deserts. It also occurs in all western states, central states, and some northeastern states.

Giant horsetail [*Equisetum telmateia* Ehrh. ssp. *braunii* (Milde) R.L. Hauke, synonym: *Equisetum braunii* Milde][EQUTE] is a widespread native *perennial* with extensive rhizomes. Giant horsetail has thick, regularly branched, pale green vegetative stems to 3 m tall and short-lived, unbranched, pale tan fertile stems to about 0.6 m tall. Giant horsetail resembles *field horsetail*, but is typically larger in all respects. Unlike field horsetail, giant horsetail has *4–6 angled stem branches* with the *lowest internode shorter than the leaf sheath immediately below*, main stems with *20–40 ribs*, and leaf sheaths about 7–20 mm long with *14–28 teeth*. In addition, fertile stems have *spikes mostly 4–8 cm long* and leaf sheaths with about *20–30 teeth*. Giant horsetail inhabits marshes and stream banks and is sometimes weedy in wet ditches and other disturbed wet places throughout the Northwestern, Central-western, and Southwestern regions, to 1000 m. It also occurs in Idaho, Oregon, and Washington. Giant horsetail is protected in Michigan and may be extirpated there, but it is a state-listed noxious weed in Oregon (class B).

Turkey mullein [*Croton setigerus* Hook.][ERMSE]

SYNONYMS: doveweed; grayweed; woolly white drouth-weed; yerba del pescado; *Eremocarpus setigerus* (Hook.) Benth. in many California floras and elsewhere

GENERAL INFORMATION: Native summer *annual* with *densely hairy, pale gray-green to yellowish green foliage*. Mature individuals often have a *mounded shape* to 0.2 m tall and 0.8 m in diameter. Turkey mullein is a ruderal native that is generally considered a desirable component of the vegetation in natural areas. The seeds are an important food source for a variety of birds and small mammals. However, turkey mullein is sometimes troublesome in dry pastures, vineyards, orchards, and other disturbed places. Livestock that ingest turkey mullein can develop digestive tract blockage from fibrous hairy masses of indigestible plant material, but animals generally avoid eating the plant when more suitable forage is available. Most problems result when animals are fed contaminated hay. The foliage also contains a substance that is toxic to fish and other compounds that have not been evaluated for toxicity, including diterpenes and a glycoside that deters insect feeding. Native Americans are reported to have put crushed turkey mullein into streams to stun fish. Turkey mullein is an introduced noxious weed in parts of Australia.

SEEDLINGS: Cotyledons oval to ovate, covered with star-shaped (stellate) hairs. Leaves alternate. First few leaves ovate to heart-shaped, densely stellate-hairy, pale gray-green, increasingly larger than cotyledons. Crushed leaves have a strong scent.

MATURE PLANT: Foliage *densely covered* with *stiff, gray to yellowish star-shaped hairs*. Crushed foliage has a *strong unpleasantly sweet scent* and clear juice. Plants branched at the base, branches spreading to ascending, often forked,

Turkey mullein (*Croton setigerus*) population in an open field.   J. M. DiTomaso

Turkey mullein

Turkey mullein (*Croton setigerus*) flowers. J. M. DiTomaso

Turkey mullein (*Croton setigerus*) plant.
J. M. DiTomaso

Turkey mullein (*Croton setigerus*) seedling.
J. K. Clark

especially near the tips. Leaves alternate, ovate to heart-shaped, 1–6 cm long, base tapered or rounded, with *3 main veins from the base*, on stalks 1–5 cm long. Stipules inconspicuous.

ROOTS AND UNDERGROUND STRUCTURES: Taproot often slender, usually with few fibrous lateral roots.

FLOWERS: May–October. Flowers inconspicuous, *lack petals*. Male and female flowers develop separately on the same plant (monoecious). Male flowers occur in clusters at the stem tips and consist of a 5–6-lobed calyx ± 2 mm long and 6–10 stamens on a hairy receptacle. Female flowers develop singly or in groups of 2–3 in the lower leaf axils, below the male flowers. Female flowers also lack a calyx and consist of a 1-chambered ovary with 4–5 small glands at the base and 1 simple style.

FRUITS AND SEEDS: Capsules oblong, ± 4 mm long, hairy, *1-chambered, contain 1 seed*. Seeds oblong-ovoid, 3–4 mm long, 2–2.5 mm wide, back rounded, front weakly 2-sided, smooth, gray, black, or white to gray mottled dark brown or black, or gray and black striped.

POSTSENESCENCE CHARACTERISTICS: In dry desert and semidesert areas, dead plants are tan and can retain the mound shape for months. In regions with moister climates, such as coastal California, dead plants usually decompose rapidly.

HABITAT: Roadsides, rangeland, dry pastures, summer-fallowed fields, orchards, dry sandy washes, and other open dry, often disturbed sites. Grows poorly in wet places. Killed by frost.

DISTRIBUTION: Throughout California, except Great Basin and central and eastern deserts, to 1000 m. Oregon, Washington, Idaho, Nevada, Utah, Arizona, and northern Baja California.

PROPAGATION/PHENOLOGY: *Reproduces by seed.* Seeds fall near the parent plant or disperse to greater distances with water, mud, human activities, and possibly as a contaminant of feed or crops. Germination occurs primarily from mid-spring through early summer. Plants produce 2 types of seeds: mottled and uniform gray. Gray seeds develop as plants senesce and appear to germinate under drier conditions than mottled seeds. Birds generally avoid consuming the gray seeds, which contain an unpalatable compound in the seed coat that is lacking in the mottled seeds. Seeds do not survive ingestion by birds.

MANAGEMENT FAVORING/DISCOURAGING SURVIVAL: Manually removing plants before seed develops can control small or sparse populations. Cultivation can control larger populations of small plants. Large plants can clog cultivation implements. Often associated with overgrazed areas. Encouraging grasses and other forbs by reduced grazing will reduce populations. Turkey mullein tolerates certain herbicides, primarily because of its dense hairiness. Plants are resistant to burning.

SIMILAR SPECIES: Turkey mullein is unlikely to be confused with other weedy species.

Turkey mullein (*Croton setigerus*) seeds. Note grayish seed at top left.   J. O'Brien

Leafy spurge [*Euphorbia esula* L.] [EPHES][Cal-IPC: High Alert][CDFA list: A]

Oblong spurge [*Euphorbia oblongata* Griseb.] [Cal-IPC: Limited][CDFA list: B]

Serrate spurge [*Euphorbia serrata* L.] [EPHSR][CDFA list: A]

SYNONYMS: Leafy spurge: faitours-grass; wolf's milk; *Euphorbia discolor* Ledeb.; *Euphorbia esula* L. var. *esula*; *Euphorbia gmelinii* Steud.; *Euphorbia virgata* Wald. & Kit.; *Galarrhoeus esual* (L.) Rydb.; *Tithymalus esula* Scop.; many others

Oblong spurge: eggleaf spurge

Serrate spurge: saw-toothed spurge; toothed spurge

GENERAL INFORMATION: Erect *perennials* to 0.8 m tall, with *milky white sap* and mostly *alternate leaves*. Native to southwestern Europe.

Leafy spurge: Plants develop clonal colonies from an *extensive system of deep, creeping roots*. Leafy spurge is one of the most tenacious weeds in the United States. Large, dense colonies displace desirable vegetation, lower rangeland carrying capacity, and reduce land value. Leafy spurge is especially problematic in

Leafy spurge (*Euphorbia esula*) plant.
J. M. DiTomaso

Leafy spurge (*Euphorbia esula*) flowering stem.
J. M. DiTomaso

Leafy spurge • Oblong spurge • Serrate spurge

Leafy spurge *(Euphorbia esula)* flowers.
J. M. DiTomaso

Leafy spurge *(Euphorbia esula)* seeds and fruit.
J. O'Brien

the north-central states and adjacent region of Canada. In all, nearly 1.2 million hectares of rangeland in 29 states are infested, with about 445,000 hectares infested in North Dakota alone. The economic loss from direct and indirect impacts due to leafy spurge has been estimated at about $130 million per year. In California, leafy spurge occurs primarily in the northeastern part of the state. Besides California, leafy spurge is a state-listed noxious weed in Arizona (prohibited), Colorado, Idaho, Iowa, (primary), Kansas, Minnesota (primary) Montana (category 1), Nebraska, Nevada, New Mexico (class A), North Dakota, Oregon (class B), South Dakota, Utah, Washington (class B), Wisconsin, and Wyoming. The milky sap of many *Euphorbia* species is *toxic* to humans and live-

Oblong spurge *(Euphorbia oblongata)* plants.
J. M. DiTomaso

Oblong spurge *(Euphorbia oblongata)* flowers.
J. M. DiTomaso

Oblong spurge *(Euphorbia oblongata)* flowering stems.
J. M. DiTomaso

Oblong spurge (*Euphorbia oblongata*) seeds and fruit. J. O'Brien

stock when plant material is ingested in a sufficient amount. However, reports on leafy spurge toxicity are mixed. In some cases the milky sap irritates skin, eyes, and digestive tracts of humans and other animals. Cattle typically avoid foraging on leafy spurge whenever possible, but goats and sheep appear to be generally immune to its irritant properties and may develop a preference for the weed. Mature plants are more likely to cause toxicity problems than immature shoots. Leafy spurge litter incorporated into the soil appears to inhibit the growth of some plants, suggesting that leafy spurge has allelopathic properties. Several insects that feed on leafy spurge have been released as biocontrol agents and are established in some western states, but none are established or registered for release in California. Leafy spurge is a highly variable species, with many regional biotypes. Some botanists recognize 2 varieties in the contiguous states, of which only variety *esula* occurs in California and all other western states. It is the most widespread variety in other states as well. Variety *uralensis*

Serrate spurge (*Euphorbia serrata*) plant.
W. J. Ferlatte

Serrate spurge (*Euphorbia serrata*) leaves.
W. J. Ferlatte

Carnation spurge (*Euphorbia terracina*) plant.
J. M. DiTomaso

Carnation spurge (*Euphorbia terracina*) leaves and stem. J. M. DiTomaso

Carnation spurge (*Euphorbia terracina*) seedling.
J. M. DiTomaso

Carnation spurge (*Euphorbia terracina*) flowers and fruit. J. M. DiTomaso

(Fisch. ex Link) Dorn, sometimes known as Russian leafy spurge, also occurs in Colorado, Montana, Wyoming, and a few central and northeastern states.

Serrate spurge, oblong spurge: Both species are considered noxious weeds but are much less problematic relative to leafy spurge. The milky sap of these species may have irritant properties as well, but toxicity problems have not been reported.

SEEDLINGS: Leafy spurge: Hypocotyl dull reddish brown at the soil line, pale green above. Cotyledons linear to elliptic, 2–4 mm wide, 4–19 mm long, become leaflike with time, upper surface covered with powdery white granules, lower surface pale and conspicuously veined. Seedling leaves are similar in shape to mature leaves except smaller. First leaves opposite, subsequent leaves generally alternate but initiated close together so as to appear subopposite. Leaf pairs folded or rolled longitudinally in bud, with one leaf enclosing the other. Seedlings develop adventitious buds near the soil line at an early stage.

MATURE PLANT: Stems often somewhat woody at the base. Leaves sessile, mostly *alternate,* but some leaves may be opposite or whorled just below the inflorescence branches. Inflorescence bracts opposite, cordate to kidney-shaped, sessile, glabrous, shorter and broader than leaves. Stipules absent.

Leafy spurge: Stems erect, glabrous or hairy. Leaves linear to narrowly oblanceolate, 2–6 cm long, typically about 0.5 cm wide, tip acute or rounded, *margin smooth,* usually glabrous. Bracts yellow-green.

Oblong spurge: Stems ascending to erect and covered with fine white hairs, especially near the petiole bases and upper stems. Leaves oblong or elliptic, mostly 4–6.5 cm long, tip broadly rounded, margin *minutely serrate* under low magnification, often glabrous.

Serrate spurge: Stems erect, glabrous. Leaves linear to ovate-lanceolate, mostly 2–5 cm long, tip acute to rounded with an abrupt toothlike tip (mucronate), margin *finely serrate* (visible without magnification), glabrous, sometimes covered with a whitish bloom (glaucous).

ROOTS AND UNDERGROUND STRUCTURES: **Leafy spurge:** Develops an extensive system of vigorously long-creeping horizontal roots, short horizontal feeder roots, and short and long vertical roots. Long horizontal roots and crowns have many *pink, scaly buds* from which new shoots can develop. Most root buds develop within 30 cm of the soil surface, but a few can develop as deep as 300 cm. First-year shoots lack woody crowns and are typically more abundant around the perimeter of a clonal population. Vertical roots can reach a depth of 9 m. Under favorable conditions, root fragments as small as 1.5 cm or as deep as 2.8 m can generate a new plant. Long roots become corky and brown with age and can tolerate considerable periods of drought. The root system stores abundant food reserves, enabling roots to produce new shoots for many years under adverse conditions such as continual grazing or mowing.

Oblong spurge, serrate spurge: Taproot woody, branched. Stems branch slightly above or below ground. Crown buds develop at the bases of stems and can produce new shoots or roots.

FLOWERS: Male and female flowers develop separately in specialized flower clusters on the same plant (monoecious). Flower clusters umbel-like at the stem tips, with the central clusters maturing first. What appears to be one flower is actually a specialized cluster of reduced unisexual flowers (cyathium) that is unique to the spurge family. A cyathium consists of several male flowers, each consisting of 1 stamen, inserted on a bell-shaped hypanthium (receptacle extension) and one female flower, consisting of an ovary, situated above the male flowers on a stalk from the center of the hypanthium. The 3-chambered ovary has 3 styles fused together at the bases and branched at the tips. The rim of the hypanthium typically has 4 *glands that lack petal-like appendages.* Insect-pollinated.

Leafy spurge: June–September. Cyathium 1.5–2.5 mm long. *Glands crescent-shaped with a horn at each end*, 1.5–2 mm long, flattened. Style branches one-half the style length. Stamens (staminate flowers) 11–21 per cyathium.

Oblong spurge: Spring/summer. Cyathium 1.5–2 mm long. Glands often 2–3 per cyathium, *oval to oblong*, ± 1 mm long, flattened, outer margin smooth. Style branches less than one-half the style length. Stamens (staminate flowers) 20–40 per cyathium.

Serrate spurge: Spring/summer. Cyathium 2–4 mm long. Glands usually *oval, cupped*, ± 1 mm long, outer margin smooth to truncate or sometimes 2-horned. Style branches less than one-half the style length. Stamens (staminate flowers) 5–20 per cyathium.

FRUITS AND SEEDS: Capsules nearly round, 3-chambered, with 1 seed per chamber. Seeds ovoid to oblong, round in cross-section, 2–3 mm long, with a yellowish caruncle (small appendage) near the end of attachment.

Leafy spurge: Capsules 3–5 mm long, surface smooth to granular. Seeds yellow-brown to gray or mottled, smooth.

Oblong spurge: Capsules 3–4.5 mm long, sparsely covered with minute tubercles. Seeds brown, smooth.

Serrate spurge: Capsules 4–6 mm long, surface smooth, glabrous. Seeds gray, smooth or minutely pitted.

POSTSENESCENCE CHARACTERISTICS: Leafy spurge shoots die back with the onset of the cold season. Leaves often turn reddish just before dropping. About 42 days of chilling are required to release plants from postsenescent dormancy. Roots survive cold winter climates.

Carnation spurge *(Euphorbia terracina)* inflorescence base.   J. M. DiTomaso

Carnation spurge (*Euphorbia terracina*) seeds. J. O'BRIEN

HABITAT: Waste areas, disturbed sites, roadsides, and fields.

Leafy spurge: Also, pastures, rangeland, alfalfa fields, riparian areas. Plants tolerate subtropic to subarctic climates, xeric to mesic conditions, and flooding for at least 4–5 months if shoots can grow above the water surface. Grows best on coarse-textured soils, although most soil types are tolerated.

Oblong spurge: Also, yards, gardens.

DISTRIBUTION: Leafy spurge: Uncommon in California. Modoc Plateau (sw Modoc and se Lassen Cos.) and South Coast (sw Los Angeles Co.), to 1400 m. Infestations that have been eradicated occurred in the Cascade Range (ne Siskiyou Co.), eastern Klamath Ranges (cw Siskiyou Co.), and southern North Coast Ranges (nw Sonoma Co.). All western states, most central states (especially the northernmost), and northeastern states.

Oblong spurge: Uncommon in California, but expanding range. Central Valley, San Francisco Bay region, southern Cascade Range, southern North Coast and North Coast Ranges, South Coast Ranges, and northern and central Sierra Nevada foothills, to 200 m. Oregon and Washington.

Serrate spurge: Known infestations that have been eradicated occurred in the San Francisco Bay region, to 200 m.

PROPAGATION/PHENOLOGY: Mature capsules of many spurges rupture and forcefully eject seeds some distance from the parent plant.

Leafy spurge: *Reproduces by seed* and *vegetatively from extensively creeping roots*. New infestations are usually initiated by seed, but population expansion is mostly vegetative. Seeds mature about 30 days after the female flower appears. Mature capsules eject seeds up to 5 m from the parent plant, but some seeds disperse to greater distances with human activities, water, mud, insects (eg., ants), animals, and as a hay or seed contaminent. Some seeds survive ingestion by sheep and goats, but rarely survive ingestion by mourning doves (*Zenaida macroura*). Seeds can germinate soon after reaching maturity. Seeds can float on water for several days and germinate on the surface. Most seeds germinate in early spring, but germination may occur throughout the growing season under favorable moisture conditions. Optimal germination occurs at temperatures fluctuating between 20°–30°C. Light and scarification appear to reduce germination. A cold, moist stratification period followed by warmer temperatures increases germination of some biotypes. Seedling emergence is optimal at depths of 1.5–5 cm, but can occur from depths to about 15 cm. Under field conditions, some seeds can remain viable for at least 8 years. Most seedlings do not flower the first season. Individual shoots typically produce about 10–50 capsules. Flower production and shoot survival decrease with increasing shade. Dormant buds on roots can tolerate a temperature as low as –13°C, and dormant crown buds can tolerate a temperature lower than –20°C.

Oblong spurge, serrate spurge: The biology of these species is poorly understood.

MANAGEMENT FAVORING/DISCOURAGING SURVIVAL: Wearing gloves while handling plants can prevent the sap from irritating the skin.

Leafy spurge: Mowing, burning, and grazing do not significantly affect roots and typically stimulate the production of new shoots from root buds. Continuous grazing by sheep or goats can reduce the soil seedbank by preventing seed production, but roots usually continue to produce new shoots for about 8 years or more. Livestock, especially sheep and goats, that may have consumed plants in fruit should be quarantined for 5–6 days before transport to an uninfested area to prevent the introduction of viable seed. Cultivation or disking can spread leafy spurge root fragments and seeds.

Oblong spurge, serrate spurge: Manually removing plants before seed develops can control these species.

SIMILAR SPECIES: There are many weedy spurges, several of which appear similar. All have milky white sap. Refer to table 32 (p. 726) for a comparison of weedy spurges. The following species are most likely to be confused with leafy spurge, oblong spurge, and serrate spurge.

Carnation spurge [*Euphorbia terracina* L., synonym: Geraldton carnationweed] [Cal-IPC: Moderate Alert][CDFA list: Q] is an uncommon but potentially noxious *perennial or biennial* to about 1 m tall, with a vertical taproot and *alternate*, linear-lanceolate to oblong-elliptic leaves that have *minutely serrate margins* under low magnification. Cyathia are yellowish green and have *glands with 2 very long, slender horns*. Seeds are ovoid, *pale gray*, and have a *smooth surface* and a prominent pale yellow boat-shaped appendage (caruncle) at the base. Carnation spurge is a recent introduction to California and is not included in California floras to date. It often forms dense patches and generally flowers from March through August. Carnation spurge inhabits disturbed places, grassland, coastal bluffs (particularly near Malibu), dunes, salt marsh, riparian areas, and oak woodlands in the South Coast (Los Angeles Co.). It also occurs in Pennsylvania. Like many other spurges, Carnation spurge is reported to have *toxic* sap. Native to southern Europe and the Mediterranean.

Caper spurge [*Euphorbia lathyris* L., synonym: gopher spurge] is a potentially noxious *annual or biennial* to 1 m tall, with *opposite leaves* and cyathia with *crescent-shaped glands*. Leaves are lanceolate, mostly 5–15 cm long, and have smooth margins. Refer to the entry for **Caper spurge and Petty spurge** (p. 710) for more information. Caper spurge inhabits coastal scrub, marshes, dunes, gardens, roadsides, and disturbed places in the North Coast, Central Coast, South Coast, and Central Valley, to 200 m, and is expanding its range. It also occurs in Oregon, Washington, Arizona, and many eastern states. Caper spurge can cause mild to severe digestive tract irritation in humans and livestock when ingested. Native to Europe.

Sun spurge [*Euphorbia helioscopia* L.][EPHHE] is an *annual* to 0.5 m tall, with broadly obovate leaves that have *finely toothed margins* and cyathia with 4 *oval glands*. Unlike oblong spurge and serrate spurge, sun spurge has seeds with a *coarsely reticulate or netlike surface*. Refer to the "Similar Species" section of the entry for **Caper spurge and Petty spurge** (p. 710) for more information. Sun spurge primarily inhabits waste places in the North Coast, western North Coast Ranges, San Francisco Bay region, and South Coast, to 200 m. It also occurs in Montana, Oregon, Washington, Wyoming, and most northern, eastern, and southern states. Ingestion of sun spurge can cause digestive tract irritation in humans and livestock, and it has been reported to be fatally *toxic* on rare occasions. Native to Europe.

Caper spurge [*Euphorbia lathyris* L.]
Petty spurge [*Euphorbia peplus* L.][EPHPE]

SYNONYMS: Caper spurge: gopher plant; gopher purge; gopher's bane; mole's bane; moleplant; petroleum plant; sassy jack; *Galarhoeus lathyris* (L.) Haw.; *Tithymalus lathyris* (L.) Hill

Petty spurge: *Galarhoeus peplus* (L.) Rydb.; *Tithymalus peplus* (L.) Hill

GENERAL INFORMATION: Both species have *milky sap, glabrous foliage, forked upper stems,* and *opposite leaflike bracts below the characteristic flower clusters.* Refer to table 32 (p. 726) for a comparison of important differences. Both species are native to Europe.

Caper spurge: Coarse, erect *biennial* to 1.5 m tall. Like many species in the genus, the milky sap of caper spurge can be *toxic* to humans and livestock when the foliage and/or fruits are ingested in a sufficient amount. Toxicity symptoms consist of mild to severe digestive tract irritation, with nausea, vomiting, and diarrhea. However, animals generally avoid consuming caper spurge if more palatable forage is available. Skin contact with the milky sap can also cause contact dermatitis in sensitive individuals. The root of this species is purported to repel gophers when planted in gardens, although there is no evidence to support this assertion.

Petty spurge: Erect winter or summer *annual* to 0.5 m tall. Petty spurge is a common minor weed of yards, landscaped areas, and urban places. Like caper spurge, ingestion of petty spurge can cause digestive tract irritation. On rare

Caper spurge (*Euphorbia lathyris*) plant.   J. M. DiTomaso

Caper spurge (*Euphorbia lathyris*) flowers and fruit.   J. M. DiTomaso

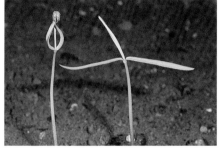

Caper spurge (*Euphorbia lathyris*) seedlings with seed coat attached.   J. M. DiTomaso

Caper spurge (*Euphorbia lathyris*) young plant with decussate leaves.   J. M. DiTomaso

occasions, petty spurge has been reported to be fatally *toxic* to humans and livestock. However, problems due to the ingestion of this species appear to be rare.

SEEDLINGS: Caper spurge: Glabrous. Cotyledons narrowly elliptic, 18–26 mm long, sessile, bases fused. Leaf pairs opposite and alternately perpendicular to one another. First and subsequent leaves narrowly elliptic to linear, about 30–35 mm long, sessile, bases fused, abruptly short-pointed at the tips (cuspidate).

Caper spurge (*Euphorbia lathyris*) seeds.   J. O'Brien

Petty spurge (*Euphorbia peplus*) seedling.
J. M. DiTomaso

Petty spurge (*Euphorbia peplus*) plant.
J. M. DiTomaso

Petty spurge (*Euphorbia peplus*) flowering stem.
J. M. DiTomaso

Petty spurge (*Euphorbia peplus*) seeds and fruit.
J. O'Brien

Petty spurge: Glabrous. Cotyledons oval, 4–7 mm long, on a stalk ± 1 mm long. First leaf pair opposite or alternate. First leaves obovate, mostly 4–7 mm long, tapered to a stalk ± 1 mm long.

MATURE PLANT: Caper spurge: Foliage glabrous. Main stems coarse, fibrous, erect, usually simple below, branched near the middle, upper stems forked, ± covered with a whitish film (glaucous). Leaves sessile, *opposite, with pairs alternately perpendicular to one another* (decussate), *linear to elliptic or lanceolate, 5–15 cm long*, 0.5–2.5 cm wide, margin smooth, upper surface dark green with a pale midvein, lower surface paler and glaucous. Inflorescence bracts of upper stems leaflike, opposite.

Petty spurge: Foliage glabrous, light green. Main stems slender, erect, simple or branched, lower branches opposite or alternate, upper stems forked. *Lower* leaves alternate to subopposite, ovate to obovate, 1–3.5 cm long, tip rounded to slightly indented, on a short stalk. Inflorescence bracts of upper stems leaflike, opposite.

ROOTS AND UNDERGROUND STRUCTURES: Caper spurge: Taproot coarse, fibrous, with fibrous lateral roots.

Petty spurge: Taproot slender, generally with few fibrous lateral roots.

FLOWERS: Male and female flowers develop separately in specialized flower clusters on the same plant (monoecious). Flower clusters umbel-like at the stem tips, with the central clusters maturing first. What appears to be one flower is actually a specialized cluster of reduced unisexual flowers (cyathium)

Sun spurge (*Euphorbia helioscopia*) flowers and fruit.   J. M. DiTomaso

Sun spurge (*Euphorbia helioscopia*) plant.
   J. M. DiTomaso

Toothed spurge (*Euphorbia dentata*) foliage.
   J. Neal

Sun spurge (*Euphorbia helioscopia*) seedling.
   J. M. DiTomaso

that is unique to the spurge family. A cyathium consists of several male flowers, each consisting of 1 stamen, inserted on a bell-shaped hypanthium (receptacle extension) and one female flower consisting of an ovary situated above the male flowers on a stalk from the center of the hypanthium. The 3-chambered ovary has 3 styles fused together at the bases and branched at the tips. The rim of the hypanthium has *4 flattened, crescent-shaped glands with a short horn at each end*. The glands of these species lack petal-like appendages. Insect-pollinated.

Caper spurge: Most of the year. Glands ± 2 mm long. Stamens (male flowers) 15–40 per cyathium.

Petty spurge: February to August. Glands less than 0.5 mm long. Stamens (male flowers) 10–15 per cyathium.

FRUITS AND SEEDS: Capsules glabrous, nearly spherical, 3-lobed, surface smooth, contain 3 seeds.

Caper spurge: Capsules *8–15 mm in diameter*. Seeds oblong, round in cross-section, 4–5 mm long, brown, surface shallowly reticulate with magnification. Seeds have a yellowish caruncle (small hatlike appendage) near the end of attachment.

Petty spurge: Capsules *1.5–3 mm in diameter*, 3-angled, lobes longitudinally 2-ridged. Seeds oblong, 4-angled, 1–1.5 mm long, white to gray, minutely pitted with magnification. Seeds lack a caruncle.

HABITAT: Caper spurge: Coastal scrub, marshes, and dunes, waste places, disturbed sites, roadsides, gardens.

Sun spurge (*Euphorbia helioscopia*) leaves at base of inflorescence. J. M. DiTomaso

Petty spurge: Landscaped areas, gardens, yards, urban areas, waste places, crop fields. Often grows on fertile soils and/or in moist, partly shaded places.

DISTRIBUTION: Caper spurge: North Coast, Central Coast, South Coast, and Central Valley, to 200 m. Expanding range. Oregon, Washington, Arizona, and many eastern states.

Petty spurge: Throughout California, except deserts and Great Basin, to 300 m. Western states except Nevada and Wyoming, northeastern states and adjacent central states, and a few southern states.

PROPAGATION/PHENOLOGY: *Reproduce by seed.* Capsules burst open and eject seed up to several meters from the parent plant. Seeds disperse to greater distances with water, mud, soil movement, vehicle tires, and human activities such as landscape maintenance.

MANAGEMENT FAVORING/DISCOURAGING SURVIVAL: Repeated cultivation, especially before seeds mature, can control both species.

SIMILAR SPECIES: Toothed spurge [*Euphorbia dentata* Michx., synonyms: toothedleaf poinsettia; *Anisophyllum dentatum* (Michx.) Haw.; *Euphorbia herronii* Riddell, *Poinsettia dentata* (Michx.) Klotzsch & Garcke][EPHDE] is an ascending to erect *annual* to 0.5 m tall, with *milky sap, hairy stems* and *opposite upper and middle leaves. Some lower leaves are usually alternate.* Leaves are stalked, lanceolate to ovate, 1–6 cm long, and have a *conspicuously toothed to lobed margin.* The cyathia are in cymelike clusters at the stem tips, and each cyathium has *1–2 cupped, oval glands* less than 1 mm long. Capsules are nearly round, 3-lobed, 4–5 mm in diameter, glabrous, and smooth. Seeds are ovoid, 3-angled, 2–3 mm long, gray to brown, and covered with tubercles. Toothed spurge is generally uncommon. It inhabits disturbed sites and waste places in the San Joaquin Valley,

Sun spurge (*Euphorbia helioscopia*) seeds and fruit. J. O'BRIEN

Central Coast, and San Francisco Bay region, to 300 m. Toothed spurge also occurs in Arizona, Colorado, most southern states, and many central and eastern states, except the northernmost. Toothed spurge has milky sap with *toxic* properties similar to those of caper spurge. Toothed spurge is a state-listed noxious weed in Idaho. Native to the eastern United States and Mexico.

Sun spurge [*Euphorbia helioscopia* L., synonyms: madwoman's milk; wartweed; *Galarhoeus helioscopius* (L.) Haw.; *Tithymalus helioscopius* (L.) Hill] [EPHHE] is an ascending to erect *annual* to 0.5 m tall, with *milky sap, sparsely hairy stems,* and *alternate leaves* that are glabrous and sessile or short-stalked. Sun spurge is further distinguished by having *broadly obovate leaves 1–3 cm long with finely toothed margins,* cyathia with *4 flat, oval glands* less than 1 mm long, and dark gray to brown nearly spherical seeds 2–2.5 mm in diameter that have a *coarsely reticulate or netlike surface.* Cyathia are in umbel-like clusters near the stem tips and in cymelike clusters below. Capsules are nearly round, 3-lobed, 2.5–3 mm in diameter, glabrous, and smooth. Plants typically flower March through August. Sun spurge primarily inhabits waste places in the North Coast, western North Coast Ranges, San Francisco Bay region, and South Coast, to 200 m. It also occurs in Montana, Oregon, Washington, Wyoming, and most northern, eastern, and southern states. Ingestion of sun spurge can cause digestive tract irritation in humans and livestock, and it has been reported to be fatally *toxic* on rare occasions. Native to Europe.

Spotted spurge [*Euphorbia maculata* L.][EPHMA]
Ground spurge [*Euphorbia prostrata* Ait.][EPHPT]
Creeping spurge [*Euphorbia serpens* Kunth][EPHSN]
Nodding spurge [*Euphorbia nutans* Lag.] [EPHNU]

SYNONYMS: Recent molecular data shows that including the genus *Chamaesyce* within the genus *Euphorbia* makes the entire group monophyletic.

Spotted spurge: milk purslane; milk spurge; prostrate spurge; spotted pusley; spotted sandmat; *Chamaesyce maculata* (L.) Small; *Chamaesyce mathewsii* Small; *Chamaesyce supina* (Raf.) Moldenke; *Chamaesyce tracyi* Small; *Euphorbia depressa* Torr.; *Euphorbia supina* Raf.

Ground spurge: prostrate sandmat; prostrate spurge; *Chamaesyce prostrata* (Ait.) Small; *Euphorbia chamaesyce* auct. non L.

Creeping spurge: matted sandmat; *Chamaesyce serpens* (Kunth) Small

Nodding spurge: eyebane; large spurge; upright spotted spurge; *Chamaesyce nutans* (Lag.) Small; *Chamaesyce preslii* (Guss.) Arthur; *Euphorbia preslii* Guss. American authors previously referred to nodding spurge as *Chamaesyce maculata* (L.) Small (synonym: *Euphorbia maculata* L.), which is a different species.

GENERAL INFORMATION: Summer *annuals*, with *opposite leaves that have an*

Red form of spotted spurge *(Euphorbia maculata)* plant. J. K. Clark

Spotted spurge • Ground spurge • Creeping spurge • Nodding spurge

Spotted spurge (*Euphorbia maculata*) leaves and white milky sap.   J. K. CLARK

Spotted spurge (*Euphorbia maculata*) fruiting stem.   J. M. DITOMASO

Green form of spotted spurge (*Euphorbia maculata*) seedling.   C. ELMORE

asymmetrically oblique base and *milky sap*. Spotted spurge, ground spurge, and creeping spurge typically have *prostrate stems* to about 0.5 m long. Nodding spurge usually has *ascending to erect stems* to about 0.5 m tall and is sometimes a biennial or short-lived perennial under favorable conditions. Refer to table 32 (p. 726)for a comparison of important distinguishing features of spurges. The milky sap of spurge species contains various diterpene esters that can cause contact dermatitis in humans and animals. Ingestion of some species can also cause mild to severe digestive tract irritation and on rare occasions can be fatally *toxic*. Spotted spurge has been responsible for the deaths of lambs

Red form of spotted spurge (*Euphorbia maculata*) seedling.   J. K. CLARK

Spotted spurge (*Euphorbia maculata*) seeds and fruit.   J. O'BRIEN

Spotted spurge • Ground spurge • Creeping spurge • Nodding spurge

Ground spurge *(Euphorbia prostrata)* plant.
J. M. DiTomaso

Ground spurge *(Euphorbia prostrata)* fruiting stem.
J. M. DiTomaso

Ground spurge *(Euphorbia prostrata)* seedling.
J. M. DiTomaso

grazing on heavily infested pasture in the southeastern United States. Spotted spurge is the most commonly encountered weedy spurge in California. It can harbor insect pests and fungal diseases that affect crops. Spotted spurge and ground spurge are native to the eastern United States. Nodding spurge is native

Ground spurge *(Euphorbia prostrata)* seeds and fruit.
J. O'Brien

to South America and the eastern United States. Creeping spurge is becoming a more serious problem in landscapes and nursery production. It can overwinter, thus acting as a weak perennial, in milder climatic areas of California. Creeping spurge is native to South America.

SEEDLINGS: Leaves opposite.

Spotted spurge: Cotyledons oval to oblong, tip rounded to slightly truncate, 1–3 mm long, 0.5–1.5 mm wide, glabrous, lower surface often maroon. First leaf pair obovate to nearly round, tip rounded and ± minutely toothed, base equally tapered, sometimes maroon-tinged or with a maroon central spot, foliage usually ± sparsely hairy. Subsequent leaves resemble mature leaves with obliquely asymmetric bases.

Nodding spurge: Cotyledons oblong, tip truncate and barely indented, base ± truncate, 3–5 mm long, 2–3 mm wide, glabrous. First few leaf pairs oblanceolate to obovate, about 5–7 mm long, ± 4 mm wide, tip acute to slightly

Creeping spurge (*Euphorbia serpens*) plant.　　　　　　　　　　J. M. DiTomaso

Creeping spurge (*Euphorbia serpens*) fruiting stem.　　　　J. M. DiTomaso

Creeping spurge (*Euphorbia serpens*) seedling.　　　　J. M. DiTomaso

Spotted spurge • Ground spurge • Creeping spurge • Nodding spurge

Creeping spurge *(Euphorbia serpens)* seeds and fruit. J. O'Brien

rounded and finely toothed, base equally tapered, glabrous to sparsely hairy, midvein sometimes reddish.

MATURE PLANT: Stem branches alternate. Leaves opposite, base usually obliquely asymmetric, on a short stalk.

Spotted spurge: Stems usually *prostrate*, sometimes ascending when competing with other plants for light, sometimes reddish. New foliage typically *hairy*, especially stems and lower leaf surfaces. Leaves oblong to ovate, 4–17 mm long, margins minutely toothed with magnification, often have a maroon central spot. Stipules separate, ± awl-shaped, often fringed.

Spotted spurge *(Euphorbia maculata)* (bottom) and creeping spurge *(E. serpens)* (top). C. Elmore

Spotted spurge • Ground spurge • Creeping spurge • Nodding spurge

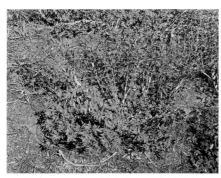
Nodding spurge *(Euphorbia nutans)* plant.
J. M. DiTomaso

Nodding spurge *(Euphorbia nutans)* flowering stem.
J. M. DiTomaso

Nodding spurge *(Euphorbia nutans)* young plant.
J. M. DiTomaso

Nodding spurge *(Euphorbia nutans)* seedlings.
J. M. DiTomaso

Nodding spurge *(Euphorbia nutans)* fruiting stem.
J. M. DiTomaso

Nodding spurge *(Euphorbia nutans)* seeds and fruit.
J. O'Brien

Ground spurge: Stems *prostrate,* sometimes slightly upturned near the tip (decumbent). New foliage usually *hairy,* especially stems and lower leaf surfaces. Leaves (ob)ovate to elliptic, 3–11 mm long, margins usually minutely toothed with magnification. Stipules separate, linear.

Creeping spurge: Stems *prostrate, often rooting at the nodes in moisture areas. Foliage glabrous.* Leaves ovate or obovate to slightly oblong, 2–7 mm long, *margins smooth.* Stipules fused, triangular, scalelike.

Nodding spurge: Stems *ascending to erect,* typically *slightly drooping or nodding near the tip.* Foliage sparsely soft-hairy to glabrous. Leaves oblong, 8–35 mm long, often with a maroon central spot, margin finely toothed without magnification and sometimes maroon-tinged. Stipules fused near the base, triangular.

ROOTS AND UNDERGROUND STRUCTURES: Taproots slender, fibrous. Spotted spurge roots can grow to about 0.5 m deep.

FLOWERS: Male and female flowers develop separately in specialized flower clusters on the same plant (monoecious). Flower clusters umbel-like at the stem tips, with the central clusters maturing first. What appears to be one flower is actually a specialized cluster of reduced unisexual flowers (cyathium) that is unique to the spurge family. A cyathium consists of several male flowers, each consisting of 1 stamen, inserted on a bell-shaped hypanthium (receptacle extension) and one female flower consisting of an ovary situated above the male flowers on a stalk from the center of the hypanthium. The 3-chambered ovary has 3 styles fused together at the bases and branched at the tips. The rim of the receptacle extension typically *has 4 flattened, oval, or oblong glands* to 0.5 mm long, each with a *petal-like appendage.* Insect-pollinated.

Spotted spurge: May–October. Flower clusters (cyathia) 1–2 per leaf axil (1–4 per node), distributed along the length of the stem. Glands oblong. *Gland appendage width ± equal to gland width, margin shallow-scalloped,* white to pink. Stamens (male flowers) *4–5 per cyathium.*

Ground spurge: August–October. Flower clusters (cyathia) 1–2 per leaf axil (1–4 per node), distributed along the length of the stem. Glands oval. *Gland appendage width equal to or greater than gland width, margin smooth or shallow-scalloped,* white. Stamens (male flowers) *4 per cyathium.*

Creeping spurge: Flower clusters (cyathia) 1–2 per leaf axil (1–4 per node), distributed along the length of the stem. Glands oblong. Gland appendage width greater than gland width, margin shallow-scalloped, white. Stamens (male flowers) *5–10 per cyathium.*

Nodding spurge: April–October. Flower clusters (cyathia) *densely clustered in the axils at the stem tips.* Glands oblong. Gland appendage width greater than gland width, margin smooth, white to red. Stamens (male flowers) *5–11 per cyathium.*

FRUITS AND SEEDS: Capsules 3-lobed, 3-chambered, with 1 seed per chamber.

Spotted spurge: Capsules ovoid, *1–1.5 mm long, evenly hairy.* Seeds

Spotted spurge • Ground spurge • Creeping spurge • Nodding spurge

Thyme-leafed spurge (*Euphorbia serpyllifolia*) fruiting stem, a native nonweedy plant resembling the weedy species. J. M. DiTomaso

Thyme-leafed spurge (*Euphorbia serpyllifolia*) seeds and fruit. J. O'Brien

oblong, ± 4-sided, truncate at 1 end, 0.8–1 mm long, ± 0.5 mm wide, light brown, *faintly transverse-wrinkled*, wrinkles rounded.

Ground spurge: Capsules nearly round, *1–1.5 mm long*, mostly *hairy on the ridge of each lobe*. Seeds oblong, 4-sided, truncate at 1 end, ± 1 mm long, 0.5 mm wide, *sharply transverse-ridged*, white, gray, or orange-brown.

Creeping spurge: Capsules nearly round, mostly *1–1.5 mm long, glabrous*. Seeds oblong, truncate at 1 end, 4-sided, ± 1 mm long, ± 0.5 mm wide, smooth, minutely granular with magnification, white to pale brown or orange-brown.

Nodding spurge: Capsules ovoid, *2–2.5 mm long, glabrous*. Seeds ovoid, truncate at 1 end, 4-sided, 1–1.5 mm long, ± 1 mm wide, shallow-wrinkled, black or brown, with a minute caruncle at the small end.

POSTSENESCENCE CHARACTERISTICS: Dead plants generally do not persist.

HABITAT: Yards, landscaped areas, walkways, roadsides, gardens, turf, waste places, vacant lots, orchards, vineyards, agronomic and vegetable crops, nursery grounds and containers, and other disturbed places.

DISTRIBUTION: Spotted spurge: Throughout California, except deserts and Great Basin, to 200 m. All contiguous states except possibly Colorado.

Ground spurge: Southwestern region, Central-western region, central Sierra Nevada foothills, probably southern Sierra Nevada foothills and San Joaquin Valley, and possibly elsewhere, to 250 m. Arizona, New Mexico, Utah, southern states, and most central and eastern states except the northernmost.

Creeping spurge: Central Valley, southern South Coast Ranges, and Southwestern region, to 200 m. Expected to expand its range in California.

Arizona, Colorado, Montana, New Mexico, Wyoming, most central and southern states, and a few northeastern states.

Nodding spurge: Throughout California, except Mojave Desert and Great Basin, to 300 m. New Mexico, central and southern states, and most eastern states.

PROPAGATION/PHENOLOGY: *Reproduce by seed.* Creeping spurge also reproduces vegetatively from creeping stems that root at the nodes. Most seeds fall near the parent plant, but some can disperse to greater distances with water, soil movement, mud, contaminated seed, foot and vehicle traffic, transported nursery containers, agricultural operations, lawn mowing, leaf blowing, and other human activities. Ants also appear to be a vector for movement of spurge seeds. Seeds germinate in spring and summer. Seedlings emerge from a soil depth to about 4 cm and can produce seed in about 2 weeks. Plants typically produce abundant long-lived seed. Buried spotted spurge seeds can survive up to ± 50 years. Like many plants in the spurge family, spotted spurge and prostrate spurge use the C4 photosynthetic pathway. The photosynthetic pathway used by nodding spurge and creeping spurge is not documented.

MANAGEMENT FAVORING/DISCOURAGING SURVIVAL: Spurges are difficult to control once a large seedbank is established. Regular cultivation or persistent manual removal of plants before seed develops can limit the quantity of seed entering the soil seedbank. When hand-pulled, the stems of established plants often break above the crown, allowing the remaining parts to grow new stems. Severing plants well below the crown can help prevent regrowth. In areas where there is an established soil seedbank, mulching with at least 4 cm of fine mulch or 7.5 cm of coarse mulch can reduce seed germination. In turf, spurge typically invades bare spots. Maintaining healthy turf can help prevent spurge from becoming established. Open spaces in turf or landscapes can lead to establishment of the spurges.

SIMILAR SPECIES: Currently, 18 native prostrate *Euphorbia* species (formerly *Chamaesyce*) are recognized in California, some of which are uncommon or rare. Native prostrate spurges typically inhabit open sites in natural areas, but some of the ruderal types can occur on the margins of landscaped areas, cultivated fields, turf, and other human-dominated landscapes. They are rarely problematic, although some resemble the weedy non-native species.

Spotted spurge • Ground spurge • Creeping spurge • Nodding spurge

Table 32. Spurge (*Euphorbia* spp.) vegetative and reproductive characteristics

Vegetative Characteristics

Euphorbia spp.	Life cycle	Form	Leaf arrangement	Leaf shape	Leaf margin	Leaf length (mm)	Sti
E. albomarginata whitemargin spurge	perennial	prostrate	opposite, leaf pairs in 1 plane		entire	3–8	triangular fused cili
E. dentata toothed spurge	annual	erect to ascending	mostly opposite		coarsely toothed	10–60	non
E. esula leafy spurge	perennial	erect	alternate (inflor bracts opp)		entire	20–60	non
E. helioscopia sun spurge	annual	ascending to erect	alternate (inflor bracts opp)		finely toothed	10–30	non
E. lathyris caper spurge	biennial	erect	opposite, leaf pairs in 2 planes (decussate)		entire	50–150	non
E. maculata spotted spurge	annual	prostrate	opposite, leaf pairs in 1 plane		minutely toothed (low magnification)	4–17	frin sep
E. nutans nodding spurge	annual	erect	opposite, leaf pairs in 1 plane		finely toothed	8–35	tria lar
E. oblongata oblong spurge	perennial	ascending to erect	alternate (inflor bracts opp)		minutely toothed (low magnification)	40–65	no
E. peplus petty spurge	annual	ascending to erect	alternate (inflor bracts opp)		entire	10–35	no
E. prostrata ground spurge	annual	prostrate	opposite, leaf pairs in 1 plane		minutely toothed (low magnification)	3–11	lin sep
E. serpens creeping spurge	annual or weak perennial	prostrate	opposite, leaf pairs in 1 plane		entire	2–7	tria lar
E. serrata serrate spurge	perennial	erect	alternate (inflor bracts opp)		finely toothed	20–50	no
E. terracina Carnation spurge	biennial, perennial	ascending to erect	alternate (inflor bracts opp)		minutely toothed (low magnification)	30–40	no

Spotted spurge • Ground spurge • Creeping spurge • Nodding spurge

Reproductive Characteristics

orbia	Inflorescence	Anther nos.	Gland shape	Gland length (mm)	Fruit surface	Fruit length (mm)	Seed surface	Seed color
bomar-a emar-purge	1/node; ± crowded on lateral branches	15–30		< 1	smooth	2–2.5	smooth	white
ntata ed ge	clusters at stem tips	< 50		< 1	smooth	4–5	tubercled	gray to brown
ula ge	umbel-like	11–21		1.5–2	granular or smooth	3–5	smooth	yellow-brown
liosco- spurge	umbel-like	8–12		< 1	smooth	2.5–3	coarsely reticulate	dark gray to brown
hyris ge	± umbel-like	15–40		~ 2	smooth	8–15	shallowly reticulate	brown
culata ed ge	1/node; ± crowded on lateral branches	4–5		< 0.5	± evenly hairy	< 1.5	transversely wrinkled	light brown
tans ing ge	dense axillary clusters (or 1/node) at stem tips	5–11		< 0.5	smooth	2–2.5	shallowly wrinkled	black or brown
ongata ig ge	umbel-like	20–40		~ 1	tubercled	3–4.5	smooth	brown
lus ge	± umbel-like	10–15		< 0.5	smooth	~ 2	dotted with shallow depressions	white to gray
strata d ge	1/node; ± crowded on lateral branches	4		< 0.5	hairy, ± margins only	1–1.5	transversely sharp-ridged	white, gray, or orange-brown
pens ing ge	1/node; ± crowded on lateral branches	5–10		< 0.5	smooth	< 1.5	smooth	white to orange-brown
rata te ge	umbel-like	5–20		~ 1	smooth	4–6	smooth or dotted with shallow depressions	gray
racina ation ge	umbel-like	(5–20)		~ 1	smooth	3–5	smooth	pale gray

Castorbean [*Ricinus communis* L.][RIICO] [Cal-IPC: Limited]

SYNONYMS: castor oil plant; palma christi; wonder plant

GENERAL INFORMATION: Summer *annual, perennial, shrub, or small tree* to 3 m tall or more, with *large, shieldlike, palmate-lobed leaves* and *soft-spiny capsules*. Some varieties of castorbean are cultivated as ornamentals and for the oil contained in the seeds. However, castorbean has escaped cultivation in California and become a noxious weed in Central and South Coast areas. Seeds and, to a lesser degree, foliage, are *highly toxic* to humans and animals when ingested. Plant material contains ricin, an extremely toxic protein. Ingestion of

Castorbean (*Ricinus communis*) plants.
J. M. DiTomaso

Castorbean (*Ricinus communis*) fruit.
J. M. DiTomaso

Castorbean (*Ricinus communis*) seedling.
J. M. DiTomaso

Castorbean (*Ricinus communis*) young plant.
J. M. DiTomaso

Castorbean

Castorbean (*Ricinus communis*) flowers and leaves.   J. M. DiTomaso

Castorbean (*Ricinus communis*) seeds.   J. O'Brien

about 4–8 seeds can kill an adult, and fewer can kill a child. Seeds must be broken or chewed to release the toxin. Castor oil derived from the seeds does not contain the water-soluble toxin. Crushed foliage has a *disagreeable odor* and is usually avoided by animals. Livestock poisonings most often occur when feed is contaminated with castorbean seeds. Handling foliage and seeds can cause severe contact dermatitis in sensitive individuals. Native to tropical Africa and Eurasia.

SEEDLINGS: Grow rapidly the first year. Cotyledons large, about 2–4 cm long, ovate to oblong, 3-veined from the base, on ± reddish stalks about as long or longer than the cotyledons, stalk below cotyledons (hypocotyl) often reddish.

MATURE PLANT: Sap clear. Stems glabrous, thick, hollow, often red-tinged. Leaves alternate, glossy dark green, *10–60 cm wide, shieldlike* (peltate), *palmate-lobed with 5–11 lanceolate lobes*, margin toothed, main veins palmate and often reddish. Leaf stalk glandular, about 10–30 cm long. Stipules fused, sheathlike, about 1–2 cm long.

ROOTS AND UNDERGROUND STRUCTURES: Roots fibrous, thick.

FLOWERS: Throughout much of the year. Male and female flowers develop on the same plant (monoecious). *Flower clusters terminal, ± panicle-like*, with female flowers above the male flowers. All flowers lack petals and have 3–5 sepals. Female flowers have a 3-chambered ovary with 3 forked, feathery, red styles. Male flowers have numerous stamens with branched filaments.

FRUITS AND SEEDS: Capsules covered with *soft spines*, usually gray-green, rarely purplish or red, mostly 1–2 cm in diameter, 3-chambered with 1 seed per chamber. Seeds slightly flattened, oblong, 10–14 mm long, smooth, glossy, mottled silver and brown, with a fleshy appendage (caruncle) 3–4 mm wide at the base.

HABITAT: Roadsides, fields, riparian areas, and disturbed waste places. Grows best when moisture is plentiful, but tolerates considerable drought. Plants are highly susceptible to frost damage and grow as an annual in cold-winter areas.

DISTRIBUTION: North Coast Ranges, Central Valley, San Francisco Bay region,

Central Coast, and South Coast, to 300 m. Expanding range in California. Arizona, Utah, most southern and eastern states, and a few central states.

PROPAGATION/PHENOLOGY: *Reproduces by seed.* Seeds disperse short distances when capsules snap open at maturity and to greater distances with human activities, soil movement, and water. Seeds germinate readily in spring. The seed caruncle absorbs water and appears to enhance germination under conditions that would otherwise be too dry. Seedlings can emerge from soil depths to 30 cm.

MANAGEMENT FAVORING/DISCOURAGING SURVIVAL: Manually removing mature plants followed by periodic removal of seedlings can readily control small colonies. Castorbean is relatively easy to control with many systemic herbicides. Removal or death of an older plant can lead to increased seed germination below the parent plant.

SIMILAR SPECIES: Castorbean is unlikely to be confused with other species.

Chinese tallowtree [*Sapium sebiferum* (L.) Roxb.] [SAQSE][Cal-IPC: Moderate Alert]

SYNONYMS: chicken tree; Chinese tallowberry; Florida aspen; popcorn tree; vegetable tallow; white wax berry; *Croton sebiferum* L.; *Excoecaria sebifera* (L.) Müll. Arg.; *Stillingia sebifera* (L.) Michx.; *Triadica sebifera* (L.) Small; *Triadica sinensis* Lour.

GENERAL INFORMATION: Fast-growing *deciduous tree* to about 16 m tall, with *milky sap* and pendent leaves. Chinese tallowtree can aggressively invade disturbed and undisturbed terrestrial, wetland, and riparian plant communities. The tree is most problematic in the southeastern United States, where large tracts of coastal prairie have been replaced by stands of Chinese tallowtree. It is a state-listed noxious weed in Florida and Louisiana. In California, Chinese tallowtree has only recently been discovered in certain riparian and wetland areas.

Chinese tallowtree *(Sapium sebiferum)* leaves and bark.   J. M. DiTomaso

Stands replace native vegetation and can significantly alter the soil nutrient status. Chinese tallowtree leaf litter decomposes rapidly and increases the level of nitrogen, phosphorus, potassium and other mineral nutrients, while decreasing the level of magnesium and sodium. The outer seed layer contains a solid fat that is used to manufacture candles, soap, cloth dressing, and fuel in Asia. Seeds also contain a quick-drying oil that is used in varnishes, paints, machine oils, as a crude lamp oil, and as a medicinal purgative and emetic. After the oil is expressed, seeds are ground into a high-protein meal used for feed and manure. The milky sap and unripe fruits are *mildly toxic* to humans and livestock when ingested. Chinese tallowtree is a popular landscape tree because it grows rapidly, adapts well to many environmental conditions, has showy fall coloration, and is nearly pest-free. It was introduced to the southeastern United States in the late 1700s as a potential oil crop tree. In the southeastern United States, seeds are often dispersed by birds, especially great-tailed grackles (*Quiscalus mexicanus*). These grackles are not found in California where Chinese tallowtree has naturalized. However, the range of this bird is spreading northward. Chinese tallowtree is native to subtropical and warm-temperate areas of eastern Asia.

SEEDLINGS: Vigorous, grow rapidly under sunny conditions, tolerate shade. Cotyledons glabrous, broadly obovate, 10–20 mm long, 8–18 mm wide, 3–5-veined from the base, tip ± slightly indented. Leaves resemble those of mature plants except much smaller (± 20 mm long, 18 mm wide). First leaf pair can appear opposite, but development is alternate.

MATURE PLANT: Canopy thin, open, generally rounded-pyramidal. Bark light gray-brown, typically with narrow longitudinal ridges, sometimes vertically peeling in narrow strips. Branches often slightly drooping. Twigs slender, waxy, many arched upward, dotted with small pores (lenticels). Leaves alternate, glabrous, *ovate-rhomboid with a long-tapered pinched-in tip* (long attenuate), mostly 3–8 cm long, 3–6 cm wide, mostly *pendent* on stalks 2–5 cm long. *Margins smooth. Stalks have 2 glands on the upper side just below the blade* and a pair of stipulelike appendages about 3 mm long at the base. Leaves emerge later in spring than most other deciduous trees and typically turn yellow, orange-red, and eventually purplish red before dropping in fall.

ROOTS AND UNDERGROUND STRUCTURES: Often extensively spreading and/or deep.

FLOWERS: April–July. *Monoecious* (male and female flowers develop separately on the same plant). Panicles terminal, narrow, dense, ± *spikelike*, 5–20 cm long, with small greenish yellow flowers that lack petals. *Male flowers develop in fascicles of 3–15 along the length of the panicle and are above the 1–5 female flowers.* Calyx 2–3-lobed. Stamens 2–3 per bract. Styles (2–)3 per female flower. Trees produce 2 types of inflorescences that help to ensure outcrossing (dichogamy). Trees with grape-type panicles have female flowers that mature first and are spirally arranged on stalks on a terminal fruiting branchlet below the male flowers. Trees with eagle-claw-type panicles initially have only male flowers on a central fruiting branchlet that mature and fall off with much of the branchlet. Then stalked female flowers and a few male flowers develop on 2 or more short stems that grow from the base of the original fruiting branchlet.

FRUITS AND SEEDS: *Capsules 3-lobed*, ± spherical, brown, ± fleshy, mostly *1–2 cm in diameter*, consists of 3 chambers each containing 1 seed. In late summer and early fall, *capsule walls fall away, leaving the large white seeds attached to a central column on the tree* (hence the name popcorn tree). *Seeds persist on trees* through winter or until removed by birds or dislodged by wind.

HABITAT: Disturbed and undisturbed riparian areas, wetlands. Tolerates shade, drought, saline, and flooded conditions and an average minimum temperature as low as −15°C. Grows best on well-drained clay-peat soil.

DISTRIBUTION: Sacramento Valley (American River, Consumnes River watershed, and Redding), San Francisco Bay Delta region, to 50 m. Southeastern United States.

PROPAGATION/PHENOLOGY: *Reproduces by seed*. Seeds disperse with animals (especially birds), water, and human activities. Germination occurs during the cool season in California. Seeds survive at least 1 year under field conditions. Trees begin to produce fruit when they reach a height of ~ 0.9 m tall. Most trees live ~ 15–25 years, but some individuals may survive much longer. Root fragments can develop new shoots under favorable conditions.

MANAGEMENT FAVORING/DISCOURAGING SURVIVAL: Manual removal of trees and seedlings can control infestations. However, cut stumps generally require treatment with a systemic herbicide to prevent resprouting from the stump and roots.

SIMILAR SPECIES: Without flowers or fruits, Chinese tallowtree superficially resembles native and nonnative poplars, aspens, and cottonwoods (*Populus* spp.). Unlike *Populus* species, Chinese tallowtree has *leaves with smooth margins and milky sap*.

Chinese tallowtree (*Sapium sebiferum*) flowering stem.   J. M. DiTomaso

Black acacia [*Acacia melanoxylon* R.Br.] [Cal-IPC: Limited]

Kangaroothorn [*Acacia paradoxa* DC.][CDFA list: B]

SYNONYMS: Black acacia: blackwood acacia

Kangaroothorn: hedge acacia; prickly acacia; prickly wattle; *Acacia armata* R.Br.

GENERAL INFORMATION: Acacias are often cultivated as woody ornamentals. Most species are not typically weedy, but a few species have escaped cultivation in some coastal regions.

Black acacia: *Tree* to 15 m tall, *usually with a single trunk* and with spherical heads of pale yellow to cream-colored flowers. Most phyllodes are simple and leaflike, but young branches often have juvenile or intermediate leaves that are pinnate-compound 2 times. Black acacia has hard dark wood that makes high-quality lumber. Clonal populations often spread from historical plantings. Native to eastern Australia.

Kangaroothorn: Spiny *shrub* to 3 m tall, 3.5 m wide, with *simple, leaflike phyllodes* (enlarged petiole) and spherical heads of bright yellow flowers. Kangaroothorn is considered a noxious weed in some areas of Australia. Native to southern and eastern Australia.

SEEDLINGS: *Juvenile leaves pinnate-compound 2 times with an even number of leaflets.*

Black acacia (*Acacia melanoxylon*) flowering branches.   J. M. DiTomaso

Black acacia (*Acacia melanoxylon*) flowering stem.   J. M. DiTomaso

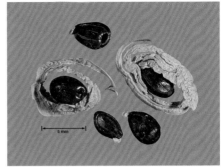
Black acacia (*Acacia melanoxylon*) seeds.   J. O'Brien

Black acacia: Intermediate leaves resemble juvenile leaves, but with *expanded, leaflike petioles*.

MATURE PLANT: Leaves are actually *simple, leaflike phyllodes* (expanded petioles). Phyllodes alternate, lanceolate to oblong, evergreen, leathery.

Black acacia: Phyllodes 6–15 cm long, *(0.5)1–3(5) cm wide*, straight to sickle-shaped, sometimes broadest near the apex, smooth, dull dark green. Main veins 2–5, parallel. Young branches or branch tips *often have leaves that are pinnate-compound 2 times with or without an expanded petiole*.

Kangaroothorn: Often bushy. Stipules modified into *spines ± 1 cm long*. Phyllodes 1–2(3) cm long, 2–8 mm wide, oblique, *margin wavy*, midvein off-center, light green, waxy. New phyllodes hairy, especially margins.

ROOTS AND UNDERGROUND STRUCTURES: Usually associate with nitrogen-fixing bacteria.

Black acacia: Can develop root suckers under certain conditions.

Kangaroothorn: Taproot branched, with numerous lateral roots.

FLOWERS: Heads spherical, 8–10 mm in diameter, consist of 30–50 flowers. *Stamens separate, exserted, more than 10 per flower.* Petals, sepals inconspicuous.

Black acacia: March–June. Heads pale yellow to cream-colored, in axillary racemes of 2–8. Racemes shorter than the length of the phyllodes.

Kangaroothorn: February–April. Heads bright yellow, solitary on stalks in leaf axils. Heads and stalks about equal to the length of the phyllodes. Partially self-incompatible.

FRUITS AND SEEDS: Pods split open along both margins. Seeds attach to pod with a long, folded stalk.

Black acacia: Pods in clusters, flattened, loosely to tightly spiraled, slightly constricted between seeds, pale to reddish brown, leathery, 3–12 cm long, 0.4–1 cm wide, with thickened margins. Seeds oval, black, 3–5 mm long, with a *long, flattened, crinkled, bright pinkish to red stalk encircling the seed in a double-fold*.

Kangaroothorn (*Acacia paradoxa*) flowering branches. J. M. DiTomaso

Kangaroothorn (*Acacia paradoxa*) flowering stem. J. M. DiTomaso

Kangaroothorn (*Acacia paradoxa*) branches with spines. J. M. DiTomaso

Kangaroothorn (*Acacia paradoxa*) seeds. J. O'Brien

Kangaroothorn: Pods aligned, short-stalked, brown, straight or slightly curved, cylindrical, 4–7 cm long, ± 0.5 cm wide, rounded over seeds, rarely narrowed between seeds, sparse to densely covered with long, soft, white hairs. Seeds dark brown to black, oblong, 3–5 mm long, with the stalk folded 2–4 times and a small aril (expanded end of seed stalk that partially covers seed).

HABITAT: Disturbed areas, usually near cultivated plants, buildings, and homestead sites.

DISTRIBUTION: San Francisco Bay region, western South Coast Ranges, South Coast, and outer North Coast ranges, to 200 m.

Black acacia: Also Santa Cruz Island and North Coast.

Kangaroothorn: s Marin, w Solano, w Contra Costa, w Alameda, nw Santa Clara, e San Mateo, sw Santa Cruz, nw Monterey, se Santa Barbara, s Los Angeles, and w San Diego Counties.

PROPAGATION/PHENOLOGY: *Reproduce by seed.* Seeds typically disperse near the parent plant or to greater distances by human activities or water. Seed is hard-coated and requires scarification or degradation to germinate. Seed is probably long-lived, but longevity is undocumented. Pods and seeds are not typically used by native wildlife.

Black acacia: In California, escaped populations appear to expand slowly from their origin. Fire stimulates germination. Trees also reproduce vegetatively by root suckers.

Kangaroothorn: Tolerates temperatures to –7°C. Produces seed at an early age.

MANAGEMENT FAVORING/DISCOURAGING SURVIVAL: Manually removing seedlings before roots are well established can prevent population expansion.

Black acacia: Cut trees appear to survive most chemical treatments and often develop new shoots from the crown and sometimes the roots. Mechanical control requires repeated visits to eliminate emergent seedlings and must include thorough root removal to eliminate regrowth.

Kangaroothorn: Cutting stems just below ground level typically kills shrubs.

SIMILAR SPECIES: There are at least 11 other *Acacia* species that have naturalized in California. For a quick comparison of distinguishing characteristics,

Cootamundra wattle (*Acacia baileyana*) flowering stem.   J. M. DiTomaso

refer to table 33 (p. 742). The following 4 species are more widespread or more likely to be problematic; unlike kangaroothorn and black acacia, these species *lack phyllodes* and have mature leaves that are *pinnate-compound 2 times with an even number of leaflets*.

Cootamundra wattle [*Acacia baileyana* F. Muell., synonym: Bailey's acacia]: *Tree to 6 m tall*, with leaves that have *3–6 pairs of pinnae* 1–3 cm long and silver-blue leaflets 5–7 mm long that are *close together and touch at the margins*. Found along roadsides and in disturbed areas in the San Francisco Bay region, western South Coast Ranges, and western Transverse Ranges, to 300 m. Native to southeastern Australia.

Silver wattle [*Acacia dealbata* Link, synonym: *Acacia decurrens* (Wendl.) Willd. var. *dealbata* (Link.) F. Muell.]: *Tree to 12 m tall*, with *8–25 pairs of overlapped pinnae* and 20–70 pairs of *overlapping leaflets* 2–5 mm long. Twigs angled, *lack winglike ridges*. Twigs (leaf and pinnae axes) *densely covered with minute, silvery hairs*. Found in disturbed sites and roadsides in the Western North Coast

Silver wattle (*Acacia dealbata*) flowering stem. J. M. DiTomaso

Silver wattle (*Acacia dealbata*) seeds. J. O'Brien

Green wattle (*Acacia decurrens*) seeds. J. O'Brien

Sweet acacia (*Acacia farnesiana*) fruit and spines.   J. M. DiTomaso

Sweet acacia (*Acacia farnesiana*) seeds.
J. O'Brien

Sweet acacia (*Acacia farnesiana*) flowering stem.
J. M. DiTomaso

Ranges, San Francisco Bay region, western South Coast Ranges, and South Coast, to 500 m. Often confused with green wattle. Native to southeastern Australia.

Green wattle [*Acacia decurrens* (Wendl.) Willd.]: *Tree* to 15 m tall, similar to silver wattle. Pairs of pinnae 5–15, *not overlapping*. Pairs leaflets 15–35, overlapping, *nearly glabrous*. *Leaf axes narrow-winged on angles*, glabrous to hairy. The occurrence of green wattle outside of cultivation in California is uncertain. Many trees reported to be green wattle are actually silver wattle. Native to eastern Australia.

Sweet acacia [*Acacia farnesiana* (L.) Willd. var. *farnesiana*, synonym: huisache]: Deciduous *Shrub* to 3 m tall, with *pairs of short, straight, stipular spines* that turn white over time and axillary flower heads of gold to dull orange flowers. Usually there are *1–3 flower heads per leaf axil*, and the *leaves are longer than the inflorescences*. Sweet acacia occurs nearly worldwide and is often a troublesome weed. It inhabits dry slopes in chaparral in the South Coast, particularly southern San Diego County, to 300 m, and is also naturalized in Arizona, New Mexico, and many southern states. Native to tropical America, Mexico, and the southern United States, including New Mexico and Arizona.

Other Australian Acacias that are naturalized to a limited extent in California include **cyclops acacia** [*Acacia cyclops* G. Don.], **cedar wattle** [*Acacia elata* A. Cunningham ex Benth.], **Sydney golden wattle** [*Acacia longifolia* (Andr.)

Black acacia • Kangaroothorn

Sydney golden wattle (*Acacia longifolia*) leaf and flower clusters. J. M. DiTomaso

Black wattle (*Acacia mearnsii*) leaves. J. M. DiTomaso

Sydney golden wattle (*Acacia longifolia*) flowering stem. J. M. DiTomaso

Willd.], **black wattle** [*Acacia mearnsii* De Wild.], **golden wattle** [*Acacia pycnantha* Benth.], **everblooming acacia** [*Acacia retinodes* Schldl.], and **star acacia** [*Acacia verticillata* (L' Her.) Willd.]. Refer to table 33 (p. 742) and the appendix, **Non-native Naturalized Plants** (p. 1601, vol. 2).

Catclaw acacia [*Acacia greggii* A. Gray] is a *native desert shrub or small tree* to 7 m tall and is a dominant shrub in some areas of the Mojave and Sonoran deserts. Catclaw acacia is *not* a weed. However, it closely resembles sweet acacia, which is a troublesome weed in some areas of southern California. Catclaw acacia is distinguished by having *curved prickles that are irregularly scattered on the stems* and *not* in pairs, and the yellow flower heads are in *spikes on short branchlets along with the leaves, which are shorter than the flower spikes*.

The mesquites (*Prosopis* spp.) are primarily distinguished from the acacias by having *10 stamens per flower,* and mesquite leaves are always *pinnate-compound 2 times with an even number of leaflets*. Refer to the entry for **Velvet mesquite and Creeping mesquite** (p. 790) for more information.

Black acacia • Kangaroothorn

Everblooming acacia (*Acacia retinoides*) plant. J. M. DiTomaso

Everblooming acacia (*Acacia retinoides*) flowering stem. J. M. DiTomaso

Catclaw acacia (*Acacia greggii*) flowering stem, the only *Acacia* species native to California. Nonweedy.
J. M. DiTomaso

Table 33. *Acacia* species
Species with innate compound leaves and spherical flower clusters

Acacia spp.	Habit	Stipular spines	Pairs of pinnae	Pairs of leaflets	Leaflet length (mm)	Pinna margins; leaflet margins	Flower heads	Other distinguishing characteristics
A. baileyana cootamundra wattle	tree	no	3–6	12–20	5–7	overlapping; touching	bright yellow	leaflets glabrous silver-blue, covered with a whitish bloom; pinnae mostly 1–3 cm long
A. dealbata silver wattle	tree	no	8–25	20–70	2–5	overlapping; overlapping	bright yellow	leaf axis angled, usually densely silver-hairy; leaves with a gland at junction of each pinna and main axis
A. decurrens green wattle	tree	no	5–15	15–35	5–15	not overlapping; touching to overlapping	bright yellow	leaf axis narrow winged, ± hairy; leaves with a gland at junction of each pinna and main axis
A. elata cedar wattle	tree	no	3–6	8–14	20–80	not touching; mostly not touching	pale yellow	leaflets lanceolate, dark green
A. farnesiana sweet acacia	shrub	yes	2–5	8–20	2–5	not overlapping; touching to overlapping	golden to dull orange	leaves deciduous; spines straight, white with age
A. mearnsii black wattle	tree	no	8–25	30–70	2–5	overlapping; overlapping	pale yellow	leaves with glands at junction of pinnae and main axis, and between pinnae on the main axis

Species with simple leaves (phyllodes):

Acacia spp.	Habit	Stipular spines	Prominent leaf veins	Leaf length (cm)	Leaf width (mm)	Flower clusters	Flower color	Other distinguishing characteristics
A. cyclops cyclops acacia	shrub	no	3–5	4-9	5–15	spherical	bright yellow	main stem usually branched at base; seed stalk yellowish orange, doubly folded around seed
A. longifolia Sydney golden wattle	shrub or tree	no	2–3	5-15	10–20	spikelike	bright yellow	pods mostly 3–5 mm wide, nearly cylindrical, narrowed between seeds
A. melanoxylon black acacia	tree	no	2–5	6-15	(5) 12–30	spherical	pale yellow to cream	main stem usually single; often some leaves 2-pinnate compound at tip; seed stalk pink to red, doubly folded around seed
A. paradoxa kangaroothorn	shrub	yes	1 off-center	1-2	2–8	spherical	bright yellow	leaf margins wavy; stipular spines straight
A. pycnantha golden wattle	tree	no	1	5-20	(5) 10–30(50)	spherical	golden	mature twigs smooth; seed stalk about one-half the length of the seed; flowers fragrant
A. retinodes everblooming acacia	tree	no	1	3-20	3–10	spherical	pale yellow	mature twigs ribbed; seed stalk doubly folded around seed, reddish-brown
A. verticillata star acacia	shrub	no	1	to 1.5	± 1	spikelike	pale yellow	leaflets whorled, needle-like, tapered to a sharp point

Camelthorn [*Alhagi maurorum* Medik.][ALHPS]
[Cal-IPC: Moderate][CDFA: A]

SYNONYMS: Caspian manna; Persian manna; *Alhagi camelorum* Fischer; *Alhagi pseudalhagi* (M. Bieb.) Desv.

GENERAL INFORMATION: Green *herbaceous perennial to shrub* to 1 to 2 m tall, with *simple leaves, many thorny branches*, and an *extensive creeping root system*. One plant can rapidly colonize an area by developing many new plants from the creeping roots. Intense eradication programs have eliminated most populations in California. Camelthorn is also a state listed weed in Arizona (restricted),

Camelthorn (*Alhagi maurorum*) plant. J.P. CLARK

Camelthorn

Camelthorn (*Alhagi maurorum*) fruit.
COURTESY OF ROSS O'CONNELL

Camelthorn (*Alhagi maurorum*) seeds.   J. O'BRIEN

Camelthorn (*Alhagi maurorum*) flowering stem.
COURTESY OF ROSS O'CONNELL

Colorado, Nevada, New Mexico (class A), Oregon (class A), and Washington (class B). Native to the Mediterranean region and western Asia.

SEEDLINGS: Lack thorns. Cotyledons ovate, thick, leathery, ~ 5–10 mm long. Often found growing in cattle manure from seed passed through digestive tract. Seedlings appear to be poor competitors in agronomic crops, including alfalfa and some clovers.

MATURE PLANT: Deciduous in cool climates. Stems nearly glabrous, greenish, longitudinally ridged, highly branched, with the leaf axil of nearly every node supporting an ascending, *leafless branchlet*, 2–5 cm long, *tipped with a thorn* about 5 mm long. *Leaves alternate, sparse, simple*, thick, leathery, elliptic or obovate, 7–20 mm long, on a stalk 1–2 mm long. Stipules about 1 mm long. Upper leaf surfaces glabrous to sparsely hairy, covered with minute red dots. Lower leaf surfaces sparse to moderately hairy. Morphology varies according to environmental conditions. In moist habitats, thorns are smaller and fewer, and leaves larger and more numerous.

ROOTS AND UNDERGROUND STRUCTURES: Root system *woody, extensively creeping*. Roots can grow more than 2 m deep and to a distance of 8(12) m or more in all directions. Horizontal roots at depths to 1.5 m produce new shoots and deep vertical roots at about 1–1.5 m intervals. In turn, each new clone develops creeping roots in all directions. Populations can spread at a rate of about 10 m in all directions per year. Roots associate with nitrogen-fixing bacteria.

FLOWERS: June–August. Each branchlet develops 2–6 alternate flowers on short stalks. *Flowers pea-like.* Petals *magenta* to pink, 8–9 mm long. *Sepals persistent, fused* and cuplike, with *small unequal teeth.* Stamens 10, 1 separate, 9 fused into a tube around the style at the base. Self-compatible. Flower production is high under hot dry conditions and low to nonexistent under moist shady conditions.

FRUITS AND SEEDS: July–August. *Pods* (loments) reddish brown at maturity, slender, often curved, *1–3 cm long, constricted between seeds,* and often tipped with a small spine. *Pods do not split open to release seeds,* but can break apart between seeds. Seeds 5–8, oval, yellowish or greenish brown with dark mottling or solid dark brown, smooth-textured, about 3 mm long, and 2.5 mm wide. Soft- and hard-coated seeds are produced. Fruits are eaten by herbivores, including cattle and horses.

POSTSENESCENCE CHARACTERISTICS: Woody stems persist after plant death.

HABITAT: Arid agricultural areas and riverbanks where roots can access water tables or other water sources during the growing season. Often grows in heavy clay soils. Tolerates some salinity. Aboveground parts can be killed by hard frost. Designated a facultative wetland indicator species.

DISTRIBUTION: Uncommon. Southeastern Sierra Nevada (sw Inyo Co.) and Mojave and Sonoran deserts (cw San Bernardino Co. and e Riverside Co.), to 500 m. Previous populations in the Central Valley have been eradicated. Arizona, Colorado, Idaho, Nevada, New Mexico, Utah, Texas, and Washington.

PROPAGATION/PHENOLOGY: *Reproduces vegetatively from creeping roots* and sometimes by seed. Often only about 20% of flowers set seed. Seeds disperse with animals that browse on fruits, water, and high winds that blow clumps of branches with fruits. Passing through the digestive tract of a herbivore or acid scarification appears to stimulate germination. Optimal temperature and soil depth for germination is near 27°C and 1 cm, respectively. Light appears to inhibit germination. Seeds can survive submersion in water for at least 8 months and can remain viable for several years in semiarid soils. Viability decreases rapidly after 1 year in cool, moist soil conditions.

MANAGEMENT FAVORING/DISCOURAGING SURVIVAL: Cattle can preferentially feed on pods. Moving livestock that has browsed on fruits can disperse seeds to new locations. Lack of soil moisture during the warm growing season discourages seedling survival. Mechanical removal can stimulate remaining roots to spread and to develop new shoots. Fire stimulates root sprouting. Several herbicides can kill plants.

SIMILAR SPECIES: Russian salttree [*Halimodendron halodendron* (L.) Voss, synonym: common salttree][CDFA list: A] is a deciduous, sprawling to erect, *thorny shrub* to about 3 m tall. Russian salttree was introduced to California as an ornamental, but quickly colonized the sites where it was planted. Because of its tendency to spread invasively by developing new shoots from lateral roots, it was placed on the noxious weed list for California as an A-listed weed. Unlike camelthorn, Russian salttree has *even-pinnate compound leaves* clustered on short spurs, typically with *4 obovate to oblanceolate leaflets* (sometimes 2 or 6), *thorn-tipped branchlets* below the spurs, and *black inflated pods 1.5–2 cm long* and ± 1 cm wide that are *not constricted between the seeds* and that eventually open to release the seeds. In California, known infestations have been eradicated in the Central Valley (UC Davis Arboretum, Kern County Park) and central South Coast (Los Angeles basin), to 200 m. It occurs in Montana, Utah, and Wyoming. Native to southwestern Asia.

Loco and milkvetch [*Astragalus* spp.]

SYNONYMS: crazyweed; locoweed; rattleweed

GENERAL INFORMATION: Native *annuals* and *perennials,* with *axillary racemes of slender pea-like flowers* and slender to inflated pods. Most species have *odd-pinnate compound leaves,* but a few species sometimes have 3 leaflets. There are between 340 and 400 *Astragalus* species in North America and about 94 species in California, of which about one-third are now rare or endangered. Locos and milkvetches are an important component of the vegetation in natural areas, and a few species, such as **Nuttall milkvetch** [*Astragalus nuttallianus* A. DC.][ASANT] are useful forage plants. However, many *Astragalus* species can be fatally *toxic* to livestock when ingested over a short to long period of time. Poisoning by *Astragalus* is a serious hazard to livestock foraging on rangeland of the western states. Most animals generally avoid consuming toxic *Astragalus* species if more palatable forage is available, but some animals develop a preference for it. There are 3 types of toxicity problems associated with *Astragalus.* Certain species contain indolizidine alkaloids, which can cause a neurological syndrome known as locoism. Varieties of **spotted loco** [*Astragalus lentiginosus* Hook., synonym: freckled milkvetch][ASALE] are widespread in the western states, including California, and are often implicated with outbreaks of locoism. Other species occasionally associated with locoism in California are **Douglas' loco** [*Astragalus douglasii* (Torr. & Gray) Gray, synonym: jacumba milkvetch],

Shorttooth Canadian milkvetch (*Astragalus canadensis* var. *brevidens*) plant.   J. M. DiTomaso

Loco and milkvetch

Locoweed (*Astragalus oxyphysus*) plant.
 J. M. DiTomaso

Locoweed (*Astragalus trichopodus* var. *lonchus*) flowers and fruit.
 L. C. Wheeler

San Joaquin milkvetch [*Astragalus asymmetricus* Sheldon, synonym: horse loco], and few less common species. Symptoms of locoism include behavioral changes, poor coordination, difficulty in eating and drinking, weight loss, failure of the right side of the heart at high elevations, abortion, and deformed and/or weak offspring. Horses are at greatest risk to developing locoism. Toxicity problems usually occur from late winter to early summer. Many *Astragalus* species in California and other western states contain aliphatic nitro compounds, also called nitrotoxins, which can cause another neurological syndrome known as cracker heels. Cattle, especially lactating cows, are most frequently affected. Symptoms can include staggering, excessive salivation, tremors, abnormal movement of the hind legs that results in clicking the feet together, wheezing, dazed appearance, hind leg paralysis, and collapse. Cracker heels most often occurs in mid-summer after animals have ingested a toxic species for about 1 week. One of the more toxic California species known to cause cracker heels is **shorttooth Canadian milkvetch** [*Astragalus canadensis* L. var. *brevidens* (Gand.) Barneby]. Besides locoism and cracker heels, some *Astragalus* species accumulate toxic levels of selenium, which can cause cardiac failure, blind staggers, and alkali disease syndrome in livestock. Selenium toxicity resulting from the ingestion of *Astragalus* is rare in California and only documented with *A. crotalariae* (Benth.) A. Gray and *A. preussii* A. Gray.

MATURE PLANT: Conspicuous stems prostrate to erect or lacking. Foliage glabrous to hairy, hairs sometimes forked and propeller-like in some species, such as shorttooth Canadian milkvetch (requires magnification). Leaves alternate, odd-pinnate compound, leaflets (3)5–numerous. Leaflets linear to elliptic, (ob)lanceolate, or (ob)ovate, 2–35 mm long. Stipules membranous, sometimes fused around the stem near the stem base.

ROOTS AND UNDERGROUND STRUCTURES: Annual plants have a taproot. Perennial plants generally have an oblique, somewhat woody rhizome and stem base.

FLOWERS: Year-round. Specific months depend on the species. Racemes axillary, loosely elongate to dense and headlike, on long stalks. Flowers slender, pea-like, vary in color from white or cream to shades of pale yellow, pink, lavender, or violet. Stamens 10, with 9 filaments fused and 1 separate. Style 1, glabrous, stigma minute. Spotted loco varieties have white, cream, violet, or white to yellow and violet flowers. Shorttooth Canadian milkvetch has cream-colored flowers that are occasionally tinged dull violet.

FRUITS AND SEEDS: Pods vary according to species from slender to inflated, 1-chambered or partially divided into 2 chambers by a membrane from the lower suture, glabrous or hairy, often mottled, sessile, stalked, or jointed on a stalk above the calyx. Seeds 2–numerous, smooth, flattened, with a slight notch at the attachment point, often rattle about in dried pods. Pods of spotted loco varieties are slightly inflated, papery, ovoid to nearly spherical, 10–48 mm long, more than 5 mm wide, and have a sunken groove on the top and bottom and a *flattened triangular tip*. Pods appear *2-chambered below* the triangular tip. Shorttooth Canadian milkvetch pods are erect, minutely hairy, cylindrical, 3–4 mm wide, and 12–20 mm long, including a curved beak ± 3 mm long.

HABITAT: Many plant communities, including desert, alpine, and wetland areas. Spotted loco inhabits dry open areas. Shorttooth Canadian milkvetch inhabits moist to wet places, such as lake margins.

DISTRIBUTION: *Astragalus* species are distributed throughout much of North America and all of California, including high elevation areas. Spotted loco varieties occur throughout California, to 3600 m, and in most other western states. Shorttooth Canadian milkvetch occurs in the Modoc Plateau and adjacent areas of the Cascade Range and the east side of the northern and central Sierra Nevada and adjacent areas to the east, mostly at 1500–2450 m. It occurs in all western states, except Arizona and New Mexico.

PROPAGATION/PHENOLOGY: *Reproduce by seed.* Seeds are probably long-lived under field conditions. Populations of some species, especially annuals, tend to be cyclic in response to yearly weather patterns. These species can be abundant in wet years and scarce in dry years.

MANAGEMENT FAVORING/DISCOURAGING SURVIVAL: Excluding animals from rangeland areas with abundant loco or milkvetch is an effective way to prevent toxicity problems in livestock.

SIMILAR SPECIES: Swainsonpea [*Sphaerophysa salsula* (Pall.) DC.] [SWASA] [CDFA list: A] is an uncommon *perennial* with pinnate-compound leaves and inflated pods that resemble those of some locos and milkvetches. Swainsonpea is distinguished by having a *vigorous creeping root system* that produces new shoots under favorable conditions and *brick-red to orange-red flowers with a finely hairy stigma*. Swainsonpea inhabits roadsides, ditches, crop fields, and other disturbed sites in the southern San Joaquin Valley to 500 m. It also occurs in all other western states. Refer to the entry for **Swainsonpea** (p. 802) for more information.

Scotch broom [*Cytisus scoparius* (L.) Link] [SAOSC][Cal-IPC: High][CDFA list: C]

French broom [*Genista monspessulana* (L.) L. Johnson] [Cal-IPC: High][CDFA list: C]

Spanish broom [*Spartium junceum* L.] [Cal-IPC: High][CDFA list: C]

SYNONYMS: Relationships among the genera of brooms are complex and weakly differentiated. Consequently, taxonomists have shuffled names back and forth among genera.

Scotch broom: English broom; *Cytisus scoparius* (L.) Link var. *andreanus*

Scotch broom *(Cytisus scoparius)* plant. J. M. DiTomaso

Scotch broom • French broom • Spanish broom

Scotch broom (*Cytisus scoparius*) flowers.
J. M. DiTomaso

Scotch broom (*Cytisus scoparius*) fruit.
J. M. DiTomaso

Scotch broom (*Cytisus scoparius*) stem with leaves.
J. M. DiTomaso

Scotch broom (*Cytisus scoparius*) seedlings.
Courtesy of Jim McHenry

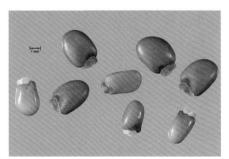

Scotch broom (*Cytisus scoparius*) seeds.
J. O'Brien

(Puiss.) Dippel and var. *scoparius*; *Sarothamnus scoparius* (L.) Wim. ex Koch; *Spartium scoparium* L.

French broom: Cape broom; Montpellier broom; *Cytisus monspessulanus* L.; *Teline monspessulanus* (L.) Koch

Spanish broom: weavers broom; *Genista juncea* Lam.

GENERAL INFORMATION: *Shrubs to 3(5) m tall, with photosynthetically active green stems* (older stems and branches brown) *and yellow, pea-like flowers*. Refer to table 34 (p. 762) for a quick comparison of distinguishing characteristics.

Scotch broom • French broom • Spanish broom

French broom *(Genista monspessulana)* infestation. J. M. DiTomaso

French broom *(Genista monspessulana)* flowering stem. J. M. DiTomaso

French broom *(Genista monspessulana)* fruit. J. M. DiTomaso

French broom *(Genista monspessulana)* seedling. Courtesy of Jim McHenry

French broom *(Genista monspessulana)* seeds. J. O'Brien

The brooms were originally introduced as landscape ornamentals. Scotch and Spanish broom were also extensively planted along highways in some areas to prevent soil erosion in the 1900s. Brooms have escaped cultivation and have aggressively invaded many natural areas. Scotch and French broom often form dense stands that displace native vegetation and wildlife. Infestations in forested areas increase the fire hazard and diminish the use of rangeland. Flowers and seeds of brooms contain quinolizidine alkaloids and can be *toxic* to humans and livestock when ingested. However, toxicity problems due to the brooms are uncommon. Toxicity symptoms include digestive tract disturbance and/or neurological effects, such as trembling or uncoordinated gait. Broom foliage may be mildly toxic, but is generally unpalatable to most livestock, except goats. Scotch and Spanish broom are used medicinally in Europe. However, Scotch broom is considered to be a herb unsafe for human consumption by the United States Food and Drug Administration.

Scotch broom: Deciduous, late summer to early spring. Seedlings can outcompete conifer seedlings and prevent reforestation. California lists Scotch broom as a C-listed noxious weed, while Idaho, Oregon, and Washington list the species as a class B noxious weed. Some botanists have recognized 2 varieties of Scotch broom in the United States: *C. scoparius* var. *scoparius* occurs in all states where Scotch broom is known to exist, and var. *andreanus* (Puiss.) Dippel occurs only in California. Three biocontrol agents are established in California, the Scotch broom seed weevil (*Apion fuscirostre*), twig miner (*Leucoptera spartifoliella*), and the accidentally introduced gorse or broom tip moth (*Agonopterix nervosa*). The seed weevil and the twig miner are specific to Scotch broom. The broom tip moth also attacks **gorse** [*Ulex europaeus* L.]. A number of native insects also feed on Scotch broom. However, insect damage has not yet been severe enough to control problematic populations. Native to Europe and northern Africa.

French broom: Evergreen. Although widespread and aggressive in California, the biology of this species remains poorly understood. In California, many plants identified as French broom may actually be hybrids involving two very similar species, **Canary Island broom** [*Genista canariensis* L.] and **leafy broom** [*Genista stenopetala* Webb & Berth, synonym: narrow-petaled broom]. In addition to its California listing, French broom is a state-listed noxious weed in Oregon (class B). Native to the Mediterranean region and Azores Islands.

Spanish broom: Deciduous, summer to early spring. This species appears to be somewhat less troublesome, to date, than French or Scotch brooms. Spanish broom is a state-listed noxious weed in Oregon (class B). In its native range, stems and fibers are used to make baskets, mats, ropes, and paper. Flowers provide a yellow dye and an essential oil that is sometimes used in perfumery. Native to the Mediterranean region and the Canary, Madeira, and Azores Islands.

MATURE PLANT: Stems erect, dense, *green*. Leaves alternate (to subopposite in Spanish broom) and typically drought-deciduous.

Scotch broom: Most stems *sharply 5-angled or -ridged*, ± star-shaped in cross-

Scotch broom • French broom • Spanish broom

Spanish broom (*Spartium junceum*) flowering stem. J. M. DiTomaso

Spanish broom (*Spartium junceum*) stem with leaves. J. M. DiTomaso

Spanish broom (*Spartium junceum*) plant. J. M. DiTomaso

section, *often with few leaves*. New twigs covered with wavy hairs, becoming glabrous over time. *Leaves compound with 3 leaflets, sometimes with 1 leaflet on new twigs.* Leaflets 5–20 mm long, oblong to obovate, upper surface glabrous or nearly glabrous, lower surface sparse to densely covered with flattened (appressed), short hairs. Stipules leaflike, mostly 5–10 mm long, deciduous.

Spanish broom (*Spartium junceum*) seeds.
J. O'Brien

Spanish broom (*Spartium junceum*) fruit.
J. M. DiTomaso

French broom: Stems typically *leafy*, often strongly 8–10-ridged (generally apparent only on young branches), *round in cross-section*, usually *moderate to densely covered with silky, silvery hairs. Leaves compound with 3 leaflets*. Leaflets oblong to obovate, variable in size, mostly 10–20 mm long, upper and lower surface sparse to densely covered with appressed, short, silvery hairs. Stipules ± awl-shaped, to 3 mm long, hairy, deciduous.

Spanish broom: Stems *often leafless, round in cross-section, finely ridged, glabrous. Leaves simple, sparse, ephemeral*, linear to elliptic or lanceolate, 10–30 mm long, upper surface glabrous, lower surface covered with flattened, short hairs. Stipules lacking.

ROOTS AND UNDERGROUND STRUCTURES: Taproots deep, branched, especially in *Cytisus* and *Spartium* species. Fine roots associate with nitrogen-fixing bacteria. Nitrogen fixation occurs year-round where winters are mild. Roots do not produce new shoots, but plants cut above the crown can grow new shoots from the crown, especially during the rainy season.

FLOWERS: Pea-like, usually *bright yellow* (pale yellow to maroon-red in *Cytisus*). Stamen filaments fused. Insect-pollinated. Mostly out-crossing.

Scotch broom: March–June. Flowers mostly 1–2.5 cm long, *single or paired in leaf axils*. Petals of some cultivars are partially or entirely dark red, or pale yellow. *Calyx glabrous, usually less than 5 mm long, 2-lipped*, top lip minutely toothed. *Style curved at the middle or for the entire length.*

French broom: March–May. Some plants in large populations flower most of the year in mild climates. Slightly fragrant. *Flower clusters headlike, 4–10-flowered, at the ends of short axillary branchlets.* Flowers ± 1 cm long. *Calyx pubescent, 5–7 mm long, ± 2-lipped, top lip 2-lobed to near middle, lower lip with (1–)3 shallow lobes. Style abruptly bent or curved near the tip.*

Spanish broom: April–June. Fragrant. Flowers 2–2.5 cm long, on *open, leafless racemes at the stem tips. Calyx glabrous*, mostly 5–8 mm long, *usually 1-lipped, parted to base on the top*, lip minutely 5-toothed. *Style abruptly bent or curved near the tip.*

FRUITS AND SEEDS: Seeds have a small cream-colored to yellowish appendage (strophiole) at the attachment scar (hilum).

Scotch broom: Pods typically mature June–July. Pods dark brown to black at maturity, *flattened, mostly 2–5 cm long, ± 1 cm wide, glabrous with margins densely lined with long, silky, golden to silvery hairs.* Seeds about 5–9 per pod on average, ovoid, compressed, ± 2 mm long, shiny, greenish brown to black, with an appendage that is attractive to certain species of ants.

French broom: Pods brown at maturity, slightly flattened, *1–3 cm long, ± 0.5 cm wide, densely covered with appressed, long, silky, silvery to reddish gold hairs.* Seeds about 5–8 per pod on average, ovoid, shiny, smooth, black, about 2–3 mm long.

Spanish broom: Pods brown at maturity, slightly flattened, mostly *4–11 cm long, ± 0.5 cm wide, densely covered with appressed, long, silky, silvery hairs.* Seeds about 10–18 seeds per pod, ovoid to rectangular, compressed, smooth, reddish brown, ± 4 mm long.

POSTSENESCENCE CHARACTERISTICS: Plants tolerate frost, but typically die back after severe cold winter conditions; however, roots or lower stems may survive and generate new growth. Some branch death can occur during seasonal drought. Natural decline and senescence occur over a period of years. Symptoms of decline in Scotch broom include an increase in the ratio of woody to green plant material, a thinning of stems, and a decrease in pod production. Eventually old plants die and topple over, opening the canopy for seedling establishment.

Portuguese broom (*Cytisus striatus*) plant.   J. M. DiTomaso

Scotch broom • French broom • Spanish broom

Portuguese broom *(Cytisus striatus)* flower.
J. M. DiTomaso

Portuguese broom *(Cytisus striatus)* fruit.
J. M. DiTomaso

HABITAT: Open disturbed sites, such as logged or burned sites, roadsides, and pastures; also relatively undisturbed grasslands, coastal scrub, oak woodlands, riparian corridors, and open forests. Brooms do not tolerate heavy shade, but they can tolerate minimal shade along the edges of forest canopies. Drought-resistant.

Scotch broom: Typically inhabits mountainous regions and cool coastal areas with dry summers. Grows best on sandy, high-phosphorus soils with an acidic to neutral pH. Rarely grows on limestone soils. Tolerates a high concentration of boron.

French broom: Inhabits humid and subhumid areas in its native range.

Spanish broom: Grows on poor, dry, stony soils and limestone soils in its native range. Tolerates urban pollution, salt-laden winds near the coast, and temperatures as low as –10°C.

DISTRIBUTION: Scotch broom: North Coast, North Coast Ranges, Klamath Ranges, Cascade Range foothills, north and central Sierra Nevada foothills, Central Valley, San Francisco Bay region, and South Coast, to 1200 m. Idaho, Montana, Oregon, Utah, Washington, and many eastern and southeastern states.

French broom: North Coast Ranges except high-elevation areas, San Francisco Bay region, western South Coast Ranges, Sierra Nevada foothills, western Transverse Ranges, Peninsular Ranges, and Channel Islands, to 500 m. Oregon and Washington.

Spanish broom: Western North Coast Ranges, San Francisco Bay region, Sacramento Valley, western South Coast Ranges, western Transverse Ranges, South Coast, and southern Channel Islands, to 600 m. Oregon and Washington.

PROPAGATION/PHENOLOGY: *Reproduce by seed.* Pods typically burst apart into spiral halves, ejecting seeds a short distance from the parent plant. Seeds disperse to greater distances with water, soil movement, vehicle tires, human activities, and animals. Seeds are hard-coated and long-lived under field conditions. Brooms can resprout from the crown when cut above.

Scotch broom: Several species of ants are attracted to the seed appendages

Scotch broom • French broom • Spanish broom

and disperse seeds while foraging. Seeds can survive 30 years or more under field conditions. About 50% of seeds produced in a season remain dormant. Scarification and/or a period of cool, moist vernalization enhances germination. Seeds heated to temperatures above 150°C for 2 minutes are killed, and seeds heated to 100°C for 1 minute appear to be more susceptible to fungal pathogens. Yet burning in the field often appears to stimulate germination. Seeds germinate after the first rains in fall through early summer in inland areas, and winter through mid-summer in coastal regions. Seedlings can emerge from soil depths up to 8 cm, with optimal emergence from about 2 cm. Seedlings are subject to high summer mortality and browsing by wildlife during the first several years. Plants typically grow rapidly during years 3–4, often reaching a height of 2 m. Seed production begins during years 2 or 3, usually when plants reach a height of about 1 m. Shrubs are fully reproductive from years 3–5 through about 9. A mature shrub can produce about 2000–3500 pods per season. Seed production often decreases significantly during drought years. Individual shrubs can live about 10–15(20) years.

French broom: In Australia, seeds germinate in spring and fall. Fire appears to stimulate germination. Where seeds are present in the soil, a large flush of seedlings may emerge on newly burned sites. Development is similar to that of Scotch broom. Pods are often copiously produced.

Spanish broom: Many seeds germinate readily without treatment, but scarification can increase germination.

Portuguese broom (*Cytisus striatus*) seeds. J. O'Brien

MANAGEMENT FAVORING/DISCOURAGING SURVIVAL: Established populations are difficult to eliminate because large, long-lived seedbanks typically accumulate. Minimizing soil disturbances, monitoring, and manually pulling young plants prior to fruit production, when discovered, can help prevent new infestations. Machines and tools used to remove stands may inadvertently transport seed to new sites. Cutting Scotch broom shrubs to ground level at the end of the dry season can help reduce resprouting from the root crown. Planting native shrubs and trees within and around broom stands can eventually help to minimize infestations by shading. Goats confined to a small area can help control stands of young shrubs or young crown sprouts from cut shrubs. Prescribed burns can eliminate aboveground growth, but do not prevent resprouting from the root crown and may stimulate a flush of seed germination. Seedlings can be killed with the use of high-temperature flaming, as from a blowtorch, during nonfire season. Herbicide treatments (cut-stem, foliar applications) are effective but costly, as recovering plants often need repeat treatments for several years. In addition, herbicides can damage nontarget native species.

SIMILAR SPECIES: Several other brooms with yellow or white flowers have naturalized locally in some areas. The following species are the most important of these.

Portuguese broom [*Cytisus striatus* (Hill) Rothm.] [Cal-IPC: Moderate] has *yellow flowers* and is often confused with Scotch broom. Unlike Scotch broom, Portuguese broom has *stems that are 8–10-angled,* flowers with a *pubescent calyx,* and *slightly inflated pods that are densely covered with white hairs.* Portuguese broom occurs in the San Francisco Bay region, South Coast, and Peninsular Ranges, to 300 m. It is expected to expand its range in California. Portuguese broom is a state-listed noxious weed in Oregon (class B). Native to Spain and Portugal.

Bridal broom [*Retama monosperma* (L.) Boiss., synonyms: *Genista monosperma* Lam.; *Lygos monosperma* L.] [Cal-IPC: Moderate Alert] has *white flowers with a purple calyx* and *nearly round pods that contain 1 or 2 seeds. Pods do not open to release the seed(s).* Bridal broom has slender, nearly leafless branches that typically *droop.* Leaves are usually *simple* or occasionally compound with 3 leaflets, and the blades are ± linear. A population of bridal broom was first discovered at Fallbrook Naval Weapons Station in San Diego County, where it spread from a patchy population on less than 4 hectares to occupy more than 800 hectares in about 6 years. Native to Spain and North Africa.

Unlike the brooms, **gorse** [*Ulex europaea* L.] has *thorny stems.* Refer to the entry for **Gorse** (p. 820) for more information.

Table 34. Brooms (*Cytisus*, *Genista*, *Spartium*, and *Retama* spp.)

	Cytisus scoparius Scotch broom	*Cytisus striatus* Portuguese broom	*Genista monspessulana* French broom	*Spartium junceum* Spanish broom	*Retama monosperma* bridal broom
Stems	usually 5-angled; ± star-shaped in cross-section	8–10-ridged; round in cross-section	8–10-ridged; round in cross-section	finely ribbed; round in cross-section	finely ribbed; round in cross-section slender, lax
Leaves	compound, leaflets 3, sometimes single on new twigs; deciduous	compound, leaflets 3, sometimes single on new twigs; deciduous	compound, leaflets 3, usually dense; evergreen	simple, sparse; deciduous	simple, sparse; deciduous
Flowers	single or paired in leaf axils	single or paired in leaf axils	4–10 in head-like clusters at ends of short axillary branchlets	several in open racemes at stem tips	2–20 in short axillary racemes or clusters
Petals	yellow or partially to entirely dark red	yellow	yellow	yellow	white
Calyx	2-lipped, top lip minutely toothed; glabrous	2-lipped, top lip minutely toothed; pubescent	2-lipped, top lip 2-lobed to near middle, lower lip shallow (1-)3-lobed; pubescent	1-lipped, parted to base on top (rarely 2-lipped and 5-lobed); glabrous	2-lipped, purplish, top lip 2-lobed to near middle, lower lip shallow 3-lobed; glabrous
Pods	flattened; glabrous with margins densely lined with long hairs; 2-5 cm long, ± 1 cm wide	slightly inflated; densely covered with long hairs; 1.5–5 cm long, ± 1 cm wide	slightly flattened; densely covered with long hairs; 1–3 cm long, ± 0.5 cm wide	slightly flattened; densely covered with long hairs; 4–11 cm long, ± 0.5 cm wide	spherical to ovoid; glabrous; do not split in half; ± 1 cm long, nearly as wide

Wild licorice [*Glycyrrhiza lepidota* Nutt. ex Pursh] [GYCLE]

SYNONYMS: American licorice; Nuttall's licorice; sweet root; *Glycyrrhiza glutinosa* Nutt. in Torr. & A. Gray; *Glycyrrhiza lepidota* Nutt. ex Pursh var. *glutinosa* (Nutt.) S. Wats.

GENERAL INFORMATION: Native *perennial* to 1.2 m tall, with *odd-pinnate-compound leaves, whitish pea-like flowers,* and *deep, creeping roots*. Wild licorice is usually not considered a weed in natural areas, but it is occasionally troublesome in moist agricultural fields, orchards, pastures, gardens, and ditches. Under favorable conditions, plants develop colonies from the extensive creeping roots. Wild licorice is a state-listed noxious weed in Nevada. The roots contain glycyrrhizin, the sweet compound that is extracted from **cultivated licorice** [*Glycyrrhiza glabra* L.], and was consumed by Native Americans. The foliage is generally unpalatable to cattle.

Wild licorice (*Glycyrrhiza lepidota*) flowering stem.   J. M. DiTomaso

Wild licorice

SEEDLINGS: Cotyledons oblong, slightly curved and asymmetrical, about 8–11 mm long, 4–5 mm wide, glabrous. First and second leaves simple, elliptic, 8–15 mm long, 5–10 mm wide, sparsely covered with minute sessile glandular dots or scales, especially lower surface. Margins ± minutely ciliate.

Wild licorice *(Glycyrrhiza lepidota)* flower cluster. J. M. DiTomaso

Wild licorice *(Glycyrrhiza lepidota)* fruiting stem. J. M. DiTomaso

Wild licorice *(Glycyrrhiza lepidota)* fruit cluster. J. M. DiTomaso

Wild licorice

Wild licorice (*Glycyrrhiza lepidota*) seedling.
J. M. DiTomaso

Wild licorice (*Glycyrrhiza lepidota*) seeds.
J. O'Brien

MATURE PLANT: Stems round or slightly angled in cross-section. Foliage ± sparsely pubescent. Leaves *odd-pinnate-compound*, 8–24 cm long, with 9–19 sessile leaflets. Leaflets lanceolate to elliptic, 2–5 cm long, tip often tapered to a minute nipplelike point (mucronulate). *Leaf axis (rachis) and leaflets, especially the lower surface, evenly covered with sessile glandular dots* (visible with 10× magnification). Stipules deciduous.

ROOTS AND UNDERGROUND STRUCTURES: Creeping roots tough, thick, long, deep.

FLOWERS: April–July. Racemes axillary, dense, stalked. Corolla dingy white to greenish white, sometimes yellowish or faintly tinged purple, mostly 9–15 mm long. Calyx 5-lobed, lobes unequal. Stamens 10, 1 separate and 9 fused into a tube.

FRUITS AND SEEDS: Pods oblong-ellipsoid, mostly 10–15 mm long, *bur-like, covered with minutely hooked bristles ± 3 mm long, persistent. Pods do not open to release seeds.* Seeds usually 2–4 per pod, kidney-shaped, mostly 2.5–3 mm long, flattened, dull brown to purplish brown.

POSTSENESCENCE CHARACTERISTICS: Foliage senesces in late summer and early fall, but stems with fruits can persist into winter.

HABITAT: Disturbed sites, roadsides, fields, pastures, cropland, orchards, riparian areas, lake margins, and seasonally moist places in many plant communities. Tolerates seasonal drought. Grows best on fertile or deep soils.

DISTRIBUTION: Throughout California, to 2400 m. Western and central states, many northeastern states.

PROPAGATION/PHENOLOGY: *Reproduces by seed* and *vegetatively from creeping roots*. Fruits probably disperse primarily by clinging to animals.

MANAGEMENT FAVORING/DISCOURAGING SURVIVAL: Frequent mowing or clean cultivation can eventually control wild licorice in agricultural fields.

SIMILAR SPECIES: The bur-like fruits, gland-dotted foliage, and odd-pinnate-compound leaves of wild licorice make it unlikely to be confused with other weedy legumes.

Everlasting peavine [*Lathyrus latifolius* L.]

SYNONYMS: perennial pea; perennial sweet pea; weedy sweet pea

GENERAL INFORMATION: *Glabrous perennial vine*, with *winged stems* to about 2 m long and *showy pink, purple, red-violet, or white flowers*. Everlasting peavine is a garden ornamental that has escaped cultivation in many areas of California, particularly in the northern part of the state. Plants are often vigorous and can establish dense colonies that exclude desirable vegetation. The fruits and, to a lesser degree, the foliage of everlasting peavine and other *Lathyrus* species contain unusual amino acids that are *toxic* to livestock and humans when ingested in quantity over time, resulting in a degenerative motor neuron disorder known as neurolathyrism. In the early stages, symptoms can be reversed or arrested, but the effects are permanent in later stages. Horses, humans, and other animals with a single stomach are more susceptible than ruminants such as cattle and sheep. Cattle and especially sheep can often ingest large quantities of everlasting peavine foliage without ill effect. Symptoms vary with the species involved. Symptoms in horses include stiff gait, poor coordination, weakness, flaccid muscles, hyperexcitability, difficulty in getting up, paralysis of the larynx, and death. In humans, symptoms include stiffness, weakness, muscle spasms and rigidity, especially in the legs, numbness, and loss of muscular control leading to incapacitation. Young males are most often affected. Since ancient times, neu-

Everlasting peavine (*Lathyrus latifolius*) infestation.   J. M. DiTomaso

Everlasting peavine (*Lathyrus latifolius*) seeds.   J. O'Brien

Everlasting peavine (*Lathyrus latifolius*) flowering stem.   J. M. DiTomaso

Everlasting peavine

Common sweetpea (*Lathyrus odoratus*) flowers. J. M. DiTomaso

Common sweetpea (*Lathyrus odoratus*) stem and leaf with two leaflets. J. M. DiTomaso

rolathyrism has been associated with people living under subsistence or famine conditions. It continues to be a problem in some poor areas of the world. Native to Europe.

SEEDLINGS: Cotyledons remain below the soil surface. Leaves alternate. First few true leaves scalelike, inconspicuous. First conspicuous leaf consists of 2 elliptic-lanceolate to ovate leaflets about 1–2 cm long on a stalk to ± 1 cm long. Stalk and leaflets sometimes have red glands. Stipules linear-lanceolate.

MATURE PLANT: *Foliage glabrous. Stem and leaf stalk wings usually more than 2 mm wide.* Leaves consist of *2 leaflets*, with the *main axis terminating in a branched, coiled tendril.* Leaflets elliptic-lanceolate to ovate, mostly 5–14 cm long, 3–5 veined from the base. Stipules lanceolate, 2–5 cm long.

ROOTS AND UNDERGROUND STRUCTURES: Root system deep, with rhizomelike roots near the soil surface.

FLOWERS: May–September. *Flowers 4–15 per cluster. Corolla 2–2.5 cm long,* pink to red-violet, white, or colored and white. Calyx unequally 5-lobed, lobes shorter than the tube. Stamens 10, 9 with fused filaments and 1 separate. Insect-pollinated. Primarily out-crossing.

FRUITS AND SEEDS: Pods glabrous, linear, 6–10 cm long, 0.7–1 cm wide, flat to ± round in cross-section, dark at maturity. Seeds ± round, hard, gray to black at maturity.

POSTSENESCENCE CHARACTERISTICS: Dead stems can persist into winter.

HABITAT: Roadsides, riparian areas, cereal crops, orchards, vineyards, and other disturbed places.

Everlasting peavine

DISTRIBUTION: Scattered throughout California, except deserts and Great Basin, to 2000 m. Most contiguous states, except possibly Florida, Iowa, and North Dakota.

PROPAGATION/PHENOLOGY: *Reproduces by seed and vegetatively from rhizomelike roots.* Seeds fall near the parent plant and disperse to greater distances with human activities and possibly water, soil movement, mud, and animals. Most seedlings emerge from seeds buried about 2–4 cm deep.

MANAGEMENT FAVORING/DISCOURAGING SURVIVAL: Cultivation, close mowing, or manual removal of stems before seeds mature can help control everlasting peavine.

SIMILAR SPECIES: The following *annual* species have *flowers 2–3 cm long*. All sometimes escape cultivation as garden ornamentals but are encountered less frequently than everlasting peavine and are less likely to persist.

Common sweetpea [*Lathyrus odoratus* L.] is an *annual* with *short, stiff-hairy foliage* and pink, purple, or white flowers *2–3 cm long*. In addition, common sweetpea has *flowers mostly in clusters of 2–4*. Stems and leaf veins on the lower surfaces are especially rough to touch. Stem wings are often less than 2 mm wide, but this characteristic is not reliable. Leaves consist of 2 ovate to narrow-elliptic leaflets 3–6 cm long. Pods are hairy and less than 6 cm long. Common sweetpea inhabits disturbed places sporadically throughout much of California, except the deserts and Great Basin, to 1000 m. It is most common in coastal areas. Common sweetpea also occurs in Utah and some eastern states. Like everlasting

Tangier pea *(Lathyrus tingitanus)* flowering stem with fruit.　　　　　J. M. DiTomaso

Common sweetpea (*Lathyrus odoratus*) seeds.   J. O'Brien

peavine, common sweetpea foliage and fruits contain unusual amino acids and can be *toxic* to humans and livestock when large amounts are ingested, although toxicity problems associated with common sweetpea ingestion appear to be less severe than neurolathyrism.

Tangier pea [*Lathyrus tingitanus* L.] is a *glabrous annual* with maroon to crimson flowers. Like common sweetpea, Tangier pea has *flowers 2–3 cm long in clusters of 2–4*, and stem wings are often less than 2 mm wide. It is distinguished from sweet pea by having *glabrous foliage and pods*. Tangier pea inhabits disturbed sites throughout California, especially in coastal areas and except deserts and Great Basin, to 500 m. It also occurs in Oregon.

A few other uncommon ornamental *Lathyrus* species have escaped cultivation in certain areas, particularly near the coast. All are annuals that have leaves with 2 leaflets and variously colored flowers 0.8–1.5 cm long. Refer to the appendix, **Non-native Naturalized Plants** (p. 1601, vol. 2), for more information.

Tangier pea (*Lathyrus tingitanus*) seeds.   J. K. Clark

Birdsfoot trefoil [*Lotus corniculatus* L.][LOTCO]

SYNONYMS: babies' slippers; birdfoot deervetch; bloomfell; broadleaf birdsfoot trefoil; cat's clover; crowtoes; ground honeysuckle; hop o' my thumb; sheepfoot; *Lotus corniculatus* L. var. *arvensis* (Schk.) Ser. ex DC.

GENERAL INFORMATION: *Glabrous to sparsely hairy perennial*, with ascending to trailing stems to about 0.5 m long and yellow flowers in headlike umbels. Cultivars of birdsfoot trefoil are used for pasture forage and hay, especially on poorly drained, low-fertility soils, and to control erosion in some areas. It generally provides excellent green forage for livestock and wildlife. On rare occasions, birdsfoot treefoil has been implicated in cyanide poisoning of livestock in other areas of the world. Plants contain variable amounts of glucosides that can generate *toxic* hydrocyanic acid under certain conditions, particularly when there is an increase in moisture and temperature. However, the risk of toxicity problems developing in livestock that graze birdsfoot trefoil pasture in the United States is very low. Birdsfoot trefoil has escaped cultivation in most states, including California, where it often grows in turf and is sometimes a weed of crop fields, orchards, vineyards, landscaped areas, managed forests, and natural areas, most notably in swales immediately inland of coastal dunes. Native to Eurasia.

SEEDLINGS: Cotyledons oval, about 3–7 mm long, glabrous, dark green. Leaves alternate, compound. First leaf consists of 3 leaflets, with the central leaflet slightly larger than the lateral leaflets. Foliage ± glabrous to sparsely hairy.

MATURE PLANT: Highly variable, with many regional types and cultivars. *Foliage a distinct blue-green color, glabrous to sparsely covered with stiff, appressed hairs* (strigose). Stems usually *solid*, branched, slightly angled near the top.

Birdsfoot trefoil (*Lotus corniculatus*) plant. J. M. DiTomaso

Birdsfoot trefoil

Birdsfoot trefoil *(Lotus corniculatus)* leaves, stipules, and stem. J. M. DiTomaso

Birdsfoot trefoil *(Lotus corniculatus)* seeds. J. O'Brien

Birdsfoot trefoil *(Lotus corniculatus)* flower and fruit clusters. J. M. DiTomaso

Leaves alternate, *pinnate-compound with 5 leaflets. The lowest 2 leaflets resemble leaflike stipules. The upper 3 leaflets are at the tip of the main axis well above the lower pair.* Leaflets narrowly elliptic to obovate, 0.5–2 mm long, tip acute to rounded with a slight indentation. *Stipules glandlike, inconspicuous.*

ROOTS AND UNDERGROUND STRUCTURES: Taproot well-developed, usually woody, branched, to 1 m deep, with numerous fibrous lateral roots that sometimes form a mat near the soil surface. Conspicuous rhizomes are lacking, but new shoots grow from the roots in spring or when the crown is damaged. Roots typically develop nodules in association with nitrogen-fixing bacteria. Carbohydrate reserves are low throughout the growing season.

FLOWERS: (February–) May–August. Flowers pea-like, ± sessile, arranged in *headlike umbels of 3–8 flowers* on a long stalk (peduncle) with a leaflike bract. Petals bright yellow, often aging orangish red, 8–14 mm long. Calyx 2–4 mm long, 5-lobed, lobes ± equal to the tube and *straight in bud*, glabrous to hairy. Stamens 10, filaments of 9 fused, 1 separate. Insect-pollinated. Primarily outcrossing.

FRUITS AND SEEDS: Pods occur in headlike clusters, each one linear, straight, ± round in cross-section, 1.5–3.5 cm long, 2–4 mm in diameter, glabrous, faintly net-veined, brown at maturity, with a persistent style at the apex, contain up to 49 seeds per pod. Pod halves twist into a spiral upon opening. Seeds irregularly round-ovoid, ± 1–1.5 mm in diameter, smooth, greenish to brown, sometimes with darker mottling.

Birdsfoot trefoil

POSTSENESCENCE CHARACTERISTICS: Frost kills the foliage. Dead foliage does not persist.

HABITAT: Turf, pastures, roadsides, crop fields, ditches, orchards, vineyards, managed forests, disturbed grassland, and wetland and riparian sites. Grows year-round in warm climate areas. Winter-dormant in cold-winter regions. Tolerates drought and infertile, dry, wet, saline, acidic, or limestone-based soils.

DISTRIBUTION: Throughout California, except deserts, to 1000 m. Most contiguous states, except a few southeastern states.

PROPAGATION/PHENOLOGY: *Reproduces by seed* and vegetatively from roots and occasionally stems. Pods snap open and eject seeds a short distance from the parent plant. Some seeds disperse to greater distances with water, mud, soil movement, animals, landscape maintenance, agricultural operations and

Spanish clover *(Lotus purshianus)* immature plant.   J. M. DiTomaso

Spanish clover *(Lotus purshianus)* seedling.   J. K. Clark

Spanish clover *(Lotus purshianus)* flowering stem.   J. M. DiTomaso

Spanish clover *(Lotus purshianus)* seeds.   J. O'Brien

other human activities, and as a seed and feed contaminant. Seeds germinate primarily in spring, but some germination can occur in fall. Some seeds are hard-coated and require scarification or decomposition of the seed coat before they can germinate. Hard-coated seeds are probably long-lived under field conditions and may survive ingestion by animals. Most seedlings emerge from a soil depth to 1.5 cm. Seedlings grow slowly and do not compete well with other vegetation. Root fragments can develop into new plants under favorable conditions. Mature stems can root at the nodes under favorable conditions.

MANAGEMENT FAVORING/DISCOURAGING SURVIVAL: Repeated clipping to near the ground can prevent seed production and weaken the root system, but plants can survive and even thrive under mowing.

SIMILAR SPECIES: Big trefoil [*Lotus uliginosus* Schk., synonyms: greater lotus; marsh lotus; *Lotus corniculatus* L. var. *major* Ser.; *Lotus pedunculatus* Cav.] is an uncommon glabrous to sparsely hairy *perennial* that closely resembles birdsfoot trefoil. Unlike birdsfoot trefoil, big trefoil has *conspicuous stolons or rhizomes, 8–12 flowers per cluster, sepal lobes that often curve outward in bud,* and *spreading hairs* when present. Usually the *stems are hollow.* Big trefoil inhabits roadsides, fields, ditches, and wetlands in the North Coast, western North Coast Ranges, northern San Francisco Bay region, and probably elsewhere, to 800 m. It has been collected in Nevada County and also occurs in Idaho, Oregon, Washington, and Florida. Native to Europe.

Spanish clover [*Lotus purshianus* (Benth.) Clements & E. G. Clements var. *purshianus,* synonyms: American bird's-foot trefoil; Pursh's lotus; Pursh's trefoil; *Hosackia americana* Benth.; *Lotus unifoliolatus* (Hook.) Benth.; *Lotus unifoliolatus* (Hook.) Benth. var. *unifoliolatus*; *Lotus americanus* Bisch.; others] is a widespread, variable, ruderal native *annual,* with erect to ± prostrate stems to 0.6 m long or tall and *white to pale pink* or sometimes yellow-tinged flowers mostly 0.5–1 cm long. Spanish clover is a desirable component of the vegetation in natural areas. It is an important source of food for wildlife, and the foliage provides good forage for livestock on rangeland. Like most other legumes, it is a nitrogen-fixing species. It is also a useful species for habitat restoration. However, because plants typically colonize open, disturbed places, Spanish clover is occasionally weedy in agronomic crops, orchards, vineyards, and landscaped areas. Spanish clover has *gray-green foliage that is covered with short, soft, spreading hairs,* although some biotypes are nearly glabrous. It is further distinguished by having most leaves consisting of *3 leaflets* (some upper leaves can consist of only 1 leaflet), *straight pods* 1.5–3 cm long that *spring open* to eject seeds, *one flower per leaf axil,* and *inconspicuous, glandlike stipules.* In addition, leaflets are ± elliptic, and *sepal lobes are much longer than the calyx tube.* Spanish clover inhabits roadsides, rangeland, crop fields, orchards, and other open, often disturbed places, including those in chaparral, forests, riparian and coastal areas throughout California except the Sonoran desert, to 2400 m. It also occurs in Arizona, Idaho, Montana, Nevada, Oregon, Washington, Wyoming, central states, and a few scattered southern and northeastern states.

Yellow bush lupine [*Lupinus arboreus* Sims]
[Cal-IPC: Native Invasive]

SYNONYMS: yellow-flowered bush lupine; tree lupine; *Lupinus propinquus* E. Greene

GENERAL INFORMATION: Fast-growing *shrub or subshrub*, with *erect stems* mostly 0.5–2 m tall and racemes of *fragrant, often yellow, pea-like flowers*. Some native populations have lilac to purple flowers, especially north of the central North Coast. Yellow bush lupine is a desirable species in its natural native range in the Central Coast and in parts of the North Coast. However, a yellow-flowered type was introduced to the Humboldt Bay area in the early 1900s as a landscape ornamental and for soil stabilization near railroad tracks. Since then, yellow bush lupine has invaded the native coastal dune community in some areas of Humboldt Bay, where it has negatively impacted the native dune community by changing the vegetation structure, species composition, and ecology. Yellow bush lupine increases soil nitrogen levels for many years and can facilitate invasion by European annual grasses. The exact demarcation between the native and introduced ranges in the southern North Coast region is uncertain, but it likely coincides with the range of native dune scrub communities that occur as far north as Sonoma County. The foliage and seeds of many lupines contain alkaloids that can be toxic to humans and animals when

Yellow bush lupine (*Lupinus arboreus*) along the northern coast.   J. M. DiTomaso

Yellow bush lupine (*Lupinus arboreus*) seeds.
   J. O'Brien

Yellow bush lupine (*Lupinus arboreus*) flowering stem.   J. M. DiTomaso

ingested, but toxicity problems due to the ingestion of yellow bush lupine have not been reported. Introduced yellow bush lupine on the North Coast hybridizes with two native lupine species, *Lupinus rivularis* and *Lupinus littoralis*, producing plants with flowers that vary from yellow to purple, even on the same inflorescence.

MATURE PLANT: Woody stems *erect*. Foliage glabrous and green to covered with silvery hairs. Leaves palmate-compound, consist of 5–12 leaflets 2–6 cm long. Leaf stalks 2–6 cm long. Stipules 8–12 mm long.

ROOTS AND UNDERGROUND STRUCTURES: Large shrubs have a deep, extensively branched taproot. Roots typically associate with nitrogen-fixing bacteria.

FLOWERS: April–June. Sweetly fragrant in most populations. Racemes about 10–30 cm long, with *yellow, sometimes lilac to purple, pea-like flowers* 14–18 mm long scattered on the stem or in whorls and deciduous bracts 8–10 mm long. Upper petal (banner) back *glabrous*. Inner lower petals (keel) *minutely ciliate along the upper margin from the middle to the tip.* Calyx base *lacks a bulge*.

FRUITS AND SEEDS: Pods brown-black, *4–7 cm long*, moderately covered with long, light-colored hairs. Seeds mostly 8–12 per pod, ± ovoid, base truncate with a bulge to one side, slightly flattened, 4–5 mm long, black to yellowish brown, often striped.

HABITAT: Coastal dunes and bluffs, sometimes inland to about 8 km (5 mi). Typically grows on sand or loamy sand. Tolerates some salinity.

DISTRIBUTION: Central Coast and North Coast, to 100 m. Sporadic along the coast of Oregon and Washington.

PROPAGATION/PHENOLOGY: *Reproduces by seed.* Seed germination and seedling emergence is greatest when seeds are shallowly buried in exposed soil of open sites. Disturbance stimulates germination. Salinity appears to enhance germination. Some seeds survive at least 15 years and probably much longer under field conditions. Plants typically do not live for more than 5 to 10 years.

MANAGEMENT FAVORING/DISCOURAGING SURVIVAL: Manual removal of yellow bush lupine, litter, and duff for at least 3 years can control infestations and help restore native dune vegetation. In addition, yearly removal can stimulate seed germination and reduce the soil seedbank. One effective approach is to use heavy equipment to scrap the duff layer then cover the scraped area with a weed mat for 2 years.

SIMILAR SPECIES: There are numerous native lupines in California that are considered a desirable component of the vegetation in natural communities. **Varied lupine** [*Lupinus variicolor* Steud.] is a North and Central Coast *perennial or subshrub* with a yellow-flowered form that could be mistaken for yellow bush lupine. Unlike yellow bush lupine, varied lupine usually has *prostrate to decumbent, weakly woody stems less than 0.5 m tall* and densely long-hairy pods *3–4 cm long*. Only yellow bush lupine and its hybrids are considered to be an invasive weed in the Humboldt area of the North Coast.

Black medic [*Medicago lupulina* L.][MEDLU]

California burclover [*Medicago polymorpha* L.] [MEDPO][Cal-IPC: Limited]

SYNONYMS: Black medic: black medick; black clover; hop clover; hop medic; nonesuch; yellow trefoil; *Medicago lupulina* L. var. *cupaniana* (Guss.) Boiss.

California burclover: barrel medic; burclover; burr medic; medic; *Medicago apiculata* auct. non Willd.; *Medicago denticulata* Willd.; *Medicago hispida* Gaertn.; *Medicago nigra* Krock.; *Medicago polycarpa* Willd.; others

GENERAL INFORMATION: Low-growing legumes to about 0.4 m tall, with *headlike racemes of yellow flowers*, and *compound leaves that consist of 3 leaflets*. Both species are good livestock forage and are sometimes cultivated for pasture or as a cover crop. However, the prickly fruits of California burclover can entangle the wool of sheep and lower its value.

Black medic: Winter or summer *annual to short-lived perennial*. Black medic seeds are a common impurity in commercial alfalfa and clover seed. Native to Europe.

California burclover: Winter or summer *annual*. California burclover is susceptible to alfalfa mosaic virus. It is a state-listed noxious weed in Arizona.

Black medic (*Medicago lupulina*) flowers and fruit.   J. M. DiTomaso

Black medic (*Medicago lupulina*) flowering stem.   J. M. DiTomaso

Black medic (*Medicago lupulina*) seedling.   J. K. Clark

Black medic • California burclover

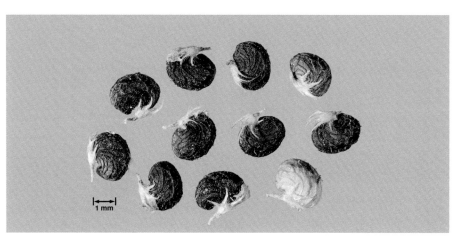

Black medic (*Medicago lupulina*) fruit.   J. O'Brien

California burclover (*Medicago polymorpha*) flowering stem.   J. M. DiTomaso

Despite its common name, California burclover is native to the Mediterranean region.

SEEDLINGS: Cotyledons oblong, about 4–9 mm long, glabrous. Leaves alternate. First leaf simple, horizontally oval, broader than long, about 4–8 mm long. Subsequent leaves compound with 3 leaflets, resemble leaves of mature plants.

MATURE PLANT: Main stem branched near base. Stems prostrate to decumbent, to about 0.4 m long. Leaves alternate. Leaflets obovate to ± heart-shaped, about 1–2 cm long, *finely toothed near the tip,* apex often tipped with a minute slender tooth. *Stalk of middle leaflet about 2–5 mm long, longer than lateral leaflet stalks.*

Black medic • California burclover

California burclover (*Medicago polymorpha*) stipules and fruit.   J. M. DiTomaso

California burclover (*Medicago polymorpha*) burred and unburred fruit.   C. Elmore

California burclover (*Medicago polymorpha*) seedling.   J. K. Clark

California burclover (*Medicago polymorpha*) seeds.   J. O'Brien

Black medic: *Foliage hairy, especially stems. Stipule margins smooth or with a few, ± shallow teeth.*

California burclover: *Foliage nearly glabrous. Stipule margins deeply lobed*, lobes slender and curved.

ROOTS AND UNDERGROUND STRUCTURES: Taproots slender, usually with many fine laterals that form a dense mat. Roots associate with nitrogen-fixing bacteria.

FLOWERS: Flowers *yellow*, slender, pea-like, in dense clusters. Calyx 5-lobed, persistent in fruit, lobes nearly equal or unequal. Stamens 10, 9 with fused filaments, 1 separate. Self-pollinating and outcrossing.

Black medic: April–July. Flower clusters ± short, *spikelike*, mostly 10–20 per cluster. *Flowers 1.5–4 mm long.* Calyx 1–1.5 mm long.

California burclover: March–June. Flower clusters *headlike*, 2–5 per cluster. *Flowers usually 3.5–6 mm long.* Calyx tubes ± 3 mm long.

FRUITS AND SEEDS: Pods do not open to release seeds.

Black medic: Pods *kidney-shaped, about 2–3 mm long*, glabrous, strongly veined, *black at maturity*, contain 1 seed. Seeds oval to barely kidney-shaped, ± 2 mm long, 1 mm wide, with a small sharp point on the concave face, smooth, yellowish to olive-green.

California burclover: Pods usually *tightly coiled 2–6 times*, mostly 4–6 mm in diameter, glabrous, brown, *smooth or with 2–3 rows of prickles on the outer face*, contain a few seeds. Prickles often minutely hooked. Seeds kidney-shaped, ± 3 mm long, smooth, yellowish brown.

POSTSENESCENCE CHARACTERISTICS: Dead stems with fruits can persist for a few months.

HABITAT: Common in turf, roadsides, fields, grassland, pastures, agronomic crops (especially alfalfa), vegetable crops, orchards, vineyards, gardens, and other disturbed places.

Black medic: Also forests, mountainous areas. Grows well on dry, high-calcium (calcareous) soils.

California burclover: Often associated with agricultural land.

DISTRIBUTION: Black medic: Throughout California, except deserts, to 2500 m. All contiguous states.

California burclover: Throughout California, except Great Basin and deserts, to 1500 m. All western states except possibly Colorado, southern states, and many northeastern states.

Spotted burclover (*Medicago arabica*) flowers, stem, and leaves with spots.   J. M. DiTomaso

Spotted burclover (*Medicago arabica*) fruit.
                                          J. M. DiTomaso

Spotted burclover (*Medicago arabica*) nitrogen-fixing nodules on roots.   J. M. DiTomaso

PROPAGATION/PHENOLOGY: *Reproduce by seed.* Flowers and fruits develop continuously throughout the growing season. Fruits disperse with water, soil movement, animals, and human activities. Seeds also disperse as commercial seed contaminants. Seeds are hard-coated and can survive many years under

Spotted burclover *(Medicago arabica)* seeds.   J. O'Brien

White-flowered form of alfalfa *(Medicago sativa).*   J. M. DiTomaso

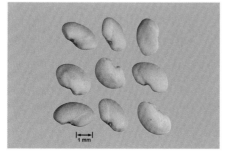

Alfalfa *(Medicago sativa)* seeds.   J. O'Brien

Purple-flowered form of alfalfa *(Medicago sativa).*   J. M. DiTomaso

Black medic • California burclover

Foxtail restharrow (*Ononis alopecuroides*) flowering stem. J. M. DiTomaso

Foxtail restharrow (*Ononis alopecuroides*) fruiting cluster. J. M. DiTomaso

Foxtail restharrow (*Ononis alopecuroides*) seedling. J. M. DiTomaso

Foxtail restharrow (*Ononis alopecuroides*) flowering cluster. J. M. DiTomaso

field conditions. In California, most germination usually occurs in fall after the first significant rain.

Black medic: Newly matured seeds can readily germinate for a variable period if field conditions are favorable. After this period, seeds become dormant. Under optimal conditions, seedlings mature to flower in ~ 6 weeks and fruit at ~ 9 weeks.

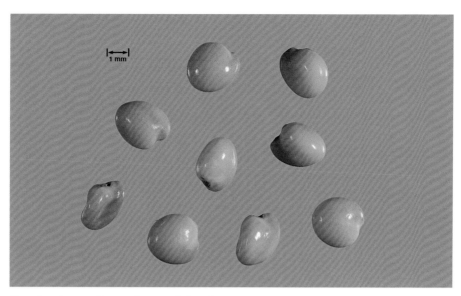

Foxtail restharrow (*Ononis alopecuroides*) seeds. J. O'Brien

MANAGEMENT FAVORING/DISCOURAGING SURVIVAL: Both species tolerate frequent mowing. Cultivation or manual removal of plants before seed production can provide control.

SIMILAR SPECIES: Spotted burclover [*Medicago arabica* (L.) Huds., synonym: *Medicago maculata* Sibth.] [MEDAB] is a low *annual* with yellow flowers that resembles California burclover. Unlike California burclover, spotted burclover typically has many *broad leaflets that have a dark purplish spot in the center*. Leaflets are usually *broadly heart-shaped with the lobes at the tip*, and the *leaflet length is about equal to the width*. Spotted burclover inhabits disturbed places and agricultural lands in the North Coast Ranges, North Coast, northern Sierra Nevada, and Central-western region, to 2000 m. It also occurs in Oregon, Washington, southern states, and a few eastern-central and northeastern states. Probably native to the Mediterranean region.

Alfalfa [*Medicago sativa* L.] [MEDSA] is a variable, *erect perennial* that is cultivated for hay and other animal feed products. Alfalfa is readily distinguished from black medic and the burclovers by its *erect habit, perennial life cycle*, and *purple flowers 6–11 mm long*. Occasionally, flowers are white to greenish yellow. In addition, flower clusters are spikelike, leaflets are narrowly lanceolate to obovate, and fruits are smooth and coiled or sickle-shaped. Alfalfa occurs along roadsides where hay trucks frequently travel. It has escaped cultivation throughout California, except deserts, to 1500 m. Native to Eurasia.

Foxtail restharrow [*Ononis alopecuroides* L., synonym: *Onosis salzmanniana* Boiss. & Reut. non sensu Ivimey-Cook in *Flora Europaea*] [CDFA list: Q] [Cal-IPC: Limited] is a weakly woody *annual* to about 1 m tall that has recently been discovered in California and is not included in California floras to date. Foxtail restharrow has *upper stems that are glandular and hairy* and *compound leaves that consist of a broadly oval central leaflet 2–5 cm long on a stalk to 15 mm long and 2 small lateral leaflets that are sessile*. Leaflet margins are finely toothed. Flowers are about 13–16 mm long and arranged in *dense, terminal spikes of pink flowers* with whitish wings. In addition, *pods are straight or nearly straight, oblong, ± 1 cm long, and open to release a few glossy, orangish brown seeds 2–3 mm long*. Foxtail restharrow generally inhabits disturbed fields, pastures, and agricultural areas. It is currently limited to one location in the western South Coast Ranges (west end of Los Oso Valley, San Luis Obispo Co.). Although it is nearly eradicated at this location, it may spread rapidly if it becomes established elsewhere in California. Native to the western Mediterranean region and southern Europe.

White sweetclover [*Melilotus albus* Medik.][MEUAL]
Indian sweetclover [*Melilotus indicus* (L.) All.][MEUIN]
Yellow sweetclover [*Melilotus officinalis* (L.) Pall.] [MEUOF]

SYNONYMS: Under the current rules of the International Code of Botanical Nomenclature, the genus *Melilotus* is to be treated as masculine, even though it has been primarily treated as feminine.

White sweetclover: honey clover; tree clover; white melilot; *Melilotus alba* Medik.; *Melilotus albus* Medik. var. *annuus* Coe; *Melilotus leucanthus* W. D. J. Koch ex DC.; *Melilotus officinalis* (L.) Pall. var. *albus* (Medik.) H. Ohashi & Y. Tateishi; *Sertula alba* (Medik.) Kuntze

Indian sweetclover: annual yellow sweetclover; bitter clover; Indian melilot;

White sweetclover (*Melilotus albus*) along edge of pond.　　J. M. DiTomaso

White sweetclover • Indian sweetclover • Yellow sweetclover

White sweetclover (*Melilotus albus*) flowering stem. J. M. DiTomaso

White sweetclover (*Melilotus albus*) inflorescence and leaf. J. M. DiTomaso

White sweetclover (*Melilotus albus*) seedling. J. K. Clark

White sweetclover (*Melilotus albus*) seeds and pods. J. O'Brien

small melilot; sour clover; yellow melilot; *Melilotus indica* (L.) All.; *Melilotus parviflora* Desf.; *Trifolium melilotus* var. *indica* L.

Yellow sweetclover: field melilot; ribbed melilot; yellow melilot; *Melilotus arvensis* Wallr.; *Melilotus graveolens* Bunge; *Melilotus officinalis* (L.) Lam. var. *micranthus* O. E. Schulz; *Melilotus suaveolens* Ledeb.; *Melilotus vulgaris* Hill; *Trifolium melilotus* var. *officinalis* L.; *Trifolium officinale* L.

GENERAL INFORMATION: Legumes with slender, ± elongate racemes of small, fragrant flowers and compound leaves that consist of 3 leaflets. Sweetclovers are sometimes cultivated for livestock forage and as cover crops. They are useful honey bee plants, and the foliage and seeds are consumed by wildlife. Sweetclover foliage and flower buds contain coumarin, a compound of low

White sweetclover • Indian sweetclover • Yellow sweetclover

Indian sweetclover (*Melilotus indicus*) plant.
J. M. DiTomaso

Indian sweetclover (*Melilotus indicus*) flowering stem.
J. M. DiTomaso

Indian sweetclover (*Melilotus indicus*) seedling.
J. M. DiTomaso

Indian sweetclover (*Melilotus indicus*) seeds and pods.
J. O'Brien

toxicity that is generally safe for livestock to ingest in moderation. However, poorly cured hay or silage promotes the growth of certain molds that convert coumarin to the *toxic* compound dicoumarin, also known as dicumerol and dicumarin. Dicoumarin interfers with vitamin K metabolism and blood clotting and is the cause of sweetclover poisoning, a severe hemorrhagic often fatal disorder primarily affecting cattle that have ingested moldy sweetclover hay. Sweetclover hay can appear to be dry and free of mold, but stems may still contain moisture and mold inside, which tend to retain moisture. Native to Eurasia.

White sweetclover, yellow sweetclover: Erect *biennials,* sometimes annuals or short-lived perennials, *to 2 m tall or more,* with white and yellow flowers, respectively. White sweetclover is generally the most weedy and widespread of the sweetclovers in California.

Indian sweetclover: Erect to spreading *annual to about 0.8 m tall,* with yellow flowers.

SEEDLINGS: First leaf simple. Subsequent leaves pinnate-compound with 3 leaflets.

White sweetclover, yellow sweetclover: Cotyledons oblong, about 5–8 mm long, 3–4 mm wide, glabrous. First leaf ± obovate, tip ± squared with a minute nipplelike point (cuspidate), about 2–5 mm long and wide, glabrous or sparsely

White sweetclover • Indian sweetclover • Yellow sweetclover

Yellow sweetclover *(Melilotus officinalis)* plant. J. M. DiTomaso

hairy, margin minutely toothed near apex. Subsequent leaves have leaflets that resemble the first leaf. Terminal leaflet short-stalked.

Indian sweetclover: Cotyledons resemble those of white and yellow sweetclover. First leaf often broadly heart-shaped and lobed at the tip, about 3–4 mm long, 5–8 mm wide.

MATURE PLANT: Foliage glabrous to sparsely minute-hairy. Leaves alternate, *pinnate-compound with 3 leaflets, terminal leaflet short-stalked.*

White sweetclover, yellow sweetclover: Stems ± woody at the base, especially in the second year. Leaflets (ob)ovate to ± oblong, mostly 1–2.5 cm long, tip rounded or weakly squared, margin smooth to weakly toothed. Stipules ± narrowly triangular, mostly 0.5–1 cm long.

Indian sweetclover: Stems generally not woody at the base. Foliage resembles that of white and yellow sweetclover. Leaflets often obovate, with tips slightly indented, squared, or rounded.

ROOTS AND UNDERGROUND STRUCTURES: Roots associate with nitrogen-fixing bacteria.

White sweetclover, yellow sweetclover: Taproot tough or woody, slender to thick, typically deep, with fibrous lateral roots.

Indian sweetclover: Taproot tough, but usually not woody, usually slender.

FLOWERS: Racemes slender, axillary and terminal, elongate as flowers mature. Flowers pea-like, slender, sweetly fragrant, on short stalks. Petals deciduous in fruit. Insect-pollinated.

White sweetclover: (March) April–November (December). Racemes mostly

White sweetclover • Indian sweetclover • Yellow sweetclover

Yellow sweetclover (*Melilotus officinalis*) leaf.
J. M. DiTomaso

Inflorescence of yellow sweetclover (*Melilotus officinalis*) (right) and indian sweetclover (*Melilotus indicus*) (left). J. M. DiTomaso

Yellow sweetclover (*Melilotus officinalis*) flowering stem. J. M. DiTomaso

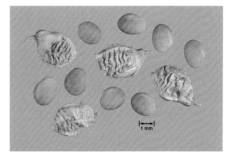

Yellow sweetclover (*Melilotus officinalis*) seeds and pods. J. O'Brien

3–12 cm long. *Flowers white, mostly 4–6 mm long.* Self-compatible.

Indian sweetclover: April–November. Racemes 2–10 cm long, often compact when flowers open. *Flowers yellow, 2–3 mm long.*

Yellow sweetclover: April–September. Racemes mostly 3–12 cm long. *Flowers yellow, 4–7 mm long.* Generally out-crossing.

FRUITS AND SEEDS: *Pods ovoid, glabrous, on stalks that bend downward (reflexed), contain 1 seed, rarely 2. Pods do not open to release seeds.*

White sweetclover: *Pods 3–5 mm long,* usually dark brown to black, *surface weakly net-veined.* Seeds ovoid, obliquely angled at the base, ± 2–2.5 mm long, ± 1.5 mm wide, smooth, dull yellowish green to orange-tan.

Indian sweetclover: *Pods 2–3 mm long*, light brown, *surface bumpy or weakly net-veined*. Seeds ± 2 mm long, 1.5 mm wide, resemble those of yellow sweetclover.

Yellow sweetclover: *Pods 3–5 mm long*, light brown, *surface sharply net-veined or ridged, most veins horizontal*. Seeds ovoid, with an oblique angle or shallow notch at the base, ± 2–2.5 mm long, 1–1.5 mm wide, smooth, dull yellowish green to orange-tan, sometimes with brown or purplish spots.

HABITAT: All inhabit roadsides, open fields, pastures, agronomic crops, and other open disturbed places. White sweetclover often grows in moist places, such as riparian areas, ditches, and other moist, open, disturbed sites in many plant communities, but it does not tolerate extended periods of standing water. Yellow sweetclover and Indian sweetclover typically inhabit dry places.

DISTRIBUTION: All occur throughout California, except low deserts, to 1500 m and are locally abundant in some areas. White sweetclover is especially common in the southern portion of the state. Yellow sweetclover also occurs in all contiguous states. Indian sweetclover occurs in Arizona, Idaho, Montana, Nevada, New Mexico, Oregon, Utah, Washington, and most southern and eastern states.

PROPAGATION/PHENOLOGY: *Reproduce by seed*. Seeds fall near the parent plant and disperse to greater distances with water, mud, human activities, as a seed or feed contaminant, and by clinging to the shoes and clothing of humans, vehicle tires, and possibly to animals. Most mature seeds are hard-coated and can remain viable for up to ± 20 years or more under field conditions. Seeds typically germinate fall through spring in California. Seedlings compete poorly with perennial species.

MANAGEMENT FAVORING/DISCOURAGING SURVIVAL: Manual removal, cultivation, or cutting plants to prevent seed production can control sweetclover. Burning can kill existing plants, but often stimulates seed germination when moisture conditions become favorable. Establishing a dense cover of desirable perennial species can eliminate sweetclover in ~ 2 years. Mowing roadsides and waste places early in the season reduces vegetative competition and can favor the survival of immature sweetclover. Cutting or mowing immature white and yellow sweetclover when root and crown carbohydrate reserves are at their lowest, usually in late August or early September, can weaken plants and reduce their growth the following season.

SIMILAR SPECIES: Burclovers (*Medicago* spp.) and some clovers (*Trifolium* spp.) have leaves consisting of 3 leaflets, with the central leaflet on a stalk. Without flowers or fruits, theses species can be confused with immature sweetclovers. Buclovers are primarily distinguished by their *coiled or kidney-shaped fruits*. In California, all yellow-flowered burclover species, except yellow-flowered forms of alfalfa [*Medicago sativa* L.], are *low, ± spreading plants* with *short racemes that consist of only a few flowers*. Clovers are also generally *low-growing plants* and are readily distinguished by having *dense, globular to elongate flower heads* and *petals that persist in fruit*.

Velvet mesquite [*Prosopis velutina* Wooton][PRCJV] [Federal Noxious Weed]

Creeping mesquite [*Prosopis strombulifera* (Lam.) Benth][CDFA list: A][Federal Noxious Weed]

SYNONYMS: Velvet mesquite: Arizona mesquite; *Netuma velutina* (Woot.) Britt. & Rose; *Prosopis articulata* S. Wats.; *Prosopis chilensis* (Mol.) Stuntz var. *velutina* (Woot.) Standl.; *Prosopis juliflora* (Sw.) DC. var. *velutina* (Woot.) Sarg.; others

Creeping mesquite: Argentine screwbean; espinilla; retortuño; *Strombocarpa strombulifera* (Lam.) A. Gray

GENERAL INFORMATION: Winter deciduous *shrubs or small trees*, with *thorns or spines* and *twice-pinnate-compound leaves*. Mesquite pods are sweet, fairly nutritious, and relished by livestock, but heavy consumption can cause digestive problems and foundering in horses. Mesquites in general are often considered rangeland weeds because they are prolific and highly competitive for moisture with herbaceous grassland species.

Velvet mesquite: Besides its designation as a federal noxious weed, velvet mesquite is a state-listed noxious weed in Florida and North Carolina (class A). Native to the southwestern United States, but not California.

Creeping mesquite: This species accidentally escaped cultivation as an experimental desert rangeland plant at Bard, California (Imperial Co.). To date, creeping mesquite has been eradicated at the Bard location and does not appear to have

Velvet mesquite (*Prosopis velutina*) plant. J. M. DiTomaso

Velvet mesquite (*Prosopis velutina*) flowering branch.   J. M. DiTomaso

Velvet mesquite (*Prosopis velutina*) seeds.   J. O'Brien

Velvet mesquite (*Prosopis velutina*) fruiting branch.   J. M. DiTomaso

spread beyond that locale. Besides California, creeping mesquite is a state-listed noxious weed in Florida and North Carolina (class A). Native to Argentina.

SEEDLINGS: Cotyledons oval, somewhat fleshy. Fast growing.

MATURE PLANT: Small branches often zig-zag slightly. Leaves alternate (sometimes appearing fascicled on short shoots), *twice-pinnate-compound* with *even numbers of leaflets and pinnae* (primary divisions of leaves). Leaflets opposite, oblong.

Velvet mesquite: Shrubs or small trees, with spreading, rounded crowns, to 15 m tall. Trunk single or multiple. Foliage and fruits *densely covered with short velvety hairs*, especially new leaves and immature fruits. Pinnae 1–2 pairs, 2–9 cm long. *Leaflets usually 15–30 pairs, 4–15 cm long, closely spaced along the pinnae.* Thorns 1–2 cm long, from axillary buds in a stipular position.

Creeping mesquite: Shrubs to 3 m tall, often forming dense stands. Pinnae 1 pair, 1–3 cm long. *Leaflets 2–8 pairs, 1 cm long or less, usually glabrous or sparsely short-hairy, widely spaced on the pinnae*, covered with a whitish bloom (glaucous). Spines 1–2 cm long, thin, white, from modified stipules.

ROOTS AND UNDERGROUND STRUCTURES: Plants develop *spreading lateral roots* that can extend several meters outward in all directions in the upper soil layers and deep taproots to depths of 4 m or more. Roots associate with nitrogen-fixing bacteria.

Velvet mesquite: Under certain conditions, trees may occasionally develop roots to nearly 50 m deep. Plants cut to ground level produce new shoots from dormant stem buds on the crown at or just below soil level, but not from roots.

Creeping mesquite: *Lateral roots long, creeping,* can develop new shoots.

FLOWERS: Flower heads axillary. Flowers small, numerous, *radially symmetrical*, each with a bell-shaped calyx. Petals 5. *Stamens 10, extended beyond petals.* Insect-pollinated. Species hybridize freely.

Velvet mesquite: April–July. Flower heads *spikelike*, sessile, 5–15 cm long, mostly drooping, and developing only on 3- or 4-year-old branches. Petals cream-colored, separate.

Creeping mesquite: Flower heads *spherical*, ± 1.5 cm in diameter, on a stalk 1–5 cm long. Petals yellow, fused at base. Stamen filaments red.

FRUITS AND SEEDS: Prolific, especially in dry years. Pods (loments) flattened, leathery, slightly constricted between seeds, and not opening at maturity. Seeds numerous, separated by spongy pulp.

Velvet mesquite: Pods tan or yellow, often with pink streaks, *linear*, 8–15 cm long, *covered with short, velvety hairs*, especially while immature. Seeds oval, brown, 5–10 mm long.

Creeping mesquite: Pods irregularly *tight-coiled*, lemon yellow, 1.5–5 cm long, glabrous. Seeds oval, gray-green, 4.5–5.5 mm long.

HABITAT: Typically grows on sandy, rocky, medium to fine-textured soils in semiarid and arid regions.

Velvet mesquite: Washes, canyons.

Creeping mesquite: Disturbed sites.

DISTRIBUTION: Velvet mesquite: Uncommon. San Joaquin Valley and Centralwest and Southwest regions, to 1700 m. Arizona, New Mexico.

Creeping mesquite: Not known to exist in California to date. The population

Creeping mesquite (*Prosopis strombulifera*) fruit.   J. O'BRIEN

Honey mesquite *(Prosopis glandulosa* var. *torreyana)*, a native nonweedy species similar to velvet mesquite.　　　　　　　　　　　　　　　　　　　　　　　　　　　　　J. M. DiTomaso

that was eradicated occurred in the southeastern Sonoran Desert (se Imperial Co.), to 50 m.

PROPAGATION/PHENOLOGY: *Reproduce by seed.* Creeping mesquite also reproduces vegetatively from creeping lateral roots. Fruits are consumed and dispersed by animals. Rodents often plant seeds. A proportion of seeds remains viable after ingestion by livestock. These seeds typically have higher germination rates. Seeds can germinate under considerable moisture stress. Most seeds germinate between 20°–40°C, and light is not required. Seeds retained within intact pods can remain viable for up to 40 years, but exposed seeds dry out or decay more rapidly. Seedlings typically emerge from soil depths of 1–2 cm. Seedling root growth can be up to 10 times more rapid than shoot growth. Plants can reach reproductive maturity after 3 years.

MANAGEMENT FAVORING/DISCOURAGING SURVIVAL: Livestock browsing on fruits can disperse seeds to new sites. Quarantining livestock for about 6 days before moving can help prevent introducing mesquite seed into new areas.

SIMILAR SPECIES: Honey mesquite [*Prosopis glandulosa* Torr. var. *torreyana* (L.Bens.) M.Johnst.] and **screw bean** [*Prosopis pubescens* Benth.] are native *shrubs or small trees* that are important components of desert communities and are not weeds. Both species are included here for identification purposes only. Honey mesquite has pods similar to those of velvet mesquite, but unlike velvet mesquite, honey mesquite has *glabrous to sparsely hairy leaflets* mostly *greater than 15 mm long with the length about 7–9 times greater than the width.*

Like creeping mesquite, screwbean has tightly coiled pods. Screw bean is distinguished by having *spikelike flower heads* and *leaflets densely covered with hairs.*

Acacia species are distinguished from the mesquites by having more than 10 stamens per flower.

Black locust [*Robinia pseudoacacia* L.][ROBPS] [Cal-IPC: Limited]

SYNONYMS: false acacia; yellow locust; *Robinia pringlei* Rose; *Robinia pseudoacacia* L. var. *rectissima* (L.) Raber

GENERAL INFORMATION: Fast-growing deciduous *tree* to about 25 m tall, with *pinnate-compound leaves* and highly *fragrant, white, pea-like flowers*. Historically, black locust was widely planted in the United States as a landscape ornamental and for its hard, rot-resistant wood. In some areas of California and elsewhere, black locust has escaped cultivation and become invasive. Trees can produce numerous suckers from the roots, and dense clonal colonies that exclude native vegetation can develop. Foliage, seeds, and especially bark are *toxic* to humans and livestock when ingested, but rarely fatal. Flowers are reported to be nontoxic to slightly narcotic. Horses that chew on the bark while tied to black locust trees for prolonged periods are most frequently affected. Toxicity symptoms include rapid onset of colic, diarrhea, vomiting (in humans), weakness, and in severe cases, labored breathing, weak pulse, and irregular heartbeat. Horses are also subject to laminitis. Black locust flowers attract bees, but the seeds and foliage are poorly consumed by wildlife. Some cultivars lack spines on the twigs or have pinkish flowers. Native to the eastern United States.

SEEDLINGS: Cotyledons oblong, about 13–18 mm long, glabrous, tips rounded. Leaves alternate. First leaf simple, broadly ovate, 10–20 mm long, base slightly lobed, ± short-hairy. Second leaf compound, leaflets 3, elliptic to ovate, ~ 7–16 mm long, ± short-hairy.

Black locust (*Robinia pseudoacacia*) flowering branches. J. M. DiTomaso

Black locust

Black locust (*Robinia pseudoacacia*) fruit pods.   J. M. DiTomaso

MATURE PLANT: Bark light gray-brown, rough, deeply furrowed. *Twigs glabrous*. Axillary buds inconspicuous. Some trees may have a pair of stout stipular spines about 1–2.5 cm long at the base of the leaf stalks. Leaves deciduous, *alternate, pinnate-compound with an odd number of leaflets*, mostly 20–36 cm long. *Leaflets 9–19*, ovate to oblong with a *smooth margin* and a minute bristle at the tip, 2–5 cm long, glabrous, on a short stalk. New foliage is often covered with very short hairs.

ROOTS AND UNDERGROUND STRUCTURES: Roots spreading, can produce new shoots, and associate with nitrogen-fixing bacteria.

FLOWERS: April–June. Racemes axillary, showy, *pendent*, about 10–20 cm long. *Flowers fragrant, pea-like, white,* 15–20 mm long. Stamens 10, 9 with fused filaments and 1 separate. Insect-pollinated.

FRUITS AND SEEDS: Fruits mature July–November. *Pods flat, glabrous,* narrowly winged along lower margin, light brown to nearly black, 5–10 cm long, mostly 1–1.5 cm wide, usually contain (1)2–8 seeds. Pods *eventually split open* on both sides during fall through winter. Seeds kidney-shaped, 4–6 mm long, smooth, reddish brown and purplish black, often finely mottled.

HABITAT: Roadsides, sites near old habitations, riparian areas, canyon slopes, disturbed woodlands, and floodplain forests. Grows best on well-drained, neutral to mildly acidic, limestone soils. Tolerates some drought, fire, and temperatures as low as –36°C. Does not tolerate shade, salinity, or extended periods of excessive soil moisture.

DISTRIBUTION: Throughout California, except deserts, to 1900 m. All contiguous states, except possibly Arizona.

PROPAGATION/PHENOLOGY: *Reproduces vegetatively from root sprouts* and *by seed.* Fruits and seeds fall near the parent tree and disperse to greater distances with water, soil movement, human activities, and possibly animals. Seeds are

Black locust

Black locust (*Robinia pseudoacacia*) thorns on stem. J. M. DiTomaso

Black locust (*Robinia pseudoacacia*) seeds. J. O'Brien

hard-coated and can remain dormant for 10 or more years under field conditions. Scarification or decomposition of the seed coat stimulates germination. Seeds typically germinate in spring. Seedlings do not tolerate shade and compete poorly with established vegetation. Seedlings that establish on favorable sites can grow up to about 1 m per year. Seedling establishment appears to be poor in many areas of California. Saplings generally begin to fruit at ± 6 years of age. Trees 15–40 years old typically produce the largest quantities of seeds.

MANAGEMENT FAVORING/DISCOURAGING SURVIVAL: Cutting or ringing trees close to ground level is likely to stimulate root sprouting. Repeatedly removing new shoots at least two times per year for several years can eventually kill trees. Immediately treating newly cut stumps with systemic herbicide is more effective than cutting alone. Drill-and-fill or hack-and-squirt treatments are also effective. Digging out small trees and the main roots can help reduce root sprouting. Fire stimulates root sprouting and possibly seed germination, and it can create conditions that favor seedling establishment. Monitoring sites and removing seedlings for about 10 years can prevent new trees from becoming established.

SIMILAR SPECIES: **Black locust** is unlikely to be confused with any other weedy tree species. **Tree-of-heaven** [*Ailanthus altissima*] and **black walnut** [*Juglans californica* S. Wats] also have pinnately compound leaves, but the leaflets of these species are pointed at the tips.

Red sesbania [*Sesbania punicea* (Cav.) Benth.] [SEBPU][Cal-IPC: High Alert]

SYNONYMS: bladderpod; coffeeweed; purple sesban; rattlebox; rattlebush; scarlet wisteria tree; *Daubentonia punicea* (Cav.) DC.; *Emerus puniceus* (Cav.) Kuntze; *Piscidia punicea* Cav.; *Piscidia ovalifolia* Larrañaga; *Sesbania tripetii* (Poit.) hort ex Hubb.

GENERAL INFORMATION: Deciduous *shrub or small tree* to about 4 m tall, with *pinnate-compound leaves that have an even number of leaflets* and *red to orange-red, pea-like flowers*. Red sesbania is cultivated as an ornamental in many countries, but it has escaped cultivation and invaded riparian areas and other moist habitats in South Africa, the southeastern United States (particularly Georgia and Florida), and certain localities in California. In South Africa, where red sesbania is especially noxious, three weevil species (*Trichapion lativentre*, *Rhyssomatus marginatus*, and *Neodiplogrammus quadrivittatus*) that feed specifically on red sesbania were released as biocontrol agents in the 1980s. Recent studies indicate that red sesbania populations there have been significantly reduced where two or three of the weevil species are well established. Foliage, flowers, and especially immature seeds contain sesbanimides and saponins

Red sesbania (*Sesbania punicea*) along bank of American River. J. M. DiTomaso

Red sesbania (*Sesbania punicea*) plants in flower. J. M. DiTomaso

Red sesbania (*Sesbania punicea*) inflorescence. J. M. DiTomaso

Red sesbania (*Sesbania punicea*) fruit pods. J. M. DiTomaso

Red sesbania

Red sesbania (*Sesbania punicea*) seedling. J. M. DiTomaso

Red sesbania (*Sesbania punicea*) seeds. J. O'Brien

and are *toxic* to humans and animals when ingested. The primary toxicity symptom is severe digestive tract disturbance. A dose of less than 0.1% of body weight in seeds ingested over a period of days can be lethal to livestock. Native to Argentina, Brazil, Paraguay, and Uruguay.

SEEDLINGS: Cotyledons oblong, sessile, about 1.5–2.5 cm long, ± 1 cm wide, stalk below (hypocotyl) often reddish. First leaf simple, elliptic to oval, tip tapered to rounded, with an abrupt point (mucro), larger than the cotyledons, short-stalked. Subsequent 2 leaves pinnate-compound with 6 leaflets, slightly larger than first leaf. Leaflets oblong to obovate, usually smaller than cotyledons, tip rounded to slightly truncate and abruptly pointed.

MATURE PLANT: Foliage sparsely hairy to glabrous. Leaves alternate, *pinnate-compound with an even number of leaflets, 8–20 cm long*, typically drooping. Leaflets mostly 10–40 per leaf, ± oblong, 1–3 cm long, nearly sessile, tip rounded to slightly truncate and with a minute, abrupt point (mucro).

ROOTS AND UNDERGROUND STRUCTURES: Main roots woody. Fine roots associate with nitrogen-fixing bacteria.

FLOWERS: Mostly June–August in California. *Flowers showy, pea-like, red to orange-red, about 2–3 cm long*, in ± drooping axillary racemes about 8–25 cm long. Calyx cuplike, ± weakly lobed, usually tinged dull dark red.

FRUITS AND SEEDS: Pods brown to dark brown, glabrous, *oblong* with a pointed tip, about 6–12 cm long, *1.5–2.5 cm wide*, with *4 lengthwise wings mostly 5–10 mm wide*. Seeds usually 4–10 per pod, separated by partitions, oblong, mostly 5–9 mm long, smooth, dull brown to tan.

HABITAT: Riparian areas, disturbed moist places, and margins of ponds, ditches, and canals.

DISTRIBUTION: Uncommon. Southern Sacramento Valley (American River Parkway), San Joaquin Valley (Suisun Marsh, San Joaquin River Parkway), southern North Coast Ranges, and possibly elsewhere, to about 50 m. Washington and southern states, from Texas to Virginia.

Red sesbania

PROPAGATION/PHENOLOGY: *Reproduces by seed.* Pods open slowly and do not eject seeds. Seeds appear to disperse primarily with water. Seeds are hard-coated and require scarification or decomposition of the seed coat before they can germinate. After water is imbibed, germination can occur at any time when the temperature is 10–35°C. Seedlings emerge from soil depths to about 12 cm. Plants typically attain reproductive maturity at 2–3 years of age. Individual trees can survive for up to ± 15 years.

SIMILAR SPECIES: Hemp sesbania [*Sesbania exaltata* (Raf.) Cory, synonyms: bigpod sesbania; coffee weed; Colorado River hemp; *Sesbania herbacea* (P. Mill.) McVaugh; *Sesbania macrocarpa* Muhl.][SEBEX] is an erect, native summer *annual* to about 3 m tall, with leaves that resemble those of red sesbania. Although sometimes weedy in crop fields and associated irrigation ditches, hemp sesbania is generally not considered a weed in natural riparian areas throughout its native range in the southwestern to southeastern United States. Unlike red sesbania, hemp sesbania has *yellow to pale orange, pea-like flowers mostly 1–1.5 cm*

Hemp sesbania *(Sesbania exaltata)* leaf and flower.
J. M. DiTomaso

Hemp sesbania *(Sesbania exaltata)* seedling.
J. K. Clark

Hemp sesbania *(Sesbania exaltata)* seeds.
J. O'Brien

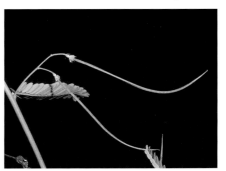

Hemp sesbania *(Sesbania exaltata)* fruit pod.
J. M. DiTomaso

Sicklepod (*Senna obtusifolia*) flowering and fruiting stem. J. M. DiTomaso

Sicklepod (*Senna obtusifolia*) flower. J. M. DiTomaso

Sicklepod (*Senna obtusifolia*) immature plant. C. Elmore

Sicklepod (*Senna obtusifolia*) seeds. J. O'Brien

long that are often maroon speckled on the upper petal (banner) and *linear pods mostly 15–20 cm long and 0.3–0.4 cm wide*. Flowers typically open in the evening. Seeds are cylindrical, rounded to slightly truncate at the ends, 3–5 mm long, glossy dark brown to tan with dark spots. Seedlings have glabrous, oblong cotyledons about 10–25 mm long and 4–10 mm wide. First leaves are pinnate-compound, with ± 15 lanceolate to oblong leaflets mostly 5–10 mm long. Leaflet tips have a small abrupt point (mucro). Hemp sesbania is sometimes cultivated for green manure or as a cover crop. In California, hemp sesbania inhabits riparian areas and is sometimes weedy along irrigation ditches and in moist agronomic fields in the eastern Sonoran Desert, primarily in the Imperial Valley, to 500 m. It may occasionally occur as an agricultural weed elsewhere in the state. Hemp sesbania also occurs in Arizona, a few northeastern states, southern states, and adjacent central states. It is a common weed of soybeans, rice, and cotton in the southern states. Like red sesbania, the foliage, flowers, and seeds of hemp sesbania are *toxic* to animals and humans when ingested, but to a lesser degree compared to red sesbania.

Sicklepod [*Senna obtusifolia* (L.) H. S. Irwin & Barneby, synonyms: java bean; arsenic weed; coffee-pod; *Cassia obtusifolia* L.] [CASOB] is a semiwoody *annual* or sometimes short-lived perennial to about 2.5 m tall, with *evenly pinnate-*

compound leaves and *slightly asymmetrical yellow flowers*. Sicklepod is a recent introduction to California and is not included in California floras to date. It is distinguished by having even-pinnate-compound leaves with 2–3 pairs of obovate leaflets 2–3 cm long and 1.5–2 cm wide, *a single elongated gland on the main leaf stalk (rachis) between the lowest pair of leaves*, flowers mostly 1–1.5 cm in diameter with *5 spreading unequal petals* that are narrowed at the base (clawed), and *4-sided, linear pods that are curved*, especially while immature. In addition, sepals are separate. Sicklepod inhabits agronomic fields, particularly cotton fields, and nearby roadsides and other areas in the Imperial Valley of the Sonoran Desert. A single plant was collected from a roadside in Fresno County, San Joaquin Valley. Sicklepod also occurs in most eastern and southern states, where it is a weed of crop fields, and it is a noxious weed in Australia. Seeds can remain viable for several years under field conditions, and some survive ingestion by livestock. Germination occurs whenever conditions are favorable. Seeds and, to a lesser degree, the foliage contain several *toxic* anthraquinones that affect numerous tissues. Ingestion of a large amount of plant material can be lethal to livestock within a few days. Ingestion of smaller amounts of plant material typically results in digestive tract disturbance. Later symptoms of more severe poisoning can include muscle weakness, inability to stand, cardiac insufficiency, and dark urine. Native to tropical and some warm-temperate regions of America, including Mexico and the southern United States.

Swainsonpea [*Sphaerophysa salsula* (Pall.) DC.] [SWASA][CDFA list: A]

SYNONYMS: alkali swainsonpea; Austrian peaweed; swainsona; *Phaca salsula* Pallas; *Swainsona salsula* (Pallas) Taubert

GENERAL INFORMATION: Herbaceous *perennial* to 1.5 m tall, with *pinnate-compound leaves, reddish flowers,* and *vigorously creeping roots.* Plants are often long-lived. Although uncommon and perhaps even eradicated in California, swainsonpea is potentially noxious wherever alfalfa [*Medicago sativa* L.] is grown for seed. A close resemblance of the two seeds makes it difficult to separate in seed production systems. Native to Asia.

SEEDLINGS: Less vigorous than alfalfa seedlings.

MATURE PLANT: Stems erect to ascending, covered with short white hairs. *Leaves odd-pinnately-compound.* Leaflets 15–23, oblong to ovate, 0.5–2 cm long, upper surface mostly glabrous, lower surface covered with short, white hairs. Stipules lacking or awl-like and are early-deciduous, to 5 mm long, fused at the base.

ROOTS AND UNDERGROUND STRUCTURES: Plants often develop an extensive system of vigorous, woody, creeping, horizontal roots that develop new shoots. Roots associate with nitrogen-fixing bacteria.

FLOWERS: May–July. Racemes axillary, near the stem tips. Flowers pea-like, 12–14 mm long, with *brick-red to orange-red petals* and bractlets at the base of the calyx, on a stalk 3–7 mm long. Style 1, with a *finely hairy stigma.* Stamens 10, 9 with fused filaments and 1 separate (diadelphous).

Swainsonpea *(Sphaerophysa salsula)* flowers. W. J. Ferlatte

Swainsonpea (*Sphaerophysa salsula*) foliage.
W. J. Ferlatte

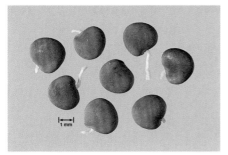

Swainsonpea (*Sphaerophysa salsula*) seeds.
J. O'Brien

Swainsonpea (*Sphaerophysa salsula*) plants.
W. J. Ferlatte

FRUITS AND SEEDS: *Pods inflated at maturity, ovoid to spherical, grooved on top, 1.5–3.5 cm long, with a stalklike base 4–7 mm long that protrudes beyond the persistent calyx tube, glabrous, membranous to papery, translucent, often mottled. Pods indehiscent.* Seeds several to numerous per pod, flattened, semicircular to triangular with a slight notch, ± 2 mm long, surface smooth, dull dark brown to black.

HABITAT: Disturbed sites, roadsides, irrigation ditches, and cultivated crops.

DISTRIBUTION: Considered eradicated in California. Infestations previously found in southern San Joaquin Valley (ce Kings and c Kern Cos.) and perhaps in northeastern California, to 500 m. All western states, Kansas, and Texas.

PROPAGATION/PHENOLOGY: *Reproduces by seed* and spreads *vegetatively from creeping roots*. Entire pods disperse as units with seeds enclosed. A large proportion of the seed is hard-coated and requires scarification to germinate. Under favorable conditions root fragments can regenerate into a new plant.

MANAGEMENT FAVORING/DISCOURAGING SURVIVAL: Manually removing individual plants, including as much of the root system as possible, followed by frequent removal of root sprouts and seedlings can eventually control small populations of swainsonpea.

SIMILAR SPECIES: Other weedy perennial members of the pea family that can superficially resemble swainsonpea include **camelthorn** [*Alhagi maurorum* Medik., synonym: *Alhagi pseudalhagi* (M.Bieb.) Desv.] and **wild licorice** [*Glycyrrhiza lepidota* Pursh.]. Unlike swainsonpea, camelthorn has *axillary thorns, simple leaves,* and *slender pods that are constricted between the seeds and not inflated*. Wild licorice is distinguished by having pinnate-compound leaves *dotted with sessile glands,* dense racemes of *dingy-white flowers,* and *bur-like fruits covered with hooked prickles*. Refer to the entries for **Camelthorn** (p. 744) and **Wild licorice** (p. 763) for more information.

In addition, many native **locos** and **milkvetch** species (*Astragalus* spp.) have leaves and inflated pods that somewhat resemble those of swainsonpea. However, locos and milkvetches *do not* develop a vigorous creeping root system that produces new shoots and most *do not have red flowers*, with the exception of scarlet milkvetch [*Astragalus coccineus* Brandegee]. In addition, all locos and milkvetches have a *glabrous style and stigma*. Scarlet milkvetch is a desirable native species that occurs only on the western edge of the deserts and east of the Sierra Nevada in sagebrush communities or pinyon woodlands. Refer to the entry for **Loco and Milkvetch** (p. 748) for more information.

Small hop clover [*Trifolium dubium* Sibth.][TRFDU]
Strawberry clover [*Trifolium fragiferum* L.][TRFFR]
Rose clover [*Trifolium hirtum* All.][Cal-IPC: Moderate]
Crimson clover [*Trifolium incarnatum* L.][TRFIN]
White clover [*Trifolium repens* L.][TRFRE]

SYNONYMS: Small hop clover: lesser trefoil; little hop clover; shamrock; suckling clover; yellow trefoil; *Chrysaspis dubia* Greene; *Trifolium minus* Relhan

Strawberry clover: *Amoria fragifera* (L.) Yu. R. Roskov; *Trifolium fragiferum* L. ssp. *bonannii* (K. Presl) Soják; *Trifolium neglectum* Fisch. & Mey.

Rose clover: hairy clover; *Trifolium hispidum* Desf.; *Trifolium oxypetasum* Heldr; *Trifolium pictum* Roth.

Crimson clover: French clover; Italian clover; *Trifolium incarnatum* L. var. *elatius* Gibelli & Belli

White clover: Dutch clover; ladina clover; wild white clover; *Amoria repens* (L.) Kuntze

GENERAL INFORMATION: Cool-season clovers with *dense, round to spikelike flower heads* and compound leaves of *3 leaflets*. Refer to table 35 (p. 818) for a comparison of *Trifolium* species. The following species are some of the frequently encountered, non-native clovers that have escaped cultivation in California.

Small hop clover (*Trifolium dubium*) plants. J. M. DiTomaso

Small hop clover • Strawberry clover • Rose clover • Crimson clover • White clover

Small hop clover (*Trifolium dubium*) flowering stem. J. M. DiTomaso

Small hop clover (*Trifolium dubium*) seeds. J. O'Brien

There are many commercial cultivars of important forage clovers.

Small hop clover: *Annual* with spreading to erect stems to about 0.5 m tall and small heads of *bright yellow flowers*. Small hop clover is sometimes a component of turf seed mixtures. Native to Europe.

Strawberry clover: Low *perennial* with *fuzzy heads* of *pink flowers*. Plants are tufted or have *creeping stems* to 0.3 m long that root at the nodes. Strawberry

Strawberry clover (*Trifolium fragiferum*) patch in turf. J. M. DiTomaso

Strawberry clover (*Trifolium fragiferum*) flowers and leaves. J. M. DiTomaso

Strawberry clover (*Trifolium fragiferum*) fruiting cluster. J. M. DiTomaso

Strawberry clover (*Trifolium fragiferum*) seeds. J. O'Brien

Small hop clover • Strawberry clover • Rose clover • Crimson clover • White clover

Rose clover (*Trifolium hirtum*) flowering stems.   J. M. DiTomaso

Rose clover (*Trifolium hirtum*) population.
J. M. DiTomaso

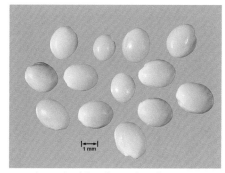

Rose clover (*Trifolium hirtum*) seeds.   J. O'Brien

clover is cultivated as pasture forage, especially on wet soil. Native to Europe, western Asia, and northern Africa.

Rose clover: *Annual* with ascending to erect stems to about 0.4 m tall and *fuzzy heads* of *pink flowers*. Rose clover readily naturalizes and is consumed as rangeland forage. Native to Eurasia and northern Africa.

Crimson clover: *Annual* with erect stems to about 1 m tall and ovoid-oblong to cylindrical heads of *crimson flowers*. Crimson clover is cultivated as a cover crop and for forage and hay. It sometimes escapes cultivation but rarely persists. Native to southern Europe.

White clover: Low *perennial* with *creeping stems* to 0.3 m long that *root at the nodes* and ± round heads of *white to pale pink flowers*. White clover is an important forage crop and a common turf weed. White clover is variably estrogenic and can cause fertility problems in livestock that are maintained on extensive

Small hop clover • Strawberry clover • Rose clover • Crimson clover • White clover

Crimson clover (*Trifolium incarnatum*) plant.
J. M. DiTomaso

Crimson clover (*Trifolium incarnatum*) seeds.
J. O'Brien

Crimson clover (*Trifolium incarnatum*) flowering stem.
J. M. DiTomaso

quantities of clover pasture, hay, or silage. Sheep are most often affected. Some varieties are weakly cyanogenic under certain conditions, but toxicity problems in livestock are extremely rare. White clover is also occasionally associated with photosensitization in horses and other species of young animals. Native to Eurasia.

SEEDLINGS: Cotyledons oval-oblong, glabrous, stalked. Leaves alternate. First leaf simple. Subsequent leaves resemble mature leaves except smaller.

Small hop clover: Cotyledons about 2–5 mm long. First leaf nearly round with a slightly indented tip, about 2–6 mm long, glabrous or nearly glabrous.

Strawberry clover: Cotyledons about 2–5 mm long. First leaf ovate, tip usually rounded, base truncate, 4–7 mm long, nearly glabrous to hairy.

Crimson clover: Cotyledons 6–12 mm long. First leaf ovate, broader than long, tip rounded to slightly truncate, base ± slightly heart-shaped, 9–16 mm long, hairy.

White clover: Cotyledons 2–4 mm long. First leaf nearly round to ovate, often broader than long, tip ± rounded, base truncate, 2–5 mm long, glabrous or nearly glabrous.

MATURE PLANT: Stems ± hairy. Stem leaves alternate.

Small hop clover: Basal tuft of leaves lacking. Stems spreading to erect. Leaves at the *middle* of stems are usually *longer than the stalks* (petioles). Leaflets *nearly* *glabrous*, obovate, tip often indented, mostly 0.6–1.2 cm long, *central leaflet on a short stalk*. Stipules ovate.

Strawberry clover: Tufted or with creeping stems that root at the nodes. Leaflets glabrous to sparsely hairy, obovate to elliptic, tip often rounded, mostly 0.5–2 cm long, central leaflet sessile. Stipules ovate to lanceolate, often overlapping.

Rose clover: Basal tuft of leaves lacking. Stems ascending to erect. Leaflets hairy, obovate, tip slightly indented, truncate, or rounded, mostly 1–2.5 cm long, central leaflet sessile. Stipules narrow with a long bristlelike tip.

Crimson clover: Basal tuft of leaves lacking. Stems erect. Leaflets hairy, obovate, tip rounded or indented, mostly 1–2.5 cm long, central leaflet sessile. Stipules ovate to oblong, with conspicuous veins.

White clover: Stems prostrate, *creeping, rooting at the nodes*. Leaves can appear to form a basal tuft when stem growth is slow. Leaflets *nearly glabrous*, obovate, tip indented or rounded, mostly 0.5–2.5 cm long, often marked with a white crescent near the center, central leaflet sessile. Stipules membranous, whitish.

ROOTS AND UNDERGROUND STRUCTURES: Roots associate with nitrogen-fixing bacteria.

Small hop clover, rose clover, crimson clover: Taproots slender, branched, with fibrous lateral roots.

Strawberry clover, white clover: Plants from seed initially have a slender

White clover (*Trifolium repens*) plant.
J. M. DiTomaso

White clover (*Trifolium repens*) flower and fruit head.
J. M. DiTomaso

White clover (*Trifolium repens*) population.
J. M. DiTomaso

White clover (*Trifolium repens*) leaf.
J. M. DiTomaso

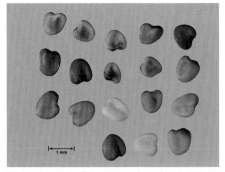

White clover (*Trifolium repens*) seeds.   J. O'Brien

branched taproot with lateral roots. Later the root system is entirely fibrous. Adventitious roots from the nodes are fibrous.

FLOWERS: These species lack bracts at the base of the flower heads. Insect-pollinated.

Small hop clover: April–July. Heads spherical to oblong, 5–10 mm long, 4–8 mm wide. Flowers 5–20 per head, initially bright yellow with *faint ± translucent striations*, aging to brown, 2.5–4 mm long, some or all *reflex downward* on stalks 1–4 mm long at maturity. Calyx 1.5–2 mm long, glabrous to sparsely hairy.

Strawberry clover: May–November. Heads spherical to ovoid, 5–15 mm

Narrowleaved crimson clover (*Trifolium angustifolium*) seeds and calyx.   J. O'Brien

Immature inflorescences of rabbitfoot clover (*Trifolium arvense*).   J. M. DiTomaso

Rabbitfoot clover (*Trifolium arvense*) flowering stem.   R. Uva

Small hop clover • Strawberry clover • Rose clover • Crimson clover • White clover

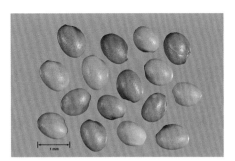

Rabbitfoot clover (*Trifolium arvense*) seeds.
J. O'Brien

Alsike clover (*Trifolium hybridum*) seeds.
J. O'Brien

Narrowleaved crimson clover (*Trifolium angustifolium*) flowering stem.
J. M. DiTomaso

long, ± 20 mm wide, sometimes with tiny slender bracts at the base, *appear fuzzy. Flowers pink,* 5–6 mm long, initially erect to spreading, mostly reflexed downward in fruit. Calyx hairy, mostly 4–6 mm long, *inflated and conspicuously veiny in fruit.* Calyx lobes linear, lowest 2 bristlelike.

Rose clover: April–May. Heads spherical to slightly ovoid, 10–25 mm long and wide, sessile, *appear fuzzy.* Flowers pink to violet-pink, 11–15 mm long, erect to spreading, numerous. Calyx densely hairy, 7–9 mm long, *not* inflated in fruit. Calyx lobes bristlelike, often dark, lowest one equal to or slightly longer than the others.

Crimson clover: April–August. *Heads oblong-ovoid to cylindrical,* 20–60 mm long, 10–15 mm wide. *Flowers crimson,* 10–14 mm long, erect to spreading, numerous. Calyx long-hairy, 7–10 mm long, strongly 10-veined, lobes bristlelike.

White clover: March–December. Heads spherical to slightly ovoid, 1–2.5 cm long and wide, with 1–2 leaves just below. *Flowers white to pale pink,* 7–11 mm long, some or all *reflex downward on stalks 2–6 mm long at maturity.* Calyx glabrous, 3–6 mm long. Sometimes self-incompatible.

FRUITS AND SEEDS: Pods round to oblong, remain within the calyx, usually do not open to release seeds, but sometimes break apart.

Small hop clover: Fruiting heads small, mostly glabrous, tan. Seeds oval to nearly round, slightly compressed with rounded sides, flat to barely indented at

Small hop clover • Strawberry clover • Rose clover • Crimson clover • White clover

Large hop clover (*Trifolium campestre*) flowering and fruiting stem. J. M. DiTomaso

Large hop clover (*Trifolium campestre*) plants. J. M. DiTomaso

Large hop clover (*Trifolium campestre*) seeds. J. O'Brien

the point of attachment, usually 0.8–1.5 mm long, nearly as wide, surface smooth, glossy, yellow to purplish.

Strawberry clover: Fruiting heads spherical, sparsely hairy, tan to reddish brown. *Calyx inflated, conspicuously veiny, reflex downward.* Seeds broadly oval to ± heart-shaped, with a notch at the point of attachment, slightly compressed, 1.5–2 mm long, 1–1.5 mm wide, surface smooth, ± dull, yellow to reddish, black-dotted.

Rose clover: Fruiting heads densely fuzzy-hairy, ± reddish brown. Calyx *not* inflated or reflexed downward. Seeds nearly spherical, with a shallow notch at the point of attachment, slightly compressed, 2–3 mm long and wide, surface smooth except a few slight wrinkles at the end near the notch, ± dull, pale yellow.

Crimson clover: Fruiting heads hairy, oblong to cylindrical. Seeds oval, with a very shallow indentation at the point of attachment, slightly compressed, about 1.5–3 mm long, 1–2 mm wide, surface smooth, highly glossy, yellowish.

White clover: Fruiting heads brown, glabrous. Seeds irregularly triangular-ovate with a shallow indentation at the point of attachment, slightly compressed, ± 1 mm long and wide, surface smooth, ± dull, yellow to orange-brown.

POSTSENESCENCE CHARACTERISTICS: Dead foliage generally decomposes rapidly, but some remnants of the dried flower heads can remain attached to stems for a few months after senescence.

Small hop clover • Strawberry clover • Rose clover • Crimson clover • White clover

Clustered clover (*Trifolium glomeratum*) plant.
J. M. DiTomaso

Clustered clover (*Trifolium glomeratum*) seeds.
J. O'Brien

HABITAT: Small hop clover: Turf, fields, grassland, open slopes, roadsides, waste places, agricultural lands, vernal pool margins, and other disturbed places.

Strawberry clover: Turf, fields, roadsides, saline sites, and other disturbed, usually moist places. Requires summer moisture.

Rose clover: Roadsides, fields, grassland, open slopes, and other disturbed places.

Crimson clover: Roadsides, fields, agricultural lands, and other disturbed places.

White clover: Turf, urban areas, agricultural lands, riparian areas, flood plains, forest clearings, mountain meadows, and other disturbed, usually moist places.

DISTRIBUTION: Small hop clover: Throughout California, except possibly deserts, to 500 m. Most common in the northern half of the state and coastal regions. Arizona, Idaho, Montana, Oregon, Washington, Wyoming, and most central, eastern, and southern states.

Strawberry clover: North Coast Ranges, Sacramento Valley, Modoc Plateau, South Coast Ranges, Southwestern region, to 1500 m. Most western states, except possibly Arizona, and a few central, eastern, and southern states.

Rose clover: Throughout California, except deserts and Great Basin, to 2060 m. Oregon and some southeastern states.

Crimson clover: Throughout California, except deserts and Great Basin, to 300 m. Oregon, Washington, Idaho, and most states in the eastern half of the United States.

White clover: Throughout California, except deserts, mostly to 1500 m. All contiguous states.

PROPAGATION/PHENOLOGY: All *reproduce by seed*. Fruits fall near the parent plants and disperse to greater distances with turf maintenance, agricultural activities, mud, soil movement, as seed contaminants, and by clinging to animals, the shoes and clothing of humans, and vehicle tires. A variable proportion of seeds are hard-coated and impermeable to water. Hard-coated seeds require

Small hop clover • Strawberry clover • Rose clover • Crimson clover • White clover

scarification or decomposition of the seed coat to germinate. Depending on the region and climate, seeds germinate in fall, winter, or early spring. Rose clover develops a high proportion of hard-coated seeds, which facilitates its persistence on dry slopes and fields. In contrast, crimson clover generally produces a low proportion of hard-coated seeds and usually does not persist for more than a couple of years outside of cultivation. Hard-coated seeds can generally survive for several years under field conditions. Buried white clover seeds have been documented to survive for up to 30 years.

Strawberry clover, white clover: Under favorable conditions, these species also *reproduce vegetatively from creeping stems that root at the nodes*. In turf, vegetative reproduction is often more important than seed reproduction.

MANAGEMENT FAVORING/DISCOURAGING SURVIVAL: Grazing generally favors the survival of all species, except crimson clover. Cultivation before seeds are produced can control annual clovers on agricultural lands. In turf and gardens, regular manual removal before seed develops can control clover. Improving turf grass health and nitrogen fertilization can help to prevent invasion and limit the spread of white clover and strawberry clover.

SIMILAR SPECIES: Refer to table 35 (p. 818) for a comparison of important distinguishing clover characteristics.

Narrowleaved crimson clover [*Trifolium angustifolium* L., synonym: narrowleaf crimson clover] is an erect *annual* to 0.4 m tall, with hairy foliage, ovoid to

Red clover (*Trifolium pratense*) flowering stem.
J. M. DiTomaso

Red clover (*Trifolium pratense*) plant.
J. M. DiTomaso

Red clover (*Trifolium pratense*) seeds.
J. O'Brien

Subterranean clover (*Trifolium subterraneum*) plant. J. M. DiTomaso

oblong heads of *pink flowers*, and *linear-elliptic to narrow lanceolate leaflets 2–5 cm long*. It is further distinguished by having flowers with a *corolla 10–12 mm long that is about equal to the calyx*. Narrowleaf crimson clover is less common than the clovers of this entry. It inhabits disturbed places along the Central Coast and in the South Coast Ranges, to 200 m. It also occurs in Oregon, Alabama, and South Carolina. Native to the Mediterranean region.

Rabbitfoot clover [*Trifolium arvense* L.] [TRFAR] is an erect *annual* to 0.4 m tall, with hairy foliage, ovoid to oblong flower heads of *pale pink to white flowers*, and *narrowly oblong to oblanceolate leaflets 0.5–2 cm long*. Unlike narrowleaf crimson clover, rabbitfoot clover has flowers with a *corolla 3–4 mm long that is shorter than the calyx*. Rabbitfoot clover occurs infrequently on disturbed sites throughout California, except deserts and Great Basin, to 300 m. It also occurs in Idaho, Montana, Oregon, Washington and most states of the eastern half of the United States. Native to Europe.

Alsike clover [*Trifolium hybridum* L., synonym: *Trifolium elegans* Savi] is a ± spreading *annual to perennial* with glabrous foliage, thick, succulent, leafy stems to about 0.6 m long, and broadly spherical heads of pink to white flowers. Alsike clover is distinguished by having *glabrous flower heads 1.5–3 cm wide* that are usually broader than long and *flowers 6–11 mm long that reflex downward on stalks 2–6 mm long*. Each flower has a *small, linear, deciduous bract to 1 mm long* just below the stalk (easiest to detect with immature flowers). Alsike clover grows best in a cool, moist climate. It is associated with 2 types of photosensitization problems in horses grazing extensively on alsike clover pasture. However, the actual causes of these conditions are uncertain. Outside of cultivation, alsike clover grows on disturbed sites primarily in the Northwestern region and occurs sporadically elsewhere, to 1500 m. Alsike clover occurs in most contiguous states. Native to Europe.

Subterranean clover (*Trifolium subterraneum*) flowering stem.   J. M. DiTomaso

Large hop clover [*Trifolium campestre* Schreb., synonyms: field clover; low hop clover; *Trifolium procumbens* L. 1755, non 1753][TRFCA] is a yellow-flowered winter *annual* that resembles small hop clover. Unlike small hop clover, large hop clover has petals that are *conspicuously striated with ± translucent veins*, flower heads that are mostly *8–13 mm wide*, often with *more than 20 flowers per head*, and *lower and middle leaves with a stalk that is longer than the leaflets*. Large hop clover inhabits roadsides, turf, moist sites, and other disturbed places throughout California, except deserts and Great Basin, to 300 m. It is most common in the northern part of the state and also occurs in Idaho, Montana, New Mexico, Oregon, Washington, and throughout most of the eastern half of the United States. Native to Europe.

Clustered clover [*Trifolium glomeratum* L.] is an uncommon, ± spreading *annual* to 0.4 m tall, with *nearly glabrous foliage* and *spherical heads of pink flowers in a series of leaf axils*. Flower heads are *7–10 mm in diameter* and *sessile*. Flowers are 5–6 mm long. In addition, the *triangular calyx lobes reflex backward in fruit*. Clustered clover inhabits roadsides, fields, and other disturbed places throughout California, except deserts and Great Basin, to 300 m. It also occurs in a few southeastern states. Native to Europe.

Red clover [*Trifolium pratense* L., synonym: purple clover][TRFPR] is a ± tufted *biennial to short-lived perennial* to 0.6 m tall, with pubescent foliage and hairy, spherical to ovoid heads of *pale red-purple to purplish pink flowers*. It most closely resembles rose clover. Red clover is distinguished primarily by having flowers with a *sparsely hairy calyx that has the lowest lobe about twice as long as the other lobes*. In addition, leaflets are 1.5–3.5 cm long, generally dark green, and sometimes marked with a pale inverted "V" or triangular mark. Red clover is an important forage crop. Under cool, moist conditions, it is particularly susceptible to black patch disease, a fungus that can cause a condition known as slobbers in livestock that ingest large amounts of infected foliage or hay over a period of at least several days. Slobbers is characterized by excessive salivation and digestive tract disturbance. Symptoms disappear with a change of feed. Black patch disease appears as bronze to black spots or rings on red clover foliage. Red clover inhabits disturbed places throughout California, except deserts, mostly to 1000 m. It is most common in the northern region of the state and grows best in areas that receive regular moisture. It has escaped cultivation in all contiguous states. Native to Europe.

Subterranean clover [*Trifolium subterraneum* L., synonym: sub clover] is a *prostrate annual* with *creeping stems that root at the nodes* and unusual *bur-like fruiting heads* on stalks (peduncles) that curve downward to implant the heads in the soil as they mature. Flower heads are composed of 2–8 fertile outer flowers with yellow and purple petals or sometimes white petals and several sterile inner flowers that are bristly and lack petals. The sterile flowers are on stalks (pedicels) that curve outward and down as they elongate, such that the bristly sterile flowers enclose the fertile flowers in a bur. Subterranean clover is cultivated for pasture forage but is becoming more invasive in natural areas. Like white clover, subterranean clover is variably weakly estrogenic and can cause fertility problems in animals that ingest a large amount of foliage. Subterranean clover inhabits roadsides, grasslands, fields, urban areas, waste places, and other disturbed sites in the North Coast Ranges, northern Sierra Nevada, Central Valley, South Coast Ranges, and probably elsewhere, to 1000 m. It also occurs in Oregon, Washington, New Jersey, Massachusetts, and many southern states. Native to southern Europe and northern Africa.

A number of other non-native clovers have escaped cultivation in California. These are less frequently encountered than the species described here. See the appendix, **Non-native Naturalized Plants** (p. 1601, vol. 2), for more information. There are also at least 30 native species of *Trifolium* found in California.

Subterranean clover (*Trifolium subterraneum*) seeds.   J. O'BRIEN

Small hop clover • Strawberry clover • Rose clover • Crimson clover • White clover

Table 35. Clovers (*Trifolium* spp.)

Trifolium spp.	Life cycle	Petal color	Flower head shape	Flower head length × width (mm)	Calyx	Mature flowers	Other
T. angustifolium narrowleaved crimson clover	annual	pink	ovoid to cylindrical	10–50 × 10–20	densely hairy	erect to spreading	leaflets very narrow, 2–5 cm long; petals 10–12 mm long
T. arvense rabbitfoot clover	annual	pale pink to white	ovoid to oblong	10–30 × ± 10	densely hairy	erect to spreading	leaflets very narrow, 0.5–2 cm long; petals 3–4 mm long
T. campestre large hop clover	annual, biennial	yellow with readily visible translucent lines	spherical to ovoid	8–15 × 8–15	± glabrous	reflexed on down-curved stalks	central leaflet stalked
T. dubium small hop clover	annual	yellow with faint translucent lines	spherical to ovoid	3–8 × 4–8	± glabrous	reflexed on down-curved stalks	central leaflet stalked; common in turf
T. fragiferum strawberry clover	perennial, tufted or with creeping stems	pink	spherical to ovoid	5–15 × ± 20	hairy	erect to spreading, reflexed in fruit	calyx inflated, conspicuously veiny in fruit; heads sometimes with linear bracts at the base, common in turf
T. glomeratum cluster clover	annual	pink	spherical	5–10 × 5–10	glabrous	erect to spreading	calyx lobes triangular, curved backward in fruit; heads sessile in sequential leaf axils
T. hirtum rose clover	annual	pink to purple-pink	spherical to ovoid	10–25 × ± 15	densely hairy	erect to spreading	lowest calyx lobe equal to or slightly longer than others; leaflets often light green
T. hybridum alsike clover	annual, biennial, perennial	pink	spherical to flattened-spherical	10–20 × 15–30	glabrous	reflexed on down-curved stalks	deciduous bract 1 mm long below each flower stalk
T. incarnatum crimson clover	annual	crimson	ovoid-oblong to cylindrical	20–60 × 10–15	hairy	erect to spreading	—

Small hop clover • Strawberry clover • Rose clover • Crimson clover • White clover

Trifolium spp.	Life cycle	Petal color	Flower head shape	Flower head length × width (mm)	Calyx	Mature flowers	Other
T. pratense red clover	biennial, perennial	purplish pink to light red-purple	spherical to ovoid	15–30 × 20–30	sparsely hairy	erect to spreading	lowest calyx lobe about twice as long as others; leaflets often dark green, sometimes marked
T. repens white clover	perennial with creeping stems	white to pale pink	spherical	10–25 × 10–25	± glabrous	reflexed on down-curved stalks	central leaflet sessile; common in turf
T. subterraneum subterranean clover	annual with creeping stems	white, or yellow and purple (fertile flowers)	spherical to ovoid	± 10 × 10	sparsely hairy	erect to spreading, reflexed in fruit	fruiting heads bur-like, implanted in soil on down-curved stalks; fertile flowers 2–8; sterile flowers many, lack petals

Note: All species lack bracts at the base of flower heads except where noted in "other" column.

Gorse [*Ulex europaeus* L.][ULEEU][Cal-IPC: High] [CDFA list: B]

SYNONYMS: common gorse; furze; prickly broom; thorn broom; whin; *Ulex europaea* L. in some California floras

GENERAL INFORMATION: *Spiny, evergreen shrub* to 3.5(5) m tall, with *yellow, pea-like flowers*. Gorse often forms dense, impenetrable thickets that exclude desirable vegetation and increase fire risk. Mature plants contain about 2–4% flammable oils. Older shrubs develop a central mass of dead material that is very flammable. Gorse also produces abundant leaf litter that can acidify the upper soil layers. The gorse seed weevil (*Apion ulicis*) and spider mite (*Tetranychus lintearius*) are introduced biocontrol agents that have become established in California. The seed weevil reduces seed production but cannot kill established stands. Heavy mite infestations can kill branches; these infestations are apparent by the dense webbing that covers the foliage. As in California, gorse is a state-listed noxious weed in Oregon (class B) and Washington (class B). Native to Western Europe and introduced as an ornamental or hedge shrub.

SEEDLINGS: Cotyledons sessile, oblong, 5–7 mm long, base slightly tapered, tip rounded, leathery, glabrous. Stalk below cotyledons (hypocotyl) often weakly woody, glabrous. First leaves mostly simple, alternate, sessile, lanceolate, 2–5 mm long, tip acute, leathery, covered with stiff, unicellular hairs. Subsequent leaves compound usually with (2)3 leaflets, about 5–10 mm long, otherwise similar to first leaves. Juvenile plants typically lack lateral branches and have compound leaves with 3 leaflets.

Gorse (*Ulex europaeus*) infestation. J. M. DiTomaso

Gorse

Gorse *(Ulex europaeus)* shrub. J. M. DiTomaso

Gorse *(Ulex europaeus)* flowering stem.
J. M. DiTomaso

Gorse *(Ulex europaeus)* seedling. J. M. DiTomaso

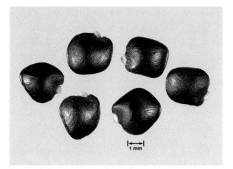
Gorse *(Ulex europaeus)* seeds. J. O'Brien

MATURE PLANT: *Stems highly branched, interwoven,* stiff, spreading, longitudinally ridged. New stems green, covered with hairs. *Branchlets terminate in a thorn.* Leaves alternate. Unlike juveniles, *mature shrubs have simple leaves modified into stiff, curved, awl-like spines,* 5–30 mm long. Stipules lacking.

ROOTS AND UNDERGROUND STRUCTURES: Taproot typically poorly developed. Lateral roots extensive, branched, mostly shallow, but a few penetrate the soil more deeply. Lower branches in contact with soil may develop adventitious roots. Roots associate with nitrogen-fixing bacteria.

FLOWERS: November–July. Flowers axillary, solitary or in few-flowered clusters. *Petals yellow,* 15–20 mm long, persistent. *Calyx yellow,* membranous, *deeply 2-lipped,* 10–15 mm long, upper lip 2-toothed, lower lip 3-toothed, covered with short hairs. Stamen filaments fused into a tube.

FRUITS AND SEEDS: *Pods ovoid to oblong, 1–2 cm long,* slightly flattened, partially enclosed by the calyx, dark brown, *covered with tawny, spreading, wavy hairs.* Seeds 2–6, shiny green to brown, smooth, ± triangular, flattened, ± 3 mm long, with a straw-colored appendage at the point of attachment.

HABITAT: Disturbed sites, sand dunes, coastal bluffs (especially where erosion is prevalent), fields, pastures, riparian corridors, logged areas, and burned sites, particularly in coastal areas where winters are mild and some moisture is available year-round. Plants tolerate shade. Frost-damaged plants can resprout from the crown. Does not survive severely cold winters or arid climates. Grows best on acidic (pH 4.5–5.0) soil. Tolerates many soil types, including serpentine, but seldom grows in high-calcium (calcareous) soils.

DISTRIBUTION: North Coast, western North Coast Ranges, northern and central Sierra Nevada foothills, San Francisco Bay region, and very rarely in the eastern South Coast Ranges, South Coast, and Peninsular Ranges, to 400 m. Oregon, Washington, Hawaii, and some northeastern states.

PROPAGATION/PHENOLOGY: *Reproduces by seed.* Plants produce abundant quantities of seed. Most seeds are ejected to within 5 m of the parent shrub when pods snap open at maturity. Some seeds may disperse to greater distances with human activities, water, soil movement, animals, and ants. Seeds have a hard coat and can survive for more than 30 years under field conditions. Large soil seedbanks often accumulate. Scarification or heating, which often occurs following fire, stimulates germination when moisture is available. Optimal temperature for germination is 15°–19°C. Germination can occur year-round when conditions are favorable. Shrubs may live for up to about 30 years.

MANAGEMENT FAVORING/DISCOURAGING SURVIVAL: Plants cut or burned to ground level may resprout from the crown or roots. Burning often stimulates a flush of seedling germination after the first rain. Browsing and trampling by goats can greatly reduce seedling establishment and crown regrowth. Successful management requires repeated removal and monitoring to completely purge the seedbank. Cut stump herbicide applications are effective in reducing resprouts. Management must typically be intensive and continue over many years to be successful.

SIMILAR SPECIES: Unlike gorse, the brooms (*Cytisus, Genista, Spartium,* and *Retama* spp.) *lack spiny foliage.*

Common vetch [*Vicia sativa* L. ssp. *sativa*][VICAN]
Narrowleaf vetch [*Vicia sativa* L. ssp. *nigra* (L.) Erhart] [VICSA]
Winter vetch [*Vicia villosa* Roth ssp. *varia* (Host) Corb.] [VICVV]
Hairy vetch [*Vicia villosa* Roth ssp. *villosa*][VICVI]

SYNONYMS: Common vetch: garden vetch; Oregon vetch; spring vetch; tare; *Vicia sativa* L. var. *linearis* Lange

Narrowleaf vetch: blackpod vetch; common vetch; garden vetch; smaller common vetch; spring vetch; *Vicia angustifolia* L.; *Vicia angustifolia* L. var. *segetalis* (Thuill.) W. D. J. Koch; *Vicia angustifolia* L. var. *uncinata* (Desv.) Rouy; *Vicia nigra* Steud.; *Vicia sativa* L. var. *angustifolia* (L.) Ser.; *Vicia sativa* L. var. *nigra* L.; *Vicia sativa* L. var. *segetalis* (Thuill.) W.D.J. Koch

Winter vetch: hairy vetch; lana vetch; sand vetch; smooth vetch; tare; woollypod vetch; *Vicia dasycarpa* Ten.; *Vicia villosa* Roth ssp. *dasycarpa* (Ten.) Cavillier; *Vicia villosa* Roth var. *glabrescens* W. D. J. Koch

Hairy vetch: sand vetch; tare; winter vetch

GENERAL INFORMATION: Cool-season legumes with viny stems to about 1 m long, *even-pinnate-compound leaves that terminate in a branched tendril*, and slender violet to white pea-like flowers. Plants climb over other vegetation and objects or

Narrowleaf vetch (*Vicia sativa* ssp. *nigra*) flowering stems. J. M. DiTomaso

form spreading tangled mats. Common vetch, winter vetch, and hairy vetch have widely escaped cultivation as cover crops, livestock forage, and green manure. Narrowleaf vetch is also cultivated to a limited extent, primarily in the southern United States, but occurs more commonly as a weed. Vetch seeds are *toxic* to livestock when ingested in quantity and can cause a neurological disorder with symptoms similar to those of rabies in cattle. Poultry food contaminated with the seeds of either species can retard the growth of chickens and cause death. Also, vetch pasture is occasionally associated with a type of dermatitis that mostly affects black cattle breeds and horses, but the actual cause of the disorder is unknown. Introduced from Europe.

Common vetch, narrowleaf vetch: Glabrous to short-hairy *annual* with nearly sessile clusters of 1–4 flowers. Hybridization between common vetch and narrowleaf vetch is possible but appears to be uncommon.

Narrowleaf vetch (*Vicia sativa* ssp. *nigra*) seedling. J. M. DiTomaso

Narrowleaf vetch (*Vicia sativa* ssp. *nigra*) fruit. J. M. DiTomaso

Common vetch (*Vicia sativa* ssp. *sativa*) flowering stem. J. M. DiTomaso

Narrowleaf vetch (*Vicia sativa* ssp. *nigra*) seeds. J. O'Brien

Common vetch • Narrowleaf vetch • Winter vetch • Hairy vetch

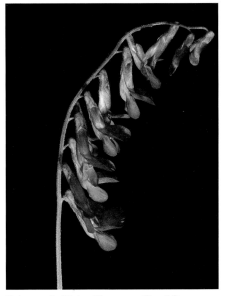

Hairy vetch (*Vicia villosa* ssp. *villosa*) flowering stem.　　　　　　　　　　J. M. DiTomaso

Hairy vetch (*Vicia villosa* ssp. *villosa*) flower cluster.　　　　　　　　　　J. M. DiTomaso

FLOWERS: Style round in cross-section, tip hairy all around or mostly on convex side.

Common vetch, narrowleaf vetch: April–July. Flower clusters *nearly sessile* in leaf axils, consist of *1–4 flowers*. Upper petal (banner) glabrous. Common vetch: Flowers usually violet, 1.8–3 cm long. Calyx 1–1.5 cm long. Narrowleaf vetch: Flowers violet to white, 1–2 cm long. Calyx 0.7–1.2 cm long.

Winter vetch, hairy vetch: March–July. Racemes stalked, *one-sided*, typically *longer than the leaf just below*, usually comprised of *10–40 ± closely spaced blue-violet, lavender, or rarely white flowers 1–2 cm long*. Pressed flower length usually 4–6 times the width. Calyx 0.5–0.6 cm long, base *obliquely attached, swollen and sac-like on one side*. Hairy vetch: Often more than 19 flowers per raceme. Lowest calyx lobe linear, 2–4 mm long. Winter vetch: Usually 10–20 flowers per raceme. Lowest calyx lobe narrowly lanceolate, 1–2.5 mm long.

FRUITS AND SEEDS: Pods ± oblong, open to release few to several dark seeds. Pod halves typically form a spiral upon opening.

Common vetch, narrowleaf vetch: Pods 2.5–7 cm long, 0.3–0.8 cm wide, nearly round to slightly flattened in cross-section, initially inconspicuously covered with short appressed hairs, but usually glabrous at maturity, lack a stalklike base above the calyx, contain 4–12 smooth seeds. Mature common vetch pods tan to black, seeds ± lens-shaped, mostly 5–8 mm in diameter, greenish gray to maroon or black. Mature narrowleaf vetch pods black, seeds nearly round, 2.5–5 mm in diameter, dark.

Winter vetch, hairy vetch: Pods 1.5–4 cm long, 0.7–1 cm wide, flattened, glabrous (hairy vetch sometimes sparsely pubescent), tan, stalklike base above the calyx 2–3 mm long, contain about 4–6 nearly round dark seeds mostly 4–5 mm in diameter.

POSTSENESCENCE CHARACTERISTICS: Leaves generally decompose rapidly, but dead stems with pod remnants can persist for months.

HABITAT: Roadsides, fields, pastures, meadows, crop fields and margins, orchards, vineyards, fence and hedge rows, and waste places. Narrowleaf vetch also inhabits woodland borders and open areas.

DISTRIBUTION: Common vetch, narrowleaf vetch: Throughout California, except deserts and Great Basin, to 1500 m. Common vetch is encountered more frequently below 500 m. Common vetch also occurs in Oregon, Washington, Idaho, most eastern and southern states, and a few central states. Narrowleaf vetch occurs in most contiguous states.

Winter vetch, hairy vetch: Throughout California, except deserts and Great Basin, to 1000 m. Hairy vetch occurs in all contiguous states. Winter vetch also occurs in Oregon, Washington, and most southern and eastern states.

PROPAGATION/PHENOLOGY: *Reproduce by seed.* Seeds are ejected a short distance from the parent plant when pods snap open. Some seeds disperse to greater distances with water, soil, animals, agricultural activities, in hay, and as a

Hairy vetch (*Vicia villosa* ssp. *villosa*) seedling.
J. M. DiTomaso

Winter vetch (*Vicia villosa* ssp. *varia*) fruit.
J. M. DiTomaso

Winter vetch (*Vicia villosa* ssp. *varia*) seeds.
J. O'Brien

Common vetch • Narrowleaf vetch • Winter vetch • Hairy vetch

Purple vetch (*Vicia benghalensis*) seedling.
J. M. DiTomaso

Purple vetch (*Vicia benghalensis*) plant.
J. M. DiTomaso

Purple vetch (*Vicia benghalensis*) flowers and fruit.
J. M. DiTomaso

Purple vetch (*Vicia benghalensis*) seeds.
J. O'Brien

Common vetch • Narrowleaf vetch • Winter vetch • Hairy vetch

Bird vetch (*Vicia cracca*) flowering stem.
J. M. DiTomaso

Bird vetch (*Vicia cracca*) flower cluster.
J. M. DiTomaso

Bird vetch (*Vicia cracca*) seedling.   J. M. DiTomaso

Bird vetch (*Vicia cracca*) seeds.   J. O'Brien

seed contaminant. Mature seeds are hard-coated, especially those of winter vetch and hairy vetch, and some survive ingestion by animals. Immature common vetch seeds have permeable seed coats and germinate readily. Germination occurs in fall and/or early spring when conditions are favorable. Buried seeds of narrowleaf vetch are reported to survive for up to ~ 18 years.

MANAGEMENT FAVORING/DISCOURAGING SURVIVAL: Cultivation, mowing, and manual removal before seeds develop can control vetch.

SIMILAR SPECIES: Refer to table 36 (p. 834) for a comparison of the more commonly encountered *Vicia* species. Less commonly encountered species are listed in the appendix, **Non-native Naturalized Plants** (p. 1601, vol. 2). The following species are native to Europe.

Tiny vetch (*Vicia hirsuta*) flowering stem.   J. M. DiTomaso

Purple vetch [*Vicia benghalensis* L., synonyms: reddish tufted vetch; *Vicia atropurpurea* Desf.] is a pubescent *annual* that is very similar to winter vetch and hairy vetch. Purple vetch is primarily distinguished by having mature racemes that that are usually *roughly equal in length to that of the leaf just below* and that consist of *3–12 flowers with rose-violet petals*, and by having *densely pubescent pods*. In addition, emerging racemes are densely pubescent. Flowers are 1.2–1.6 cm long, with calyx bases obliquely attached and 1 side slightly swollen and sac-like or not. Purple vetch inhabits roadsides, fields, and other disturbed places throughout California, except deserts and Great Basin, to 200 m. It is also naturalized in New York and sporadically escapes cultivation in many places.

Bird vetch [*Vicia cracca* L. ssp. *cracca*, synonym: *Vicia semicinta* Greene] [VICCR] is a pubescent *perennial with creeping roots* that also resembles winter vetch, hairy vetch, and purple vetch. Besides the perennial life cycle and creeping roots, bird vetch is distinguished by having flowers 1–1.5 cm long with the *length usually 3–3.5 times the width when pressed flat* (mostly 2–4 mm wide) and glabrous pods 2–3 cm long and *0.5–0.7 cm wide*. Flowers are blue-violet to lavender, or occasionally white, with calyx bases obliquely attached, but not swollen and sac-like on one side. Racemes consist of 20–50 flowers and are typically much longer than the leaf just below. Bird vetch inhabits roadsides, other disturbed places, and woodlands in the North Coast Ranges, San Joaquin Valley, Modoc Plateau, and possibly elsewhere, to 1500 m. It also occurs in most northern and eastern states and in a few southern states.

Common vetch • Narrowleaf vetch • Winter vetch • Hairy vetch

Tiny vetch *(Vicia hirsuta)* fruiting stem. J. M. DiTomaso

Tiny vetch *(Vicia hirsuta)* mature fruit. J. M. DiTomaso

Tiny vetch [*Vicia hirsuta* (L.) S.F. Gray][VICHI] is a glabrous to pubescent *annual* with stalked racemes of 2–8 *white to pale blue flowers 3–5 mm long*. Tiny vetch is further distinguished by having 2-seeded pods *0.6–1 cm long* and 3–5 mm wide that are *covered with spreading hairs*. Seeds are 1.5–3 mm in diameter. Tiny vetch inhabits roadsides, fields, open woodlands, slopes, and other disturbed places in the North Coast, North Coast Ranges, San Francisco Bay region, Southwestern region, and probably elsewhere, to 200 m. It also occurs in Oregon, Washington, and the southern and eastern United States.

Sparrow vetch [*Vicia tetrasperma* (L.) Schreb., synonyms: four-seeded vetch; lentil vetch; slender vetch][VICTE] is a nearly glabrous *annual* that is similar to tiny vetch. Unlike tiny vetch, sparrow vetch has *glabrous pods* with a *blunt or rounded terminal end*. In addition, the stalked flower clusters consist of *1–4 light violet to pale lavender flowers 4–6 mm long*. Pods are 1–1.5 cm long, 3–5 mm wide, and often contain 4 seeds 1.5–2 mm in diameter. Sparrow vetch inhabits disturbed sites and woodlands in the Northwestern region, San Francisco Bay region, Peninsular Ranges, and probably elsewhere, to 1000 m. Sparrow vetch also occurs in Oregon, Washington, Idaho, Montana, southern and eastern states, and a few east-central states.

Tiny vetch (*Vicia hirsuta*) seeds. J. O'BRIEN

Sparrow vetch (*Vicia tetrasperma*) stems with flowers and fruit. JOE NEAL

Sparrow vetch (*Vicia tetrasperma*) seedling. J. M. DITOMASO

Sparrow vetch (*Vicia tetrasperma*) seeds. J. O'BRIEN

Common vetch • Narrowleaf vetch • Winter vetch • Hairy vetch

Table 36. Vetches (*Vicia* spp.)

Vicia spp.	Life cycle	Inflorescence	Flower number	Flower length (cm)	Flower color	Mature pod length × width (cm); hairiness	Other
V. benghalensis purple vetch	annual	stalked raceme, ± equal to leaf below	3–12	1–1.6	rose-violet	2.5–3.5 × 0.8–1.2; densely pubescent	emerging raceme densely pubescent; calyx base oblique, sac-like or not
V. cracca bird vetch	perennial with creeping roots	stalked raceme, much longer than leaf below	20–50	1–1.5	blue-violet, lavender, or nearly white	2–3 × 0.5–0.7; glabrous	calyx base oblique, not sac like
V. hirsuta tiny vetch	annual	stalked cluster or raceme, ± shorter than leaf below	2–8	1–2	pale blue to dull white	0.5–1 × 0.3–0.5; spreading-hairy	free end of pod tapered
V. sativa ssp. *nigra* narrowleaf vetch	annual	nearly sessile cluster	1–4	1–2	violet, lavender, or white	2.5–7 × 0.3–0.8; glabrous (immature ± pubescent)	calyx 0.7–1.2 cm long; pods black
V. sativa ssp. *sativa* common vetch	annual	nearly sessile cluster	1–4	1.8–3	violet, rarely white	2.5–7 × 0.3–0.8; glabrous (immature ± pubescent)	calyx 1–1.5 cm long; pods black or tan
V. tetrasperma sparrow vetch	annual	stalked cluster, equal to or longer than leaf below	1–4	0.4–0.6	light violet to pale lavender	1–1.5 × 0.3–0.5; glabrous	free end of pod blunt or rounded
V. villosa ssp. *varia* winter vetch	annual, biennial, short-lived perennial	stalked raceme, much longer than leaf below	10–20	1–2	blue-violet, lavender, or white	1.5–4 × 0.6–1; glabrous (sometimes sparsely pubescent)	calyx base oblique and sac-like; foliage glabrous or with hairs ± 1 mm long
V. villosa ssp. *villosa* hairy vetch	annual, biennial, short-lived perennial	stalked raceme, much longer than leaf below	10–40	1–2	blue-violet, lavender, or white	1.5–4 × 0.6–1; glabrous	calyx base oblique and sac-like; foliage with hairs 1–2 mm long

Common vetch • Narrowleaf vetch • Winter vetch • Hairy vetch

Winter vetch, hairy vetch: Variably hairy *annual, biennial, or short-lived perennial* with one-sided racemes of mostly 10–40 flowers.

SEEDLINGS: Cotyledons nearly round, thick, remain below the soil surface. Leaves alternate. First 2 or 3 leaves inconspicuous, scalelike. First conspicuous leaves have 2–4 narrow elongate leaflets. Subsequent 1–2 leaves usually resemble the first conspicuous leaf.

Common vetch, narrowleaf vetch: First conspicuous leaf usually has 2 leaflets and often terminates with a short tendril. Narrowleaf vetch leaflets mostly

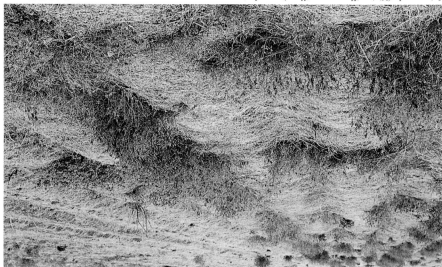

Common vetch (*Vicia sativa* ssp. *sativa*) seeds. J. O'Brien

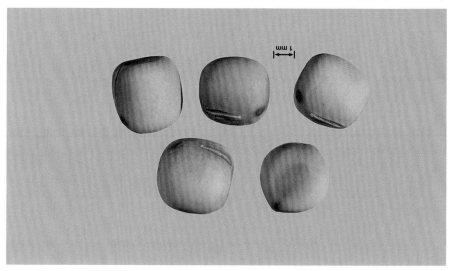

Hairy vetch (*Vicia villosa* ssp. *villosa*) population. J. M. DiTomaso

Common vetch • Narrowleaf vetch • Winter vetch • Hairy vetch

Hairy vetch (*Vicia villosa* ssp. *villosa*) flower clusters with white and purple forms. J. M. DiTomaso

10-20 mm long, 2–3 mm wide, tip often tapered to a minute point (mucro), glabrous or nearly glabrous.

Winter vetch, hairy vetch: First conspicuous leaf has 2 or 4 leaflets and lacks or terminates with a short tendril. Leaflets mostly 5–18 mm long, 2–3 mm wide, tip slightly rounded or tapered to a point, glabrous or hairy. Subsequent few leaves have 4–6 leaflets.

MATURE PLANT: Stems ridged or angled. Leaves alternate. Leaflets folded in bud, alternate to opposite.

Common vetch, narrowleaf vetch: Foliage glabrous to pubescent. Leaflets usually 8–16 per leaf, linear to oblong or wedge-shaped, 1–3.5 cm long, tips acute, truncate, or slightly indented, often with a minute abrupt point at the apex. Stipules often toothed. Common vetch leaflets 4–15 mm wide, with the length 2–6 times the width. Narrowleaf vetch leaflets 1–7 mm wide, with the length 4–10 times the width, usually glabrous or nearly glabrous.

Winter vetch, hairy vetch: Leaflets usually 10–20 per leaf, linear to narrowly oblong or elliptic, 1–3 cm long, tips acute to slightly rounded, with an abrupt minute point at the apex. Stipules often narrow, entire. Hairy vetch foliage spreading-hairy, with hairs 1–2 mm long. Winter vetch foliage nearly glabrous to pubescent, with hairs ± 1 mm long.

ROOTS AND UNDERGROUND STRUCTURES: Taproots usually branched, shallow, or can grow to nearly 1 m deep, usually associate with nitrogen-fixing bacteria.